Nanotechnology Innovations for Food Security and Sustainable Agriculture

Nanotechnology Innovations for Food Security and Sustainable Agriculture

Edited by

Zia Ur Rehman Mashwani
Pir Mehr Ali Shah Arid Agriculture University
Rawalpindi, Pakistan

Shah Fahad
Abdul Wali Khan University
Khyber Pakhtunkhwa, Pakistan

Ilyas Ahmad
Pir Mehr Ali Shah Arid Agriculture University
Rawalpindi, Pakistan

Shah Saud
Linyi University
Shandong, China

Sajid Fiaz
The University of Haripur
Pakistan

Taufiq Nawaz
South Dakota State University
South Dakota, USA

WILEY

This edition first published 2026

Registered Offices
John Wiley & Sons, Inc., 111 River Street, Hoboken, NJ 07030, USA
John Wiley & Sons Ltd, New Era House, 8 Oldlands Way, Bognor Regis, West Sussex, PO22 9NQ, UK

For details of our global editorial offices, customer services, and more information about Wiley products visit us at www.wiley.com.

The manufacturer's authorized representative according to the EU General Product Safety Regulation is Wiley-VCH GmbH, Boschstr. 12, 69469 Weinheim, Germany, e-mail: Product_Safety@wiley.com.

Library of Congress Cataloging-in-Publication Data Applied for:

Hardback ISBN: 9781394348985
ePDF ISBN: 9781394349005
ePUB ISBN: 9781394348992

Cover Design: Wiley
Cover Image: © Deyan Georgiev/stock.adobe.com

Set in 9.5/12.5pt STIXTwoText by Straive, Pondicherry, India

Contents

List of Contributors

Qamar Abbas
School of Business
Hengxing University
Qingdao, Shandong, China

Banzeer Ahsan Abbasi
Department of Botany
Rawalpindi Women University
Rawalpindi, Pakistan

Ilyas Ahmad
Department of Botany
Arid Agriculture University
Rawalpindi, Pakistan

Waqar Ahmad
Department of Botany
Government Degree College No 1
Dera Ismail Khan, Pakistan

Iftikhar Ali Ahmed
College of Resources and Environment/
Microelement Research Center/Key
Laboratory of Arable Land Conservation
(Middle and Lower Reaches of Yangtze River)
Ministry of Agriculture and Rural Affairs,
Huazhong Agricultural University
Wuhan, China

Rasheed Akbar
School of Environmental sciences and
Engineering
Jiangsu University
Zhenjiang, China

Md. Zubair Al Islam
Soil Science Division
Bangladesh Institute of Nuclear Agriculture
Mymensingh, Bangladesh

Amjad Ali
Department of Sustainable Crop Production
Università Cattolica Del Sacro Cuore
Piacenza, Italy

Ayaz Ali
College of Resources and Environment/
Microelement Research Center/Key
Laboratory of Arable Land Conservation
(Middle and Lower Reaches of Yangtze River)
Ministry of Agriculture and Rural Affairs,
Huazhong Agricultural University
Wuhan, China

Md. Yousuf Ali
Maize Breeding Division
Bangladesh Wheat and Maize Research
Institute
Nashipur, Dinajpur, Bangladesh

Nawab Ali
Biosystems and Agricultural Engineering
Michigan State University
East Lansing, USA

Rovaid Ali
Department of Horticulture
The University of Agriculture
Peshawar, Pakistan

Muhammad Anas
Department of Plant Sciences, Faculty of Biological Sciences
Quaid-I-Azam University
Islamabad, Pakistan

and

Department of Soil Science & Plant Nutrition
Selçuk University Campus
Konya, Türkiye

Sadaf Anees
Department of Clinical Studies, Faculty of Veterinary and Animal Sciences
Arid Agriculture University
Rawalpindi, Pakistan

Muhammad Mehran Anjum
AGROSCOPE, Climate Change and Agriculture Research Division
Agroecology and Environment
Zurich, Switzerland

Muhammad Anwar-ul-Haq
Institute of Soil and Environmental Sciences
University of Agriculture
Faisalabad, Pakistan

Annika Jahan Aonti
Soil Science Division
Bangladesh Wheat and Maize Research Institute
Dinajpur, Bangladesh

Sanam Ashraf
Department of Plant Sciences, Faculty of Biological Sciences
Quaid-i-Azam University
Islamabad, Pakistan

Nasir Assad
Institute of Chemistry
University of Sargodha
Sargodha, Pakistan

Asma Ayoob
Department of Clinical Studies, Faculty of Veterinary and Animal Sciences
Arid Agriculture University
Rawalpindi, Pakistan

Jalal Bayar
Department of Agronomy
The University of Agriculture
Peshawar, Pakistan

Khadija Bibi
Department of Botany, Faculty of Sciences
Ghazi University
Dera Ghazi Khan, Pakistan

Nizakat Bibi
International joint research laboratory for Global Change Ecology
School of Life Science, Henan University
Kaifeng, Henan, China

Younsuk Dong
Biosystems and Agricultural Engineering
Michigan State University
East Lansing, USA

Shah Fahad
Department of Agronomy
Abdul Wali Khan University
Mardan, Khyber Pakhtunkhwa, Pakistan

Aliza Falak
Department of Plant Sciences, Faculty of Biological Sciences
Quaid-i-Azam University
Islamabad, Pakistan

Afshan Farid
Department of Botany
PMAS Arid Agricultural University
Rawalpindi, Punjab, Pakistan

Aimen Fatima
Department of Microbiology, Faculty of biological sciences
Quaid-i-Azam University
Islamabad, Pakistan

Saba Fatima
Animal Quarantine Department
Ministry of National Food Security and Research
Islamabad, Pakistan

Zuha Fatima
Department of Botany
Government Degree College No 1
Dera Ismail Khan, KPK, Pakistan

Sharjeel Haider
College of Resources and Environment/ Microelement Research Center/Key Laboratory of Arable Land Conservation (Middle and Lower Reaches of Yangtze River) Ministry of Agriculture and Rural Affairs, Huazhong Agricultural University
Wuhan, China

Abdul Haq
Pakistan Council of Scientific and Industrial Research, Laboratories Complex
Peshawar, Pakistan

Sameel Hassan
Department of Agriculture & Forest Science (DAFNE)
University of Tuscia
Viterbo, Italy

Akbar Hossain
Soil Science Division
Bangladesh Wheat and Maize Research Institute
Dinajpur, Bangladesh

Chudary Sadam Hussain
Department of Botany
PMAS Arid Agriculture University
Rawalpindi, Pakistan

Muhammad Nasir Hussain
Department of Chemistry
Quaid-i-Azam University
Islamabad, Pakistan

Sajid Hussain
Department of Botany
PMAS Arid Agriculture University
Rawalpindi, Pakistan

Tahmina Iasmin
Adaptive Research and Extension Division
Bangladesh Institute of Nuclear Agriculture
Mymensingh, Bangladesh

Naila Ijaz
Nano composite lab, Department of Physics
University of Agriculture
Faisalabad, Pakistan

Shumaila Ijaz
Department of Burn and Plastic Surgery Shenzhen Institute of Translational Medicine, The First Affiliated Hospital of Shenzhen University, School of Biomedical Engineering, Shenzhen University Medical School, Shenzhen University
Shenzhen, China

Javed Iqbal
Department of Botany
Bacha Khan University
Charsadda, Khyber Pakhtunkhwa, Pakistan

Majid Iqbal
Key Laboratory of Ecosystem Network Observation and Modeling
Institute of Geographic Sciences and Natural Resources
Research, Chinese Academy of Sciences
Beijing, China

and

University of Chinese Academy of Sciences (UCAS)
Beijing, China

Parvez Kabir
Soil Science Division
Bangladesh Wheat and Maize Research Institute
Dinajpur, Bangladesh

Alevcan Kaplan
Department of Crop and Animal Production
Sason Vocational School, Batman University
Batman, Turkey

Amir Abdullah Khan
School of Environmental sciences and Engineering
Jiangsu University
Zhenjiang, China

Amjid Khan
Department of Plant Sciences
Faculty of Biological Sciences
Quaid-i-Azam University
Islamabad, Pakistan

UNESCO UNISA Africa Chair in Nanoscience and Nanotechnology
College of Graduate Studies
University of South Africa
Pretoria, South Africa

Asghar Khan
Department of Clinical Studies, Faculty of Veterinary and Animal Sciences
Arid Agriculture University
Rawalpindi, Pakistan

Gul Roz Khan
Pakistan Council of Scientific and Industrial Research, Laboratories Complex
Peshawar, Pakistan

Muhammad Nauman Khan
Department of Botany, Biology Laboratory
Islamia College Peshawar, University Public School, University of Peshawar
Peshawar, Pakistan

Ramzan Khan
Department of Clinical Studies, Faculty of Veterinary and Animal Sciences
Arid Agriculture University
Rawalpindi, Pakistan

Waseem Ahmed Khattak
Department of Plant Sciences, Faculty of Biological Sciences
Quaid-I-Azam University
Islamabad, Pakistan

Marzia Batool Laila
Institute of Chemistry
University of Sargodha
Sargodha, Pakistan

Tariq Mahmood
Department of Plant Sciences, Faculty of Biological Sciences
Quaid-i-Azam University
Islamabad, Pakistan

Muhammad Majeed
Department of Botany
University of Gujrat
Punjab, Pakistan

Zia ur Rehman Mashwani
Department of Botany
Arid Agriculture University
Rawalpindi, Pakistan

Nosheen Mirza
Department of Soil and Environmental Sciences
Ghazi University
Dera Ghazi Khan, Pakistan

Nirma Mubeen
Department of Botany, Faculty of Plant Science
University of Narowal
Punjab, Pakistan

Akhtar Munir
Department of Chemistry
Quaid-i-Azam University
Islamabad, Pakistan

Most Sirajum Munira
Entomology Division
Bangladesh Wheat and Maize Research Institute
Nashipur, Dinajpur, Bangladesh

Venuste Munyaneza
College of Resources and Environment/
Microelement Research Center/Key
Laboratory of Arable Land Conservation
(Middle and Lower Reaches of Yangtze River)
Ministry of Agriculture and Rural Affairs,
Huazhong Agricultural University
Wuhan, China

Nazia
Department of Botany
Government Degree College No 1
Dera Ismail Khan, KPK, Pakistan

Nipa Rani Paul
Plant Pathology Division
Bangladesh Institute of Nuclear Agriculture
Mymensingh, Bangladesh

Umar Masood Qurashi
Department of Plant Sciences, Faculty of Biological Sciences
Quaid-I-Azam University
Islamabad, Pakistan

Naveed Iqbal Raja
Department of Botany
PMAS Arid Agriculture University
Rawalpindi, Pakistan

Muhammad Ramzan
School of Finance and Economics
Jiangsu University
Zhenjiang, China

Sara Rauf
College of Resources and Environment/
Microelement Research Center/Key
Laboratory of Arable Land Conservation
(Middle and Lower Reaches of Yangtze River)
Ministry of Agriculture and Rural Affairs,
Huazhong Agricultural University
Wuhan, China

Sadia Riaz
Department of Plant Sciences, Faculty of Biological Sciences
Quaid-I-Azam University
Islamabad, Pakistan

Syed Mukaram Shah
Centre of Plant Biodiversity
University of Peshawar
Peshawar, Pakistan

Sidra Sohail
State Key Lab of Soil Erosion and Dryland Farming on the Loess Plateau
Institute of Soil and Water Conservation,
Northwest A&F University
Yangling, China

Sudenaz Soylu
Department of Field Crops, Faculty of Agriculture
Selcuk University
Konya, Türkiye

Taharat Al Tauhid
Plant Pathology Division
Bangladesh Wheat and Maize Research Institute
Dinajpur, Bangladesh

Tooba
Institute of Chemistry
University of Sargodha
Sargodha, Pakistan

Nasir Uddin
Sustainable Agricultural Systems
International Maize and Wheat Improvement Center
Dhaka, Bangladesh

Barkat Ullah
Department of Botany
Islamia College Peshawar
Peshawar, Pakistan

Faiz Ullah
Department of Botany
PMAS Arid Agriculture University
Rawalpindi, Pakistan

Zakir Ullah
Department of Plant Sciences, Faculty of Biological Sciences
Quaid-i-Azam University
Islamabad, Pakistan

and

Center for Pharmaceutical Engineering Science and Institute of Cancer Therapeutics
School of Pharmacy and Medical Sciences, Faculty of Life Sciences, University of Bradford
Bradford, United Kingdom

Ayesha Unzila
Department of Botany
Government Degree College No 1
Dera Ismail Khan, KPK, Pakistan

Sana Wahab
Department of Plant Sciences
Quaid-i-Azam University
Islamabad, Pakistan

Rooma Waqar
Department of Plant Sciences, Faculty of Biological Sciences
Quaid-i-Azam University
Islamabad, Khyber Pakhtunkhwa, Pakistan

Tabassum Yaseen
Department of Botany
Bacha Khan University
Charsadda, Khyber Pakhtunkhwa, Pakistan

Zohaib Younas
Department of Botany
PMAS Arid Agriculture University
Rawalpindi, Pakistan

Arfan Yousaf
Department of Clinical Studies, Faculty of Veterinary and Animal Sciences
Arid Agriculture University
Rawalpindi, Pakistan

Tayyaba Yousaf
Department of Botany
Arid Agriculture University
Rawalpindi, Pakistan

Asma Zafar
College of Natural Resources and Environment
Northwest A&F University
Yangling, Shaanxi, China

Syeda Anber Zahra
Institute of Biomaterials and biomedical, School of Food and Pharmacy
Shanghai Zhongqiao Vocational and Technical University
Shanghai, China

Dua e Zainab
Key Laboratory of Green and Low Carbon Agriculture on Dryland in Northwest China
Ministry of Agriculture and Rural Affairs, College of Natural Resource and Environment, Northwest A & F University
Yangling, China

Foreword

In an increasingly interconnected world, the grand challenges of feeding a growing population and ensuring the health of our planet are intrinsically linked. Agriculture, the bedrock of human civilization, faces a complex array of pressures, from climate volatility and resource scarcity to pest resistance and diminishing arable land. Addressing these multifaceted issues demands not just incremental improvements, but truly transformative innovations that can fundamentally alter our approach to food production and sustainability.

This book, "Nanotechnology Innovations for Food Security and Sustainable Agriculture," arrives at a crucial juncture, offering a profound and timely exploration of a technology poised to revolutionize these critical sectors. It is clear that the nanoscale holds immense promise, offering precision tools and novel materials that can overcome limitations inherent in traditional agricultural practices.

The editors and contributing authors of this volume have meticulously compiled a comprehensive body of knowledge, meticulously detailing how nanotechnology can enhance every facet of the agricultural value chain. From the intelligent delivery of nutrients and targeted pest control to advanced biosensing for disease detection and innovative food preservation techniques, the breadth of topics covered here underscores the pervasive impact of nanotechnology. What makes this book particularly compelling is its balanced perspective, not only highlighting the exciting opportunities but also rigorously addressing the crucial considerations of safety, ethics, environmental impact, and economic viability.

This work serves as a testament to the power of interdisciplinary research, bringing together insights from materials science, biology, engineering, environmental science, and policy. It provides an indispensable resource for scientists striving to push the boundaries of knowledge, for policymakers grappling with regulatory frameworks, for industry leaders seeking sustainable solutions, and for students eager to understand the future of food.

I commend the editors for their vision in curating such a vital collection. This book is more than just a compendium of current research; it is a roadmap for a more resilient, efficient, and sustainable agricultural future. Its pages will undoubtedly inspire new collaborations, spark further innovations, and guide the responsible development of nanotechnology for the benefit of humanity and the planet.

Shah Fahad
Assistant Professor
Abdul Wali Khan University, Pakistan

Preface

In an era defined by unprecedented challenges to global food security and environmental sustainability, the imperative to innovate within agriculture has never been more critical. A rapidly growing world population, coupled with the escalating impacts of climate change, dwindling natural resources, and persistent crop losses, demands revolutionary approaches to ensure nutritious food for all while safeguarding our planet for future generations.

It is within this transformative landscape that nanotechnology emerges as a beacon of hope and a frontier of scientific exploration. Operating at the nanoscale, this powerful discipline offers precise, efficient, and often more environmentally benign solutions to long-standing agricultural dilemmas. From enhancing nutrient delivery and protecting crops from pests and diseases to monitoring plant health in real time and extending the shelf life of produce, nanotechnology promises to redefine the paradigms of food production and preservation.

This book, *Nanotechnology Innovations for Food Security and Sustainable Agriculture*, is meticulously crafted to provide a comprehensive and insightful exploration into the myriad ways nanotechnology is reshaping the agricultural sector. Across its seventeen chapters, we investigate into the fundamental concepts, cutting-edge applications, and critical considerations surrounding this exciting field.

Beginning with an **Overview of Nanotechnology in Food Security and Agriculture: Introduction, Current Status, and Concerns**, the book establishes a foundational understanding, discussing its current status, applications, and initial concerns (Chapter 1). We then transition to **Integrating Nanobased Engineering Advances to Enhance Crop Yields and Optimize Food Production Systems**, showcasing how engineering principles at the nanoscale can revolutionize farming practices (Chapter 2). The **Development of Nanofertilizers for Soil Health and Fertility** is explored in depth, highlighting their mechanisms, benefits, and future directions for sustainable nutrient management (Chapter 3).

Further chapters illuminate specific applications, including the fascinating realm of **Nano-Biofortification of Cereal Crops**, which promises to address hidden hunger by enhancing the nutritional content of staple foods (Chapter 4). The **Role of Nanomaterials in Improving Oil Yield and Quality of Oilseeds Crops** is detailed, revealing their potential to boost both quantity and nutritional value (Chapter 5). The evolution toward **Next-Generation Agriculture with Nanopesticides and Fungicides** demonstrates how targeted, efficient solutions can minimize environmental impact (Chapter 6), complemented by a dedicated examination of **The Role of Nanomaterials in Modern Plant Disease Management** (Chapter 7).

The journey continues by exploring **Revolutionizing Fruit and Vegetable Farming: Nanotechnology from Soil to Shelf**, covering everything from soil improvement to post-harvest processing (Chapter 8). A specific focus on the **Role of Nanotechnology in Increasing the Shelf Life of Fruits** highlights its significance in reducing food waste (Chapter 9). The adoption of **Harnessing Nanoherbicides and Insecticides for Eco-Friendly Solutions: Advancing Green Agriculture** underscores the move toward green agriculture (Chapter 10), while the **Development and Application of Nanobiosensors for Better Stress Management in Crops** offer advanced tools for real-time monitoring (Chapter 11).

Beyond plant agriculture, the book broadens its scope to include **Nano-Based Applications for Veterinary and Dairy Production**, showcasing the versatility of nanotechnology in animal health and food production (Chapter 12). Comprehensive discussions on **Role of Nanotechnology for Better Food Preservation** (Chapter 13) and **Nanoencapsulation Approaches in Food Processing and Packaging** (Chapter 14) emphasize its pivotal role in ensuring food quality and safety across the supply chain.

Crucially, this volume does not shy away from the challenges and responsibilities associated with this powerful technology. Dedicated chapters address the **Potential Health Risks Associated with Nanoparticles** (Chapter 15), providing a balanced perspective on safety considerations. A **Cost–Benefit Analysis of Nanoagricultural Technologies** offers economic insights into their adoption (Chapter 16), culminating in a thorough examination of the **Ethical Safety and Regulatory Considerations in Nanotechnology for Agriculture** (Chapter 17), which champions responsible innovation and sustainable deployment.

This book is an invaluable resource for researchers, academics, policymakers, industry professionals, and students across disciplines such as agricultural science, nanotechnology, environmental science, food science, and biotechnology. It aims to foster a deeper understanding of nanotechnology's potential while encouraging thoughtful deliberation on its societal and environmental implications.

We believe that the collective knowledge presented within these pages will serve as a foundational text, inspiring further research, informing policy decisions, and ultimately contributing to a more food-secure and agriculturally sustainable future for all.

The Editors

Zia Ur Rehman Mashwani, Shah Fahad, Ilyas Ahmad,
Shah Saud, Sajid Fiaz, and Taufiq Nawaz

1

Overview of Nanotechnology in Food Security and Agriculture: Introduction, Current Status, and Concerns

Muhammad Anas[1,2], Waseem Ahmed Khattak[1], Aliza Falak[1], Afshan Farid[3], Muhammad Majeed[4], Sadia Riaz[1], Shah Fahad[5], and Umar Masood Quraishi[1]

[1] *Department of Plant Sciences, Faculty of Biological Sciences, Quaid-I-Azam University, Islamabad, Pakistan*
[2] *Department of Soil Science & Plant Nutrition, Selçuk University Campus, Konya, Türkiye*
[3] *Department of Botany, PMAS Arid Agricultural University, Rawalpindi, Punjab, Pakistan*
[4] *Department of Botany, University of Gujrat, Punjab, Pakistan*
[5] *Department of Agronomy, Abdul Wali Khan University Mardan, Khyber Pakh-tunkhwa, Pakistan*

1.1 Introduction

The agricultural sector is the backbone of the economy, as it has always been important in ensuring food security for a growing population. There is a growing need for the environment in agriculture to be more resource-efficient, economically viable, and sustainable. However, the industry struggles with many problems like declining crop production, climate change, soil nutrient depletion, water shortage, lower fertility of soil, crop diseases, and insufficient workforce. Scientists have been forced to look for other potential technologies that could help ease the looming food supply crisis through enhanced supply chain and logistics. It has easily been observed that the potential scope of agriculture productivity improvement from application of efficient pesticides, herbicides, and nanofertilizers along with soil nutrition control and wastewater treatment lies within the arms of nanotechnology. It, alongside industrial agriculture, has made great strides in food processing contributing greatly to the surge of food productivity and improvement of food quality through nutrition and powerful antimicrobial action. Progress in agriculture results from changes in climate, functions of ecosystems, social factors, and the availability of resources (Eswaran et al., 2024). Hence, there is a great need to make further progress in agricultural development to overcome problems like poverty and famine. The UN has stressed the importance of sustainable agriculture and has even adopted it as one of its goals. Most people are aware and worried about sustainability since it has always been closely linked with national economics and food production. Sustainable agriculture includes the optimal application of soil nutrients, water resources, agro wastes handling, minimum fertilizer usage, plant pest control, and modern farming practices. Assessment can only be made under certain conditions like the level of technological advancement, funding accessibility to farmers, and the

Nanotechnology Innovations for Food Security and Sustainable Agriculture, First Edition. Edited by Zia Ur Rehman Mashwani, Shah Fahad, Ilyas Ahmad, Shah Saud, Sajid Fiaz, and Taufiq Nawaz.

degree of participation in research and development for sophisticated methodologies. There exists a broader application of nanotechnology in achieving sustainability in agriculture by improving the efficiency of agricultural inputs and hence increasing food production and crop yield (Anas et al., 2025). This would lead to sustaining the environment, preserving ecological balance, and stabilizing the economy simultaneously.

Nanotechnology offers possible solutions to food security issues through the enhancement of agriculturally related products and services in the sectors of water, environment, medicine, and health. Some of the emerging areas of nanotechnology should be profoundly useful in solving food challenges such as delivery systems, protein bioseparation, rapid sampling of chemical and biological contaminants, nutraceuticals, nanoencapsulation, and solubilization (Zhu, 2025). By manipulating processes at the nanolevel, food nanotechnology focuses on reinforcing the bioactive constituents with the goal of altering their biological accessibility and protecting them from environmental or chemical fluctuations. There isn't a single method or practice that nanotechnology makes easy. There are few, if any, available studies regarding the constraints and toxicity of these technologies, which makes legislative control very complex and leads to the absence of approval. Moreover, the existence of complex nanomaterials like metallic nanoparticles, carbon nanotubes, nanowires, and nanocomposites confuses nonexperts. Therefore, there is a strong need for research on the possible harmful effects of these structures on the human body and the environment in relation to their bioaccumulation, degradation, assimilation, and toxicity. Food nanotechnology has the potential to achieve immense advances in the efficiency of the food production cycle, ranging from processing and storage to even the transportation of food products (Chauhan et al., 2024) (Figure 1.1).

The impacts that these nanomaterials could have on animals, plants, and ecosystems are still largely undocumented, particularly regarding side effects. Nevertheless, these situations can be rectified with proper disposal mechanisms engineered by future scholars, state bodies, and agro-industrial enterprises. Sustaining agricultural practices has incorporated various measures like the use of nanobiotechnology and genome editing. However, the use of nanotechnology is becoming well-acknowledged in the areas of humanitarian drug development, medicine, and food security (Erdoğar et al., 2019). The structure, shape, and size features of the nanoparticles vary. Their shape and size vary from 1 to 100 nm and can be spherical, cylindrical, tubular, conical, hollow core, flat, or irregular. There might be asymmetric or homogeneous surface variations. Some nanoparticles are crystalline or amorphous and possess one or more clusters of loosely or tightly packed crystals (Li et al., 2024). There is an increase in the number of production processes that are being or have been developed to improve the product quality and decrease the price of production. Special nanoparticles that have high optical, mechanical, physical, and chemical capabilities are processed using different techniques to enhance these capabilities (Bolledla & Bakshi, 2024). Progress in instrumentation has enabled characterization of nanoparticles, which can now be done in detail with subsequent use. Three broad classifications of nanomaterials include carbon-based, organic, and inorganic. Organic nanoparticles or polymers are frequently referred to as dendrimers, micelles, liposomes, and ferritin. They are nontoxic and biodegradable, and some of them, like micelles and liposomes, have nanosized hollow cores known as nanocapsules, which can be altered by direct energy like heat and light (Arora et al., 2024). Due to these unique properties, they are ideal for use in drug

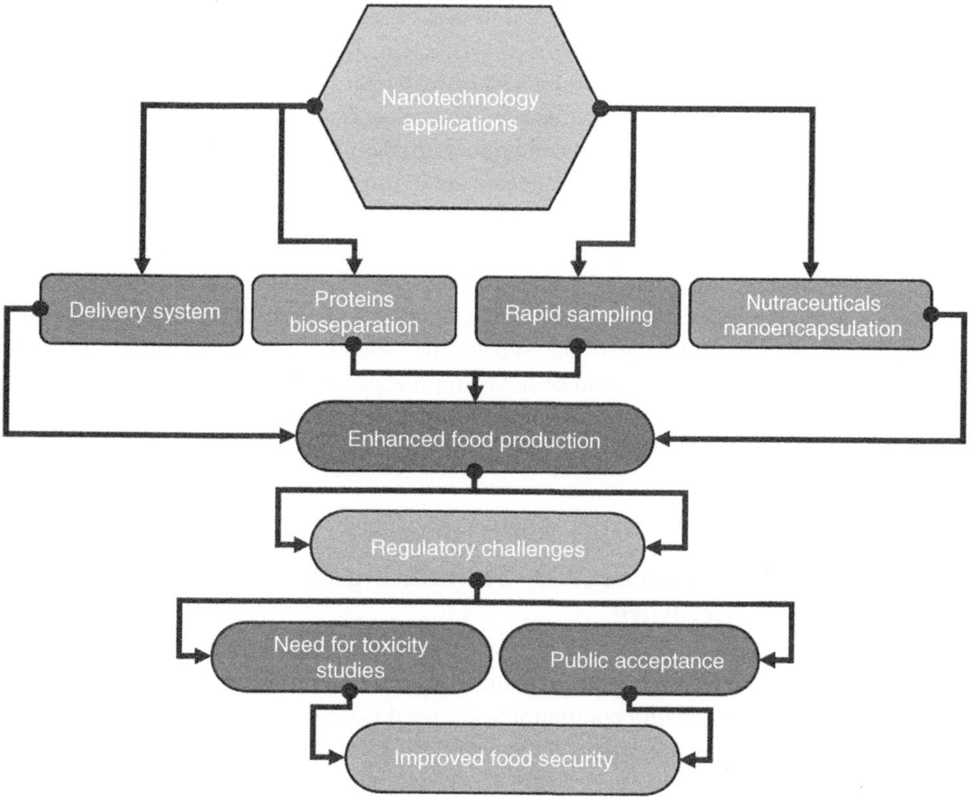

Figure 1.1 Nanotechnology in food security.

delivery systems. Inorganic nanoparticles are non-carbon-containing and are generally defined as those composed of a metal or metal oxide.

From fully destructing or constructing metals to a nanometric size, metal-based nanoparticles can be produced (Sune et al., 2024). The most common metals that are used in the manufacturing of particles include aluminum (Al), cadmium (Cd), cobalt (Co), copper (Cu), gold (Au), iron (Fe), lead (Pb), silver (Ag), and lead (Zn). Nanoparticles with their corresponding metal oxides are produced in order to modify the features of metal-based nanoparticles. For example, iron nanoparticles oxidize quickly into iron oxide (Fe_2O_3), which makes their reactivity greater than iron brand nanoparticles. This enhances reactivity and efficiency which makes it favorable to produce aluminum oxide nanoparticles (Al_2O_3). Carbon-based nanoparticles include all the nanoparticles made of carbon. They are classified into fullerenes, graphene, carbon nanotubes (CNTs), carbon nanofibers, carbon black, and in some cases activated carbon at the nanoscale (Yu & Feng, 2024). A significant step was taken with the emergence of biologically based methods for synthesizing nanomaterials, such as plant extracts, which aim to reduce waste production and help preserve the ecosystem. It has been reported that plant extracts can serve as a more cost-effective option compared to chemical and physical techniques for nanomaterial construction, as plant extracts are solely composed of secondary metabolites and highly reducing and stabilizing agents.

Clustered regularly interspaced short palindromic repeats (CRISPR) technologies, when paired with nanotechnology, can organically enhance agricultural productivity at a low price point. For example, silver and titanium nanoparticles are primarily used in food packaging and storage. Zinc oxide nanoparticles also aid in improving the nutritional quality of food. In addition, food analysis makes use of gold and platinum nanowire-based biosensors. Recently, zeolite-based nanosensors have been developed and can also be employed as catalyst carriers. Nanobiotechnology has been primarily focused on creating new biosensors as immobilization platforms, catalytic tools, or optical labels in food industries due to their elevated performance regarding sensitivity, selectivity, and stability (Thakur et al., 2022). Clearly, the development of novel approaches for food detection using these sensors has garnered a lot of attention. Hence, this review seeks to facilitate other researchers in utilizing nanomaterials to tackle issues surrounding food security and sustainable agriculture.

1.2 Nanoparticles as a Growth-Stimulating Element of Sustainable Agriculture

Sustainable developments in agricultural technology are of utmost importance to developing nations, which have growing populations and need innovative research. In plant technology, some nanoparticles are proved to be useful for increasing both the stress tolerance and the growth of plants (Zhao et al., 2020). Lowered impact of biotic and abiotic environmental stresses is achieved through usage of biosynthesized nanoparticles in plants, which is known as nano-agriculture. Plants are negatively stressed, and their homeostatic equilibrium is disrupted, which leads to a decline in their productivity. Nanoparticle biosynthesis has spectacular effects on plants, like improving antioxidant status and lowering the level of reactive oxygen species (ROS), which, in turn, affects many molecular and biochemical signaling pathways, resulting in a more favorable plant yield. The biologically active role of nanoparticles mostly depends on how such particles were produced, the dosage used, and the physicochemical features. Therefore, in this part we attempt to outline the impact of various types of nanoparticles on some abiotic (such as heat, water deficit, and heavy metals) and biotic (pathogens) plant stresses and plant growth (El-Saadony et al., 2022) (Figure 1.2).

There are reports that nanoparticles enhance the growth of plants by improving root expansion, seed sprouting, shoot formation, and biomass yield, which will surely encourage prospective researchers and agencies to use these suitable nanoparticles toward agricultural sustainability and development. In the near future, agriculture will likely have to confront ecological issues that threaten the food supply to the rapidly increasing global population. One potential avenue to investigate is the modification of conventional fertilization methods to improve plant biomass, plant productivity, and ultimately grain yield. The over-application of chemical fertilizers can be detrimental to human health, animal health, environmental health, and plant/crop health (Mubarik et al., 2022). These challenges can be difficult to solve, but nanofertilizers (NFs) may offer a promising solution.

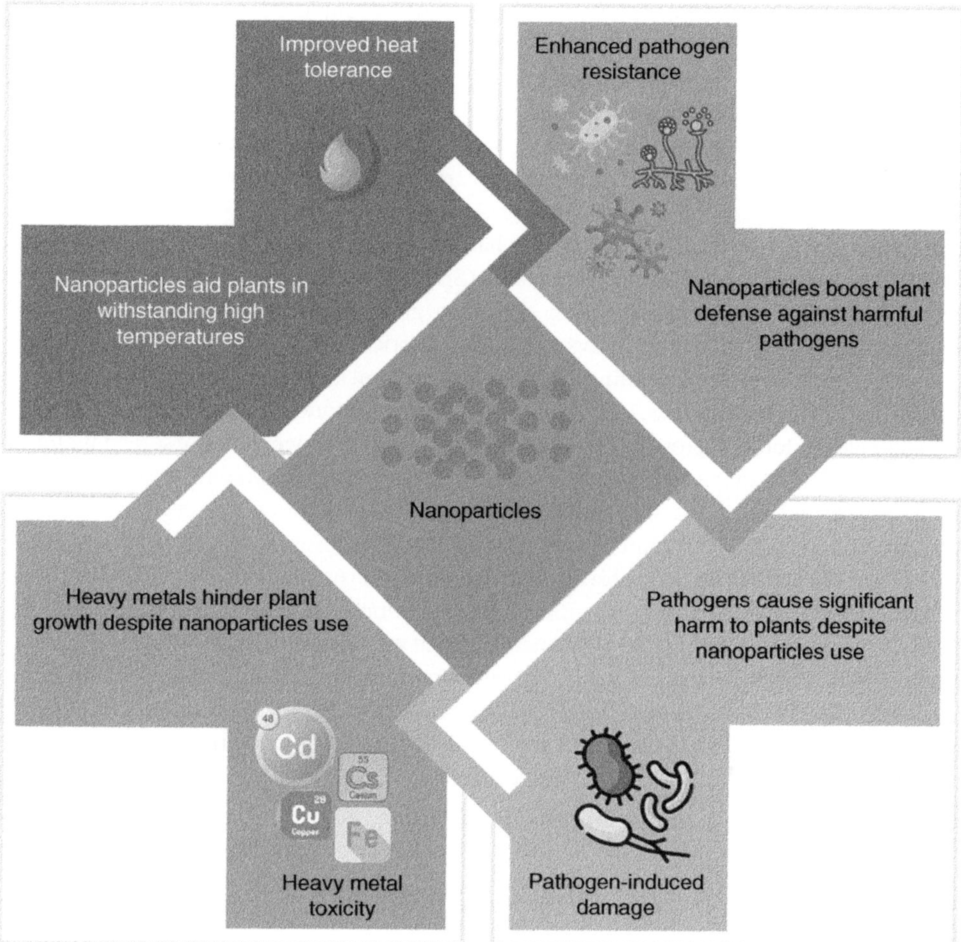

Figure 1.2 Impact of nanoparticles on plant growth and stress.

The application of NFs has been shown to be one of the most effective ways to improve resource productivity and plant production and to decrease pollution levels. Therefore, nanofertilizers can be used instead of conventional fertilizers to increase agricultural products (Darbre, 2023).

In plants, photosynthetic activity is essential for metabolism, but it remains susceptible to a variety of stresses such as nutrient deficiency, drought, salt concentration, and temperature. Silicon dioxide (SiO_2) nanoparticles have been applied to pumpkin plants to elicit defense responses to abiotic stress by enhancing modern transpiration (water use efficiency) as well as carbonic anhydrase activities and photosynthetic pigment contents (Javed et al., 2022). Titanium dioxide (TiO_2) has interfered with the photo-reduction potential and linolenic acid blocking in the electron transport chain (ETC) in chloroplasts, where oxygen evolution occurs (Thapa, 2021). As a result of abiotic stress, the primary symptom exhibited by plant cell organelles is the overproduction of reactive oxygen species (ROS). It is well established that, normally, plants possess exquisitely designed enzymatic machinery for nullifying this oxidative stress induced by the environment. However, in conditions of abiotic stress, crops are exposed to detrimental effects of these situations, which occur when the plant defense mechanisms fail. Nanomaterials can provide this relief by accumulating osmolytes, switching on specific stress-inducible genes, and triggering the supply of free amino acids and other nutrients. Cationic rather than anionic nanoparticles are readily propenetrated by the negatively charged plant cell walls. Therefore, negatively charged nanoparticles possess greater translocational efficiency. For example, positively charged cerium oxide (CeO) nanoparticles are adsorbed onto negatively charged root surfaces (Spielman-Sun et al., 2017); however, negatively charged CeO_2 nanoparticles have shown to greatly limit root accumulation while enhancing shoot uptake by overcoming electrostatic barriers.

The use of nanofertilizers has improved both biochemical and physiological parameters of crop plants. In sunflower plants, it was noted that magnetic nanofluid had a beneficial effect on the content of total chlorophyll. In *Zea mays*, improved crop yield was observed with the foliar spray of TiO_2 nanoparticles due to increased photosynthetic pigments. During stressful conditions such as lack of nutrients, salinity, or drought, plants express certain types of organic substances known as antioxidants (secondary metabolites). Nanofertilizers such as $n\text{TiO}_2$ enhance photoassimilation and subsequently, the yield of leaves and grains, which results in better nutrient supply for improved antioxidant regulation in plant cells. More than 95% of plants that were subjected to $n\text{TiO}_2$ treatment showed an increase in fresh and dry mass due to improved photosynthetic activity and nitrogen metabolism, resulting from greater pigment biosynthesis and transformation of light into biochemical energy through photophosphorylation (Li et al., 2023), which increased biological carbon sequestration through the light-independent reactions known as the Calvin cycle.

1.3 Current Status on Food and Agriculture Nanotechnology

The field of nanotechnology deals with materials that have at least one dimension ranging from 1 to 100 nm. Various nanomaterials have been explored in the food sector, including inorganic (metal and metal oxide nanoparticles), organic (mainly natural product nanoparticles), and composite materials (such as clay). Among the metal nanoparticles, silver is the most commonly produced and used due to its antimicrobial qualities, while gold nanoparticles are extensively studied for their potential in sensing and detection applications. Titanium dioxide nanoparticles are thoroughly examined for their disinfectant properties and their application as a food additive for enhancing whiteness and flavor (Wasilewska et al., 2023). Biological products-based nanoparticles are typically engineered as delivery systems and are integrated into food products as ingredients or supplements.

1.3.1 Current Status on Food Nanotechnology

Food nanotechnology has progressed in areas including food preservation, packaging, and additives, as well as in the use of cutting-edge technology for food processing, storage, and security. Nanometer-scale voids are known to exist in a large number of common compounds used as food additives or packaging materials. Food-grade titanium dioxide nanoparticles (TiO_2 NPs), for instance, are thought to be present at concentrations of around 40%. Titanium dioxide nanoparticles (TiO_2 NPs) are thought to be nontoxic in comparison to other nanomaterials, but persistent exposure to the environment can turn even innocuous materials dangerous (Rashid et al., 2021). The use of novel nanotechnology in food science, in combination with the presence of chemical entities at the nanoscale, can be equally harmful and is therefore controversial. In this chapter, we report on the developments in food nanotechnology.

1.3.1.1 Food Processing

In the food industry, engineered nanomaterials are increasingly being used as preservatives, color or taste enhancers, and even as transporters for nutritional supplements, such as animal feed. From processing to packaging, the food sector benefits greatly from their exceptional qualities. The U.S. FDA has allowed the use of some inorganic oxide compounds, including SiO_2 (E551), MgO (E530), and TiO_2 (E171), as food taste enhancers, anti-caking agents, and food color preservatives (Siddiqui, & Alrumman, 2021). For instance, TiO_2 is widely used in gums, white sauces, cake sponge, candies, and puddings. With the exception of titanium nitride and carbon black, which are no longer permitted by the U.S. FDA but were approved by EC 10/2011, the current approval for the compounds is based on traditional particle sizes. These ingredients are often found in food items at the nanoscale. The effective and safe production of edible animal products is made possible by animal feeds, which are vital to the global food system even beyond human use. The FDA in the United States has declared copper oxide, iron oxide, and zinc oxide to be "generally recognized as safe" (GRAS) for use in animal feeds as nutritional supplements. Additionally, the use of copper oxide (Cu_2O) as a dietary supplement for all animal species does not pose any health risks to consumers, according to the European Food Safety Authority's (EFSA's) Panel on Additives and Products or Substances Used in Animal Feed (FEEDAP Panel) (Forouzandeh, 2023).

1.3.1.2 Food Packaging

Materials designed for direct contact with food during production, transport, and storage are known as food contact materials. In the food industry, nanotechnology has emerged as an innovative solution for food packaging, with extensive research and development in this area. Compared to traditional packaging materials, nanomaterials engineered for food packaging offer numerous benefits. Among the various novel nanomaterials, nanoclay stands out as one of the most extensively studied and utilized options for food packaging, owing to its mechanical, thermal, and barrier properties, as well as its cost-effectiveness (Padil et al., 2022). The selectivity of a nanocomposite membrane containing 1 wt% bentonite clay in poly(vinyl alcohol) was 46, and it penetrated water at a rate of 6500 GPU. In another work, the glass transition temperature rose by around 6 C, and interlaminar fracture toughness was significantly improved in woven carbon fiber/compatibilized polypropylene nanocomposites loaded with 3% nanoclay (Franco-Urquiza, 2021). In addition, these epoxy/clay nanocomposites were claimed to give better performance in corrosion resistance.

Nanoclays are further separated into montmorillonite, bentonite, kaolinite, hectorite, and halloysite based on their physicochemical characteristics. The U.S. FDA approved bentonite and montmorillonite as GRAS and included them in notifications for Effective Food Contact Substances (FCS) (Siddiqui & Alrumman, 2021). The FCS notifications represent the U.S. FDA's grant of premarket approval of food contact substances that were found to be safe for the intended use. But recent research has pointed out the possible migration concerns of using nanoclay packaging. The ability to preserve food has been greatly enhanced with the utilization of nanomaterial edible coatings. Active respiration in fresh products such as vegetables and fruits is emphasized with controlled coatings during transport as well as the storage facility. When respiration processes are active, the rate at which the product is cosmetically and nutritionally tainted post harvesting is alarming and if uncontrolled, renders the product useless. For fresh food products to have an extended shelf life, their weight and the nutrients must be monitored closely. The most important factors are the relative humidity and the temperature within the storage environment. These two factors are the primary ones that affect the respiration of fresh food and mold the microbial activities of the products (Preetha & Narayanan, 2020). To increase the storage life of fruits and vegetables while also enhancing their mechanical properties, nano-edible film, which is usually less than 100 µm thick, can be used that serves as a barrier against gas and moisture. Highly decomposable food products are most effectively preserved with chitosan-quinoa protein, lemongrass essential oil, and apple pectin due to their edible coating properties.

Additional uses for food contact packaging nanomaterials involve the detection of harmful substances, like pesticides, pathogens, and toxins, which are the subject of active research and development due to the extensive sensitivity of nanohybrid structures (Lin et al., 2024). Atrazine, glyphosate, tetradifon, and aldrin were among the herbicides that ZnO quantum dots (QDs) could detect. This occurred as a result of the insecticides' chlorine bond, which worked with QDs rapidly (binding affinity $k = 107$ M-1). Additionally, when ZnO QDs interact with pesticides, they may photocatalyze. Additionally, "smart packaging" and "active packaging," which automatically use sensors based on nanomaterials, have become quite popular and significantly improve food quality (Shivakumar et al., 2018).

1.3.2 Current Status on Agriculture Nanotechnology

The objective of adopting and researching nanotechnology in agriculture is to enhance food production while ensuring the nutritional, qualitative, and safety aspects of the food are not compromised. The modern agricultural production systems make optimal use of plant growth factors/regulators, fertilizers, insecticides, and herbicides. It is possible to increase the efficiency of pesticide, herbicide, and plant growth regulator delivery by employing nanocarriers (He et al., 2019). For example, poly (ε-caprolactone) nanocapsules were developed recently to serve as the carrier of atrazine herbicide. In comparison to commercial atrazine, the herbicidal activity of mustard plants (*Brassica juncea*) treated with poly(ε -caprolactone) nanocapsules containing atrazine was increased. Significant drops in stomatal conductance and net photosynthetic rates, as well as an increase in oxidative stress and ultimately weight loss and growth inhibition of the studied plants, demonstrated this. Moreover, S polymeric NPs and S and C silica NPs are being developed as modified-release systems in the form of nanocarriers. Such nanoscale carriers allow efficient targeting and controlled release of the entities (Yalcinkaya et al., 2024). These approaches can be categorized as "precision farming," as they enhance yield without causing damage to soil and water resources.

Most critically, utilizing nanoencapsulation technology allows for extended intervals between herbicide applications while still effectively managing a particular group of plants, something that is beneficial to the environment. Also, alongside the use of nanocarriers, the development of insect-resistant varieties through plant transformation via gene or DNA material was conducted using gold nanoparticles. Additional information is provided in the previous reviews. Furthermore, some nanoparticles may have significant lethality and specificity that qualifies them as pesticides. Among the least and most studied metal oxide nanomaterials in agriculture are ZnO, TiO_2, and CuO, which because of their poisonous features may aid in the control of plant pathogens. For instance, we take ZnO NPs. ZnO NPs have been shown to possess strong antifungal, antibacterial, and antimycotic activity against *Fusarium graminearum, Aspergillus flavus, Aspergillus niger, Aspergillus fumigatus, Fusarium culmorum,* and *Fusarium oxysporum*.

Traditional mineral fertilizers do not perform well in recovering nutrients, resulting in significant losses, and the use of fertilizers over long periods results in severe economic wastage. This is where the application of nanofertilizers may greatly help. Nanofertilizers can help with improving incorporation of nutrients by minimizing loss and further include crops and soil microorganisms to improve utilization. The nanofertilizers that have been commercialized so far are primarily nanoscale micronutrients. Nanoparticles containing essential nutrients, such as manganese, copper, iron, zinc, molybdenum, boron, and nitrogen, have been developed as nano-fertilizers to enhance nutrient uptake, improve crop productivity, and reduce environmental losses. Moreover, other forms of nanomaterials, for example, carbon nano-onions and chitosan NPs may also improve the growth and quality of crops better than standard fertilizers do. Now it is expected that the introduction of novel types of nanofertilizers will alter the landscape of fertilizer production industries in the next 10 years.

1.4 Toxicological Fundamentals and Risk Assessment

1.4.1 Exposure Routes and Interactions

The growing use of nanotechnology in the food and agricultural industries is a matter of public concern in the past 10 years. Many foods and agricultural products contain nanomaterials that are intentionally added as food additives or, in many instances, such materials are added via migration. This expansion of nano-products, as well as their inherently unique physicochemical characteristics, raises concern for both environmental and human health. The concern for environmental health can be postulated to arise from the use of nanomaterials for nanofertilizers, nanopesticides, and nanoherbicides; though, if not seldom, immobilized nanosensors do pose a lack of concern. The properties and impacts that nanomaterials have on the environment are mostly dependent on the physicochemical properties of those selected nanomaterials. Furthermore, the nanomaterials' behavior and footprints in the environment are difficult to predict due to the complex state of environmental conditions. Nomadic distribution of nanomaterials and the difficulty in tracing and monitoring them make the spontaneous interactions between them and nano-bio-eco systems difficult. While an integrative method has been suggested for understanding the interactions that nanomaterials have with biotic and abiotic organisms in such ecosystems, nanotoxicity must be assessed through singular, targeted experimentation.

Aluminum NP exposure pathways serve as an example of the complexity and possible experimental workflow to estimate nano-product risks. The processes associated with the exposure routes of aluminum NPs in the human digestion system are rather complex and require elaborate steps to be dissected accurately. In the first phases of digestion, the samples remain intact inside artificial saliva. Aluminum NPs partially dissolve releasing aluminum ions in stomach fluid. During the entire digestion process, particle agglomeration has the potential to occur; however, it is most significant in stomach fluid. During the deagglomeration process, agglomerates consisting of more than one nanoparticle are broken down into their primary forms. In addition, free ions can also undergo the physical change of evaporation to create new nanoparticulate structures. Nanoparticles of a different kind are assumed to follow the same procedure, in this case, silver NPs. Silver colloids have been known to exist in all mediums as complex mixtures with varying species that are absorbed. This leads to the release of Ag^+ ions, which, along with toxicity, possess other features like being antimicrobial for drug-resistant microbes. Additionally, gold NPs can also discharge Au [I, III] atoms under normal condition ayatas and participate in reactions.

1.4.2 Nanotoxicology Mechanisms

The concern over the use of nanomaterials in food and agricultural products as additives or other ingredients is just as worrisome regarding health risks. Employing nanomaterials as food additives or as functional or nutritional constituents may pose direct health risks to humans. One of the foremost toxicological mechanisms that leads to cell death and damage is the generation of ROS. Damage includes over-autophagy, which can severely damage neurons and DNA, and has potential mutagenic, carcinogenic, and aging-associated disease impacts on human beings. Release from nanomaterials may cause allergic reactions or damage due to metal ions, which is another potential adverse effect due to the consumption of food with nano-ingredients. Furthermore, the incorporation of nanomaterials into the edible portions of plants (seeds) and the human body has the potential to culminate in more serious problems at higher concentrations coupled with prolonged exposure.

1.4.3 Data Generating and Analysis

A comprehensive evaluation of nanotoxicology is necessary to guarantee the engineering, processing, and use of nanomaterials in food and agricultural goods. Furthermore, current methodologies used for toxicology are mostly uninformative toward improving sustainable design for industrial application by nature. In addition to several advances concerning in vitro and in vivo cellular damage, toxicological information remains vastly insufficient to construct any conclusive generalizations regarding the standard pattern of nanomaterials exposure and toxic effects on human health (Pietroiusti et al., 2018). The intricacy of nano-bio-eco interactions, as we have elaborated upon in other instances, accounts for this narrowing phenomenon. Traditional in vitro and in vivo assessments focused on certain endpoints and processes, such as ROS generation, DNA damage, and immunological responses, yield little information, and the costs are high. Model organisms and cell lines, such as *Escherichia coli* and the human lung adenocarcinoma A549 cell line, respectively, are used to generate omics data, indicating the expected path of progress in the field of nanotoxicity research. Concurrently, a machine learning framework must be developed in order to mine the growing repository of data.

1.5 Frontier Topics

Although it has been hypothesized that the introduction of nanomaterials has a small toxic effect on food and agricultural products, it is conceivable that the toxicity within continuous exposure to the nanomaterials will change. With limited knowledge on the bioavailability and biodistribution of nanomaterials (as well as the acute and chronic toxicity after exposure to nanomaterials), caution may be appropriate. Recently, France became the first country to assume the global lead in legislative action on reassessing the safety of TiO_2 (E171) as a food additive (Younes et al., 2021). In 2017, animals fed TiO_2 NPs for 100 days have, in a group of rats led by the French Agricultural Research Institute (INRA), been found with nonobstructive, pre-tumorous lesions in the colon at doses found in ice cream and powders used in toothpastes and consumer products. Thereafter, the French food safety agency (ANSES) looked at the French INRA study and drew conclusions with regard to the probable carcinogenicity of TiO_2 to the European Chemicals Agency (ECHA). One of the amendments of the French Farm and Food bill in the latter half of 2018, which passed the National Assembly (not final), was aimed to temporarily stop the import and marketing of products of food containing TiO_2 additives by 2020. For instance, several French food manufacturers, including Mars Chocolate France, the French branch of the Mars Inc., promised to comply with the TiO_2 ban. Dunkin' Donuts' powdered sugar donuts caused the US company Dunkin' Brands to state that they will stop using their purported TiO_2 NPs as well. Another example is the lifting of the ban on E171, in a way similar in its influence on the use and marketing of nanofood products. Regardless of the astounding marketed applications of nanotechnology in the food sector, the future of nanotechnology in the food sector depends on the societal ethnoscape and perception, advancements in science and technology, and government policy and legislation due to the said factors. The controversy is still ongoing as to the possible dangers of conventional nanomaterials, which are still being investigated actively. One is bound to need more evidence on risk evaluation. Additionally, there have been several efforts to prevent engineered nanomaterials from being toxic and simultaneously to make the material whose actions are highly selective and reliable. For instance, functionality of surface, doping, and morphology (size and shape) control are some of the manipulations proven to make nanomaterials more sustainable and less toxic. In the next section, guidance and options of bioinspired or biosynthesized (or "green synthesized") nanomaterials for future research and exploration are outlined (Jha & Shimpi, 2018).

1.5.1 Perspectives on Biosynthesized and Bioinspired Nanomaterials

1.5.1.1 Biosynthesized Nanomaterials

Biosynthesis has become a leading method for producing various nanomaterials, offering a potentially efficient and eco-friendly alternative to traditional manufacturing techniques (Figure 1.3). Researchers have highlighted three key benefits of this approach: First, biological systems used as production hosts can act as capping, stabilizing, and reducing agents, reducing the need for harmful substances in engineering processes. Second, biosynthesis typically takes place at room temperature, normal pressure, and neutral pH, which helps conserve energy and minimizes the use of dangerous chemicals. Third, the surface functionalization of most nanomaterials produced through biosynthesis improves their biocompatibility and nontoxicity, making them particularly well-suited for medical uses. This section will provide a brief overview of these three aspects of biosynthesis in relation to nanomaterial design and engineering (Dhanker et al., 2021).

In place of chemical capping, stabilizing, or reducing, biological systems can function on their own. Functional amide and carboxyl groups found in some macro-biomolecules, such as proteins and lipids, can be adsorbed onto the surface of the nanoparticles and contribute to the stability of Au NPs and the decrease of Au^{3+}. Extracellular protein sodium dodecyl sulfate–polyacrylamide gel electrophoresis

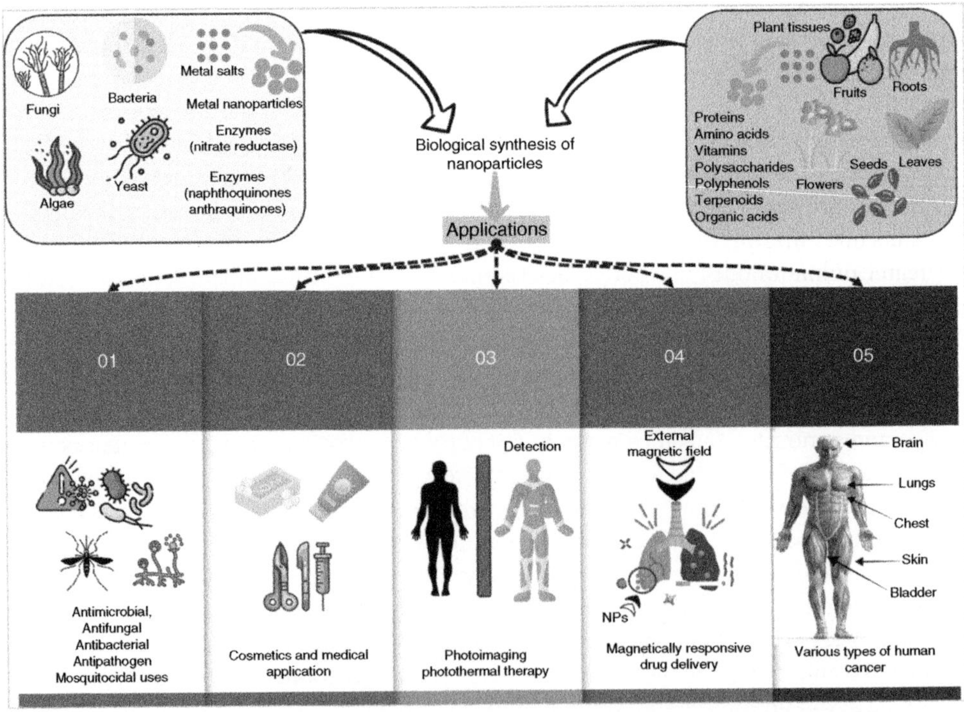

Figure 1.3 Biological synthesis and applications of nanoparticles.

(SDS-PAGE) profiles revealed biosynthesis, which stabilizes silver nanoparticles in colloidal solutions (Zaki et al., 2022). In addition, a number of other studies have found other classes of biomolecules that serve as capping agents in the synthesis of composite nanomaterials. Furthermore, biological systems generally have neutral pH, temperature, and pressure as their default conditions. The chemical synthesis of nanomaterials is routinely done at elevated temperatures, pressures, and extremes of pH. For example, the fabrication of TiO_2 NPs involves the use of solution containing the precursors, which is heated to between 100 and 400 °C for some hours, while the extreme pH conditions are used in the synthesis of ZnO and, most importantly, the surface modification of biosynthesized nanomaterials is very easily verified using Fourier transform infrared (FT-IR) spectroscopy. For instance, the FT-IR measurement results demonstrated that the biosynthesized iron NPs had limits of iron and oxygen in addition to a range of functional groups such as hydroxyl, carboxyl, carbonyl, and amine. In addition, surface engineering of nanomaterials for biocompatibility is one of the most common approaches in nanobiotechnology, which aims to imitate numerous natural bioprocesses (Omran & Omran, 2020). As a result, a biological organism often serves as a natural source for the surface engineering of nanomaterials.

Nevertheless, the biosynthesis of nanomaterials faces challenges such as minimal output and severely automated engineering processes. This is mainly due to the fact that the biological mechanisms and processes that govern nanoparticle biosynthesis are still not completely understood. It has been suggested that a multitude of biomolecules and metabolites (Panda et al., 2021), proteins, amino acids, vitamins, flavonoids, alkaloids, and polysaccharides, among others, take part in the capping, stabilizing, and metal reduction processes. The precise paths, however, are still unclear. It is also highly tough to produce the exact geometric and dimensional properties of nanostructures in biological systems. This is especially important for the biological engineering of these nanomaterials, since they need a certain morphological shape in order to act as nanocarriers. A handful of publications have sought to alter the morphology of nanomaterials in different biological systems (Kinnear et al., 2017). For example, orange extract produced silver NPs that were round and nearly spherical, while pineapple extract produced angled nanoparticles. While the results showed some degree of manufacturing control, understanding the processes at a more mechanistic level is essential. However, further work in this area with respect to biosynthesized nanomaterials is plausible.

1.5.1.2 Bioinspired Nanomaterials

It is rather common to confuse bioinspired approaches with the biosynthesis approach, where the term bioinspired approach is misused to define biosynthesis, and at times, it is referred to as bioinspired synthesis. Broadly speaking, a bioinspired approach is usually associated with the sculpturing of novel nanomaterials that have intertwined functions and shapes, for instance skin of mussel or cilia of a cell or tentacles of an insect. This approach is often taken after considering its broad application bioinspired approach in almost all fields, including medicine and engineering. Many modern structures have been designed with mimicking the the fluent bioactive multiscale interface, the same principles that let plants and animals survive throughout the years (Harun-Ur-Rashid et al., 2023). Some of those inventions include, but are not limited to, spider-web membranes, artificial compounds eyes, and gecko-inspired adhesives. For example, in biology, gecko is an interesting case study because nature designed a complex hierarchical adhesion system consisting of nanoscale fibers that produce adhesion and attachment through van der Waals forces, where it can also detach by altering the angle of load. This class of gecko adhesives is derived from those elongate nanoscale features acquired during evolution, which acted as a selective pressure over millions of years.

Like naturally occurring biological systems, novel bioinspired nanomaterials have a biotically existing function that allows them to shift their interfacial physicochemical properties in response to the changes in surrounding conditions with environmental stimuli (Chakraborty et al., 2023). Achieving that level of precision, where the environment is altered at a moment's notice in response to a change, is very complex. Replicating the dynamic nature of biological systems is indeed a challenging task. The exploration of bioinspired nanomaterials for agricultural and food applications is still relatively underdeveloped. Nonetheless, some examples of implemented artificial bioinspired nano-devices are available, including pesticide delivery systems and devices for chemical and pathogen detection and environmental monitoring.

To showcase the application of bioinspired nanostructures in food and agriculture systems, three captivating topics are examined. (1) Pesticide delivery: A bioinspired mussel ivermectin nanoparticle [P(St–MAA)–Av–Cat] was developed, offering enhanced adhesion to crop leaves, reducing soil contamination, and enabling more precise pesticide dosing. This nanostructure exhibited high avermectin retention, good storage stability, and controlled release. The toxicity of HighP (St–MAA)–Av–Cat was significantly increased under moderate conditions. (2) Environmental sensing: A novel microelectromechanical flow sensor was created at the microscale, mimicking hair cell anatomy and physiology. The bioinspired cactus $ZnSnO_3$ nanoneedles, with tip areas of 1.5–2.5 nm, can absorb and desorb water molecules based on environmental humidity. The device demonstrated rapid response and recovery times of ~2.5s and ~3s respectively, making it suitable for detecting minor moisture fluctuations. The sensor comprises tall pillars simulating vertebral microscopes and 54 short pillars for body regulation. Flow velocity and direction are determined by measuring voltage output from tensile stress applied to nanofibers at contact pads. This self-powered device shows high biocompatibility and sensitivity to flow characteristics. (3) Pathogen monitoring: In 2012, Mannoor and colleagues successfully fabricated a silk-inspired, graphene wireless pathogen sensor on tooth enamel. Silk channels filled with graphene formed a structure with a wireless coil on the tooth surface, capable of detecting single bacterial cells. While not directly applied to food and agriculture, this technology holds promise for real-time pathogen monitoring in these sectors. It is anticipated that more intricate natural bioregions will be discovered and analyzed, leading to the development of synthetic biomimetic nanomaterials that replicate these bioregions using cost-effective, high-throughput techniques yet to be perfected.

1.5.2 Regulation and Legislation

Rules and laws have paramount importance in integrating nanotechnology and marketing products because they cover both regulatory scopes. They form the basis of the public's knowledge and understanding. Legislation generally has three primary steps: (1) First, proposals are made by intellectual stakeholders, which can be a research center, an institution, governmental body, or even an individual. (2) A suggestion that is considered plausible is picked up by a government body (sometimes known as the agency of guidance) and put forth by the head of a state/country or legislative authority, to be acted upon in the form of legislation. It is then voted on by a parliament of a country or other legislative body of a government and gets converted into laws. (3) A particular legislation or law is bound to cover generalities associated with a particular sector. Therefore, when such legislation is implemented by the relevant authority, there emerges a need for regulations, which attend to specific and precise instruction, which can be quite diverse, and has a narrower focus than the original regulations (Figure 1.4).

1.5.2.1 Recent Updates

Similar protocols are employed for regulating nanotechnology in food production. Appropriate governments worldwide established or appointed responsible agencies in 2003 when the concept was first proposed. In the European Union (EU), the primary regulation is the Registration, Evaluation, Authorisation and Restriction of Chemicals (REACH), which is the most proactive regarding nanotechnology legislation in the food industry, closely followed by the United States Food and Drug Administration (FDA) (Malik, & Patil, 2020). Significant progress has been made in legislation over the past decades, as extensively covered in previous articles. While various agencies differ in their identification and expression methods, they share a common core. For instance, nanomaterials are universally defined as substances with at least one dimension measuring within the 100 nm range.

1.5.2.2 Limitation and Urgent Need for Legislation

Nanomaterials have distinct features and behaviors, which differ from bulk materials on the "standard protocol" is troublingly close to incomprehensible. This characteristic includes lack of control as well as understanding and manipulation of nanoscale functions and interactions as well as charges, which dictate the useful application, harmful effect, transformation, and the eventual fate of nanomaterials in the environment ecosystem (Zhang et al., 2018). Therefore, it is quite impossible to make any recommendations, because there exists no database or record of proven scientific results to build suggestions on. Thus, there is no way for any regulatory bodies to act or create guidelines. Each type of

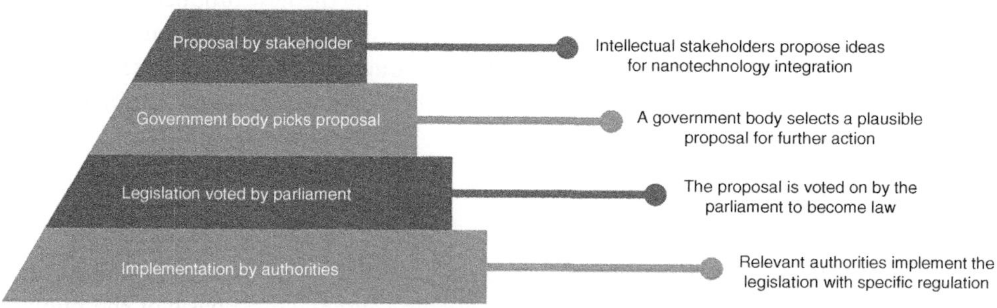

Figure 1.4 The legislative journey from proposal to regulation.

nanomaterials has its distinct behaviors, which does not help the matter as it necessitates a "case-by-case" or "product-by-product" approach for evaluation and action. With all these restrictions highlighted, current governance is still at its peak. It is possible that there exists a preliminary structure that attempts to encompass all the facets of use, threat, food safety, and waste. At present, it has coverage for incredibly vast and general situations concerning nanomaterials and nanotechnology. It requires an assessment or a fresh assessment when it comes to a specific nanomaterial. In the majority of cases, the requirements and techniques to evaluate the criterion are still unclear. The FDA does not categorically evaluate all goods containing nanomaterials or involving the application of nanotechnology as innately benign or dangerous, for example, according to the final guidelines for industry that the agency recommended. This leaves the evaluation criteria and methods open-ended (He et al., 2019).

Nevertheless, food industry stakeholders have marketed and incorporated the use of nanotechnology in food contacts such as materials for packages, sensors, and detectors. This encourages and aids the marketing of nanomaterials. In implants, these nanomaterials are used as food additives and pigments, apart from titanium dioxide and iron oxide, which are already used as food colorants. This fact is established by the food safety code. The highest level of safety is given to food products intended for human consumption. Such products undergo rigorous control and evaluation, whereas the so-called "nonedible" products (such as food contact materials) have lower requirements in terms of safety and are easier to obtain approval and be sold in the market (Precup et al., 2024). Primarily, safety (primarily from the perspective of toxicity) is the foremost consideration of these products to humans and their living and health conditions. Foods that are directly eaten have the highest level of safety scrutiny and control, whereas "side" products have lower level of safety scrutiny and control. On the other hand, the impact those products may impose on the environment and natural ecosystems is an important issue that needs to be addressed. Food manufacturers and innovative product developers do not currently prioritize comprehending the environmental fate of nanomaterials and their effects on natural ecosystems. The likelihood of this information being relevant is concerned with one interested academic research group which, therefore, makes it very precise. Improvement is possible if the data is needed to sell a certain item, as the EU commission suggested. Otherwise, the chances are that extensive use of nanomaterials would yield another catastrophe, akin to plastic waste in the oceans, which would ultimately become a tragedy for the planet. Currently, with the focus on applying nanotechnology to create new products for the food industry, folders containing templates marketing the new products increase on an annual basis (Zhu, 2025). This trend makes it more important to introduce regulations and laws into the market. Whether or not these new products will be able to capture a share in the food market also relies on the sentiments of the populace and their willingness to accept them.

1.5.3 Public Awareness and Acceptance

Some silver NPs were incorporated into some package materials because of their antimicrobial properties and these materials have become extensively used in a number of food packages (e.g. milk). The public consumed all these products without even knowing that they had added nanoparticles. This could be a moral, social, and legislative dilemma. Today, it is mandatory for manufacturers to label information that informs consumers about what they are buying. It's the manufacturer's duty to make the relevant information known and easily accessible to ensure that the public knows what they are consuming. Acknowledgment and awareness from the public is a factor that is significant yet often overlooked by many food producers. In a way, food producers prefer to keep their product innovation activities "under cover" and restrict access to proprietary information (which may in part be due to competition and trade secrets). This poses a challenge because the public generally wants to know the reasons for which a food manufacturer is launching a new product (Garcia-Garcia et al., 2021). An illustrative example is a case study in Singapore, which showed that lack of knowledge concerning nanotechnology, and its negative consequences lead to growing public negativity. When it comes to the agricultural and food organizations that are considered stakeholders, it is even more alarming that they have, according to a survey conducted in Ireland, very little knowledge about the technology of nanotechnology.

Regarding the application of nanotechnology in the food industry, two viewpoints can be observed: agreement and disagreement, or optimism and doubt. The backing for technology often depends on its specific use. Endorsement for nanotechnology in food decreased from 49% in 2005 to 32% in 2008, which remains one of the lowest levels of support for any nanotechnology application. In contrast, 73% of participants in 2008 favored food packaging that monitors and tracks food conditions (McCarron, 2016). The use of nanotechnology in food garnered the highest level of concern at 28%. A nationwide online survey in the United States revealed that consumers are unwilling to pay extra for canola oil modified or packaged using nanotechnology, including oil with nanoenhanced health-promoting droplets. While people seem to have a positive attitude towards nano-engineered canola oil, it appears that not all nano-foods are viewed equally. Currently, "organic" is likely the most common standard for healthy food. Conventional and "organic" foods are preferred (Vigar et al., 2019), particularly when compared to genetically modified foods. There is sometimes confusion between genetically modified and nano-engineered foods. This confusion primarily stems from a lack of public information and resources about food nanotechnology. The technology has not yet reached a level of maturity suitable for use in the food industry. The scarcity of scientific evidence makes it challenging, and it is considered unethical to market this ambiguous technology to the public.

1.6 Future Perspective

Nanotechnology is believed to have multifarious applications in the food industry, although this is speculation supported by little information available from laboratory sources. The practical use of nanotechnology is highly improbable, especially when considering the poor control of the nanoscale phenomena, the unknown environmental impact, and the virtually absent toxicity information. Alongside these reasons, the undeveloped and poorly definable limitations for legislation and regulation also serve as barriers for the introduction of innovative products. People don't seem to know much about food nanotechnology, and their attitude can change based on how it is presented or advocated for. Conflict arises when the public wants to be kept up to date on food nanotechnology (especially the novel products under development) while food companies try to keep this information secret (Yang, & Duncan, 2021). As for the public and food manufacturers, a comprehensive database and evidence base must be established to meet the needs of both parties, which is greatly overdue. The bioinspired approach is gaining traction in biological research and in other areas because the implementation of environmentally friendly practices has become more crucial to achieving success in the modern biotechnology industry. At present, however, the development and research of bioinspired nanomaterials for application in agriculture and food industries is very low compared to that in the biomedical fields (Chaudhry et al., 2018). The food industry, being a multi-trillion-dollar economy, has driven the global marketing of numerous products containing novel nanotechnology, especially in food contact materials and technologies (packaging materials, monitors). Due to regulatory safety control from the government, it remains a contested area for the manufacturer.

1.7 Conclusions

The over-dependence on agrochemicals has severely destroyed crop yield, which resulted in profound negative impacts on our ecosystem. The increasing population calls for more food production, which is eventually linked to food quality. As a result, nanobiotechnology has provided a possible answer to this complex problem not only in increasing growth yield but also the survival rate of plants. Due to their wide range of distinctive properties, nanomaterials can be utilized as a safe delivery system and therefore nanofertilizers, nanosensors, and nanopesticides can be used in agriculture for other purposes that may assist in supporting sustainability. On the other hand, great efforts should be made in determining the sustainability of the environment and economy when using nanomaterials on a large scale. Furthermore, nanotechnology is essential for processing, modifying, detecting, and preventing food loss through sophisticated packaging in addition to food stability. Common nanoparticles, such as Zn, Ag, TiO_2, Au, ZnO, MgO, and SiO_2, pose significant health hazards because of their ease of cellular penetration in both humans and animals. Only by using more environmentally friendly methods, such as using plant extracts to synthesize nanoparticles for increased productivity and long-term sustainability in agriculture and the ecosystem, can such health hazards be reduced.

References

Anas, M., Falak, A., Hassan, S., Khattak, W. A., Saleem, M. H., Khan, K. A., Khalid, A., & Fahad, S. (2025). Microbial interactions and bacterial responses to metal stress in plants: Mechanisms, adaptations, and applications for sustainable agriculture. *Journal of Crop Health*, *77*, 36. https://doi.org/10.1007/s10343-025-01108-w

Arora, D., Kumar, M., Bhatt, S., Gautam, R. K., & Taneja, Y. (2024). *Biodegradability and sustainability of biobased nanomaterials* (pp. 509–535). https://doi.org/10.1007/978-981-97-0542-9_16

Bolledla, N., & Bakshi, V. (2024). A discussion of the properties of nanoparticles. *International Journal of Drug Delivery Technology, 14*(03), 1814–1817. https://doi.org/10.25258/ijddt.14.3.79

Chakraborty, S., Bera, D., Roy, L., & Ghosh, C. K. (2023). Biomimetic and bioinspired nanostructures: Recent developments and applications. In *Bioinspired and green synthesis of nanostructures: A sustainable approach* (pp. 353–404). Wiley.

Chaudhry, N., Dwivedi, S., Chaudhry, V., Singh, A., Saquib, Q., Azam, A., & Musarrat, J. (2018). Bio-inspired nanomaterials in agriculture and food: Current status, foreseen applications and challenges. *Microbial Pathogenesis, 123*, 196–200.

Chauhan, P., Chauhan, E. S., & Singh, P. (2024). A review of nanotechnology applications in food processing, packaging, and preservation. *Advances in Research, 25*(6), 427–441. https://doi.org/10.9734/air/2024/v25i61215

Darbre, P. D. (2023). *Nanomaterials* (pp. 67–78). Elsevier BV. https://doi.org/10.1016/b978-0-323-99684-6.00009-4

Dhanker, R., Hussain, T., Tyagi, P., Singh, K. J., & Kamble, S. S. (2021). The emerging trend of bio-engineering approaches microbial nanomaterial synthesis and its applications. *Frontiers in Microbiology, 12*, 638003.

El-Saadony, M. T., Saad, A. M., Soliman, S. M., Salem, H. M., Desoky, E. S. M., Babalghith, A. O., El-Tahan, A. M., Ibrahim, O. M., Ebrahim, A. A. M., Abd El-Mageed, T. A., Elrys, A. S., Elbadawi, A. A., El-Tarabily, K. A., & AbuQamar, S. F. (2022). Role of nanoparticles in enhancing crop tolerance to abiotic stress: A comprehensive review. *Frontiers in Plant Science, 13*, 946717.

Erdoğar, N., Akkın, S., & Bilensoy, E. (2019). Nanocapsules for drug delivery: An updated review of the last decade. *Recent Patents on Drug Delivery & Formulation, 12*(4), 252–266. https://doi.org/10.2174/1872211313666190123153711

Eswaran, S., Anand, A., Lairenjam, G., Mohan, G., Sharma, N., Khare, A., & Bhargavi, A. (2024). Climate change impacts on agricultural systems mitigation and adaptation strategies: A review. *Journal of Experimental Agriculture International, 46*(11), 1–12. https://doi.org/10.9734/jeai/2024/v46i113021

Forouzandeh, A. (2023). Effects of high copper supplementation from different sources on the productive performance and intestinal microbiota of broilers and pigs, and its environmental dimension.

Franco-Urquiza, E. A. (2021). Clay-based polymer nanocomposites: Essential work of fracture. *Polymers, 13*(15), 2399.

Garcia-Garcia, G., Garcia-Garcia, G., Azanedo, L., & Rahimifard, S. (2021). Embedding sustainability analysis in new food product development. *Trends in Food Science and Technology, 108*, 236–244. https://doi.org/10.1016/J.TIFS.2020.12.018

Harun-Ur-Rashid, M., Jahan, I., Foyez, T., & Imran, A. B. (2023). Bio-inspired nanomaterials for micro/nanodevices: A new era in biomedical applications. *Micromachines, 14*(9), 1786.

He, X., Deng, H., & Hwang, H. M. (2019). The current application of nanotechnology in food and agriculture. *Journal of Food and Drug Analysis, 27*(1), 1–21.

Javed, T., Shabbir, R., Hussain, S., Naseer, M. A., Ejaz, I., Ali, M. M., Ahmar, S., & Yousef, A. F. (2022). Nanotechnology for endorsing abiotic stresses: A review on the role of nanoparticles and nanocompositions. *Functional Plant Biology, 50*(11), 831–849.

Jha, M., & Shimpi, N. G. (2018). Spherical nanosilver: Bio-inspired green synthesis, characterizations, and catalytic applications. *Nano-Structures & Nano-Objects, 16*, 234–249.

Kinnear, C., Moore, T. L., Rodriguez-Lorenzo, L., Rothen-Rutishauser, B., & Petri-Fink, A. (2017). Form follows function: Nanoparticle shape and its implications for nanomedicine. *Chemical Reviews, 117*(17), 11476–11521.

Li, X., Ghanizadeh, H., Han, Z., Wang, Q., Li, F., Qiu, Y., Zhang, Y., Chen, X., Liu, J., & Wang, A. (2023). Metabolic profile and physiological mechanisms underlying the promoting effects of TiO_2NPs on the photosynthesis and growth of tomato. *Scientia Horticulturae, 322*, 112394.

Li, L., Zhao, H., Wang, N., & Wang, L. (2024). Recent advances of amorphous nanomaterials: Synthesis and applications. *Chinese Journal of Chemistry*. https://doi.org/10.1002/cjoc.202400388

Lin, X., Yu, W., Tong, X., Li, C., Duan, N., Wang, Z., & Wu, S. (2024). Application of nanomaterials for coping with mycotoxin contamination in food safety: From detection to control. *Critical Reviews in Analytical Chemistry, 54*(2), 355–388.

Malik, R., & Patil, S. (2020). Nanotechnology: Regulatory outlook on nanomaterials and nanomedicines in United States, Europe and India. *Applied Clinical Research, Clinical Trials and Regulatory Affairs, 7*(3), 225–236.

McCarron, E. (2016). *Nanotechnology and food: investigating consumers' acceptance of foods produced using nanotechnology* (Doctoral dissertation, Dublin Business School).

Mubarik, N., Gulelala, G., Iqbal, S., Shahmeel, M., Hussain, A. A., Razzaq, K., & Akram, M. (2022). Different methods, novel tools towards the synthesis of nanoparticles and applications in engineering, chemical. *Physical Sciences and Technology. Scholars Bulletin, 8*(2), 71–74. https://doi.org/10.36348/sb.2022.v08i02.004

Omran, B. A., & Omran, B. A. (2020). Fundamentals of nanotechnology and nanobiotechnology. In *Nanobiotechnology: A multidisciplinary field of science* (pp. 1–36). Springer.

Padil, V. V., Kumar, K. A., Murugesan, S., Torres-Mendieta, R., Wacławek, S., Cheong, J. Y., Černík, M., & Varma, R. S. (2022). Sustainable and safer nanoclay composites for multifaceted applications. *Green Chemistry, 24*(8), 3081–3114.

Panda, P. K., Verma, S. K., & Suar, M. (2021). Nanoparticle–biological interactions: the renaissance of bionomics in the myriad nanomedical technologies. *Nanomedicine, 16*(25), 2249–2254.

Pietroiusti, A., Stockmann-Juvala, H., Lucaroni, F., & Savolainen, K. (2018). Nanomaterial exposure, toxicity, and impact on human health. *Wiley Interdisciplinary Reviews: Nanomedicine and Nanobiotechnology, 10*(5), e1513.

Precup, G., Marini, E., Zakidou, P., Beneventi, E., Consuelo, C., Fernández-Fraguas, C., Garciarena, I. N., Fernandez, P. R., Torres, R. R., Rossi, A., Ruggeri, L., Suriano, F., Ververis, E., Liu, Y., Smeraldi, C., & Germini, A. (2024). Novel foods, food enzymes, and food additives derived from food by-products of plant or animal origin: Principles and overview of the EFSA safety assessment. *Frontiers in Nutrition, 11*, 1390734.

Preetha, S. S., & Narayanan, R. (2020). Factors influencing the development of microbes in food. *Shanlax International Journal of Arts, Science and Humanities, 7*(3), 57–77.

Rashid, M. M., Forte Tavčer, P., & Tomšič, B. (2021). Influence of titanium dioxide nanoparticles on human health and the environment. *Nanomaterials, 11*(9), 2354.

Shivakumar, N., Madhusudan, P., & Daniel, S. K. (2018). Nanomaterials for smart food packaging. In *Handbook of nanomaterials for industrial applications* (pp. 260–270).

Siddiqui, S., & Alrumman, S. A. (2021). Influence of nanoparticles on food: An analytical assessment. *Journal of King Saud University, Science, 33*(6), 101530.

Spielman-Sun, E., Lombi, E., Donner, E., Howard, D., Unrine, J. M., & Lowry, G. V. (2017). Impact of surface charge on cerium oxide nanoparticle uptake and translocation by wheat (*Triticum aestivum*). *Environmental Science & Technology, 51*(13), 7361–7368.

Sune, P. R., Jumde, K. S., Hatwar, P. R., Bakal, R. L., More, D. S., & Korde, A. V. (2024). Nanoparticles: Classification, types and applications: A comprehensive review. *GSC Biological and Pharmaceutical Sciences, 29*(3), 190–197. https://doi.org/10.30574/gscbps.2024.29.3.046

Thakur, M., Wang, B., & Verma, M. L. (2022). Development and applications of nanobiosensors for sustainable agricultural and food industries: Recent developments, challenges and perspectives. *Environmental Technology & Innovation, 26*, 102371.

Thapa, M. (2021). Zinc Sulphide nanoparticle as an alternative source of Zinc micronutrient for crop growth and yield: Application and evaluation.

Vigar, V., Myers, S., Oliver, C., Arellano, J., Robinson, S., & Leifert, C. (2019). A systematic review of organic versus conventional food consumption: Is there a measurable benefit on human health? *Nutrients, 12*(1), 7.

Wasilewska, A., Bielicka, M., Klekotka, U., & Kalska-Szostko, B. (2023). Nanoparticle applications in food–A review. *Food & Function, 14*(6), 2544–2567.

Yalcinkaya, D. B., Dogan, S., Özalp, V. C., & Tuna, B. G. (2024). *Targeted nanocarriers*. In *Characterisation of drug nanocarriers* (pp. 297–326). Royal Society of Chemistry. https://doi.org/10.1039/9781837672981-00297

Yang, T., & Duncan, T. V. (2021). Challenges and potential solutions for nanosensors intended for use with foods. *Nature Nanotechnology, 16*(3), 251–265.

Younes, M., Aquilina, G., Castle, L., Engel, K. H., Fowler, P., Fernandez, M. J. F., Fürst, P., Gundert-Remy, U., Gürtler, R., Husøy, T., Manco, M., Mennes, W., Moldeus, P., Passamonti, S., Shah, R., Waalkens-Berendsen, I., Wölfle, D., Corsini, E., Cubadda, F., De Groot, D., ... Wright, M. (2021). EFSA panel on food additives and flavourings (FAF), safety assessment of titanium dioxide (E171) as a food additive. *EFSA Journal, 19*(5), e06585.

Yu, H., & Feng, W. (2024). Carbon-based multifunctional nanomaterials. *Nanomaterials, 14*(19), 1600. https://doi.org/10.3390/nano14191600

Zaki, A., Aziz, M. N., Ahmad, R., Ahamad, I., Ali, M. S., Yasin, D., Afzal, B., Ali, S. M., Chopra, A., Hadda, V., Srivastav, P., Kumar, R., & Fatma, T. (2022). Synthesis, purification and characterization of Plectonema derived AgNPs with elucidation of the role of protein in nanoparticle stabilization. *RSC Advances, 12*(4), 2497–2510.

Zhang, J., Guo, W., Li, Q., Wang, Z., & Liu, S. (2018). The effects and the potential mechanism of environmental transformation of metal nanoparticles on their toxicity in organisms. *Environmental Science: Nano, 5*(11), 2482–2499.

Zhao, L., Lu, L., Wang, A., Zhang, H., Huang, M., Wu, H., Xing, B., Wang, Z., & Ji, R. (2020). Nano-biotechnology in agriculture: Use of nanomaterials to promote plant growth and stress tolerance. *Journal of Agricultural and Food Chemistry, 68*(7), 1935–1947.

Zhu, S. (2025). Nanotechnology on the food industry: Applications and challenges. *Applied and Computational Engineering, 122*(1), 124–129. https://doi.org/10.54254/2755-2721/2025.19707

2

Integrating Nanobased Engineering Advances to Enhance Crop Yields and Optimize Food Production Systems

Nawab Ali[1], *Younsuk Dong*[1], *Jalal Bayar*[2], *Muhammad Mehran Anjum*[3], *Abdul Haq*[4], *Gul Roz Khan*[4], *and Rovaid Ali*[5]

[1] *Biosystems and Agricultural Engineering, Michigan State University, East Lansing, USA*
[2] *Department of Agronomy, The University of Agriculture, Peshawar, Pakistan*
[3] *AGROSCOPE, Climate Change and Agriculture Research Division, Agroecology and Environment, Zurich, Switzerland*
[4] *Pakistan Council of Scientific and Industrial Research, Laboratories Complex, Peshawar, Pakistan*
[5] *Department of Horticulture, The University of Agriculture, Peshawar, Pakistan*

2.1 Introduction to Nanobased Engineering in Agriculture

Agriculture has long been a cornerstone of human civilization, providing the food and resources necessary for survival and economic development. However, as the world's population continues to rise, there is an increasing need for food production. To meet this demand, traditional agricultural practices must evolve, integrating more advanced, efficient, and sustainable technologies. One such technological approach is nanobased engineering, the manipulation and application of nanotechnology to improve various agricultural processes. Nanotechnology, the science of designing, producing, and utilizing structures, devices, and systems by controlling material properties at the nanometer scale (1–100 nm), has shown immense potential in revolutionizing agriculture. Agricultural practices utilizing engineered nanomaterials can adapt to changing environmental and climatic conditions (Mittal et al., 2020). Nanobased engineering encompasses a broad spectrum of applications, including crop management, pest control, soil remediation, and the enhancement of plant growth. The goal of these applications is to increase agricultural productivity and sustainability by using the special qualities of nanomaterials, such as their enormous surface area, and capacity to connect to biological systems at the molecular level. The integration of nanotechnology in agriculture addresses many present-day challenges, such as reducing environmental pollution, optimizing water usage, and improving crop resilience. For example, nanomaterials can be used in fertilizers and pesticides to enhance their efficiency and diminish their environmental impact (Rana et al., 2024). Additionally, nanotechnology has the potential to transform precision farming by allowing more precise observing of soil health, crop growth, and water usage (Sharma and Kumar, 2024).

Nanotechnology Innovations for Food Security and Sustainable Agriculture, First Edition. Edited by Zia Ur Rehman Mashwani, Shah Fahad, Ilyas Ahmad, Shah Saud, Sajid Fiaz, and Taufiq Nawaz.

In agriculture, nanotechnology can be applied across various stages, including crops' vegetative stage, production, harvesting, storage, packaging, and transportation of cultivated products (Joshi et al., 2019). Nanotechnology plays a significant role in enhancing crop yields by improving plant growth and increasing resistance to stress factors such as drought, pests, and diseases. By making nanoparticles with specific characteristics, it is possible to influence plant biology at the molecular and cellular levels. These materials can be applied to seeds, soil, or plants, promoting more efficient nutrient uptake, improving photosynthesis, and enhancing plant immunity. For example, nanomaterials facilitate the slow release of nutrients, ensuring that plants receive the optimal amount precisely when needed. This controlled release prevents nutrient leaching, a common problem with conventional fertilizers, and improves fertilizer efficiency, leading to increased crop yields while reducing environmental influence. Nanoparticles can also increase the surface area of fertilizers, making them more bioavailable to plants. Furthermore, nanotechnology enables the development of nano-encapsulated pesticides, which can target particular pests or diseases without damaging beneficial organisms. This targeted method lessens the need for broad-spectrum insecticides, which frequently harm the environment and encourage the rise of pest resistance. Nanobased solutions can be more efficient, using smaller quantities to achieve the same or better results, leading to a more sustainable approach to pest control (Table 2.1).

Nanotechnology in agriculture aims to improve crop yields by enhancing nutrient uptake, pest control, and stress tolerance, ultimately leading to higher productivity. Additionally, it focuses on optimizing resource use through efficient water and nutrient management, ensuring sustainable farming practices. Moreover, nanotechnology helps lessen environmental impact by reducing the use of chemical inputs and decreasing runoff, contributing to a more eco-friendly and resilient agricultural system.

Table 2.1 Different nanomaterials used in agriculture.

Category	Details	References
Fertilizers	Urea nanoparticles for rice and maize	Singh et al. (2023)
Micronutrient supply	Iron nanoparticles for wheat growth	Feng et al. (2022)
Insect management	Nano-chlorpyrifos against *Rhynchophorus ferrugineus* (RPW)	Abd El-Fattah et al. (2019)
Nano-fungicides	Nano-silica (nano-SiO_2) and nano-selenium (nano-Se) in common bean (*Phaseolus vulgaris* L.) against the leaf spot disease caused by *Alternaria alternata*	Taha et al. (2023)
Nano-herbicides	Zinc oxide against *Cucuminus sativus*	Zhao et al. (2014)
Biosensors	Graphene-based biosensors for pesticide detection	Xu et al. (2020)

2.1.1 Importance of Nanobased Engineering in Modern Agriculture

The global population is estimated to extent nearly 10 billion by 2050, leading to a 50% increase in agricultural demand compared to 2013 (FAO, 2017). Traditional agricultural practices are often unsustainable, leading to soil degradation, water scarcity, and environmental pollution. Nanobased engineering provides a pathway to address these issues by enabling precise and efficient management of agricultural resources (Table 2.2). The combination of nanobased methods into agricultural systems has the potential to increase economic growth by up to 3 trillion USD in recent years (Yadav et al., 2023).

2.1.2 Key Concepts and Definitions

- **Nanomaterials**: Materials with minimum one dimension in the nanoscale, exhibiting unique physical, chemical, and biological properties.

Table 2.2 Applications of nanotechnology in agriculture and their benefits.

Category	Key Elements	Description
Nanotechnology in smart agriculture	• Nanotechnology-based materials	• Utilization of nanotechnology for enhanced agricultural productivity and sustainability
Enrichment of soil	• Nanoparticles and nanofertilizers • Soil application and foliar application	• Enhances plant growth and development • Enriches soil profile • Improves microbial diversity • Increases bioavailability of nutrients • Supports sustainability
Precision farming	• Nanosensors • Soil, crop, and yield mapping	• Detects soil quality • Early detection of plant diseases • Real-time monitoring of nutrients
Protected crop	• Nanopesticides	• Uses nanoparticles to target pests • Reduces disease in plants • Promotes healthy plant growth
Crop growth and promotion	• Nanotechnology-based fertilizers • Enhanced plant productivity	• Boosts plant health and productivity
Crop improvement	• Nano-biotechnology	• Genetically enhances crop traits • Improves resistance to environmental stress
Stress mitigation	• Protection against heat, cold, drought • Prevents chilling injury	• Helps plants tolerate extreme weather conditions
Precision breeding	• Transgenic plants	• Uses nanotechnology for genetic modifications to improve crop traits

- **Nanofertilizers**: Fertilizers engineered at the nanoscale to increase nutrient uptake and lessen environmental influence.
- **Nanopesticides**: Pesticides formulated with nanomaterials to enhance efficacy and reduce toxicity.
- **Nanosensors**: Devices that detect and respond to physical, chemical, or biological signals at the nanoscale.

2.2 Nanomaterials for Enhanced Soil Health

The health of soil is one of the most crucial factors in determining crop productivity. Healthy soils provide essential nutrients, water retention, and a balanced microbial environment for plant growth. However, over time, soil degradation has become a widespread issue, with factors such as erosion, overuse of chemical fertilizers, and contamination leading to a decline in soil quality. In this context, nanomaterials offer an innovative solution to enhance soil health and restore fertility.

Nanomaterials can be used to improve the physical, chemical, and biological properties of soil. For example, platinum nanoparticles increase the strength of soil structure and the saturated hydraulic conductivity by reducing the space among soil units (Suazo-Hernández et al., 2023). Additionally, nanomaterials can increase the amount of soil organic matter by fortifying the link among soil units or by lowering bulk density (Bayat et al., 2018; Kubayat et al., 2020). Nanoclays, for example, have been found to increase soil structure by improving its water-holding capacity, reducing erosion, and enhancing aeration. These properties are specifically valuable in regions experiencing water scarcity or in areas with sandy or loamy soils disposed to erosion. Nanoclays can be integrated into soil to increase its ability to retain water, promoting better crop growth in arid conditions.

In addition to improving soil structure, nanomaterials can also enhance soil nutrient availability. It has been demonstrated that nano-sized biochar, a type of charcoal made by heating organic material without oxygen, enhances soil nutrient retention and lessens nutrient leaching. Nano-biochar has become a promising alternative for soil amendment, effective enzyme support, and a suitable adsorbent due to its physical and chemical properties as well as its permeable nature (Chausali et al., 2021). This material has the unique ability to adsorb and hold onto crucial nutrients such as nitrogen, phosphorus, and potassium, making them more readily accessible to plants. Furthermore, biochar can enhance soil pH, reducing acidity or alkalinity and improving the overall health of the soil ecosystem.

Nanotechnology also has significant implications for soil remediation. Contaminants like heavy metals, pesticides, and industrial chemicals can build up in the soil, lessening its fertility and posing risks to both human and animal health. Nanomaterials, particularly nanoparticles of metals and metal oxides, have shown promise in removing these contaminants through processes such as adsorption, chemical reactions, and photodegradation. According to Sharma and Kumar (2024), sorption–desorption, complexation with organic ligands, ion exchange, and surface precipitation interactions with soil constituents are the main mechanisms that control the activity and toxicity of soil pollutants, which include both organic and inorganic compounds. Nanoparticles of iron oxide can be used to remove heavy metals like lead aqueous solution and wastewater (Feng et al., 2022), while nanosilver particles can help break down organic pollutants.

Another critical aspect of soil health is the microbial community that resides within the soil. Nanotechnology can also help improve soil and plant health by improving the microbiome, enabling sophisticated crop management through the integration of nanosensors into plants, and reducing losses through increased crop efficiency and resilience. Beneficial soil microbes play a vital role in nutrient cycling, decomposition, and disease suppression. Some nanomaterials exhibit growth promotion and positive effects, while others are microbicidal in nature and can specifically target a particular class of microorganisms (Khan et al., 2018). Beneficial microorganisms like nitrogen-fixing bacteria or mycorrhizal fungi are frequently included as nanofertilizers in nanoscale carriers (Rezaei and Khanmirzaei 2024). The work of Saberi-Rise and Moradi-Pour (2020), which used sodium alginate, starch, and bentonite to make a nano-biofertilizer with a suspension of plant growth promoting rhizobacteria (PGPR), is a relevant example. Nanoparticles of essential elements such as copper and zinc can stimulate microbial growth and improve their nutrient cycling capacity. At the same time, nanoparticles can be engineered to minimize any negative impact on non-target organisms, ensuring that only the beneficial microbes are encouraged.

Nanofertilizers represent a cutting-edge advancement in modern agriculture, offering a favorable solution to improve crop productivity while reducing environmental influences. These fertilizers are engineered at the nanoscale, allowing for better solubility, higher absorption rates, and reduced wastage. The application of nanotechnology in fertilizers has gained significant consideration due to its potential to revolutionize plant nutrition. Nanofertilizers increase soil fertility and plant growth while decreasing nutrient losses from leaching, runoff, and volatilization. They have the potential to completely transform agriculture (Saurabh et al., 2024) (Figure 2.1). Nanofertilizers can be categorized into three main types: nanoparticle-based fertilizers, nanoscale coatings on conventional fertilizers, and nanosized nutrient carriers. These formulations enhance nutrient uptake, reduce fertilizer runoff, and promote

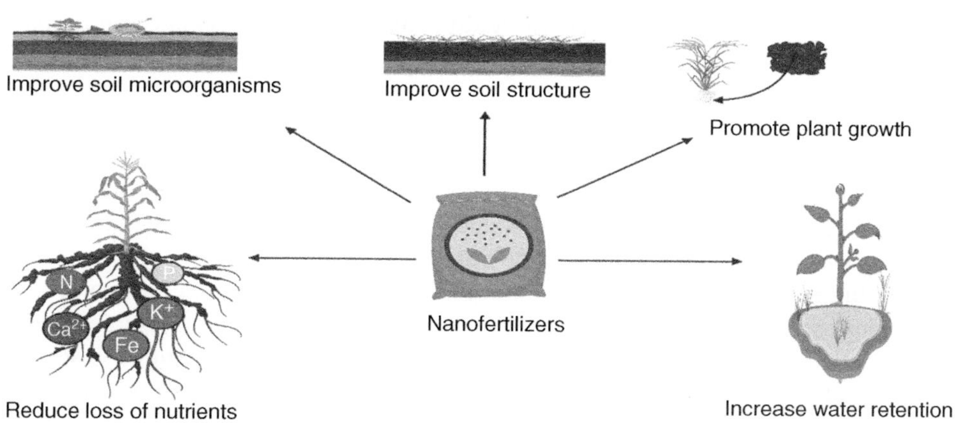

Figure 2.1 Potential applications of nanomaterials in agriculture for improving soil and plant health.

sustainable agricultural practices. Furthermore, these fertilizers can be engineered to release nutrients in response to specific environmental triggers, enhancing their effectiveness and minimizing reliance on chemical inputs.

As global agriculture grapples with climate change, soil degradation, and increasing food demand, nanofertilizers emerge as a sustainable alternative to conventional fertilizers. Their precise nutrient delivery mechanisms and reduced environmental footprint make them a key component of future agricultural strategies aimed at improving crop yields while maintaining an ecological balance.

2.3 Nano-Enhanced Crop Protection

Crop protection is one of the cornerstones of modern agriculture. The effectiveness of conventional pesticides in controlling crop-damaging pests, diseases, and weeds is often compromised by issues such as the development of resistance, toxicity to non-target organisms, and environmental pollution. Nanotechnology offers advanced solutions to these issues by enabling the development of nanobased pesticides, herbicides, and fungicides that are more proficient, targeted, and environmentally friendly.

Nanoparticles can be engineered to deliver active ingredients more effectively, reducing the need for large quantities of chemicals. Nanocarriers such as liposomes, dendrimers, and nanoparticles have been utilized to encapsulate pesticides, allowing for controlled and slow release over extended periods. This controlled discharge guarantees that the active ingredients are more effective at killing pests while minimizing their impact on the surrounding ecosystem. Savi et al. (2015) examined the efficiency of zinc compounds in reducing the development of deoxynivalenol and wheat Fusarium head blight. Wheat seedlings infected with *Pseudocercosporella herpotrichoides* had a defensive reaction when metal nanoparticles (Zn, Ag, Fe, Mn, and Cu) were applied to the seed prior to planting (Panyuta et al., 2016). Because of their unique shape and characteristics, nanosized particles are believed to investigate pesticide actions in novel nanocarrier formulations based on a variety of materials, including silica, lipids, polymers, copolymers, ceramic, metal, carbon, and others (Agostini et al., 2012). Additionally, these nano-enhanced formulations can reduce the leaching of pesticides into soil and water systems, decreasing the risk of contamination.

Furthermore, nanoparticles can also be used in combination with natural compounds such as plant extracts and biological agents to create biopesticides. By delivering these agents in nanoparticle form, their stability, efficacy, and bioavailability can be significantly improved. This leads to more sustainable pest management practices that reduce reliance on harmful chemicals and support integrated pest management (IPM) strategies. In addition to pest control, nanotechnology can be used to improve crop protection against abiotic stresses such as salinity, temperature and extreme drought. Nano-enabled stress mitigation strategies, such as nanomaterials that enhance water retention in soil or improve plant antioxidant systems, can help crops thrive in challenging growing conditions.

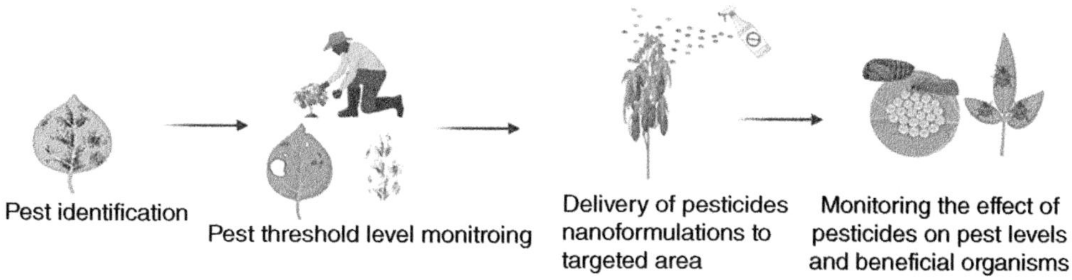

Figure 2.2 Application of nanotechnology-based pesticides for sustainable pest control and their impact on crop protection.

2.3.1 Types of Nano-Enhanced Crop Protection

Crop protection is essential to ensure food security. Traditional pesticides often have negative environmental impacts and can lead to resistance in pests. Nano-enhanced crop protection offers a more targeted and efficient approach (Figure 2.2).

1) **Nanopesticides**: Nanopesticides are a novel class of pesticides that utilize nanotechnology to enhance the efficacy, stability, and targeted supply of active ingredients while minimizing environmental influence. These pesticides are formulated using nanocarriers such as nanoparticles, nanoemulsions, nanogels, and nanocapsules, which improve solubility, controlled release, and bioavailability of active compounds. Unlike conventional herbicides, which frequently suffer from problems such as leaching, volatilization, and nontarget effects, nanopesticides offer better solubility, targeted delivery, and prolonged activity, thereby reducing overall herbicide application rates and mitigating environmental contamination. Nanopesticides offer solutions to these challenges by ensuring precise targeting of pests, reducing the total chemical burden, and enhancing the sustainability of pest control strategies. Their controlled-release mechanisms allow for prolonged activity, reducing the frequency of applications and minimizing resistance development in pests.
2) **Nanoherbicides**: Nanoherbicides involve the application of nanoparticles or nanoformulated herbicidal compounds that enhance the precision, efficacy, and controlled release of active ingredients. The incorporation of nanotechnology in herbicide formulations can take various forms, including nanoencapsulation, nanocarriers, and nanomaterial-based herbicidal agents. These advanced formulations improve the bioavailability of active compounds, enable site-specific action, and enhance weed suppression while minimizing toxicity to non-target organisms. Furthermore, nanoherbicides have the potential to combat herbicide resistance, a growing concern in modern agriculture, by improving the mechanism of herbicide uptake and action.
3) **Nanofungicides**: Nanofungicides represent an innovative advancement in the field of agriculture, combining the efficacy of traditional fungicides with the unique properties of nanotechnology. Nanofungicides are designed to target fungal pathogens more precisely, providing effective disease control with potentially lower chemical inputs, thereby reducing environmental and health risks. Their ability to penetrate plant tissues and deliver active ingredients in a controlled manner makes them an attractive solution

for IPM strategies. With the growing concern over the environmental impact of traditional chemical pesticides, nanofungicides are emerging as a favorable and sustainable substitute for protecting crops and ensuring food security.

2.4 Nanotechnology in Irrigation Systems

Water scarcity is a growing global concern, particularly in regions that face frequent droughts or limited access to freshwater resources. Agriculture is a major consumer of water, and inefficient irrigation practices are one of the foremost reasons for water wastage in farming systems. To optimize water use, reduce consumption, and ensure crops receive adequate hydration, nanotechnology is being integrated into irrigation systems to improve their efficiency and sustainability (Figure 2.3).

One of the key innovations in this area is the development of nanomaterials for water purification. Three key applications of nanotechnology in water treatment are pollution monitoring using nanosensors and detectors specific to certain contaminants, pollution prevention, and remediation by eliminating or partially removing pollutants (Nishu and Kumar, 2023). Nanomaterials have emerged as a promising solution due to their unique physicochemical properties, such as high surface area, enhanced reactivity, and tunable porosity. These materials, which include nanoparticles, nanocomposites, carbon-based nanomaterials, and membrane-based nanostructures, offer superior performance in removing contaminants such as heavy metals, organic pollutants, and microbial pathogens. By leveraging adsorption, catalysis, and membrane filtration mechanisms,

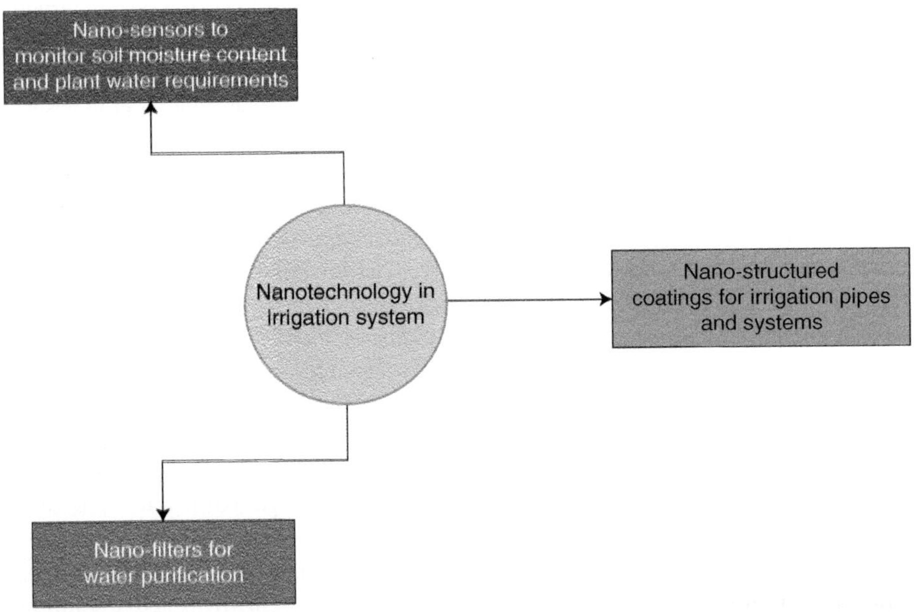

Figure 2.3 Applications of nanotechnology in irrigation systems for precision water management.

nanomaterials significantly improve water treatment efficiency, providing sustainable and cost-effective solutions for clean water access. This ensures that irrigation water is of high quality and reduces the risks of soil contamination or crop damage from polluted water. Zhang (2011) found that nitrates can be reduced and eliminated from soils, sediments, and water by using nanoparticulate zero-valent iron. At the nanoscale, zero-valent iron is frequently selected for nanoremediation because of its huge surface area, greater number of reactive sites than microsized particles, and property of both adsorption and reduction (Rajan, 2011).

Nanotechnology-based sensors have emerged as a revolutionary tool for detecting, analyzing, and monitoring various physical, chemical, and biological parameters with high precision and sensitivity. Sensors based on nanotechnology have been created to measure soil moisture in real time, permitting for effective irrigation control and water conservation (Parameswari et al., 2024). These sensors influence the distinctive properties of nanomaterials, such as a high surface-to-volume ratio, enhanced electrical conductivity, and superior mechanical strength, to achieve unparalleled performance compared to conventional sensors. Plant nanosensors can facilitate communication between farmers and plants, thereby guaranteeing plant sustainability and protection (Khandelwal et al., 2019). When used appropriately, nanosensors can increase crop yields, which will benefit farmers (Muthumalai et al., 2022). The integration of nanotechnology in sensor development has led to significant advancements in multiple fields, including agriculture, environmental monitoring, healthcare, and industrial applications. Among, the nanobased systems used for automated evaluation of the nutritional balance, stress levels, and health condition of plants are nano-biosensors, global positioning systems, and global navigation satellite systems (Ijaz et al., 2023). In agriculture, nanotechnology-based sensors are used for actual soil nutrient valuation, precision irrigation, and early disease detection in crops. This introduction explores the fundamental principles, types, and applications of nanotechnology-based sensors, highlighting their role in advancing modern scientific and technological innovations.

Another promising application is the use **of nano-structured coatings for irrigation** pipes and systems. These coatings, made from hydrophobic or superhydrophobic materials, can stop the accumulation of dirt, algae, and scale within irrigation infrastructure. This guarantees that water flows efficiently through the system without clogging, reducing the need for frequent maintenance and ensuring optimal water delivery.

2.5 Advancements in Crop Improvement

As global agricultural challenges intensify due to climate change, soil degradation, and increasing food demand, traditional methods of crop breeding and genetic modification are being supplemented by innovative approaches. Among these, nanotechnology has emerged as a transformative approach in modern agriculture, offering novel solutions to enhance crop productivity, resilience, and sustainability. It plays a vital role in advancing crop improvement by improving the efficiency of genetic engineering, facilitating precise nutrient supply, and enhancing plant growth and development. The use of nanoparticles can boost seed germination, improve water and nutrient uptake, and increase photosynthetic efficiency, eventually leading to higher crop yields. Additionally, nanomaterials can be engineered to protect plants from biotic and abiotic stresses, such as pathogens, drought, and extreme temperatures, thereby strengthening crop resilience in challenging environments.

Furthermore, nanoscale innovations contribute to sustainable farming by decreasing excessive fertilizer and pesticide use, minimizing environmental contamination, and promoting eco-friendly agricultural practices. As research in agricultural nanotechnology continues to expand, its potential to transform food production and ensure global food security becomes increasingly evident. One of the most exciting areas of research is the use of nanomaterials for gene delivery. The delivery of genetic material into plant cells is a crucial step in genetically modifying crops. Traditional methods such as Agrobacterium-mediated transformation and particle bombardment can be inefficient and expensive. However, nanotechnology provides a promising alternative by using nanoparticles as carriers to deliver DNA, RNA, or other genetic material directly into plant cells. Recently, nanomaterials like layer double hydroxides (LDHs), carbon nanotubes (CNTs), mesoporous silica particles (MSNs), and AuNPs have shown great promise as vectors for efficiently delivering genome engineering tools (DNA, RNA, proteins, and RNPs) to plants in a species-independent manner (Zhi et al., 2022). This process is not only more efficient but can also be more precise, reducing the risk of off-target effects and unintended genetic changes. Several nanoparticles, as well as polymers, liposomes, and inorganic nanoparticles, function as non-viral nanocarriers and have demonstrated exceptional gene delivery capabilities (Jat et al., 2020). Silica nanoparticles (SiNPs) upregulated the expression of basic genes associated with soluble sugars (SPS, SUS, and α-glucosidase), chlorophyll (CHLH, CAO, and POR), auxin (TIR1), and abscisic acid (PYR/PYL, PP2C, SnRK2, and ABF) (Li et al., 2023).

Moreover, nanotechnology has the potential to enhance plant growth and productivity through the use of nanofertilizers and nanonutrients. These nanomaterials can supply nutrients to plants more efficiently by improving their bioavailability and providing a controlled discharge of crucial elements. This leads to better nutrient uptake, reduced nutrient loss to the environment, and improved crop yields. At every stage of agriculture, from seed sowing to post-harvesting, a range of nanotechnology applications will improve agricultural productivity and quality (Singh et al., 2021). In addition to macronutrients such as NPK, nanotechnology can also help deliver micronutrients like zinc, iron, and copper, which are often deficient in soils and critical for plant health. Silicon nanoparticles elevate auxin (IAA) and fructose levels, which directly or indirectly stimulate wheat growth (Li et al., 2023).

Nanotechnology can also be used to improve crop resilience to biotic and abiotic stresses. For example, the application of nanoparticles can enhance plant natural defense mechanisms against pathogens or pests. Nanotechnology has been incorporated into several scientific disciplines to develop advanced products and technologies that enhance crop plant traits (Fraceto et al., 2016). By delivering nutrients precisely, developing cost-effective resource recovery plans, turning plants into smart sensors that can anticipate impending stress, genetically modifying more resilient cultivars, improving crop management to prevent crop losses, and boosting the microbiome to restore soil quality and boost plant growth, nanomaterials can improve nutrient utilization (Pal et al., 2023). Additionally, nanomaterials such as silica nanoparticles improve drought tolerance by promoting water retention in plant tissues. This makes crops more resilient to fluctuating environmental conditions, reducing the impact of climate change on agricultural productivity.

Finally, nanobased diagnostic tools are being developed to monitor plant health and detect diseases at initial stages. These diagnostic tools, which use nanosensors and nanobiosensors, can deliver actual data on the presence of pathogens, pests, or environmental stresses, allowing for timely interventions and improving crop management practices.

2.6 Integration of Nanotechnology with Precision Agriculture

Precision agriculture refers to the application of technology and data analysis to enhance farming practices, improving crop yields, reducing costs, and minimizing environmental impact. Nanotechnology, which includes the operation of material at the molecular or atomic scale, can offer novel solutions to address critical challenges in modern farming, including resource management, pest control, and crop protection. Real-time plant health monitoring makes it possible to manage resources effectively before they are lost, which significantly increases plant growth and survival. With increased control over resource management, soil health, crop development through genetic engineering, and the strategic provision of agrochemicals like fertilizers, it has cleared the path for sustainable agriculture (Giraldo et al., 2019; Bartolucci et al., 2022). Precision farming has been transformed by nanotechnology, which makes it easier to use resources efficiently, lowers expenses, and has fewer negative effects on the environment (Zain et al., 2024). It is expected that the growth of remote sensing and the incorporation of data from remote sensing into decision support tools for modern agricultural systems will enhance crop production and management while lowering operational expenses and their negative effects on the environment (Chlingaryan et al., 2018). Over the past 50 years, models of agricultural systems have developed that address issues in the past, present, and future at field, farm, landscape, regional, and global spatial scales (Jones et al., 2017).

Precision agriculture is a paradigm change in agricultural management that seeks to balance the need to protect ecosystems, biodiversity, and natural resources with the growing demand for food (Khose et al., 2023). The use of nanosensors in precision farming has the potential to increase precision methods. According to Soutter (2021), the use of GPS-based nanosensors has shown promise in identifying soil health and crop growth. Additionally, he revealed that the use of smart sensors related to precision farming significantly enhanced field productivity. Nanofertilizers have been developed to improve photosynthetic activity, seed germination rates, and seedling growth in response to the adverse environmental effects of traditional agricultural methods (Thathsarani, 2021). Nanomaterials have the potential to enhance soil quality, improve irrigation efficiency, supply nutrients more efficiently to plants, and offer unique pest control solutions. The combination of nanotechnology with the precision agriculture framework enables farmers to observe and cope agricultural systems with unparalleled accuracy, leading to additional sustainable and proficient farming practices. This integration involves the use of various nanomaterials such as nanoparticles, nanofibers, and nanocomposites, which can interact with plant tissues, soil, and microorganisms in ways that traditional technologies cannot. Through innovative methods such as smart sensors, nanoscale fertilizers, and nanoencapsulation, nanotechnology enables precision delivery of inputs and actual monitoring of agricultural methods, ensuring that resources are used optimally. Furthermore, nanotechnology can help in producing more environmental friendly and effective pest management strategies by minimizing chemical pesticide use, which in turn contributes to sustainable agricultural practices.

2.7 Case Studies and Practical Applications

Incorporating real-world case studies into the discussion of nanotechnology in precision agriculture provides insight into the practical applications of these technologies. Several studies and field trials around the world are already showing the potential benefits of nanotechnology, from improving crop yields to enhancing pest control and optimizing irrigation. Examining these case studies helps to validate the feasibility of integrating nanotechnology into agricultural practices and highlights the impact of these technologies on crop productivity, sustainability, and overall farm management.

Some case studies have focused on using nano-enabled sensors to monitor soil moisture and nutrient levels, providing growers with real-time data on their crops' health and needs. Smart nanobiosensors make it possible to precisely perform the laborious task of routine plant monitoring (Mittal et al., 2020). Alvarado et al. (2019) have proposed combining nanobiosensors with information technology in order to help farmers in rural places.

Other studies have explored the use of nano-encapsulated fertilizers that deliver nutrients more efficiently, thereby reducing the quantity of fertilizers needed and minimizing their negative environmental impact. According to Ramirez-Rodriguez et al. (2020), a simple and environmentally friendly synthetic method of doping CaP with potassium and nitrogen can be used to create multi-nutrient nanofertilizers, which release essential plant macronutrients gradually and can lessen the quantity of nitrogen given to durum wheat by 40% compared to conventional treatment without affecting the final grain weight per plant in any way, making their application difficult. Another study by Delfani et al. (2014) examined the effects of nano and conventional forms of iron (Fe) and magnesium (Mg) as foliar applications on black-eyed peas. Three replicates were used in a factorial experiment with varying doses of Mg and Fe. Data was collected at 85 days following planting, and the treatments were supplied to the leaves at 56 and 72 days. The results showed that Fe had a substantial impact on yield, chlorophyll content, leaf iron content, stem magnesium content, and plasma membrane stability. Two treatment combinations, $0.5\,g\,L^{-1}$ standard Fe + 0.5% nano-Mg and $0.5\,g\,L^{-1}$ standard Fe + $0.5\,g\,L^{-1}$ standard Mg, showed the most significant improvements. The increased photosynthetic efficiency brought about by the foliar application of these components was responsible for the overall improvement of plant characteristics. In one study, 10-day-old wheat seedlings were exposed to low concentrations of nickel nanoparticles (NPs) at 0.01 and $0.1\,mg\,L^{-1}$, which resulted in an increase in the amount of chlorophyll a and b (Tsygvintsev et al., 2018).

Additionally, nanoparticles have been tested for pest control, offering targeted solutions that are less toxic and more effective than traditional chemical pesticides. According to Zhang et al., when cucumbers were exposed to AgNPs, their antioxidant defenses were activated. In contrast to Ag ion treatment, AgNPs caused noticeable lipid peroxidation and serious poisoning symptoms in cucumbers (Zhang et al., 2018). According to Meng et al. (2017), Bombyx mori's resistance to oxidative stress is diminished by high concentrations of AgNPs over $800\,mg\,L^{-1}$. Su et al. (2023) developed a nanodelivery platform using star polycation to load thiophanate-methyl for vascular wilt disease treatment. Root irrigation with the nanopesticide significantly reduced the disease index compared to thiophanate-methyl alone. Improved efficacy was linked to enhanced absorption and translocation in stems and leaves. According to Van et al. (2016), CuO nanoparticles at concentrations more than 10 mg/L significantly inhibited cotton growth and development, as measured by plant height, root length, number, and biomass. The practical use of nanotechnology in agriculture is still in its early stages but is rapidly proceeding. Case studies provide concrete examples of how nanotechnology can address specific agricultural challenges, paving the way for broader adoption and integration into farming practices.

2.8 Challenges and Limitations

Increasing crop yields, increasing agricultural productivity, and optimizing food production systems are all made possible by nanotechnology, which provides innovative solutions like nanopesticides, nanofertilizers, and smart nutrition and agrochemical delivery systems. However, despite these developments, several challenges and limitations delay the widespread application of nanotechnology in agriculture. These challenges range from scientific and technical difficulties to regulatory, economic, and environmental concerns.

2.8.1 Scientific and Technical Challenges

2.8.1.1 Lack of Understanding of Nanoparticle–Plant Interactions

One of the primary limitations of nanotechnology in agriculture is the insufficient understanding of how nanoparticles interact with plants, soil, and microbes. The absorption, translocation, and metabolism of nanoparticles within plant tissues remain complex processes that require further investigation. Understanding how nanoparticles influence plant physiology and biochemistry is essential for their effective application.

2.8.1.2 Controlled Release and Targeting Issues

While nanomaterials are designed for controlled release and targeted supply of agrochemicals, attaining precise control over their release kinetics remains challenging. Factors such as soil composition, moisture content, and environmental circumstances can influence the effectiveness of nano-agrochemicals, leading to unpredictable results in different agricultural sites.

2.8.1.3 Stability and Shelf Life of Nanoformulations

Nanoparticles tend to aggregate or degrade over time, affecting their stability and efficacy. The shelf life of nanobased agrochemicals is a serious factor that determines their commercial feasibility. Developing formulations that remain stable under varying storage and field conditions is a significant challenge.

2.8.1.4 Variability in Efficacy Across Different Crops and Soil Types

Nanotechnology-based agricultural inputs may not work uniformly across all crop species and soil types. Differences in soil pH, organic matter content, and microbial communities can affect the behavior and effectiveness of nanomaterials, making it difficult to standardize their application.

2.8.2 Regulatory and Safety Challenges

2.8.2.1 Lack of Standardized Regulations and Guidelines

The regulatory landscape for nanotechnology in agriculture is still evolving. There is currently no globally accepted framework for assessing the safety and efficacy of nano-agrochemicals. Differences in regulatory approaches between countries create uncertainty for manufacturers and hinder the commercialization of nanotechnology-based agricultural products.

2.8.2.2 Inadequate Risk Assessment Protocols

Existing risk assessment methods for conventional agrochemicals may not be suitable for assessing nanomaterials. Nanoparticles exhibit unique physicochemical properties that require specialized testing methodologies to evaluate their toxicity, environmental impact, and potential risks to human health.

2.8.2.3 Public Perception and Ethical Concerns

The general public and policymakers often perceive nanotechnology as an emerging and potentially risky technology. Misinformation and lack of awareness regarding the benefits and risks of nano-agriculture may lead to resistance against its adoption. Ethical concerns regarding the use of engineered nanomaterials in food production also pose challenges to widespread acceptance.

2.8.3 Economic and Commercialization Barriers

2.8.3.1 High Cost of Nanomaterial Production

The synthesis and large-scale production of high-quality nanomaterials require sophisticated equipment and processes, making them expensive compared to conventional agrochemicals. The high production cost is a significant barrier to their affordability and accessibility, particularly for smallholder farmers in developing countries.

2.8.3.2 Limited Market Availability and Adoption

Despite extensive research, only a few nano-enabled agricultural products have reached the commercial market. The slow transition from laboratory research to field application is due to regulatory hurdles, high costs, and the need for extensive field trials to validate their performance.

2.8.3.3 Compatibility with Existing Farming Practices

The integration of nanotechnology into conventional farming systems requires adjustments in application techniques, equipment, and management practices. Many farmers may lack the technical knowledge and resources to adopt nanotechnology-based solutions, creating a gap between research advancements and practical implementation.

2.9 Future Prospects and Research Directions

Nanotechnology is transforming agriculture by providing innovative solutions to boost crop productivity, enhance soil health, and promote sustainable farming practices. As worldwide food demand increases and environmental challenges intensify, nanotechnology presents a promising approach to addressing these issues through precision farming, smart delivery systems, and efficient resource utilization. Future advancements in nano-agriculture focus on developing nanofertilizers, nanopesticides, and nanosensors that optimize nutrient use efficiency, reduce agrochemical wastage, and minimize environmental pollution. Smart nanocarriers can deliver nutrients and pesticides in a controlled manner, enhancing plant uptake while reducing harmful runoff. Furthermore, Internet of Things (IoT)-enabled nanosensors can offer real-time monitoring of plant health, soil moisture and nutrient levels, and other factors, enabling farmers to make data-driven decisions.

Furthermore, nanotechnology is playing a crucial role in enhancing plant stress tolerance against drought, salinity, and pathogens. Engineered nanoparticles can trigger plant defense mechanisms, improve photosynthesis efficiency, and promote healthy root development. Emerging research also explores the use of nanomaterials for genetic engineering and precision breeding, accelerating the development of resilient crop varieties. Despite its vast potential, challenges such as toxicity concerns, regulatory frameworks, and cost-effectiveness must be addressed for large-scale adoption. However, with ongoing research and advancements, nanotechnology is poised to revolutionize agriculture into a more efficient, sustainable, and resilient industry, ensuring global food security for future generations.

As nanotechnology continues to advance, future research will prioritize enhancing its scalability, cost-effectiveness, and sustainability while addressing concerns regarding safety and environmental impact. The agricultural sector faces mounting challenges, including resource scarcity, pest infestations, soil degradation, and the growing effects of climate change. Nanotechnology offers promising solutions to these issues, but further research is necessary to refine its applications and ensure its safe integration into farming systems. One key research direction focuses on the development of safer and eco-friendly nanomaterials. The possible risks related with nanoparticle accumulation in soil and water must be mitigated by designing biodegradable and non-toxic nanomaterials that minimize ecological harm while maintaining efficacy. Scientists are actively working on bio-based nanoparticles and green synthesis methods to reduce negative environmental footprints. Another crucial area of research is the enhancement of nanomaterial production. While nanotechnology holds great promise, the high cost of production and scalability challenges have limited its widespread adoption in agriculture. Researchers are exploring more cost-effective and energy-efficient manufacturing processes that can make nano-enabled products more affordable and accessible to farmers worldwide. Innovations in mass production techniques will play a key role in linking the gap between laboratory research and field applications.

The integration of nanotechnology with advanced agricultural systems is another emerging research focus. The convergence of nanotechnology with remote sensing, artificial intelligence (AI), and automation can lead to smarter, more efficient precision farming systems. Additionally, nano-enabled smart delivery systems can enhance the efficiency of fertilizers and pesticides, reducing chemical runoff and minimizing environmental contamination. Furthermore, research efforts will aim to expand the applications of nanotechnology in climate-resilient agriculture. As climate change intensifies, innovative nanomaterials can help crops withstand extreme weather conditions such as drought, heat stress, and salinity. Nanoparticles that enhance photosynthesis, improve water retention in soil, or act as stress protectants can contribute to building resilient agricultural systems capable of sustaining food production under adverse climatic conditions. In conclusion, the future of nanotechnology in agriculture holds immense potential, but continued research is essential to unlock its full capabilities. By developing safer nanomaterials, reducing production costs, and integrating nanotechnology with other emerging technologies, researchers can pave the way for a more sustainable and efficient agricultural landscape. As these advancements progress, nanotechnology is expected to revolutionize farming practices and contribute significantly to global food security.

2.10 Conclusion

Nanotechnology has the potential to transform agriculture by increasing crop yields, optimizing resource efficiency, and minimizing environmental impact. From nanofertilizers and nanopesticides to nanosensors and nano-enabled irrigation systems, the use of nanotechnology in agriculture are vast and varied. To fully understand the potential of nanotechnology in agriculture, it is essential to address the technical, environmental, and social challenges. This requires a concerted effort from researchers, policymakers, industry stakeholders, and the farming community. By adopting innovation, promoting safety, and ensuring equitable access, nanotechnology can play a key role in building a sustainable and resilient food production system for the future.

As we stand on the threshold of a new era in agriculture, the integration of nanobased engineering presents a transformative opportunity to meet the rising global demand for food while safeguarding the health of our planet. Although this journey is accompanied by significant challenges, continuous research, collaboration, and responsible innovation will result in a more sustainable and productive agricultural future. This concluding chapter combines the key findings and insights from previous discussions, underscoring the profound potential of nanobased engineering in revolutionizing agriculture. It reiterates the importance of continued research and technological advancements to harness the power of nanotechnology in optimizing food production systems, enhancing crop yields, and promoting long-term sustainability.

Furthermore, the chapter highlights the necessity of well-balanced regulatory frameworks and active public engagement to guarantee the safe and ethical application of nanotechnology in agriculture. Addressing concerns associated with environmental impact, human health, and economic feasibility will be crucial in fostering widespread acceptance and successful implementation. Ultimately, nanobased engineering is poised to play a pivotal role in shaping the future of agriculture and food production. By offering cutting-edge solutions to some of the most persistent challenges in global food security, this innovative approach has the potential to drive a more efficient, resilient, and sustainable agricultural system for generations to come.

References

Abd El-Fattah, A. Y., El-Helaly, A. A., & El-Wahab, A. S. (2019). Testing nano-pesticides toxicity against red palm weevil *Rhynchophorus ferrugineus* (Olivier) in Egypt. *Crop Protection, 19*(1), 1559–1568.

Agostini, A., Mondragón, L., Coll, C., Aznar, E., Marcos, M. D., Martínez-Máñez, R., Sancenón, F., Soto, J., Pérez-Payá, E., & Amorós, P. (2012). Dual enzyme-triggered controlled release on capped nanometric silica mesoporous supports. *ChemistryOpen, 1*, 17–20. https://doi.org/10.1002/open.201200003

Alvarado, K., Bolaños, M., Camacho, C., & Quesada, E. (2019). Nanobiotechnology in agricultural sector: Overview and novel applications. *Journal of Biomaterials and Nanobiotechnology, 10*, 120–141.

Bartolucci, C., Scognamiglio, V., Antonacci, A., & Fraceto, L. F. (2022). What makes nanotechnologies applied to agriculture green? *Nano Today, 43*, Article 101389. https://doi.org/10.1016/J.NANTOD.2022.101389

Bayat, H., Kolahchi, Z., Valaey, S., Rastgou, M., & Mahdavi, S. (2018). Novel impacts of nanoparticles on soil properties: Tensile strength of aggregates and compression characteristics of soil. *Archives of Agronomy and Soil Science, 64*, 776–789. https://doi.org/10.1080/03650340.2017.1386952

Chausali, N., Saxena, J., & Prasad, R. (2021). Nanobiochar and biochar-based nanocomposites: Advances and applications. *Journal of Agriculture and Food Research, 5*, Article 100191. https://doi.org/10.1016/j.jafr.2021.10.191

Chlingaryan, A., Sukkarieh, S., & Whelan, B. (2018). Machine learning approaches for crop yield prediction and nitrogen status estimation in precision agriculture: A review. *Computers and Electronics in Agriculture, 151*, 61–69.

Delfani, M., Baradarn Firouzabadi, M., Farrokhi, N., & Makarian, H. (2014). Some physiological responses of black-eyed pea to iron and magnesium nanofertilizers. *Communications in Soil Science and Plant Analysis, 45*, 530–540.

Feng, Y., Kreslavski, V. D., Shmarev, A. N., Ivanov, A. A., Zharmukhamedov, S. K., Kosobryukhov, A., Yu, M., Allakhverdiev, S. I., & Shabala, S. (2022). Effects of iron oxide nanoparticles (Fe_3O_4) on growth, photosynthesis, antioxidant activity and distribution of mineral elements in wheat (*Triticum aestivum*) plants. *Plants, 11*, Article 1894

Food and Agriculture Organization (FAO). (2017). *The future of food and agriculture – Trends and challenges*. Rome: FAO.

Fraceto, L. F., Grillo, R., de Medeiros, G. A., Scognamiglio, V., Rea, G., & Bartolucci, C. (2016). Nanotechnology in agriculture: Which innovation potential does it have? *Frontiers in Environmental Science, 4*, Article 20. https://doi.org/10.3389/fenvs.2016.00020

Giraldo, J. P., Wu, H., Newkirk, G. M., & Kruss, S. (2019). Nanobiotechnology approaches for engineering smart plant sensors. *Nature Nanotechnology, 14*, 541–553. https://doi.org/10.1038/S41565-019-0470-6

Ijaz, M., Khan, F., Ahmed, T., Noman, M., Zulfiqar, F., Rizwan, M., Chen, J., Siddique, K. H. M., & Li, B. (2023). Nanobiotechnology to advance stress resilience in plants: Current opportunities and challenges. *Materials Today Bio, 22,* Article 100759. https://doi.org/10.1016/j.mtbio.2023.100759

Jat, S. K., Bhattacharya, J., & Sharma, M. K. (2020). Nanomaterial-based gene delivery: A promising method for plant genome engineering. *Journal of Materials Chemistry B, 8*(16), 4165–4175. 10.1039/D0TB00217H

Jones, J. W., et al. (2017). Toward a new generation of agricultural system data, models, and knowledge products: State of agricultural systems science. *Agricultural Systems, 155*, 269–288.

Joshi, H., Somdutt, C., & P. and Mundra, S.L. (2019). Future prospects of nanotechnology in agriculture. *International Journal of Chemical Studies, 7*(2), 957–963.

Khan, S. T., Ahmad, J., Ahamed, M., & Jousset, A. (2018). Sub-lethal doses of widespread nanoparticles promote antifungal activity in Pseudomonas protegens CHA0. *Science of the Total Environment, 627*, 658–662. https://doi.org/10.1016/j.scitotenv.2018.01.257

Khandelwal, A., Joshi, R., Mukherjee, P., Singh, S. D., & Shrivastava, M. (2019). Use of bio-based nanoparticles in agriculture. In *Nanotechnology in agriculture: Advances for sustainable agriculture* (pp. 89–100). Springer. https://doi.org/10.1007/978-981-32-9370-0_6

Khose, S. B., Shekhar, S., & Bhausaheb Dhokale, K. (2023). The role of precision farming in sustainable agriculture: Advancements and impacts. *Agriculture and Food E-newsletter, 5*(9), 115–119.

Kubavat, D., Trivedi, K., Vaghela, P., Prasad, K., Vijay Anand, G. K., Trivedi, H., Patidar, R., Chaudhari, J., Andhariya, B., & Ghosh, A. (2020). Characterization of a chitosan-based sustained release nanofertilizer formulation used as a soil conditioner while simultaneously improving biomass production of *Zea mays* L. *Land Degradation & Development, 31*(17), 2734–2745.

Li, Y., Xi, K., Liu, X., Han, S., Han, X., Li, G., Yang, L., Ma, D., Fang, Z., Gong, S., Yin, J., & Zhu, Y. (2023). Silica nanoparticles promote wheat growth by mediating hormones and sugar metabolism. *Journal of Nanobiotechnology, 21*, Article number: 2

Meng, X., Abdlli, N., Wang, N., Lu, P., Nie, Z., Dong, X., Lu, S., & Chen, K. (2017). Effects of Ag nanoparticles on growth and fat body proteins in silkworms (*Bombyx mori*). *Biological Trace Element Research, 2*, 327–337. https://doi.org/10.1007/s12011-017-10.1-7

Mittal, D., Kaur, G., Singh, P., Yadav, K., & Ali, S. A. (2020). Nanoparticle-based sustainable agriculture and food science: Recent advances and future outlook. Frontiers. *Nanotechnology, 2*, Article 579954. https://doi.org/10.3389/FNANO.2020.579954

Muthumalai, K., Gokila, N., Haldorai, Y., & Rajendra Kumar, R. T. (2022). Nanosensors for crop protection: Design and fabrication. In *Nanosensors for smart agriculture* (pp. 403–422). Elsevier. 10.1016/B978-0-12-824554-5.00041-0

Nishu, & Kumar, S. (2023). Smart and innovative nanotechnology applications for water purification. *Hybrid Advances, 3*, 100044. https://doi.org/10.1016/j.hybadv.2023.100044

Pal, A., Kaur, P., Dwivedi, N., Rookes, J., Bohidar, H. B., Yang, W., Cahill, D. M., & Manna, P. K. (2023). Clay-nanocomposite-based smart delivery systems: A promising tool for sustainable farming. *ACS Agricultural Science & Technology*, *3*(1), 3–16. https://doi.org/10.1021/acsagscitech.2c00140

Panyuta, O., Belava, V., Fomaidi, S., Kalinichenko, O., Volkogon, M., & Taran, N. (2016). The effect of pre-sowing seed treatment with metal nanoparticles on the formation of the defensive reaction of wheat seedlings infected with the eyespot causal agent. *Nanoscale Research Letters*, *11*, 1–5. https://doi.org/10.1186/s11671-016-1305-0

Parameswari, P., Belagalla, N., Singh, B. V., Abhishek, G. J., Rajesh, G. M., Katiyar, D., Hazarika, B., & Paul, S. (2024). Nanotechnology-based sensors for real-time monitoring and assessment of soil health and quality: A review. *Asian Journal of Soil Science and Plant Nutrition*, *10*(2), 157–173.

Rajan, C. S. (2011). Nanotechnology in groundwater remediation. *International Journal of Environmental Sustainability and Development*, *2*(3), 182. https://doi.org/10.7763/ijesd.2011.v2.12

Ramirez-Rodriguez, G. B., Dal Sasso, G., Carmona, F. J., Miguel-Rojas, C., Perez-de-Luque, A., Masciocchi, N., Guagliardi, A., & Delgado-Lopez, J. M. (2020). Engineering biomimetic calcium phosphate nanoparticles: A green synthesis of slow-release multinutrient (NPK) nanofertilizers. *ACS Applied Bio Materials*, *3*, 1344–1353.

Rana, L., Kumar, M., Rajput, J., Kumar, N., Sow, S., Kumar, S., Kumar, A., Singh, S. N., Jha, C. K., Singh, A. K., Ranjan, S., Sahoo, R., Samanta, D., Nath, D., Panday, R., & Raigar, B. L. (2024). Nexus between nanotechnology and agricultural production systems: Challenges and future prospects. *Discover Applied Sciences*, *6*, Article 555

Rezaei, S., & Khanmirzaei, A. (2024). Nano-biofertilizers: A new area for enhancing plant nutrition. In *Plant holobiome engineering for climate-smart agriculture* (pp. 617–632). Singapore: Springer Nat. Singapore. https://doi.org/10.1007/978-981-99-9388-8_32

Saberi-Rise, R., & Moradi-Pour, M. (2020). The effect of *Bacillus subtilis* Vru1 encapsulated in alginate–bentonite coating enriched with titanium nanoparticles against *Rhizoctonia solani* on bean. *International Journal of Biological Macromolecules*, *152*, 1089–1097.

Saurabh, K., Prakash, V., Dubey, A. K., Ghosh, S., Kumari, A., Sundaram, P. K., ... Singh, R. R. (2024). Enhancing sustainability in agriculture with nanofertilizers. *Discover Applied Sciences*, *6*(11), 559.

Savi, G. D., Piacentini, K. C., de Souza, S. R., Costa, M. E. B., Santos, C. M. R., & Scussel, V. M. (2015). Efficacy of zinc compounds in controlling *Fusarium* head blight and deoxynivalenol formation in wheat (*Triticum aestivum* L.). *International Journal of Food Microbiology*, *205*, 98–104. https://doi.org/10.1016/J.IJFOODMICRO.2015.04.001

Sharma, R., & Kumar, V. (2024). Nano-enabled agriculture for sustainable soil. *Waste Management Bulletin*, *2*(1), 152–161. https://doi.org/10.1016/j.wmb.2024.01.002

Singh, H., Sharma, A., Bhardwaj, S. K., Arya, S. K., Bhardwaj, N., & Khatri, M. (2021). Recent advances in the applications of nano-agrochemicals for sustainable agricultural development. *Environmental Science: Processes & Impacts*, *23*(2), 213–239. 10.1039/D0EM00404A

Singh, P. P., Priyam, A., Singh, J., & Gupta, N. (2023). Biologically synthesised urea-based nanomaterial shows enhanced agronomic benefits in maize and rice crops during *Kharif* season. *Scientia Horticulturae*, *315*, Article 111988. https://doi.org/10.1016/j.scienta.2023.111988

Soutter, W. (2021). Nanotechnology in agriculture. *AZoNano*. https://www.azonano.com/article.aspx?ArticleID=3141

Su, X., Yan, S., Zhao, W., Liu, H., Jiang, Q., Wei, Y., ... Cheng, H. (2023). Self-assembled thiophanate-methyl/star polycation complex prevents plant cell-wall penetration and fungal carbon utilization during cotton infection by *Verticillium dahliae*. *International Journal of Biological Macromolecules*, *239*, 124354.

Suazo-Hernández, S., et al. (2023). Impact on some soil physical and chemical properties caused by metal and metallic oxide engineered nanoparticles: A review. *Nanomaterials*, *13*(8), Article 1053. https://doi.org/10.3390/nano130810.3

Taha, N. A., Hamden, S., Bayoumi, Y. A., Elsakhawy, T., El-Ramady, H., & Solberg, S. Ø. (2023). Nanofungicides with selenium and silicon can boost the growth and yield of common bean (*Phaseolus vulgaris* L.) and control *Alternaria* leaf spot disease. *Microorganisms*, *11*(3), Article 728. https://doi.org/10.3390/microorganisms11030728

Thathsarani, S. (2021). Nanofertilizer for precision and sustainable agriculture. *Journal of Research Technology and Engineering*, *2*(1), 81–85.

Tsygvintsev, P. N., Goncharova, L. I., Manin, K. V., & Rachkova, V. M. (2018). Morphophysiological features of wheat (*Triticum aestivum* L.) seedlings upon exposure to nickel nanoparticles. *Sel'skokhozyaistvennaya Biol*, *53*, 578–586.

Van, N. L., Ma, C., Shang, J., Rui, Y., Liu, S., & Xing, B. (2016). Effects of CuO nanoparticles on insecticidal activity and phytotoxicity in conventional and transgenic cotton. *Chemosphere*, *144*, 661–670. https://doi.org/10.1016/j.chemosphere.2015.09.028

Xu, Y., Dhaouadi, Y., Stoodley, P., & Ren, D. (2020). Sensing the unreachable: Challenges and opportunities in biofilm detection. *Current Opinion in Biotechnology*, *64*, 79–84.

Yadav, N., Garg, V. K., Chhillar, A. K., & Rana, J. S. (2023). Recent advances in nanotechnology for the improvement of conventional agricultural systems: A review. *Plant Nano Biology*, *4*, Article 100032

Zain, M., Ma, H., Rahman, S. U., Nuruzzaman, M., Chaudhary, S., Azeem, I., ... Sun, C. (2024). Nanotechnology in precision agriculture: Advancing towards sustainable crop production. *Plant Physiology and Biochemistry*, *206*, 108244.

Zhang, H., Du, W., Peralta-Videa, J. R., Gardea-Torresdey, J. L., White, J. C., Keller, A., Guo, H., Ji, R., & Zhao, L. (2018). Metabolomics reveals how cucumber (*Cucumis sativus*) reprograms metabolites to cope with silver ions and silver nanoparticle-induced oxidative stress. *Environmental Science & Technology*, *14*, 8016–8026. https://doi.org/10.1021/acs.est.8b02440

Zhang, M., He, F., Zhao, D., & Hao, X. (2011). Degradation of soil-sorbed trichloroethylene by stabilized zero valent iron nanoparticles: Effects of sorption, surfactants, and natural organic matter. *Water research*, *45*(7), 2401–2414.

Zhao, L., Peralta-Videa, J. R., Rico, C. M., Hernandez-Viezcas, J. A., Sun, Y., Niu, G., ... Gardea-Torresdey, J. L. (2014). CeO_2 and ZnO nanoparticles change the nutritional qualities of cucumber (*Cucumis sativus*). *Journal of agricultural and food chemistry*, *62*(13), 2752–2759.

Zhi, H., Zhou, S., Pan, W., Shang, Y., Zeng, Z., & Zhang, H. (2022). The promising nanovectors for gene delivery in plant genome engineering. *Frontiers in Plant Science*, *13*, Article 936876. https://doi.org/10.3389/fpls.2022.936876

3

Development of Nanofertilizers for Soil Health and Fertility

Waseem Ahmed Khattak[1], Muhammad Majeed[2], Sudenaz Soylu[3], Afshan Farid[4], Sadia Riaz[1], Muhammad Anas[5], and Shah Fahad[6]

[1] *Department of Plant Sciences, Faculty of Biological Sciences, Quaid-I-Azam University, Islamabad, Pakistan*
[2] *Department of Botany, University of Gujrat, Punjab, Pakistan*
[3] *Department of Field Crops, Faculty of Agriculture, Selcuk University, Konya, Türkiye*
[4] *Department of Botany, PMAS Arid Agricultural University, Rawalpindi, Punjab, Pakistan*
[5] *Department of Soil Science & Plant Nutrition, Selçuk University. Campus, Konya 42079 Türkiye*
[6] *Department of Agronomy, Abdul Wali Khan University Mardan, Khyber Pakh-tunkhwa, Pakistan*

3.1 Introduction

Recently, the adoption of green technologies in agriculture is becoming prominent as an alternative to traditional pesticides and fertilizers (Zhang et al., 2024). Drones defeat the negative sides of conventional fertilizer applications in agriculture. Consequently, the application of nanoparticles (NPs) is gaining considerable popularity in agronomy. This work provides a fresh perspective on the latest advancements in NP architecture and design for usage as nanofertilizers and other agricultural applications. Additionally, this study covers some of the most recent research on NP–plant interactions and how they affect agricultural plants (Figure 3.1). Mostly, they play a vital role in reducing the need for inorganic fertilizers and their detrimental impact on the environment. Moreover, nanofertilizers are highly reactive and capable of penetrating the epidermis, which allows them to slowly release and disperse, thereby greatly enhancing the efficiency of nutrient utilization. Additionally, they can mitigate heavy metal toxicity and abiotic stress (Anas et al., 2023, 2024b). On the other hand, there are also findings regarding NP toxicity in some plants due to the deposition of NPs on cell surfaces and organelles, which causes oxidative stress. For instance, one study highlighted the negative influences of nanomaterials, for example, NPs stunted plant growth, inhibited the formation of chlorophyll, and decreased photosynthetic efficiency. The impact of NPs is plant-, NP type-, and concentration-specific; therefore, it can be positive or negative. CuO nanoparticles are known to have cytotoxic and genotoxic effects as they enter cells and their organelles (Sajjad et al., 2023). According to some findings, NPs may penetrate the cell and interfere with the electron-transport system (ETS) cycle in mitochondria and chloroplasts, causing

Nanotechnology Innovations for Food Security and Sustainable Agriculture, First Edition. Edited by Zia Ur Rehman Mashwani, Shah Fahad, Ilyas Ahmad, Shah Saud, Sajid Fiaz, and Taufiq Nawaz.

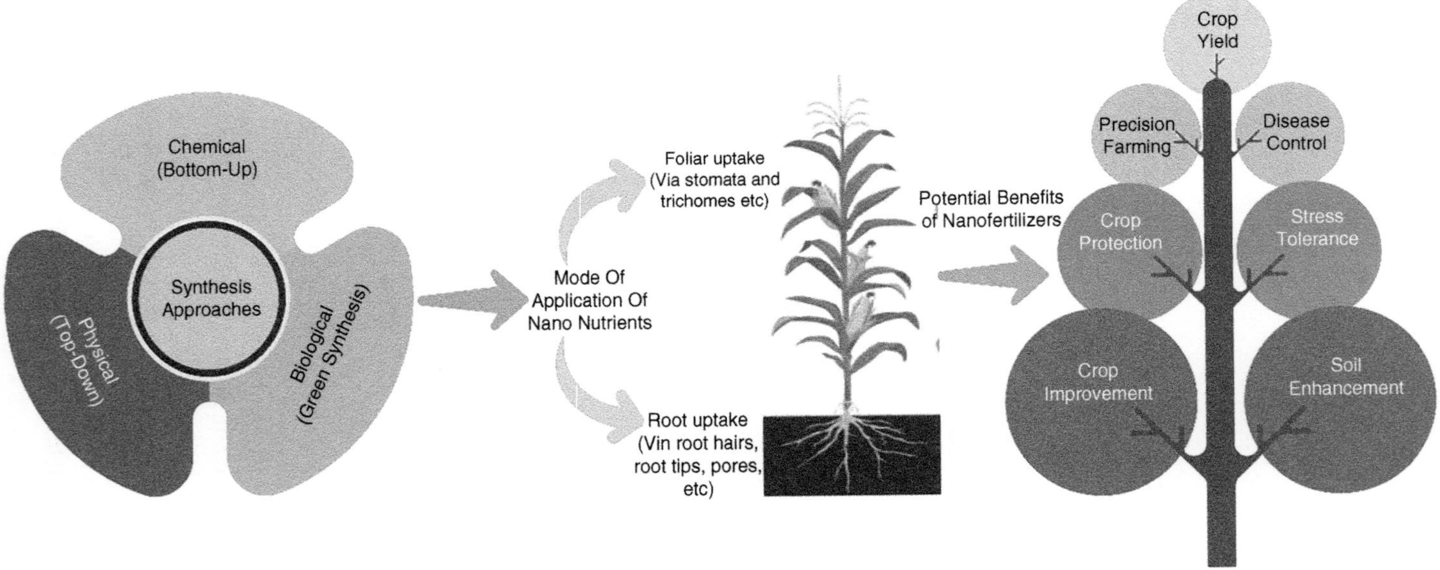

Figure 3.1 Graphical representation that emphasizes the chapter's main points.

an oxidative burst because of increased reactive oxygen species (ROS). Metalloids and metal NPs depolarize many plant species and impose oxidative stress. ROS as a result of oxidative damage to mitochondrial and chloroplast DNA lead to changes to the proteins that were encoded within them, which disable and completely inactivate the particular proteins. The treatment of *Ricinus communis* seed with silver nanoparticles (AgNPs) enhances the water-soluble hydrogen peroxide (H_2O_2) production and this is correlated with antioxidant defense mechanisms (superoxide dismutase enzyme (SOD), peroxidase (POD) activity, and phenolic acids) (Raza et al., 2025). Interestingly, some studies claim that ZnO NPs dramatically reduce ryegrass biomass and also induce root-tip swelling and vacuolation of the root epidermal and cortical cells. Single ZnO NPs were detected in the apoplast and protoplast of the root endodermis and stele and the root surface. The translocation factor of Zn from root to shoot stayed low with lower treatments of ZnO NPs. Some studies noted the unexplained accumulation of CeO_2 NPs in plants that poses a risk to human health. There are documented cases on the effects of metal oxide NPs on soil bacterial communities in the grassland soil. These values are, however, way too high, contrasted to the actual environment today. Nevertheless, it is in these contexts that evidence should be gathered regarding the impact of artificial NPs in the rhizosphere microbiome and how to mitigate their probable negative effects on agricultural soils. An intriguing case of study was found that claims to have a straightforward, low-cost, and operational method of treating wastewater using renewable super magnetic biosorbent Fe_3O_4@PW-αCFs to remove enormous amounts of engineered nanomaterials (ENMs) such as CuO and ZnO from different real-world wastewater samples. Also, the available information on the use of nanofertilizers as a substitute for conventional fertilizers, their limitations, and prospects has been analyzed (Raheman et al., 2024).

The nanoparticles that comprise nanofertilizers can include carbon, metal oxides, or other nanoporous materials, based on their specific compositional and structural properties. They can be made through biological means or chemical (bottom-up) or physical (top-down) processes. Compared with physical and chemical techniques that dominate nanoproducts fabrication, other novel approaches like green nanotechnology or biofabrication are equally clean and non-toxic and particularly eco-friendly, so they could also be used for nanofertilizer fabrication. The ability to supply different types of macro- and micronutrients in set concentrations, along with increased surface area and the ability to control concentration, are some of the many benefits that these fertilizers offer. Plants respond based on the nutrients released by the nanofertilizers into the soil. Different types of macronutrient- and micronutrient-providing nanofertilizers, as well as nanofertilizers enhanced with other nanomaterials, could be created to meet nutrient requirements. These nanofertilizers are linked to improved phytoavailability, increased nutrient solubility, the dispersion of insoluble nutrients, and tailored nutrient delivery to minimize losses. Additionally, they impact a number of vitamins, proteins, purines, coenzymes, and photosynthetic pigments for the respiratory and photosynthesis processes of plants (Bloom, & Lancaster, 2018).

This chapter offers insight into the recent progress aimed at the design and formulation of nanoparticles intended for use as nanofertilizers, their applications, and their possible agro-environmental impacts. The emphasis is particularly placed on the possibilities that nanofertilizers offer and stresses the known gaps and concerns that exist with the massive use of NPs prior to their application in the field. A recent evaluation brought to light the cost and environmental consequences of using nanofertilizers. The field data, collected over three consecutive rice-growing seasons, demonstrated that the aqueous polyacrylate-coated urea fertilizer enhanced with nano-FeIII-tannic acid had a better agronomic, environmental, and ecosystem economic impact than the widely used polyurethane-coated urea. Modified urea reduced reactive nitrogen loss by over 24% while increasing yields by 8.3%, farmer profits by nearly 10%, and net ecosystem economic benefits by nearly 11% as compared to traditional farmers' fertilizer application (Zhang et al., 2021). Additionally, a study in the field later focused on wheat and stated that using nano-FeIII-tannic acid modified waterborne polyacrylate-coated urea alone, with the nitrogen rate reduced by one-third, had the potential to maintain a high wheat grain yield along with high net ecosystem economic gains of the wheat crop in comparison to traditional nitrogen, while also eliminating 58.8% of reactive nitrogen loss (Wang et al., 2024).

3.2 Scope and Importance of Nanofertilizers

In modern farming, nanofertilizers are significant for their specificity in form and the way they are delivered to plants physiologically (Nongbet et al., 2022). The minute dimension of nanofertilizers possessing a high surface mass ratio enhances the uptake of nutrients through the roots of the plants. Furthermore, unlike bulk particles or ionic salts, nanoparticles exhibit unique physicochemical properties that influence their behavior and interactions in biological and environmental systems. These nanosized fertilizers reduce leaching, do not undergo chemical alteration, and greatly enhance the efficiency of nutrient utilization and the quality of the environment.

3.3 New Innovations in the Development of Nanofertilizers

One of the primary concerns over the next few years would be soil-fertility maintenance and nutrient management for crops because of the over-reliance on chemical fertilizers and the current farming practices. Traditional fertilizers are generally applied to crops due to higher profit margins, resulting in excessive loss of water bodies, low crop nutrient productivity, and negative impacts on the environment (Kumar et al., 2019). Therefore, an eco-friendly novel fertilization method that is better than synthetic fertilizers is needed to augment the overall production in our agriculture. Now, there has been heightened interest in the use of smart delivery systems for pesticides through the application of environmentally friendly technologies such as nanohybrid constructs in fertilizers, referred to as nanofertilizers. The introduction of engineered structures of nanohybrid systems like nanofertilizers has sparked interest of many scientists globally owing to their unique functional construction and controllable the physicochemical properties. In this context, the progress made in nanotechnology has transformed the materials design and fabrication through the modification of the properties and characteristics to fulfill a myriad of needs. There have been some reports suggesting an improvement in crop efficiency with the use of nanofertilizers in comparison to that with ordinary chemical compost (Raheman et al., 2024).

Furthermore, attempts have been made recently to enhance the response of the nanofertilizers to the environment so as to include humidity, moisture, temperature, pH, etc. In this regard, nano-biotechnology has emerged as a novel possible candidate to supplement crops with nutrients in a controlled manner and potentially change agricultural systems. Moreover, engineered NP-based nanofertilizers can assist in the nutritional management of different crops by improving their tolerance to abiotic stress and enhancing agricultural productivity (El-Saadony et al., 2022). This section presents the newest information regarding the scope of design, interactions, and effects of nanofertilizers on edible plants and their illustration in Figure 3.2.

3.3.1 Design and Formulation of Nanofertilizers

The application of nutrient-loaded NPs is an important element of nanoscale fertilizer approaches. The processes for synthesizing the nanomaterials include both physical (top-down) and chemical (bottom-up) methodologies. These methods can be divided into three categories: biological, chemical, and physical. The bottom-up NPs' chemical synthesis belongs to these approaches, while biosynthesis and top-down physical synthesis are of a different nature. Numerous investigations have already shown the production and use of nanomaterials as nanofertilizers. As nanofertilizers, basic minerals such as zeolites, copper (Cu), silver (Ag), aluminum (Al), potassium (K), carbon (C), nitrogen (N), zinc (Zn), silicon (Si), magnesium (Mg), iron (Fe), sodium (Na), manganese (Mn), and calcium (Ca) can be

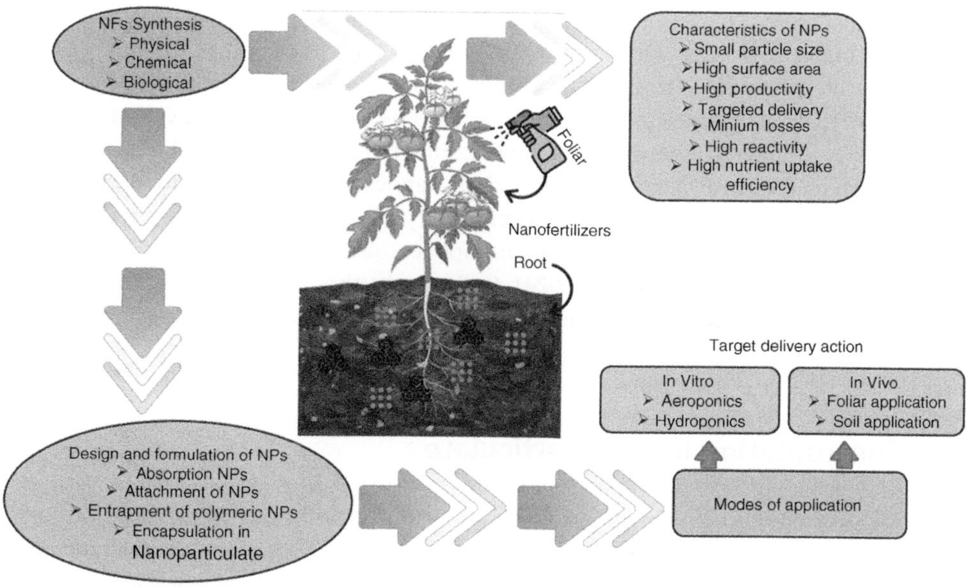

Figure 3.2 Overview of design and formulation of nanofertilizers based on the respective synthesis method and their characteristics.

nanostructured. Nanofertilizers are also made from other plant-derived materials, such as banana peels or grape plant substrates. Recently, natural zeolite (composed of over 50 minerals) has been turned into zeolite-based nanofertilizers. This is made possible because of the widespread occurrence and low cost of this material. Nanofertilizers have been developed and may be further refined to remedy a particular nutrient deficiency for the plants. This is valid because of the ability to manipulate the properties of NP surfaces by functionalization with specific atoms (Ahmad et al., 2022). The following methods can be used to prepare nanofertilizers:

3.3.1.1 Absorption of NPs

Absorption of NPs involves the adsorption of nutrient molecules onto the surface of nanoparticles via electrostatic interactions, hydrogen bonding, or van der Waals forces. This method enhances nutrient retention, prevents leaching, and facilitates a controlled release mechanism, ensuring efficient uptake by plants (Asadu et al., 2024). It is widely used for macronutrients such as nitrogen, phosphorus, and potassium, allowing for improved nutrient-use efficiency (NUE) and reduced environmental losses.

3.3.1.2 Attachment to NPs

Attachment to NPs refers to the direct binding of nutrients to the surface of nanoparticles through chemical or physical interactions. In this method, functionalized nanoparticles, such as metal oxides or carbon-based nanomaterials, provide active sites for nutrient attachment. This strong adhesion prevents nutrient washout and enhances stability, making it ideal for precision agriculture. Additionally, the attachment process allows for the modification of nanoparticles to improve their affinity for specific nutrients, ensuring targeted and sustained delivery to plants (Anas et al., 2024a).

3.3.1.3 Entrapment of Polymeric NPs

Entrapment of polymeric NPs involves encapsulating nutrients within a biodegradable polymer matrix, such as chitosan, alginate, or polylactic-*co*-glycolic acid (PLGA). This technique provides a protective barrier against environmental degradation and allows for the slow and sustained release of nutrients. Polymeric nanocarriers enhance nutrient bioavailability, minimize toxicity risks, and promote efficient plant uptake. Furthermore, the use of biodegradable polymers ensures eco-friendly applications, reducing the environmental footprint of fertilizer usage (Liyakat, 2024). Collectively, these nanofertilizer preparation methods contribute to enhanced nutrient efficiency, reduced losses, and sustainable agricultural practices.

3.4 Encapsulation in Nanoparticulate

The most accepted method of synthesizing nanofertilizers with nanomaterials is through nutrient encapsulation, which has been developing on a yearly basis. In particular, the last couple of decades have seen nutrients encapsulated in thin film polymers at the nanoscale porous structures. Encapsulation of favorable microorganisms like fungi or bacteria have been shown to enhance the availability of nitrogen, phosphorus, and potassium in the root zone. In conclusion, a nanofertilizer is a nanomaterial serving as a nutrient or a nutrient carrier containing either micro- or macronutrients for plants. Along with that, as modified synthetic fertilizers, nanofertilizers can also be synthesized via green synthesis from plant extracts, which are nanomaterials particles with biological origin encapsulated or coated to resemble the synthetic materials. Nanofertilizers can be classified into three categories according to the technique of their preparation: (1) elder known nanoscale fertilizers, which are NP like in size or are already in the form of the fertilizer; (2) the new nanoscale additive fertilizers, which are conventional fertilizers with some added nanomaterials; and (3) nanoscale coating fertilizers, which have nutrients encapsulated by nanofilms or incorporated into the nanoscale pores of a host material. Clays or other host materials that have nutrient-bearing nanopores perform the function of nanocomposite structures with the enclosed nutrients and can serve for controlled release of nutrients (Tripathy & Gupta, 2024).

3.4.1 Characteristics of Nanofertilizers

The use of the latest preparation techniques improves the distribution of insoluble nutrients to minimize losses, as well as the accessibility, bioavailability, and solubility of certain nutrients. In one study, the goal was to increase the efficiency of absorption of certain plant targets by creating nanoparticles and then encapsulating them with the necessary nutrients. Nanofertilizers must possess advanced chemical stability, elevated surface tension, enhanced absorbability, mobility, and leniency in pH and ionizing properties. The slow-release function and regulated, direct distribution of different (micro- and macro-) nutrients to plants at the right size and optimal value are made possible by these advantageous characteristics of nanofertilizers. Enhancing NUE, lowering the overall chemical load on the soil, reducing the number of treatments required, and reducing the negative effects of traditional bulky fertilizers are all possible with nanofertilizers. Nanofertilizers are essentially used to increase crop quality, soil fertility, and plant nutrient bioavailability. Nanofertilizers are most suited for usage in contemporary agriculture due to their functionalities.

Nanofertilizers, as nanomaterials, have superior retention and absorption capabilities compared to conventional synthetic chemical fertilizers due to their small size in nanometers. Nanofertilizers are more effective than chemical fertilizers because their higher surface area enables them to contain more nutrients, which are then released gradually in accordance with the crops' requirements. Just like their larger counterparts, synthetic fertilizers also emit harmful gaseous nutrients, resulting in inefficient nutrient absorption and wastage. Nevertheless, negative environmental impacts are mitigated by the nanofertilizers' slow and steady nutrient distribution. The literature suggests that plant cells use structures with pore openings of 5–20 nm in diameter as a self-defense mechanism. Only external carriers smaller than plant cells can penetrate the size exclusion limits (SELs) to capture essential nutrients normally available in soil or on the surface of leaves. Moreover, some other nanoparticles (NPs) can also be integrated into cell walls and used to improve genetic traits. Nano-encapsulated fertilizers have shown greatly enhanced mobility relative to traditional water-soluble fertilizers when injected into the plant's cells/tissues. There are reports of penetrations in the roots of *Arabidopsis thaliana* plants by SiO_2 NPs with diameters of 50–200 nm. In comparison, TiO_2 NPs between 36 and 140 nm in diameter could not pass through the parenchyma roots of wheat, but NPs of 36 nm could cross the barriers in wheat and subsequently move to other parts of the plant. When TiO_2 nanoparticles (NPs) with 30 nm radii were applied to maize, they could not translocate because their diameter was larger than the root's cell diameter of 6.6 nm. In another research, the penetration of TiO_2 NPs into wheat plants was performed, and some of the NPs were noticed to penetrate through the root cells while some others were not noticed. The NPs were polydisperse, with the particle sizes less than 20 nm penetrating through the roots while the bigger particles agglomerated in the soil and could not penetrate. This indicates that particle size is an important parameter for the translocation and uptake of nanofibers (NFs) in the plant system. Besides particle size, the surface charge of an NF also has a similar effect on the uptake and transport of NFs in plants. The roots of *Arabidopsis thaliana* excreted mucilage, a sticky viscous fluid, to facilitate the uptake of positively charged gold NPs of 12 nm radius; However, negatively charged gold NPs of the same size exhibited lower cellular uptake compared to their positively charged counterparts (Li et al., 2024).

3.4.2 Controlled Release and Targeted Delivery

The sustainable use of agriculture is supported by nanofertilizers, which improve the agricultural yield by improving NUE in fields and greenhouse crops. Nitrogenous fertilizers, whether organic or synthetic, in combination with nanofertilizers, can have a slow and constant nutrient release rate. Their release period is about 40–50 days, which is much longer than the 4–10 days of synthetic fertilizers. Therefore, when urea is chemically synthesized and applied in the field, nitrogen (N) losses through volatilization and leaching can exceed 70%, leaving only about 20% available for plant uptake. The application of nanohybrid urea, a modified form of hydroxyapatite, in rice (*Oryza sativa* L.) can result in N release up to 12 times slower than synthetic urea while also achieving half the grain production rate of conventional urea. More recently, N was synthesized in the form of NFs by enclosing urea within hydroxyapatite NPs, which delayed N release to plants. Recently, N-nanofertilizers were made through N encapsulation/coating with hydroxyl apatite NPs to achieve more gradual N release to plants. Nanofertilizers enable plants to directly absorb phosphorus, which minimizes loss of this resource. Surface modification enables targeted delivery and controlled release of nanoagrochemicals. In agriculture, nanofertilizers are crucial in enabling seed germination and growth. They easily penetrate soil and roots, increasing nutrient release, dry matter, and chlorophyll production, which subsequently enhance plant growth. The tendency of some nanomaterials to cluster together, called agglomeration, is a phenomenon that reduces their effectiveness and encourages the formation of ROS. Poly(methacrylic acid) (PMAA), poly(N-vinyl-2-pyrrolidone) (PVP), poly(methyl methacrylate) (PMMA), and polyethylene glycol (PEG) are stabilizing polymers, and a variety of surfactants, including cellulose, gluconic acid, polyallylamine, and chitosan, are employed to reduce agglomeration. (Li et al., 2024).

3.5 Modes of Application

Nanofertilizers (NFs) should be used in a manner that enables them to maintain their desirable features such as effectiveness, time-controlled release, solubility, targeted delivery, stability, and low toxicity. The application's effectiveness and impact greatly depend on how NFs are delivered and disposed of, such that no harm shall be caused. Some methods of applying nanofertilizers to plants are outlined below.

3.5.1 In Vitro Techniques

Both aeroponics and hydroponics are methods of farming that do not make use of soil. In the case of aeroponics, the roots of the plant hang in mid-air, and a nutritional solution is constantly sprayed. This method allows for the control of the entire environment around the roots. Currently, however, this method is not very popular because it requires too many nutrients in order to achieve optimal and rapid growth. Hydroponics, known as a "solution culture" method, uses inorganic salts that can be dissolved easily as a nutrient source. Special care is placed on the pH, oxygen, and volume of the solution surrounding the roots. In some commercial uses, other materials that provide support like sand or gravel are introduced. The solution is drawn out of one side and replaced with a nutrient solution from the other. One disadvantage of this method is the plants are vulnerable to pathogen attack and too much moisture can be injurious to soil-based plants, which leads to excessive wilting (Montoya-Martínez et al. 2024b).

3.5.2 In Vivo Methods

Micronutrients are provided during foliar application through the spraying of liquid fertilizers onto the leaves. This approach has an advantage with regards to the timing of nutrient absorption by plants during the exponential growth phase, as it shortens the period that elapsed between application and absorption. Compared to other soil application methods where micronutrients like iron, manganese, and copper are more leached than absorbed by soil particles, the method also increases the relative absorption of the micronutrients. Because cell intake occurs in the epidermis and stomata, there are possible agronomic benefits when using nanofertilizers, however, with regards to stomata, the time needed to reach the open position must be considered. Nitrogen-phosphorus-potassium (NPK) and NPs were tested for their foliar application. It was noted that the use of Nanochitosan-NPK fertilizer on wheat improved the mobilization index, crop index, and harvest index; therefore, those figures changed significantly. Mantling watermelon crops with a foliar nanocomposite also showed the same results. The aspect ratios of stem height, number of branches, diameter, and seed yield were all promising following the foliar application of gold nanoparticles. For the best foliar absorption, the type of plant being utilized, the working environment (gas, light, and water), and the distribution mechanisms of the nanoparticles should all be examined in addition to particle size.

Where nanoparticles are located, their size, how they relate to other particles, the form of the plant, activities in the rhizosphere, and the growth stage of a given crop are among the factors that determine the soil application or absorption of nanofertilizers at the root level. Soil-applied nanofertilizers are absorbed by the root surface and face several obstacles before they can access the plant's vascular system. The root's cuticle layer serves as the first barrier, analogous to the leaf cuticle layer. After rupturing the cuticle, the following step in nanoparticle uptake is epidermal NP permeation. When they permeate the root epidermis, they have a choice of either the apoplastic or the symplastic pathway. Nanofertilizers move into intercellular gaps via the apoplastic pathway after first penetrating cell wall porosities. There is a barrier to the apoplastic direction, the diameter of the cell wall, which is around 5–20 nm. This barrier restricts large NFs (greater than 20 nm) from going through because it is 20 nm. Notwithstanding, cells may die due to NPs, which may lead to enlarging the pore size. However, the major hindrance in apoplastic is the vascular cylinder, which directly restricts NPs flowing in, while cast off band. Meanwhile, 30 nm ZnO-bonaferts in vascular plus some mix plant root stories were noted. On the other hand, in the symplastic pathway, the NPs move from one cell to the next cell through the plasmodesmata channel. NPs enter the aerial parts of the plants via the central cylinder and subsequently the xylem's stream of transport. The work of scientists on increased yield of different crops with the use of nanofertilizers may provide new possibilities for sustainable and green agriculture (Saurabh et al. 2024a). Furthermore, additional research done in these regards is presented in this Table 3.1 based on the properties of nanofertilizers and their action methods.

3.6 Role of Nanofertilizers in Soil

The benefits of nanofertilizers (NFs) are twofold when they are applied through irrigation, as they enhance soil quality and plant growth productivity, since there may be restrictions with regards to the availability of inorganic fertilizers for the soils. In this regard, NFs may be more effective for nutrient uptake through the plants' roots. Soil and various ecosystem components have an impact on the range of elements and compounds that a given soil may contain and can modify biological species' populations and rhizic microbial life to improve soil water and fertility along with vegetation productivity. Roots play a key role in plant–soil interactions whereby water, nutrients, and other vital materials can be taken in Figure 3.3, while the development of roots is controlled by soil aeration, concentration of nutrients, pH, and soil texture. The main processes involved in the movement and storage of nutrients within the soil and into the upper part of the plants are diffusion and mass flow movement. A series of cells and their surrounding extracellular spaces experience mineral movement through the concentration gradient, termed diffusion, whereas the term bulk flow refers to the pressure-regulated movement of xylem liquid constituents and solutes driven by evaporation and soil's nutrient supply. The mechanisms of NP absorption through aerial surfaces, roots, grains, climate features, cell wall stiffness, and even the physiology, anatomy, and biochemistry of plant species/cultivars explain the accumulation of NPs. The release of nutrients from fertilizer particles is efficiently controlled by the NPs due to their tension being higher than that of a typical fertilizer.

After entering the plants, the nanoparticles (NPs) are very likely to be transported through apoplastic and symplastic ways. Movement from the outer to the inner within the stem toward the core cylinder and vascular structures and beneath, into the surfaces, is promoted. The apoplastic pathway is crucial for the remittance of the NPs to all parts of the body, but the Casparian strip confines the NPs in the root endodermis from moving radially, which can be circumvented by apoplastic becoming symplastic. Cell migration is facilitated through plasmodesmata after NPs have entered the cytoplasm. The tiniest of the

Table 3.1 Modes of nanofertilizer application and their characteristics.

Modes of application	Description	Advantages	Challenges	References
In vitro techniques—Aeroponics	Roots are suspended in air and sprayed with nutrient mist. Allows precise control of the root environment but requires high nutrient input.	Precise nutrient control, no soil-borne diseases, high growth potential.	High nutrient demand, expensive setup, limited adoption.	Montoya-Martínez et al. (2024b)
In vitro techniques—Hydroponics	Roots are submerged in a nutrient solution. Requires careful monitoring of pH, oxygen levels, and volume. Vulnerable to pathogens and excess moisture.	Efficient nutrient delivery, controlled environment, supports plant growth.	Requires constant monitoring, susceptible to root infections, not suitable for all crops.	Montoya-Martínez et al. (2024a)
In vivo methods—Foliar application	Liquid nanofertilizers are sprayed onto leaves, ensuring rapid absorption. Enhances nutrient uptake and plant growth parameters. Factors like plant type, environment, and delivery methods impact effectiveness.	Rapid uptake, reduced leaching, effective for micronutrients like iron and manganese.	Stomatal timing affects uptake, formulation and particle size influence efficiency.	Saurabh et al. (2024b)
In vivo methods—Soil application	Nanofertilizers are absorbed through roots, encountering multiple barriers such as the cuticle and vascular system. Uptake occurs via apoplastic or symplastic pathways, influencing translocation efficiency.	Efficient root absorption, controlled nutrient release, long-lasting effects.	Barriers like root cuticle and vascular cylinder restrict absorption, larger NPs face uptake challenges.	Saurabh et al. (2024b)

TiO_2 particles form in the roots of the plant and are distributed to the other parts of the organism without undergoing the changes of going to a solution or a crystal phase. With a diameter of 140 nm, the NPs do not get deposited in the roots of the wheat plant. Particles with a diameter of 36 nm or greater are indeed deposited within the parenchyma of the roots but cannot get to the stele to be passed to the upper parts of the plant. The intermolecularly separated NPs may cause an increase in root growth at early developmental stages. TiO_2 NPs at concentrations of 1–100 ppm were observed to be harmless to the soil microbial community, while CuO, ZnO, and Ag NPs were observed to be harmful. Metal NPs can be absorbed by seeds and subsequently translocated into seedlings to promote plant growth through seed priming (Djanaguiraman et al., 2024). NFs may have the capacity to provide their nutrients at a controlled slow-release rate when used in combination

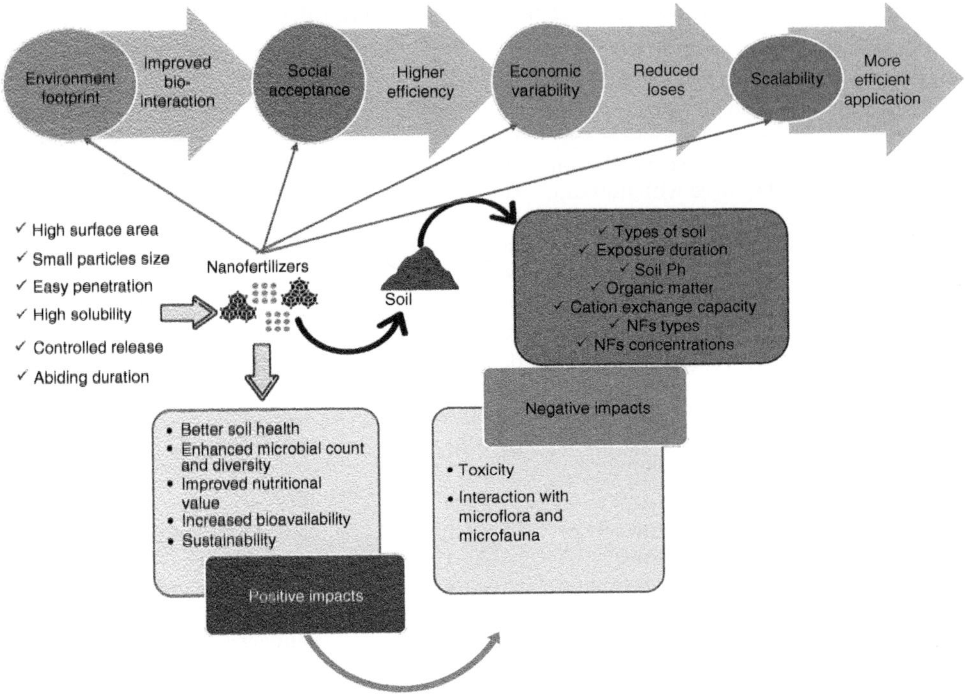

Figure 3.3 An overview of the role of nanofertilizers in soil.

with organic or synthetic fertilizers or when used alone. Controlled release may take up to 40–50 days to occur, which is substantially longer than the 4–10 days required for synthetic fertilizers.

3.7 On Soil Chemistry

Soil refers to an environmental matrix that includes elements like silicon (Si), strontium (Sr), aluminum (Al), barium (Ba), calcium (Ca), iron (Fe), potassium (K), sodium (Na), and rubidium (Rb). Numerous components make up the soil, including silicates, phosphates, sulfates, carbonates, hydroxides, clay, and iron oxides. Soil stability and NP behavior are determined by the physicochemical characteristics of the soil, including pH, moisture, clay content, temperature, texture, ionic charge, organic matter, microbiota, etc. Since nanoparticles are used in various industrial and agricultural sectors and come from a number of sources, including plants, the atmosphere, and water, soil is already recognized to be one of the primary absorbers of NPs and an eventual sinkhole. The use of nano-biofertilizers, however, only produces highly biodegradable NPs and non-toxic biomolecules. Enzyme-fabricated NPs and plant growth-promoting microorganisms (PGPM) proteins are ecologically safe and devoid of harmful chemicals. Because these NPs are colloidal, it is said that they improve the ability of soil's natural and particle organic materials to aggregate. This improves the soil's capacity to retain moisture, which in turn increases the plants' capacity to take in water (Anas et al. 2025b). While some studies have shown that NPs like TiO_2, Au, and Pt are integrated into the soil's organic matter and contribute to its stability, other studies found that Fe and Ag NPs did not affect the physicochemical properties of the soil but did increase the soil porosity. These biogenic NPs have also been found to impact one of the most significant biogeochemical cycles, namely the nitrogen fixation capacity of soil-associated microbes. In one case, it was demonstrated that carbon NPs increased the nitrogen-fixing activity and, consequently, the plant biomass (Ouyang et al., 2021). The period of exposure and the quantity of NPs in the soil commonly controlled how these components altered the soil's microbial metabolic pathways. Soil purification is another big and fascinating advantage supplied by the NPs. Fe is known to be one of the most researched and used elements for the remediation process, and the work made with the manufactured nanoparticles has been effective. Since halogen compounds and heavy metals including As, Cu, Ni, Ca, Cr, Fe, Pb, Mn, and Zn are present in extensively contaminated soil, biologically produced NPs may be used to remediate the area (Anas et al., 2024c; Anas et al., 2024d; Anas et al., 2025b). By including metal-oxidizing bacteria, nano-biofertilizers may effectively eliminate heavy metals from polluted systems. The biologically coated protein NPs of Fe and other metals have proven to be very stable under natural environmental conditions, and thus, they are able to participate in the remediation process. Stripped of contaminants, the soil-applied nanoparticles are able to aid in the decontamination of heavy metals in the fields (Ningombam et al., 2024b).

3.8 Soil Microbes

The effects of these biogenically modified NPs vary greatly depending on the kind and dose of nano-biofertilizers utilized. After being sprayed on the crops, the NPs eventually go into the soil and have a significant impact on the plant microbiomes and soil. These biogenic NPs improve the symbiotic microorganisms' capacity to cling to the plant rhizosphere. There are only a handful of studies that have documented the favorable impacts of metal NPs on soil microbes. It was noted in one study that silver NPs markedly boosted the populations of Acidobacteria, Actinobacteria, and Bacteroidetes in the soil (Jain et al., 2024a). Another study citing waste sources of NPs reported that Archaebacteria and Cyanobacteria soil populations were enhanced. Several studies have documented that in comparison with these nano-biofertilizers, multiwalled carbon nanotubes enhanced plant nitrogen fixation and subsequently plant biomass. In terms of the plant pathogenic microbes existing in the soil, these NPs are some of the most powerful growth inhibitors and therefore serve a key function for the protection of the plants. A wide variety of research has shown that antimicrobial properties of metallic NPs could inhibit the proliferation of disease-causing microorganisms within plants (Table 3.2).

3.9 Limitations Regarding the Use of Nanofertilizers

The application of new technology in the agriculture sector remains essential for curbing the growing appetite for food globally (Obaideen et al., 2022). The two most important factors in the establishment of a crop yield framework are climate change containment and the growth of agriculture sustainably. Nevertheless, over the last few decades, it is evident that excessive application of nanofertilizers is harmful to the environment and living organisms. While the conventional agriculture system benefits from the use of these fertilizers, some experts are far too pessimistic about any other potential harm that may arise. Because of the reactivity, NPs can combine with different constituent parts of an environment, and as a result get transformed through change of their physicochemical attributes (Joudeh, & Linke, 2022). Such altered nanofertilizers (NFs) can become toxic when they react with soil constituents. The other side of the equation is that the aggregation of NFs may also result in the formation of ROS, which causes growth retardation and ultimately cell death. However, consuming foods with excessive NFs might be fatal. An investigation was conducted to analyze the absorption and bioavailability translocation of ceramic oxide nanoparticles (NPs) in plant cells in plants of cucumber where 15% was reduced from cerium (Ce) IV to Ce III, and the products of the change were moved towards the phloem. These species could potentially be detrimental to human health, and therefore, their risk situation requires further assessment (Roy et al., 2018).

The overuse of nanofertilizers (NFs) in agriculture could lead to intractable and hazardous ecological problems. The production and field application of nanomaterials have drawn attention to their reactivity and unpredictability, which poses potential threats to worker safety. Hence, it is important to assess coverage and negative impacts of risk NFs, including life-cycle analysis. In addition, these fertilizers should be utilized solely at the degree that is deemed appropriate, so as not to contribute to their adverse contamination of the ecosystem (Khan et al., 2018). At the international level, the Green Revolution enhanced the production of food grains while having an imbalance and massive employment of artificial or synthetic pesticides and fertilizers, both having a profound impact on ecosystem damage. It is equally important to ensure we do not repeat the same mistakes so future studies with ellipses also needed to assess the environmental impacts of such nanofertilizers, if any. Also, for the purpose of increasing crop yield under sustainable agriculture, these green synthesized or biosynthesized nanofertilizers, nano-biofertilizer with biostimulant, should be used. In this view, the cautious and scientifically justifiable use of nanofertilizers needs to be carefully considered before reaching a commercial decision about the advertising and selling of these nanofertilizers. Research on NPs or NFs should be conducted with special attention to the possible risks of toxicity, bioavailability, and overall safety preceding the application in agricultural production (Pudhuvai et al., 2024).

3.9.1 Pros and Cons of NFs

The production of agro-industrial, physiological, and agronomical vital NPs is likely to improve with new technologies, especially essential metallic NPs for making fertilizers, which have lower nutritional losses and better NUE owing to smart delivery systems. With the advancement of agriculture technology NPs and nanobiotechnology, the NPs are likely to be used as NFs with a targeted approach by acting as a delivery system for supporting nutrients as well as the utilization of the NPs to be used in soil or on plants to enhance fertilizer uptake and utilization (Munir et al., 2024). NFs are released at controlled rates so that the health and fertility of the soil can be maintained by reducing nutrient leaching into groundwater, preventing proliferation of toxicity, and striking a balance in nutrients

Table 3.2 Role of nanofertilizers in soil and their impact on soil chemistry and microbes.

Aspect	Description	Advantages	Challenges	References
Enhancing soil quality	Nanofertilizers improve soil structure, water retention, and microbial diversity. They provide controlled nutrient release, promoting plant growth and reducing dependency on conventional fertilizers.	Improves soil fertility, enhances water retention, supports microbial life.	Potential for long-term accumulation, unknown ecological impacts.	Ningombam et al. (2024b)
Nutrient absorption and transport	NPs are absorbed through diffusion and bulk flow, enabling slow and sustained nutrient release. They enter plants via apoplastic and symplastic pathways, ensuring efficient distribution.	Increased nutrient efficiency, reduced leaching, sustained release for 40-50 days.	Large NPs may not be efficiently translocated within plants, potential toxicity at high doses.	Ningombam et al. (2024a)
Impact on soil chemistry	Nanoparticles interact with soil constituents such as clay, oxides, and organic matter, influencing soil stability and physicochemical properties like pH, moisture, and porosity.	Enhances soil aggregation, improves moisture retention, boosts soil stability.	Some NPs like Ag and Fe increase porosity but may not significantly alter soil chemistry.	Ningombam et al. (2024a)
Role in soil decontamination	Engineered NPs such as Fe, TiO_2, and biologically coated protein NPs aid in heavy metal remediation. Nano-biofertilizers remove pollutants efficiently from contaminated soils.	Effective for remediating As, Cu, Pb, Mn, Zn, and other heavy metals.	Stability and long-term effects of NPs in remediation need further study.	Ningombam et al. (2024a)
Effect on soil microbes	Biogenic NPs enhance microbial interactions, improving nitrogen fixation and plant biomass. Silver and carbon NPs have been shown to boost microbial populations, including Acidobacteria, Actinobacteria, and Cyanobacteria.	Promotes beneficial microbes, improves nitrogen fixation, inhibits plant pathogens.	High concentrations of some NPs (CuO, ZnO, Ag) may be toxic to soil microbes.	Jain et al. (2024a)
Pathogen control	Metallic NPs exhibit strong antimicrobial properties, inhibiting pathogenic microbes in the soil. This reduces plant disease occurrence and enhances overall plant health.	Protecting crops from soil-borne diseases, reduces need for chemical pesticides.	Overuse may lead to resistance in some microbial populations.	Jain et al. (2024b)

(Padhan et al., 2023). Zeolites can be slowly, uniformly, and continuously supplied to plants because of their nanoporous structure, which has great selectivity for plant minerals and large specific surface area. It was observed that the entire life cycle of the treated plants (24%) was on average about a quarter shorter than that of normal fertilized plants from sowing to maturity. Currently, there is a shift toward the adoption of new age fertilizers in place of the traditional ones that focus on balanced nutrition to cope up with different environmental factors and are also beneficial for the physiological health and functioning of the plants or crops. Synthetic fertilizers can release their nutrients in a week while nanofertilizers take about six to seven weeks. When applied in the field, synthetic urea fertilizers leak and evaporate at a rapid rate, losing approximately 70% of their nitrogen content and becoming less than 20% useful to plants (Yadav et al., 2023). Chemical fertilizers, which are known to improve the crop productivity of an area, are used extensively through many means, often resulting in pollution. Most farmers overuse fertilizers like the primary macronutrients nitrogen, phosphorus, and potassium without knowing that 40–70%, 80–90%, and 50–90% will be lost and cause water pollution. Farmers often misuse these fertilizers in hopes of reaching desired production levels, which may lead to the degradation of soil health, and the crops are also negatively impacted due to increased salt in the soil (Hossain et al., 2020).

3.10 Future Perspectives

Food security will remain a critical challenge in the near future with agriculture being impacted by climate change, population growth, and decrease in liveable land. Changes need to be made in implementing systematic crop production with the use of nano-agricultural inputs or nonconventional fertilizers (NPs) to enhance the ability of crops to grow in different ecological zones. This will facilitate the achievement of the challenges present. There is a need to be implemented. The integration of proteomics and metabolomics if the gene expression coding NPs in plants crops is executed and controlled by the system of nuclear DNA (nDNA) and organelle DNA (cpDNA, mtDNA) which enables a plant to grow, develop, function physiologically, and metabolize carbon dioxide considering light in the form of energy, photosynthesis, along with its biophysical processes using PSII and PSI, It is possible to integrate and regulate primary and secondary plant productivity. This leads to improved agricultural productivity globally though implementation of different cropping systems. NPs larger than certain sizes can only have direct access to the plant's genomic DNA during cell mitosis (Mehrian & De Lima, 2016).

There are two classes of genotoxicity processes for NPs: direct genotoxicity (where NPs either mechanically or chemically bond to the DNA and damage it) and indirect genotoxicity (which includes formation of ROS, diminished DNA repair activity, and binding with nuclear protein). The most damaging process of NPs to plants is DNA damage, which can be achieved via indirect or direct pathways. With the application of NiO NPs, direct genotoxicity was observed in the tomato plants, where the NPs encountered the cell's DNA, resulting in cell death. Co_3O_4 NPs cause indirect apoptosis inducing DNA damage in eggplants at the cellular level. Damage to DNA is caused by the breakdown of mitochondria, cristae, proliferation of peroxisomes, generation of NO, and vacuolization. In *Allium cepa* L., *Nicotiana tabacum* L., and *Vicia faba* L. NPs of ZnO caused destruction of the membrane integrity, breakage of DNA strands, and chromosomal aberrations. In regard to the specific impact of CD QD exposure on plastid (pt) and mitochondrial (mt) DNA, it may have modified the organelle genomes at the sub-stoichiometric level; however, nanoscale FeO_x and ZnS QDs constituents increased the amount of ptDNA and the amount of mtDNA by a factor of 1–3. NP CeO_2 did not change the stoichiometry of ptDNA and mtDNA (Marmiroli et al., 2022).

This indicates that changes in stoichiometry constitute a possible morpho-functional adaptive response to the exposure of nanoparticles that result from alterations of the bioenergetic redox balance in which the rate of photosynthesis or cellular respiration is lower. Interactions between plants and nanoparticles in the environment need to be carefully analyzed at the molecular level in order to reduce phytotoxic impacts so that soil health is maintained, which can increase yields and widen the environmentally friendly system by preventing the excessive use of traditional fertilizers. Therefore, the advancement of nanoformulations and their application in soil and groundwater offers innovative, safe, and cost-effective approaches to promote agroecological sustainability, supporting future generations and contributing to the development of high-quality human capital for the nation. The sustainable crop improvement strategy designed using the intrinsic capabilities of plants identified the profound need for healthy food for all future generations globally (Shah & Wu, 2019).

3.11 Conclusion

The emergence of nanofertilizers has brought with it innovative solutions to enhance soil fertility without further harming the environment, marking a modernization in agriculture. Their properties such as controlled nutrient release, better bioavailability, and lower levels of leaching make them far more effective than traditional methods. In this chapter, we covered how nanofertilizers were developed, as well as their possible applications and challenges in agricultural systems. The incorporation of nanotechnology into agriculture, especially in fertilizer manufacturing, has greatly increased crop yield and soil nutrients. They are designed at the nanoscale enabling precise delivery of elements such as macro- and micronutrients that plants require for faster growth. Their higher surface area-to-volume ratio increases nutrient absorption while restraining the overuse of fertilizers. In addition, the slow-release mechanisms greatly reduce losses of nutrients through volatilization and leaching, lowering the levels of contamination in groundwater while increasing nutrient use efficiency. It is possible to remedy soil nutrient insufficiencies with a high level of precision and control using nanofertilizer, which is their primary merit. Unlike conventional fertilizers, nanofertilizers, which overused often lead to degradation of soil, improve the soil structure and microbial activity by bringing over essential nutrients in a controlled manner. Several kinds of nanofertilizers have been created, including nanoclay-based fertilizers, polymeric nanocarriers, and metal oxide nanoparticles, for the optimal usage of plant nutrients while minimizing ecological consequences. The incorporation of biopolymeric materials such as chitosan and alginate further improve nutrients retention, gradual release to enhance soil health and plant productivity. In addition to improving nutrient efficiency, nanofertilizers are important for stress mitigation and improving plant responsiveness against some abiotic stresses like drought, salinity, and heavy metal toxicity. Some studies show that nanoparticles like zinc oxide (ZnO) and silicon dioxide (SiO_2) work at the basis of docile plant self-defense, modifying ROS and stimulating antioxidant enzymes, thus increasing their focus on sustainable crop production through enhanced growth and development. This ensures integration of the crops within the environment where they are subjected to stressors and increases farming practicality in adverse conditions.

Even with the many advantages presented by nanofertilizers, some barriers exist to their widespread use. A primary concern is the hypothesized adverse effects of nanoparticles on soil microorganisms, plants, and human health. The consequences of nanoparticle buildup in soil ecosystems require extensive research to ascertain their safe, long-term application. Furthermore, the responsible application of these types of fertilizers in agriculture requires the establishment of regulatory policies and standardization protocols for monitoring the impacts of nanofertilizers and their environmental consequences. Moreover, the profitability and volume of production of the fertilizers are additional serious issues to address. Although nanofertilizers have been proven effective in laboratory studies, their cost needs to be significantly lowered for them to be successfully synthesized on a commercial scale, which is the most efficient way to produce them. The use of plant extracts and microbial systems in nanoparticles' (Anas et al., 2025a) biosynthesis offers a better opportunity to produce ecologically friendly and economical nanofertilizers. It is essential to concentrate on improving environmentally friendly production technologies without compromising efficiency and environmental protection in future studies.

Moreover, public understanding and acceptance by farmers are instrumental in the marketing of nanofertilizers. A significant uptake of nanofertilizers can be achieved through public education campaigns directed toward farmers and training on the use of application techniques in farming. There is a need for researchers, policymakers, and industry stakeholders to work together in developing public awareness campaigns together with incentive programs and subsidizing farming activity that would facilitate the shift to using the nanofertilizers. The fusion of modern technologies like AI, precision agriculture, and smart delivery systems will dictate the future of nanofertilizers. Using AI-based models to track the condition of soil and the application of nanofertilizers would transform how resources are managed. Moreover, developing environmentally friendly formulations and biodegradable nanocarriers will improve the sustainability of nanofertilizers. In summary, enhancing soil fertility, nutrient use efficiency, and crop productivity can be optimally achieved with minimum environmental impact using nanofertilizers, the global world has emerged in the modern world of agriculture. There are still toxicity, cost, and regulatory issues that need addressing, but research efforts and current technological change are progressing in the direction of enabling practical use and broad-based development of the materials. Incorporating scientific research with environmentally friendly practices of farming would enable nanofertilizers to change the food security and ecological conservation scenario for the better. Improving agriculture will greatly depend on how these technologies are developed and implemented, as it will determine whether we can effectively protect the food production systems for the future.

References

Ahmad, F., Salem-Bekhit, M. M., Khan, F., Alshehri, S., Khan, A., Ghoneim, M. M., & Elbagory, I. (2022). Unique properties of surface-functionalized nanoparticles for bio-application: functionalization mechanisms and importance in application. *Nanomaterials, 12*(8), 1333.

Anas, M., Saeed, M., Elahi, M., Naeem, K., Shafique, M. A., & Quraishi, U. M. (2023). Histological and ionomics assessment to elucidate tolerance mechanisms of nickel-tolerant and sensitive cultivars of bread wheat (*Triticum aestivum* L.). *Plant Stress, 10*, 100277. https://doi.org/10.1016/j.stress.2023.100277

Anas, M., Khattak, W. A., Sarfraz, W., Khalid, N., Naeem, N., Farid, M., ... Fahad, S. (2024a). EDTA-enhanced phytoextraction: Leveraging *Ricinus communis* for the decontamination of Pb and Cd in agricultural soils. *Plant Growth Regulation*, 1–14.

Anas, M., Saeed, M., Naeem, K., Shafique, M. A., & Quraishi, U. M. (2024b). Anatomical and ionomics investigation of bread wheat (*Triticum aestivum* L.) to decipher tolerance mechanisms under arsenic stress. *Journal of Plant Growth Regulation, 43*(10), 3609–3625. https://doi.org/10.1007/s00344-024-11332-9

Anas, M., Bashir, M. S., Saleem, K., Noor, A., & Quraishi, U. M. (2024c). Adaptive mechanisms of wheat cultivars to lead toxicity through enhanced oxidative defense, ionomic redistribution, and anatomical modifications. *Journal of Plant Physiology, 303*, 154370.

Anas, M., Bibi, S., Farwa, U., Khana, A., & Quraishi, U. M. (2024d). Manganese oxide nanoparticles elicit cadmium tolerance in wheat seedlings by ionomic and phenomic adjustment, regulation of AsA–GSH pathway, cellular thickness recovery, and antioxidant modulation. *Environmental Science: Nano, 11*(11), 4588–4605. https://doi.org/10.1039/D4EN00490F

Anas, M., Falak, A., Hassan, S., Khattak, W. A., Saleem, M. H., Khan, K. A., Khalid, A., & Fahad, S. (2025a). Microbial interactions and bacterial responses to metal stress in plants: Mechanisms, adaptations, and applications for sustainable agriculture. *Journal of Crop Health, 77*, 36. https://doi.org/10.1007/s10343-025-01108-w

Anas, M., Khattak, W. A., Fahad, S., Alrawiq, N., Alrawiq, H. S., Abdelsalam, N. R., Jaremko, M., & Quraishi, U. M. (2025b). Mitigating heavy metal pollution in agriculture: A multi-omics and nanotechnology approach to safeguard global wheat production. *Journal of Hazardous Materials Advances, 17*, 100584. https://doi.org/10.1016/j.hazadv.2024.100584

Asadu, C. O., Ezema, C. A., Ekwueme, B. N., Onu, C. E., Onoh, I. M., Adejoh, T., ... Emmanuel, U. O. (2024). Enhanced efficiency fertilizers: Overview of production methods, materials used, nutrients release mechanisms, benefits and considerations. *Environmental Pollution and Management, 1*, 32–48.

Bloom, A. J., & Lancaster, K. M. (2018). Manganese binding to Rubisco could drive a photorespiratory pathway that increases the energy efficiency of photosynthesis. *Nature Plants, 4*(7), 414–422. https://doi.org/10.1038/S41477-018-0191-0

Djanaguiraman, M., Anbazhagan, V., Dhankher, O. P., & Prasad, P. V. (2024). Uptake, translocation, toxicity, and impact of nanoparticles on plant physiological processes. *Plants*. https://doi.org/10.3390/plants13223137

El-Saadony, M. T., Saad, A. M., Soliman, S. M., Salem, H. M., Desoky, E. S. M., Babalghith, A. O., & AbuQamar, S. F. (2022). Role of nanoparticles in enhancing crop tolerance to abiotic stress: A comprehensive review. *Frontiers in plant science, 13*, 946717.

Hossain, A., Krupnik, T. J., Timsina, J., Mahboob, M. G., Chaki, A. K., Farooq, M., ... Hasanuzzaman, M. (2020). Agricultural land degradation: processes and problems undermining future food security. In *Environment, climate, plant and vegetation growth* (pp. 17–61). Cham: Springer International Publishing.

Jain, A., Joy, L., & Reddy, U. (2024a). Environmental decontamination: Soil-microbe-plant interaction. In *Progress in Soil Microbiome Research* (pp. 219–239). Cham: Springer Nature Switzerland.

Jain, J., Yang, J., & Shi, H. (2024b). Vcoder: Versatile vision encoders for multimodal large language models. In *Proceedings of the IEEE/CVF Conference on Computer Vision and Pattern Recognition* (pp. 27992–28002).

Joudeh, N., & Linke, D. (2022). Nanoparticle classification, physicochemical properties, characterization, and applications: A comprehensive review for biologists. *Journal of Nanobiotechnology, 20*(1), 262.

Khan, M. N., Mobin, M., Abbas, Z. K., & Alamri, S. A. (2018). Fertilizers and their contaminants in soils, surface and groundwater. *Encyclopedia of the Anthropocene, 5*, 225–240.

Kumar, R., Kumar, R., & Prakash, O. (2019). Chapter-5: The impact of chemical fertilizers on our environment and ecosystem. *Chief Ed, 35*(69), 1173–1189.

Li, G., Gao, Q., Nyande, A., Dong, Z., Khan, E. H., Han, Y., & Wu, H. (2024). Cerium oxide nanoparticles promoted lateral root formation in Arabidopsis by modulating reactive oxygen species and Ca2+ level. *Functional Plant Biology, 51*(10). https://doi.org/10.1071/fp24196

Liyakat, K. K. S. (2024). Review of biopolymers in agriculture application: An eco-friendly alternative. *International Journal of Composite and Constituent Materials, 10*(1), 50–62p.

Marmiroli, M., Marmiroli, N., & Pagano, L. (2022). Nanomaterials induced genotoxicity in plant: Methods and strategies. *Nanomaterials, 12*, 1658. *The Genetic Changes Induced by Engineered Manufactured Nanomaterials (EMNs)*, 53

Mehrian, S. K., & De Lima, R. (2016). Nanoparticles cyto and genotoxicity in plants: Mechanisms and abnormalities. *Environmental Nanotechnology, Monitoring & Management, 6*, 184–193.

Montoya-Martínez, A. C., Chávez-Luzanía, R. A., Olguín-Martínez, A. I., Ruíz-Castrejón, A., Moreno-Cárdenas, J. D., Esquivel-Chávez, F., ... de los Santos-Villalobos, S. (2024a). Biological control of streptomyces species causing common scabs in potato tubers in the Yaqui valley. *Mexico. Horticulturae, 10*(8), 865.

Montoya-Martínez, A. C., Valenzuela Ruiz, V., Chávez-Luzanía, R. A., Parra-Cota, F. I., & de los Santos-Villalobos, S. (2024b). *Plant–pathogen interactions: Mechanisms involved in plant diseases* (pp. 17–25). Elsevier BV. https://doi.org/10.1016/b978.0-443-18855-8.00003-5

Munir, A., Salah, M. A., Ali, M., Ali, B., Saleem, M. H., Samarasinghe, K. G. B. A., De Silva, S. I. S., Ercisli, S., Iqbal, N., & Anas, M. (2024). Advancing agriculture: Harnessing smart nanoparticles for precision fertilization. *BioNanoScience, 14*(4), 3846–3863. https://doi.org/10.1007/s12668-024-01597-5

Ningombam, L., Hazarika, B. N., Yumkhaibam, T., Heisnam, P., & Singh, Y. D. (2024a). Heavy metal priming plant stress tolerance deciphering through physiological, biochemical, molecular and omics mechanism. *South African Journal of Botany, 168*, 16–25.

Ningombam, L., Mana, T., Apum, G., Ningthoujam, R., & Singh, Y. D. (2024b). Nano-bioremediation: A prospective approach for environmental decontamination in focus to soil, water and heavy metals. *Environmental Nanotechnology, Monitoring & Management*, 100931.

Nongbet, A., Mishra, A. K., Mohanta, Y. K., Mahanta, S., Ray, M. K., Khan, M., & Chakrabartty, I. (2022). Nanofertilizers: A smart and sustainable attribute to modern agriculture. *Plants, 11*(19), 2587.

Obaideen, K., Shehata, N., Sayed, E. T., Abdelkareem, M. A., Mahmoud, M. S., & Olabi, A. G. (2022). The role of wastewater treatment in achieving sustainable development goals (SDGs) and sustainability guideline. *Energy Nexus*, *7*, 100112.

Ouyang, B., Yilihamu, A., Liu, D., Ouyang, P., Zhang, D., Wu, X., & Yang, S. T. (2021). Toxicity and environmental impact of multi-walled carbon nanotubes to nitrogen-fixing bacterium *Azotobacter chroococcum*. *Journal of Environmental Chemical Engineering*, *9*(4), 105291.

Padhan, S. R., Kar, I., Mohanty, A., & Panigrahi, K. K. (2023). *Nanofertilizers: Types, Synthesis, Methods, and Mechanisms* (pp. 61–98). Springer International Publishing. https://doi.org/10.1007/978-3-031-41329-2_3

Pudhuvai, B., Koul, B., Das, R., & Shah, M. P. (2024). Nano-fertilizers (NFs) for resurgence in nutrient use efficiency (NUE): A sustainable agricultural strategy. *Current Pollution Reports, 11*(1), 1.

Raheman, S., Joseph, M., Hemalatha, M., Bhuvaneswari, J., Srinivasan, S., & Leninraja, D. (2024). Nano-fertilizers: The future of nutrient approaches for cereals. *Indian Journal of Agricultural Sciences, 94*(11), 1155–1164. https://doi.org/10.56093/ijas.v94i11.150587

Raza, A., Anas, M., Bhardwaj, S., Mir, R. A., Charagh, S., Elahi, M., Zhang, X., Mir, R. R., Weckwerth, W., Fernie, A. R., Siddique, K. H. M., Hu, Z., & Varshney, R. K. (2025). Harnessing metabolomics for enhanced crop drought tolerance. *The Crop Journal*. https://doi.org/10.1016/j.cj.2025.01.001

Roy, H. E., Rabitsch, W., Scalera, R., Stewart, A., Gallardo, B., Genovesi, P., & Zenetos, A. (2018). Developing a framework of minimum standards for the risk assessment of alien species. *Journal of Applied Ecology*, *55*(2), 526–538.

Sajjad, H., Sajjad, A., & Zia, M. (2023). Copper oxide nanoparticles: In vitro and in vivo toxicity, mechanisms of action and factors influencing their toxicology. *Comparative Biochemistry and Physiology C-Toxicology & Pharmacology, 271*, 109682. https://doi.org/10.1016/j.cbpc.2023.109682

Saurabh, K., Prakash, V., Dubey, A., Ghosh, S., Kumari, A., Sundaram, P., Jeet, P., Sarkar, B., Upadhyaya, A., Das, A., Kumar, S., Makarana, G., Kumar, U., Kumar, A., & Singh, R. R. (2024a). Enhancing sustainability in agriculture with nanofertilizers. *Deleted Journal*, *6*(11). https://doi.org/10.1007/s42452-024-06267-5

Saurabh, S., Bambhaniya, P., & Joshi, P. S. (2024b). Imaging ultracompact objects with radiatively inefficient accretion flows. *Astronomy & Astrophysics*, *682*, A113.

Shah, F., & Wu, W. (2019). Soil and crop management strategies to ensure higher crop productivity within sustainable environments. *Sustainability*, *11*(5), 1485.

Tripathy, D. B., & Gupta, A. (2024). Nanocomposites as sustainable smart materials: A review. *Journal of Reinforced Plastics and Composites*. https://doi.org/10.1177/07316844241233162

Wang, J., Qian, R., Li, J., Wei, F., Ma, Z., Gao, S., & Ren, X. (2024). Nitrogen reduction enhances crop productivity, decreases soil nitrogen loss and optimize its balance in wheat-maize cropping area of the Loess Plateau, China. *European Journal of Agronomy*, *161*, 127352.

Yadav, M. R., Kumar, S., Lal, M. K., Kumar, D., Kumar, R., Yadav, R. K., & Rajput, V. D. (2023). Mechanistic understanding of leakage and consequences and recent technological advances in improving nitrogen use efficiency in cereals. *Agronomy*, *13*(2), 527.

Zhang, L., Liang, Z., Hu, Y., Schmidhalter, U., Zhang, W., Ruan, S., & Chen, X. (2021). Integrated assessment of agronomic, environmental and ecosystem economic benefits of blending use of controlled-release and common urea in wheat production. *Journal of Cleaner Production*, *287*, 125572.

Zhang, J., Xie, S., Li, X., & Xia, X. (2024). Adoption of green production technologies by farmers through traditional and digital agro-technology promotion–An example of physical versus biological control technologies. *Journal of Environmental Management*, *370*, 122813. https://doi.org/10.1016/j.jenvman.2024.122813

4

Nano-Biofortification of Cereal Crops

Annika Jahan Aonti[1], *Md. Yousuf Ali*[2], *Most. Sirajum Munira*[3], *Md. Zubair Al Islam*[4], *and Akbar Hossain*[1]

[1] *Soil Science Division, Bangladesh Wheat and Maize Research Institute, Dinajpur 5200, Bangladesh*
[2] *Maize Breeding Division, Bangladesh Wheat and Maize Research Institute, Nashipur, Dinajpur 5200, Bangladesh*
[3] *Entomology Division, Bangladesh Wheat and Maize Research Institute, Nashipur, Dinajpur 5200, Bangladesh*
[4] *Soil Science Division, Bangladesh Institute of Nuclear Agriculture, Mymensingh, 2202, Bangladesh*

4.1 Introduction

Nano-biofortification of cereal crops involves the use of nanotechnology to increase micronutrient contents in crops such as rice, wheat, and maize, thereby alleviating nutritional inadequacies and enhancing food security. This technology employs nanoparticles (NPs) to more efficiently transport micronutrients to plants, resulting in increased nutrient content in edible sections. The distinct characteristics of nanomaterials, such as easy solubility, regulated release, improved bioavailability, localized distribution, and defensive encapsulation, hold significant potential for advancing nutrient delivery and strengthening agricultural resilience. Nanotechnology includes altering materials at the nanoscopic scale, with dimensions ranging from 1 to 100 nm (Mousavi & Rezaei, 2011). Materials at this minute scale have unique optical, electrical, and biological capabilities that can be harnessed in both plants and animals (Liu et al., 2019).

The integration of NPs, nanotubes, and nanofibers into agricultural practices can increase profitability and nutrient use efficiency, thereby contributing to food security. Nanofertilizers (NFs) are used to assure target-bound, delayed nutrient delivery to plants, limit nutrient volatilization and leaching, and increase nutrient bioavailability (Kushwaha et al., 2025). Biofortification uses plant breeding and biotechnology to increase micronutrient levels in cereal crops (Bouis & Saltzman, 2017). Traditional breeding procedures, on the other hand, have a limited ability to increase micronutrient levels in crops (De Steur et al., 2014). By using biofortification and antimicrobials, nanotechnology, the engineering of materials at the nanoscale level, offers creative ways to improve food safety and nutritional value (Sozer & Kokini, 2020). Encapsulated micronutrient nanodelivery systems, NP plant fortifiers, and nanoformulated fertilizers to improve beneficial

Nanotechnology Innovations for Food Security and Sustainable Agriculture, First Edition. Edited by Zia Ur Rehman Mashwani, Shah Fahad, Ilyas Ahmad, Shah Saud, Sajid Fiaz, and Taufiq Nawaz.

elements in foods are among the key advances in biofortification employing nanotechnology (Kanmani & Rhim, 2014). Compared with traditional application methods, this accuracy improves pesticide efficacy and decreases potential toxicity (Paris et al., 2020).

Nanotechnology is a new technology that can be gradually employed in agriculture to fortify crops and address the shortcomings of genetic and traditional agronomic biofortification (Thakur et al., 2018). Nanomaterials, particularly in the form of NFs, can ameliorate various stress conditions and increase crop production by increasing nitrogen metabolism, seedling growth, carbohydrate and protein synthesis, photosynthesis, and increased transport of nutrients from roots to leaves (Zulfiqar et al., 2019). Furthermore, a lower amount of NF can efficiently increase the nutritional content with less harm to the surrounding environment because of its target-bound delayed delivery (Al-Mamun et al., 2021)

There are several conflicting ways to supplement food crops, particularly cereals, with vital vitamins and minerals, including nutritional diversification. However, these solutions are no longer sustainable or affordable. Biofortification through the application of nanotechnology is a present, impending, hopeful, inexpensive, and permanent agricultural approach involving the provision of micronutrients that are needed to diverse populations of human beings who have limited access to good diets (Sahoo et al., 2023). Nanotechnology is emerging as a transformative tool in agriculture, offering new ways to improve crop productivity, resource efficiency, and sustainability (Dimkpa & Bindraban, 2016). Its main potential lies in the precise delivery of agrochemicals, enhanced crop protection, and improved food quality, but careful management and further research are needed to address potential risks (Nawaz et al., 2024).

4.2 Mechanisms of Nano-Biofortification

Nano-biofortification is an effective term in nanotechnology for the delivery of micronutrients with a significant impact on the stability of micronutrient levels to obtain a high yield status in biofortified crops. Some of the key mechanisms of nano-biofortification are discussed below:

4.2.1 Nanoparticle-Based Controlled and Slow Release: Nutrient Delivery Systems

Slow-release nutrient fertilizer (SRNF) and controlled release nutrient fertilizer (CRNF) both supply nutrients in a controlled fashion. CRNFs are mainly governed by one element: soil temperature (Lawrencia et al., 2021). These fertilizers dispense nutrients at a steady and predictable pace. The thick wall granulose form of fertilizer is crucial for nutrient release. On the other hand, SRNFs are generated by several factors, such as the soil temperature, moisture content, microbes, surface area, and pH, all of which are highly important for nutrient release (Liu et al., 2021). SRNFs come in different forms, including granules, nuggets, or packets/sachets, and provide a variety of release mechanisms. Unlike CRNFs, SRNFs do not have a specific coating material, and the dissipation of nutrients relies entirely on the properties of the fertilizers themselves.

4.2.2 Nanoencapsulation of Fertilizers

Nanoencapsulation, in the context of fertilizers, refers to the process of enclosing nutrients (macronutrients such as nitrogen, phosphorus, potassium, or micronutrients) within a nanoscale carrier or shell. Nutrients are released gradually and as needed by plants rather than all at once, which can lead to nutrient deficiency and environmental contamination (Singh et al., 2024). Nutrients are delivered directly to plant roots or specific areas where they are needed, maximizing nutrient uptake and reducing waste. Polymeric NPs and nanocapsules have the ability to encapsulate essential micronutrients such as iron, zinc, and vitamin A, ensuring their protection until they are released within plant tissues (Kanmani & Rhim, 2014). Compared with traditional soil fertilization techniques, these nanoencapsulated materials offer improved nutrient absorption. Nanocapsules also enable regulated, continuous release directly into plant cells and organs over time (Nair et al., 2010). Nanotechnology offers chances to increase the nutritional value of foods through biofortification (Nair et al., 2010).

- Chitosan: A natural polysaccharide that can form NPs and is used to encapsulate nutrients (Sharma et al., 2022).
- Biochar: A carbon-rich material that can be used to create a matrix for nutrient encapsulation (Hamed et al., 2024).

4.2.3 Nanomaterial-Mediated Improved Soil Interaction

The use of nanomaterials in the soil aims to enhance their characteristics and functions. Nanoamendments have the ability to enhance soil composition, moisture retention, and nutrient availability. NPs have tremendous potential for influencing soil properties and interactions because of their different physical and chemical properties at the nanoscale. When NPs are put into soil, they interact with soil particles, organic matter, and water components, altering the soil structure, nutrient availability, and water dynamics. These particles can impact soil composition by influencing soil aggregation, permeability, and density. For example, NPs can connect soil particles and enhance aggregate stability, which improves the soil structure and minimizes erosion (Zaib et al., 2023).

4.2.4 Nanostructured Micronutrient Fortification

Food fortification has proven to be the most safe and cost effective way to increase the nutritional value of a given diet. Nanotechnology can help crops obtain bioactive micronutrients (such as zinc, iron, and calcium) more efficiently. Wheat flour and sugar have been chosen for fortification in various developing nations to enhance foods deficient in important micronutrients and lower the incidence of detected deficiencies. Many countries have utilized rice as an important staple meal for fortification. These micronutrients can be encapsulated in NPs and given directly to plant tissues, allowing for more precise nutrient delivery (Fathima et al., 2017).

4.2.5 Improved Plant Stress Resistance

Sustainable agriculture involves less water and pesticide usage (Yadav et al., 2023). NF-treated plants often develop resilient defenses against pathogens and pests (Elbasuney et al., 2022). Some NFs, such as those containing copper or zinc oxide, possess inherent antimicrobial properties (Singh et al., 2023). These NPs can precisely target and neutralize toxic bacteria, fungi, and viruses, thus lowering the incidence of crop diseases. Additionally, NFs can increase the accumulation of antioxidants in plants (El-Bialy et al., 2023). These antioxidants are vital for removing harmful free radicals of ROS those are generated under stress, thereby providing shelter for plant cells and tissues from damage. Moreover, NFs may influence the synthesis of plant hormones such as auxins and cytokinins, which are crucial for stress resistance (Ghosh et al., 2021). These NPs can help plants adapt more successfully to difficult environmental conditions by maintaining an adequate hormonal balance (Al-Khayri et al., 2023).

4.3 Types of Nanomaterials Used in Biofortification

To improve the nutritional content of crops, especially cereals, biofortification with different types of nanomaterials is increasing daily. These materials are readily bioavailable, quickly absorbed, and stable in plants. Some commonly used nanomaterials for the biofortification of cereals are given below.

4.3.1 Nanoparticles

Nanoparticulate fertilizers are not the same as nanoparticulated systems, which provide vital bioactive components to plants directly or indirectly (Akbar et al., 2025). Fertilizers may contain NPs based on carbon, silica, or other organic polymers that are used to increase the bioavailability of nutrients to cereals such as wheat, rice, and maize.

- **Carbon nanotubes (CNTs):** CNTs as a controlled-release fertilizer and a plant growth enhancer. Their distinctive properties, including a large surface area and the capacity for functionalization, make them ideal for supplying nutrients to plants in a regulated way (Patel et al., 2020). Multiwalled CNTs promote plant growth and development in barley (*Hordeum vulgare*) and maize (*Zea mays*) (Lahiani et al., 2013).
- **Zinc (ZnO NPs):** ZnO NPs enhance a plant's resilience to salinity stress by stabilizing cell membranes and neutralizing reactive oxygen species that arise during stress. They are also involved in controlling cell division, facilitating the transport of nutrients and water, and adjusting the levels of carbohydrates, amino acids, protein metabolism, photosynthetic pigments, and osmoregulators (Zhou et al., 2024).
- **Iron Nanoparticles (Fe NPs):** Fe NPs aid in increasing the absorption and storage of iron in plants, effectively addressing iron deficiency (Xue et al., 2023).
- **Copper Nanoparticles (Cu NPs):** Cu NPs improve the uptake and distribution of copper within plants, which is vital for numerous enzymatic functions (Moreira et al., 2022).
- **Selenium Nanoparticles (Se NPs):** Se NPs increase the content of selenium, a crucial micronutrient for human health, in plants.

- **Silica nanoparticles:** Silicon (Si) ranks as the second most prevalent element on Earth, following oxygen, and has become increasingly significant in agriculture. Si is considered an essential and nonessential element for plants, as it is not solely crucial for their survival; however, its presence can offer substantial benefits (Luyckx et al., 2017). These NPs can engage with plants either directly or indirectly, leading to morphological and physiological alterations that increase stress tolerance (Babajani et al., 2019).

4.3.2 Hydroxyapatite Nanoparticles

Hydroxyapatite (HA) NPs constitute another category of substances that have gained attention for their ability to facilitate nutrient release. A promising nanoenabled fertilizer formulation for the progressive release of nitrogen was presented in 2017 by a research team headed by Amaratunga et al. that created HA nanohybrids loaded with urea (Kottegoda et al., 2017). HA $[Ca_{10}(PO4)_6(OH)_2]$ has a high surface area-to-volume ratio, making it a suitable carrier for both calcium and phosphorus. The agricultural effectiveness of these materials was determined through a field trial with rice (*Oryza sativa* L.). Compared with pure urea, the urea-HA nanohybrids resulted in an 8.2% increase in rice yield while requiring only half the nitrogen. This improvement was attributed to the slower release of urea and a 30% increase in nitrogen agronomic use efficiency (NAE) along with increased yield per unit of nutrient applied (Guo et al., 2018).

4.3.3 Nanofertilizers

These are nanoscale fertilizers; the individual sizes of the molecules should be less than 100 nm, the bulk size should be closer to 100 nm, and the nanoproduct should be stable (Akbar et al., 2025). These fertilizers result in the controlled release of nutrients, increased nutrient uptake efficiency (NUE), and reduced environmental impact compared with traditional fertilizers (Sempeho et al., 2014). A number of variables affect the size, concentration, composition, and chemical characteristics of nanomaterials (Thakur et al., 2018). For example, liquid nanourea is a liquid form of urea, a nitrogen fertilizer, in NP form, and nanourea, a foliar spray, increases crop production and provides the nutrients required for saving nitrogen fertilizers in crops to 25–34% (Velmurugan & Swarnam, 2021). The combination of nanourea and nano-Zn with 50% of the recommended amount of conventional urea is used to achieve the optimal yield (Ferdousi, 2024). DAP (liquid): An efficient source of nitrogen and phosphorus for application in the cereal crops wheat, maize, and rice with the aim of biofortification efforts. In addition to titanium dioxide, ZnS/ZnCdSe core–shell quantum dots (QDs), InP/ZnS core–shell QDs, Mn/ZnSe QDs, gold nanorods, Al_2O_3, TiO_2, CeO_2, and FeO have been developed (Sangeetha et al., 2017).

4.3.4 Macronutrient Nanofertilizers

One or more macronutrients (such Ca, Mg, K, N, and P) that have been encapsulated or intercalated with nanomaterials make up the macronutrient NFs. Since plants need large amounts of macronutrients, especially nitrogen–phosphorus–potassium (NPK), to grow and produce, it is projected that by 2050, the world's consumption of composite NPK fertilizer will reach

263 million tons (Alexandratos & Bruinsma, 2012). For example, NPK fertilizer application greatly increases wheat production per capita, but it also pollutes land and water (Basavegowda & Baek, 2021). In this context, a more environmentally friendly alternative might be offered by increasing the use of macronutrient NFs in wheat cultivation.

4.3.5 Nanoemulsion

Nanoemulsions are termed oil-in-water emulsions or nanosized droplets in water (Mittal et al., 2020). Nanoemulsions enhance the bioavailability and stability of micronutrients, including the encapsulation of micronutrients within tiny droplets, shielding them from environmental degradation, and improving their absorption in the digestive system. Additionally, insights into the different methods used to create nanoemulsions, such as high-energy and low-energy techniques, are needed. Several case studies have focused on the use of nanoemulsion technology to enrich staple grains such as rice, wheat, and maize. These examples highlight practical advantages, such as better nutritional profiles, longer shelf life, and minimal effects on sensory characteristics (Akhavan-Mahdavi et al., 2025). For example, nanoemulsions deliver water-insoluble nutrients such as vitamin A or D to fortify cereals.

4.4 Applications in Cereal Crops

Cereals are a crucial component of human diets, serving as a vital food source. However, they typically offer low-quality protein, and the micronutrients they contain, such as iron, zinc, iodine, and vitamin A, are not easily absorbed by the body (Tailor et al., 2022). On average, cereals consist of approximately 75% carbohydrates (primarily starch) and 6–15% protein in whole grains, with variations across different species, contributing to approximately 50% of worldwide energy intake. The importance of cereals is underscored by their role in global food security, with an annual availability of 2,600 million tons (Laskowski et al., 2019). The European Food Safety Authority recommends that adults consume 8–11 mg of zinc, 8–18 mg of iron, and 750 mg of calcium per day, depending on sex. These requirements are often unmet in daily diets, leading to micronutrient deficiencies (Sakellariou & Mylona, 2020). Populations that rely solely on cereal-based diets that lack essential micronutrients experience adverse health effects, ultimately leading to malnutrition. Essential micronutrients such as iron, zinc, and provitamin A are vital for maintaining the structural stability of biological systems. Biofortification aids in improving iron, zinc, vitamin A, and iodine levels in cereal crops (Bouis & Saltzman, 2017). While conventional breeding methods have faced challenges in significantly increasing the levels of these micronutrients in plants, nanoenabled delivery systems and fortificants present promising alternatives by increasing both nutrient content and bioavailability.

4.4.1 Cereals

Among the different cereal crops, rice, wheat, maize, sorghum, barley, millet, and oat are the most common for human consumption. Rice is considered the first cereal crop, which is very much deficient in Fe, Zn, and provitamin A; results in hidden hunger in rice-eating areas around the globe; and causes anemia, night blindness, and loss of vision (Majumder et al., 2019). Therefore, the biofortification of rice with vitamins A, B1, B6, and E, iron, zinc, and other compounds is increasing daily. Iron-biofortified rice ingestion is valuable in the developing world to improve iron accumulation among poor women with a dietary iron deficit. An experiment was conducted with 192 women in Philippines for nine months. These women were divided into two groups: iron-fortified rice (3.21 mg/kg Fe) and local rice (0.57 mg/kg Fe). There was an approximately 17% difference in total iron uptake compared with that of local rice varieties (Ram et al., 2015). The use of the biofortified crop golden rice (1999) is another innovation in modern science. The gene inserted from bacteria and daffodil into rice plants presented an approximately 20-fold greater beta-carotene content than did the protype of rice. The β-carotene content in golden rice parallels that of pure β-carotene in oil, which is essential for providing vitamin A to children (Tang et al., 2012). For example, foliar applications of ZnO have been found to increase grain zinc levels by 50% compared with bulk zinc applications (Pradhan et al., 2013).

Wheat is an important food crop for almost one-third of the global population. Moreover, it lacks important micronutrients such as iron and zinc, resulting in severe deficiency in several developed nations. Wheat biofortification has been performed via a variety of methods, including conventional breeding, genetic engineering, and agronomic strategies. These procedures lead to significant storage of iron and zinc in wheat grains, and people who consume biofortified wheat products have improved their health and nutritional status (Altaf et al., 2024). Conventional breeding involves the identification and selection of wheat varieties with naturally relatively high levels of essential nutrients. The application of iron NPs to soil during plant growth has resulted in a more than 25% increase in the grain iron content (Misra et al., 2016). Moreover, compared with bulk zinc applications, the use of zinc oxide (ZnO) NPs as a foliar spray led to a greater than 40% increase in the zinc content in wheat grains. A study was performed to determine the most effective Zn and Mn genotypes by comparing the nutrient usage efficiency characteristics of 14 wheat genotypes. BW-824, BWL-8855, PBW-872, PBW-869, and PBW Zn1 were the most productive genotypes in terms of crop production, whereas PBW-824 and PBW-869 were the highest ranked in terms of quality metrics. These Zn + Mn-efficient genotypes are valuable resources for improving the quality yield of wheat (Dhaliwal et al., 2023). Genetic engineering modifies the wheat genome to increase the levels of specific nutrients. The agronomic strategies involved the nanoapplication of urea (4% nitrogen), DAP (8.0% N w/v), phosphorus (16.0% P2O5 w/v), zinc, and iron to the leaves of the plants. These approaches are the most environmentally friendly and cost-effective approaches for increasing the nutrient content of wheat plants.

Maize ranks as the third most popular cereal crop, following wheat and rice, making up a significant portion of poultry feed at 49%. It is also used for human food (25%), animal feed (12%), industrial products (12%), brewing (1%), and seeds (1%). Only 30–40% of maize is consumed by people, and the bulk, 60–70%, is used as livestock feed within the country (Gwirtz & Garcia-Casal, 2014). Soil or foliar application of zinc fertilizer enhances corn output and grain concentration, leading to improved corn quality (Xue et al., 2023). In the maize endosperm, the insertion of the bacterial crtI and crtB genes, regulated by the endosperm-specific super gema zein promoter, increased vitamin A levels to as much as 9.8 μg/g of dry matter. Similarly, the overexpression of the maize Psy1 gene led to an increase in the β-carotene content. The overexpression of the dehydroascorbate reductase gene increased vitamin C levels by converting the oxidized form of ascorbic acid into its reduced form (Foyer et al., 2020). An increase in iron bioavailability was achieved through the insertion of *Aspergillus phytase*, *Aspergillus niger* phy2, and the soybean ferrite gene in maize. The overexpression of the sb401 gene from potato increased the lysine content in maize grains. Antisense dsRNA targeting α-zeins improved the accumulation of tryptophan and lysine in maize. Homogentisic acid geranylgeranyl transferase (HGGT) gene integration in maize increases tocopherol and tocotrienol levels, resulting in biofortified maize. Additionally, the overexpression of the PSY1 maize gene resulted in increased vitamin A accumulation of 59.32 μg/g DW (Hossain et al., 2019; Thanh, 2019).

Barley (*H. vulgare*) is recognized as another important key cereal crop, after rice, corn, and wheat. It is cultivated extensively across the globe, ranking fourth in both production volume and the area cultivated among grain crops (Mittal, 2022). Notably, whole-grain barley is rich in dietary fiber, containing approximately 14.8 g per 100 g of raw material, and it also provides significant amounts of bioactive compounds and minerals such as iron (6 mg/100 g), zinc (3.3 mg/100 g), and calcium (50 mg/100 g) (Langridge, 2018). Compared with nonbiofortified flour, biofortified barley flour has been shown to have better physicochemical qualities, as demonstrated by different milling processes, such as roller and hammer milling (Acar et al., 2020). With respect to agronomic practices, the combination of NPK fertilizers with organic fertilizers and micronutrient supplementation is the most effective strategy for improving barley grain quality. Specifically, fertilizing spring barley with NPK supplemented with sulfur significantly increases the content of essential micronutrients such as manganese (Mn), iron (Fe), zinc (Zn), and copper (Cu) in grain, thereby increasing its nutritional value (Barczak et al., 2019). In barley, selenium (Se) biofortification is primarily pursued through fertilization (either soil or foliar application), and the addition of either Na_2SeO_4 or Na_2SeO_3 to the soil can significantly increase the Se content in the grain (Rodrigo et al., 2013). The selenium concentration of beer brewed from biofortified barley is six times greater than that of beer brewed from unfertilized barley (Rodrigo et al., 2014). Plant growth-promoting rhizobacteria (PGPRs), another type of biofortification, increase barley yield by producing hormones (Singh et al., 2020). Inoculation of barley cultivars with the arbuscular mycorrhizal fungus *Rhizophagus irregularis* positively affects ZIP (Zrt-like, Irt-like protein) transporter genes in the roots (notably upregulating the HvZIP13 gene), leading to increased zinc concentrations in both grains and straw (Watts-Williams & Cavagnaro, 2018). Table 4.1 shows an overview of the application of nano-biofortification in cereal crops.

4.5 Benefits of Nano-Biofortification in Cereal Crops

4.5.1 Improved Nutrient Uptake and Bioavailability

Nano-biofortification enhances the accumulation and bioavailability of essential micronutrients such as iron (Fe), zinc (Zn), selenium (Se), and iodine (I) in cereal grains. Owing to their nanoscale size and high surface reactivity, nanoenabled fertilizers and supplements facilitate efficient nutrient uptake by plants and increase nutrient translocation to

Table 4.1 Nano-biofortification of cereal crops.

Targeted plant	Nano dose treatment	Nutrient forms	Growing media	Main results	References
Paddy rice	25–100 $\mu mol\ L^{-1}$ Se NPs	Nano-Se (chemical)	Pot experiment (foliar)	An appropriate dosage (50 μmol/L) of Se NPs can reduce Cd accumulation by enhancing photosynthesis and organic Se contents in rice grains	Wang et al. (2021)
Wheat	From 7.5 to 750 $mg\ L^{-1}$	Zn-EDTA, $ZnCl_2$ and ZnO-NPs (45 nm, chemical)	Hydroponic glasshouse	ZnO-NPs foliar application is capable to translocate Fe into wheat grains	Doolette et al. (2020)
Wheat	From 40 to 120 $mg\ L^{-1}$	$ZnNO_3$ and ZnO-NPs (biogenic)	Sandy loam soil in pots	80 ppm ZnO-NPs ppm dose performed superior results among different treatments and caused the highest increase in height, seed weight, yield, and biomass	Sheoran et al. (2021)
Wheat	2.17 and 4.34 $mg\ kg^{-1}$	ZnO-NPs and ZnO (>1,000 nm) ZnONPs (18 nm, chemical)	Pot experiment	ZnO-NPs may increase yield under water stress	Dimkpa et al. (2020)
Sorghum	1, 3, and 5 mg Zn kg^{-1}	ZnO-NPs (18 nm, chemical)	Pot experiment	ZnO-NP may reduce moisture stress	Dimkpa et al. (2019)
Wheat	From 20 to 1,000 $mg\ L^{-1}$	$ZnSO_4$ and ZnO-NPs (<100 nm, chemical)	Soil in pots	ZnO-NPs increased Zn content in grain than in leaf comparing to $ZnSO_4$	Du et al. (2019)

edible parts (Rizwan et al., 2019). In contrast to traditional fertilizers, which often result in poor nutrient absorption due to soil interactions, NFs facilitate direct nutrient uptake by plant roots and leaves (Dimkpa & Bindraban, 2018). Research has shown that the application of ZnO NPs in wheat leads to increased zinc accumulation in grains, which not only improves plant growth but also enhances the nutritional value of the produce (Rizwan et al., 2019). Similarly, the use of selenium NPs in rice has been linked to increased selenium content, which plays an important role in human antioxidant defense mechanisms and immune system function (Feregrino-Perez et al., 2018). Compared with conventional fertilizers, NFs provide steady and controlled nutrient release, leading to sustained crop growth (Feregrino-Perez et al., 2018). Since NFs are applied in relatively small amounts, they help prevent soil contamination caused by excessive accumulation of byproducts from chemical fertilizers, thereby reducing environmental hazards (León-Silva et al., 2018). These micronutrient-rich nanoformulations facilitate nutrient accumulation in sink organs, such as grains and edible parts of crops, thus improving food quality (Liu & Lal, 2015). Furthermore, modern nanosized formulations increase nutrient solubility and dispersion, significantly reducing nutrient losses and increasing the efficiency and sustainability of fertilization (Prasad et al., 2017).

4.5.2 Improved Crop Yield and Quality

Nano-biofortification plays a crucial role in improving both the nutrient content and overall productivity of cereal crops by increasing nutrient absorption and utilization. This is particularly important in addressing micronutrient malnutrition, commonly referred to as "hidden hunger," which affects a large portion of the world's population. In rice cultivation, the use of nanochelated iron fertilizers has demonstrated notable benefits. Researchers have indicated that applying these NFs increased the biological yield by 27%, decreased the number of hollow grains by 254%, and increased the protein content by 13%. Additionally, the concentrations of essential nutrients such as nitrogen, phosphorus, potassium, iron, and zinc in white rice increased by 46%, 43%, 41%, 25%, and 50%, respectively (Fakharzadeh et al., 2020).

4.5.3 Reduction in Fertilizer Wastage and Environmental Impact

Traditional fertilizers often lead to significant nutrient losses through runoff, volatilization, and leaching, contributing to environmental pollution and economic inefficiencies. NFs offer more controlled and sustained nutrient release, minimizing these losses and enhancing nutrient use efficiency (Parisi et al., 2015). Nanoencapsulated urea reduces nitrogen leaching into groundwater, ensuring better absorption by plants and promoting sustainable agricultural practices. Compared with conventional fertilizers, NFs require lower application rates because of their high surface area and reactivity, which improves nutrient uptake while reducing fertilizer waste. This targeted approach also decreases greenhouse gas emissions, such as nitrous oxide (N_2O), associated with conventional fertilizer application (Adisa et al., 2019; Manjunatha et al., 2016; Mathur et al., 2022). Additionally, NFs support soil health by preventing heavy metal accumulation and enhancing microbial activity, making them an environmentally sustainable alternative for modern agriculture.

4.5.4 Improved Resistance to Biotic and Abiotic Stresses

NPs are essential for strengthening plant defenses against biotic (pathogens and pests) and abiotic (drought, salt, and heavy metal toxicity) stressors. Nanoenabled fertilizers outperform conventional fertilizers in terms of transport efficiency and provide sustained nutrient release, reducing losses from leaching or soil fixation. Advanced nanocarriers target delivery through root and foliar penetration, thereby improving stress resistance (Bouhadi et al., 2025; Zulfiqar et al., 2019). Research has demonstrated that NFs significantly mitigate various abiotic stresses, including drought (Jaberzadeh et al., 2013), salinity (Siddiqui et al., 2014), heavy metal toxicity

(Tripathi et al., 2015), and extreme temperature fluctuations (Haghighi & Teixeira da Silva, 2014; Jaberzadeh et al., 2013). This multifunctional capability makes nano-biofortification a promising strategy for sustainable and climate-resilient agriculture.

4.5.5 Improving Overall Soil Health

Nano-biofortification, which focuses on enhancing the nutrient content of crops via NPs, can indirectly support soil health through improved nutrient cycling, increased plant growth, and increased microbial activity. Nano-biofortification improves the solubility and bioavailability of micronutrients such as zinc, iron, and selenium in plants. This enhanced uptake not only increases crop nutritional content but also reduces the need for excessive chemical fertilizers, which can degrade soil health over time (Prasad et al., 2017). NFs, through their controlled release mechanisms, minimize these losses, reduce environmental pollution, and preserve soil integrity. The presence of NPs, especially those used in nano-biofortification, can influence soil microbial communities. The stimulation of microbial activity can, therefore, improve soil fertility and health (Maurice 2023). The use of NFs in crops can improve root development and structure, which, in turn, enhances soil aeration, water retention, and overall soil quality (Sekhon, 2014). By reducing the frequency of fertilizer applications and improving the NUE of plants, nano-biofortification can help minimize soil contamination from excess chemicals. This is especially important for addressing concerns related to heavy metal accumulation in soils and the overall ecological health of agricultural environments (Mathur et al., 2022).

4.5.6 Potential for Sustainable Agriculture

Nano-biofortification supports sustainable agricultural practices by minimizing reliance on conventional chemical fertilizers, enhancing soil health, and promoting environmentally friendly nutrient management. The precision-targeted delivery of NFs improves nutrient use efficiency, ensuring that plants receive essential nutrients in an optimal and controlled manner, thereby reducing excessive fertilizer application and soil degradation (Sekhon, 2014). Furthermore, the gradual and controlled release of nutrients from NFs lowers the risk of nutrient runoff and leaching, mitigating environmental pollution in water bodies and reducing greenhouse gas emissions associated with traditional fertilizer production. By integrating nanotechnology into agricultural systems, long-term soil fertility and productivity can be maintained, ensuring food security for an expanding global population. The application of NFs also supports the development of climate-resilient crops by increasing plant growth under changing environmental conditions, making them valuable tools in sustainable farming.

4.6 Challenges of Nano-Bioprotection in Cereal Crops

4.6.1 Safety Concerns and Potential Toxicity of Nano-Biofortification

Despite its potential benefits, nano-biofortification has raised concerns regarding the toxicity and environmental impact of NPs. The unintended accumulation of NPs in plant, soil, and water ecosystems may pose risks to both ecological balance and human health. Certain synthetic NPs, especially those produced through chemical and physical methods, may exhibit greater toxicity than biogenic NPs do (Basavegowda & Baek, 2021). Owing to their metallic nature, many nanomaterials pose a risk of heavy metal toxicity in the soil, necessitating careful monitoring of their dosages, types, and application methods. Additionally, their nanoscale size allows easy absorption by plants, raising concerns about their potential entry into the food chain and associated health hazards (Khan et al., 2021). The bioaccumulation of NPs within the food chain presents uncertainties regarding potential long-term health effects (Maharramov et al., 2019). Nanomaterials are increasingly present in the environment and can impact microbial communities through various toxic mechanisms (Kapoor et al., 2022). While the exact pathways of toxicity are not fully understood, research highlights several direct and indirect effects on microbial structure, function, and diversity. Key mechanisms include cell membrane damage, oxidative stress, genotoxicity, and disruption of community composition and function (Fiol et al., 2021; Khanna et al., 2015; Klaine et al., 2008; Paramo et al., 2020; Pérez-de-Luque, 2017). Human exposure to NPs through dermal contact, inhalation, and ingestion is a growing concern. Some studies indicate that NPs can bypass biological barriers, including the blood–brain barrier, potentially contributing to neurological disorders such as Alzheimer's disease and Parkinson's disease (Nuthalapati et al., 2022). However, the full extent of these effects remains unclear, necessitating comprehensive toxicological evaluations (Ferdous & Nemmar, 2020). Thorough risk assessments are needed before the widespread application of nanomaterials in agriculture. Additionally, extensive studies on the behavior of NPs in the environment and their long-term impact on human health are essential for developing standardized safety guidelines for their agricultural use (Jain et al., 2018).

4.6.2 Cost and Economic Feasibility

The widespread adoption of NFs is hindered by their high production costs and limited commercial availability. Although nanoenabled fertilizers demonstrate superior efficiency in nutrient delivery and uptake, their cost effectiveness compared with that of conventional fertilizers remains a significant concern, especially for resource-limited farmers in developing countries (Tarafdar et al., 2014). The complex and expensive synthesis processes involved in NP production contribute to their elevated costs, making them less accessible for large-scale agricultural use. Additionally, knowledge gaps and market accessibility further restrict their adoption. Many small-scale farmers lack awareness of the potential benefits and proper application methods of NFs, leading them to rely on conventional fertilizers, which are more affordable and readily available (Reddy et al., 2024). The absence of well-established distribution networks and regulatory frameworks for NFs has also slowed their market expansion and integration into mainstream agricultural practices.

4.6.3 Lack of Standardized Regulations

The commercialization and widespread adoption of NFs are significantly hampered by the lack of globally accepted regulatory frameworks. A standardized regulatory framework is essential to sustain the responsible development and deployment of nanotechnology in agriculture. Establishing uniform guidelines for NF production, handling, and application would facilitate their integration into mainstream farming while addressing potential risks. Additionally, the adoption of comprehensive risk assessment methodologies, such as life cycle assessment (LCA), can provide a systematic approach for evaluating the environmental and health implications of NFs throughout their production, usage, and disposal phases (Salieri et al., 2018). LCA helps determine whether manufactured nanomaterials pose ecological risks and ensures that their benefits outweigh potential hazards.

4.6.4 Limited Field-Based Research

Most nano-biofortification studies have been conducted in controlled environments, with few large-scale field trials (Singh, 2022). Field conditions such as soil type, climate, and interactions with agricultural inputs can influence the efficacy of NF. Small-scale trials have shown benefits in terms of wheat nutrition (Munir et al., 2018), but large-scale adoption remains a challenge. Standardized field methodologies and regulatory frameworks are needed to assess scalability, environmental safety, and long-term effects. Collaborative efforts between research institutions, policymakers, and industry can facilitate real-world applications (Kiran et al., 2022).

4.6.5 Public Perception and Acceptance

The adoption of nano-biofortification in agriculture is influenced by public perception, which is often shaped by concerns regarding food safety, environmental risks, and potential long-term health effects. Many consumers are hesitant about consuming nanoenabled food products because of a lack of understanding of nanotechnology and its implications (Vandermoere et al., 2011). Strong public awareness campaigns, transparent labeling, and well-defined regulatory frameworks are vital for successful integration into agricultural and food production systems (Siegrist et al., 2007). Risk communication strategies should involve engagement with stakeholders, including farmers, policymakers, and consumers, to enhance the trust and acceptance of nano-biofortified crops (Frewer et al., 2016). Consumer education on the benefits, safety assessments, and sustainability aspects of nanotechnology in agriculture can help dispel misconceptions and promote informed decision-making.

4.7 Studies and Field Trials

Several studies have demonstrated the efficacy of nanotechnology-based fertilizer, specifically nano-Zn fertilizer, in increasing grain nutrient content and improving overall crop nutrition and productivity across cereal crops. A multilocation field experiment was conducted by Xue et al. (2023) on three corn varieties, namely, ZD958, DH605, and LD510, which provided evidence of nanotechnology-based biofortification. This study evaluated those varieties on the basis of conventional and nanoformulations. For nanoformulations, they used nano-ZnO and Zn, and this study tested both conventional zinc applications and nanoformulations such as ZnO- and Zn-chelated NPs. The other treatment (traditional) consisted of traditional treatments such as U + Zn and a cocktail approach, which results in higher Zn contents in the grain. For example, variety ZD958 presented increased Zn concentrations in grain compared with those of the control, from 13.6 to 31.2 $mg\,kg^{-1}$ at the Linyi location. Nano-Zn fertilizer also significantly increased the Zn content in maize grains. Compared with the control, nanotreatment, such as ZnO(Nano) and chelated nanozinc (U + HZnONPs and U + HZnCNP), increased the Zn enrichment to 134% in ZD958, 77% in DH605, and 74.9% in LD510. These treatments ranked third and fourth in terms of increasing the Zn concentration in the grain.

Another study was conducted by Stewart et al. (2019), who reported the effectiveness of lipid-based Fe and Zn nanoformulations in increasing the grain Zn and Fe contents of maize. In the present study, these two nanoformulations resulted in greater nutrient translocation to young leaves and grains than did the conventional and chelated sulfate formulations. These findings emphasize the delivery advantages and efficiency of lipid-based nanoformulations. Similar studies were conducted by Dimkpa et al. (2017) on sorghum. In this study, they summarized the results on the basis of treatment with $ZnSO_4$ and nano-ZnO. On the basis of the observed and analyzed data, the authors concluded that the nano-Zn treatment increased shoot uptake and grain translocation of P only at low and high NPK levels. These results suggest that nanoformulations can show superior nutrient interaction efficiency through controlled release, enhanced absorption and stimulation of P-mobilizing physiological pathways. In wheat, the multiple trials of NFs documented in Table 4.2 illustrate how fertilizers with nanotechnology outperform conventional fertilizers in increasing the nutrient concentration profile of wheat grains.

In rice, nutrients accumulate in the same way as noted in the work of Fakharzadeh et al. (2020), who reported increases in nitrogen, phosphorus, potassium, iron, and zinc contents of 46%, 43%, 41%, 25%, and 50%, respectively, after foliar application of nanochelated iron at the booting stage, which demonstrates the efficacy of nanochelated micronutrients in augmenting macro- and micronutrient levels. Wijerathna (2023) also reported advancements in nanodelivery systems in which carbon-based nanoparticles (CNPs) were used to shuttle zinc to rice. In the Zn-efficient IR36 rice variety, ZnCNPs significantly increased the mineral and protein contents of the grains, which were enriched in Zn and Fe. This change was due to the increased expression of some genes and proteins responsible for nutrient uptake, nutrient accumulation, nutrient storage, grain filling, and resorption under stress, revealing a twofold effect of CNPs on nutrient delivery and genetic manipulation.

Additionally, Parashar et al. (2023) studied the use of zinc nanocomposites in rice and reported a 55% increase in Zn accumulation in grains compared with a 35% increase under conventional zinc fertilization. These treatments also enhanced physiological functions and grain quality, reinforcing their application in biofortification strategies aimed at improving both nutritional value and yield. Finally, Table 4.3 shows findings from a two-season experiment (2018 and 2019) with the rice variety Giza179, which involved comparisons between conventional $ZnSO_4$ and foliar nano-Zn (ZnONP) applications. Compared with the control treatment ($ZnSO_4$+$ZnONP_{60}$), the $ZnSO_4$+$ZnONP_{60}$ treatment resulted in the maximum N, P, K, and Zn uptake, which resulted in the lowest nutrient uptake. Collectively, these findings demonstrate that nanozinc formulations offer a sustainable,

Table 4.2 The effects of nanofertilizers on the nutrient contents of wheat grains were compared with those of conventional or no fertilizer treatments on nutrient contents.

Application procedure	Trial condition	Medium of growth	Applied nanofertilizer	Applied amount	Nutrient amount (nano free)	Nutrient amount (with nano)	Traditional fertilizer	Applied amount	Nutrient content (conventional)	References
Leaf-applied	PVC columns	Sand	Zn-CNP	≈20 mg/g, ≈25 mL	≈15–19 μg/g DW Zn	≈21–25 μg/g DW Zn	NC	—	—	Deshpande et al. (2017)
Soil	Pot	Soil	Nanochitin	0.006 g/kg	184.3 g/kg protein, 58.6 mg/kg Fe, 42.13 mg/kg Zn	204.1 g/kg protein, 65.39 mg/kg Fe, 51.45 mg/kg Zn	NC	—	—	Xue et al. (2017)
Leaf-applied	Field	Soil	ZnO NP	2 g/L at 1.2 kg/ha	18.4 mg/kg Zn (Y1), 23.6 mg/kg Zn (Y2)	26.5 mg/kg Zn (Y1), 34.6 mg/kg Zn (Y2)	$ZnSO_4$	7 g/L at 4.2 kg/ ha	21.1 mg/kg Zn (Y1), 29.5 m g/kg Zn (Y2)	Zhang et al. (2017)
Leaf-applied	Field	Soil	Zn-CNP	40 mg/L	—	53.3 μg/g Zn	$ZnSO_4$	400 mg/L	59.4 μg/g Zn	Dapkekar et al. (2018)
Seed priming	Pot	Soil	ZnO NPs	100 mg/L	≈12 mg/kg DW Zn	≈20 mg/kg DW Zn	NC	—	—	Munir et al. (2018)
Leaf-applied	Field	Soil	Nano NPK	750:90:600 mg/L	—	13.5% protein	Traditional NPK	400 kg ha^{-1} urea, 200 kg/ha TSP, 100 kg/ha K_2SO_4	10.68% protein	Burhan and Al-Hassan et al. (2019)
Leaf-applied	Field	Soil	Nanochelated	240 kg/ha	—	17% protein	Urea	240 kg/ha	17% protein, 58 mg P, 27 mg K	Astaneh et al. (2021)

(Continued)

Table 4.2 (Continued)

Application procedure	Trial condition	Medium of growth	Applied nanofertilizer	Applied amount	Nutrient amount (nano free)	Nutrient amount (with nano)	Traditional fertilizer	Applied amount	Nutrient content (conventional)	References
Soil	Pot	Soil	ZnO NP	120 ppm	17.48 mg/g FW protein	22.71 mg/g FW protein	Chemical Zn	—	19.91 mg/g FW protein	Sheoran et al. (2021)
Soil	Field	Soil	ZnO NP	100 mg/L	≈39 mg/kg DW Zn	≈45 mg/kg DW Zn	NC	—	—	Khan et al. (2019)
Soil	Field	Soil	ZnO NPs	100 mg/L	≈30 mg/kg DW Zn	≈50 mg/kg DW Zn	NC	—	—	Rizwan et al. (2019)
Seed priming	Pot	Soil	Fe NPs	100 mg/L	≈20 mg/kg DW Fe	≈70 mg/kg DW Fe	NC	—	—	Rizwan et al. (2019)
Leaf-applied	Pot Greenhouse	Soil	Fe_2O_3 NP	600 ppm	≈30.4 mg/g FW protein	≈33.5 mg/g FW protein	NC	—	—	Sundaria et al. (2019)
Soil	Pot	Soil	Fe_2O_3 NP	100 mg/kg	≈25 mg/kg DW Fe	≈40 mg/kg DW Fe	NC	—	—	Adrees et al. (2020)

NP = Nanoparticle, NC = No comparison was performed with conventional fertilizer in this study; PVC: Polyvinyl Chloride (PVC) piping, FW: Fresh weight and DW: Dry weight.

Table 4.3 Influence of zinc application on nutrient uptake in rice cultivar Giza179 (2018 and 2019).

Treatments	N uptake (kg/ha)		P uptake (kg/ha)		K uptake (kg/ha)		Zn uptake (g/ha)	
Year	2018	2019	2018	2019	2018	2019	2018	2019
Zn_0	91.33^d	90.23^d	33.16^a	35.44^a	43.65	42.80^c	265.22^d	336.31^d
$ZnSO_4$	109.85^c	115.18^c	30.88^{ab}	34.25^b	45.01^{bc}	45.56^{bc}	349.46^c	385.32^c
$ZnSO_4$ + $ZnONP_{20}$	125.14^b	130.10^b	30.11^{ab}	32.54^b	47.84^{ab}	49.06^a	405.17^{ab}	430.16^b
$ZnSO_4$ + $ZnONP_{40}$	137.12^a	148.07^a	29.60^{ab}	29.94^b	49.15^a	50.77^a	447.13^a	462.85^a
$ZnSO_4$ + $ZnONP_{60}$	140.92^a	151.12^a	26.71^b	29.06^b	48.94^a	51.10^a	460.31^a	471.42^a
$ZnoNP_{20}$	112.22^c	124.51^c	30.23^{ab}	33.46^b	46.88^{bc}	48.75^{bc}	374.82^c	391.36^c
$ZnoNP_{40}$	121.2^{bc}	131.18^b	28.91^{ab}	32.11^b	47.26^{ab}	49.31^a	390.84^{bc}	431.61^b
$ZnoNP_{60}$	130.84^{ab}	143.28^a	28.30^b	31.74^b	49.07^a	50.41^a	435.16^{ab}	450.69^a

In column, similar superscript letter(s) indicate did not differ significantly and dissimilar superscript letter(s) indicate differed significantly.
Source: Elshayb et al., (2021)/MDPI/CC BY 4.0.

efficient and impactful solution for improving nutrient uptake and grain quality across cereals, positioning them as vital tools for combating global micronutrient deficiencies in agriculture.

4.8 Future Directions and Research Needs

Although nanotechnology holds immense promise in enhancing micronutrient uptake in cereal crops and combating malnutrition, several critical barriers must be addressed before its application can be confidently recommended for agricultural practices. One of the most urgent needs is the development of standardized methods to assess the effectiveness, safety, and long-term environmental impacts of nanoformulations across diverse agroecological landscapes (Atanda et al., 2025). While numerous successful laboratory-scale studies have been reported, validating these findings under field conditions is essential for determining their scalability, consistency, and practical feasibility in real-world farming systems.

The genotype specificity of the NF response is another important avenue to explore. Cereal genotypes can differ markedly in their efficiency of nutrient uptake, translocation, and utilization from a nanodelivery perspective, which may result in variation in their overall biofortification potential (Xue et al., 2023). To address this issue, future work needs to incorporate omics techniques, proteomics, and transcriptomics, which involve the study of genes and their product proteins, to elucidate the molecular and physiological processes involved in NP uptake and nutrient utilization.

Lin et al. (2024) emphasized the need for the development and increasing focus on biodegradable biosafe nanomaterials to mitigate environmental impact and protect food safety. Focusing on advanced multinutrient nanoformulations that can deliver important micronutrients such as Fe, Zn, and Se together should also be prioritized to increase the nutritional value of crops to improve human health (Tarafder et al., 2020). Owing to their nanoscale size and high surface area-to-volume ratio, nanomaterials possess unique features that differ from those of their bulk counterparts, posing potential toxicological concerns, as they can readily pass biological barriers and interact with tissues, cells, and even DNA (Baruah & Dutta, 2009; Dasgupta et al., 2015; De Azeredo, 2009). Therefore, it is necessary to educate farmers and promote interactions among agronomists, plant biotechnologists, toxicologists, policymakers, and other disciplines to enable the responsible use and spread of technologies. A special exciting avenue of research is to investigate plant–nanoparticle interactions at the rhizosphere level and focus on how root exudates, soil physicochemical properties, and microbiome dynamics affect the behavior, transformation, and uptake of NPs (Rajput et al., 2023). Knowing these intricate interactions may contribute to maximizing NF compositions and utilization methods for better efficiency. In addition, by assimilating precision agriculture techniques, including nanosensors and remote sensing, it becomes possible to monitor real-time crop nutrient status on a continuous basis, thus basing nutrient fertilizer application on a data-driven and targeted location (Gupta & Kumar Pal, 2025). Finally, investigating the climate resilience of nano-biofortification practices, as their potential to sustain or increase nutrient uptake under conditions of drought, salinity, or heat stress, is crucial for making this technology suitable for the objectives of climate-smart and sustainable agriculture (Zaman et al., 2025).

Together, these multidimensional research efforts will pave the way for the next generation of nanoenabled biofortification tools capable of sustainably addressing both agricultural productivity and global nutritional security.

4.9 Conclusion

In terms of nutritional content and food safety, nanotechnology has become a promising phenomenon worldwide. Different types of nanotechnology and their application approaches are now able to increase the abundance of several cereal crops with micronutrients such as iron, zinc, and vitamin A through biofortification. Nonetheless, as nanotechnology seeks to expand within the agriculture and food industries, it faces several challenges and unresolved issues. The potential toxicity and long-term effects of NP accumulation and their interactions with the environment require thorough investigation, particularly concerning prolonged exposure. Publicity and the establishment of standardized regulations are also crucial for acceptance of social content. Future research should focus on further enhancing the fineness, responsiveness, and cost effectiveness of nanoenabled systems for identification, purification, and nutrient delivery. By prioritizing safety and sustainability, nanotechnology has the potential to revolutionize nutrition and food safety, thereby bolstering global food security. However, the successful implementation of nanoinnovations for tangible benefits in a real-world context will necessitate collaboration among various stakeholders, from scientists to consumers.

References

Acar, O., Izydorczyk, M. S., Kletke, J., Yazici, M. A., Ozdemir, B., Cakmak, I., & Koksel, H. (2020). Effects of roller and hammer milling on the yield and physicochemical properties of fibre-rich fractions from biofortified and nonbiofortified hull-less barley. *Journal of Cereal Science*, *92*, 102907.

Adisa, I. O., Pullagurala, V. L. R., Peralta-Videa, J. R., Dimkpa, C. O., Elmer, W. H., Gardea-Torresdey, J. L., & White, J. C. (2019). Recent advances in nanoenabled fertilizers and pesticides: a critical review of mechanisms of action. *Environmental Science: Nano*, *6*(7), 2002–2030.

Adrees, M., Khan, Z. S., Ali, S., Hafeez, M., Khalid, S., & ur Rehman, M. Z., & Rizwan, M. (2020). Simultaneous mitigation of cadmium and drought stress in wheat by soil application of iron nanoparticles. *Chemosphere*, *238*, 124681.

Akbar, A., Yaqoob, A., Asif, F., Waqar, M., Rani, M., Hyder, Q. U. A., & Usman, H. M. (2025). Nano-Enabled Strategies for Soil Fertility and Plant Productivity: A Sustainable Approach. *Dialogue Social Science Review (DSSR)*, *3*(2), 1137–1151.

Akhavan-Mahdavi, S., Abdullahi, A., Navina, B., & Periakaruppan, R. (2025). Nanoemulsion Technology for Micronutrient Fortification in Staple Grains. In *Fortified foods* (pp. 1–28). New York, NY: Humana.

Alexandratos, N., & Bruinsma, J. (2012). World agriculture towards 2030/2050, the 2012 revision. In *2012, ESA Working Paper No. 12–03*. Rome: FAO.

Al-Khayri, J. M., Rashmi, R., Surya Ulhas, R., Sudheer, W. N., Banadka, A., Nagella, P., Aldaej, M. I., Rezk, A. A. S., Shehata, W. F., & Almaghasla, M. I. (2023). The role of nanoparticles in response of plants to abiotic stress at physiological, biochemical, and molecular levels. *Plants*, *12*(2), 292.

Al-Mamun, M. R., Hasan, M. R., Ahommed, M. S., Bacchu, M. S., Ali, M. R., & Khan, M. Z. H. (2021). Nanofertilizers towards sustainable agriculture and environment. *Environmental Technology & Innovation*, *23*, 101658.

Altaf, M. T., Liaqat, W., Bedir, M., Ali, A., Nadeem, M. A., & Baloch, F. S. (2024). Wheat biofortification: a promising approach to improve public health. In *Advances in wheat breeding: towards climate resilience and nutrient security* (pp. 623–651). Springer Nature Singapore: Singapore.

Astaneh, N., Bazrafshan, F., Zare, M., Amiri, B., & Bahrani, A. (2021). Nanofertilizer prevents environmental pollution and improves physiological traits of wheat grown under drought stress conditions. *Scientia Agropecuaria*, *12*(1), 41–47.

Atanda, S. A., Shaibu, R. O., & Agunbiade, F. O. (2025). Nanoparticles in agriculture: balancing food security and environmental sustainability. *Discover Agriculture*, *3*(1), 26.

Babajani, A., Iranbakhsh, A., Oraghi Ardebili, Z., & Eslami, B. (2019). Differential growth, nutrition, physiology, and gene expression in Melissa officinalis mediated by zinc oxide and elemental selenium nanoparticles. *Environmental Science and Pollution Research*, *26*(24), 24430–24444.

Barczak, B., Jastrzębska, M., & Kostrzewska, M. K. (2019). Biofortification of spring barley grain with microelements through sulfur fertilization. *Journal of Chemistry*, *2019*(1), 8214298.

Baruah, S., & Dutta, J. (2009). Nanotechnology applications in pollution sensing and degradation in agriculture: a review. *Environmental Chemistry Letters*, *7*, 191–204.

Basavegowda, N., & Baek, K. H. (2021). Current and future perspectives on the use of nanofertilizers for sustainable agriculture: the case of phosphorus nanofertilizer *3. Biotech*, *11*(7), 357.

Bouhadi, M., Javed, Q., Jakubus, M., Elkouali, M. H., Fougrach, H., Ansar, A., Ban, S. G., Ban, D., Heath, D., & Černe, M. (2025). Nanoparticles for sustainable agriculture: assessment of benefits and risks. *Agronomy*, *15*(5), 1131.

Bouis, H. E., & Saltzman, A. (2017). Improving nutrition through biofortification: a review of evidence from HarvestPlus, 2003 through 2016. Global Food Security, 12, 49–58.

Burhan, M. G., & Al-Hassan, S. A. (2019). Impact of Nano NPK fertilizers to correlation between productivity, quality and flag leaf of some bread wheat varieties. Iraqi Journal of. *Agricultural Sciences*, *50*.

Dapkekar, A., Deshpande, P., Oak, M. D., Paknikar, K. M., & Rajwade, J. M. (2018). Zinc use efficiency is enhanced in wheat through nanofertilization. Scientific Reports, 8(1), 6832.

Dasgupta, N., Ranjan, S., Mundekkad, D., Ramalingam, C., Shanker, R., & Kumar, A. (2015). Nanotechnology in agro-food: from field to plate. *Food Research International*, *69*, 381–400.

De Azeredo, H. M. (2009). Nanocomposites for food packaging applications. *Food research international*, *42*(9), 1240–1253.

De Steur, H., Mogendi, J. B., Blancquaert, D., Lambert, W., Van Der Straeten, D., & Gellynck, X. (2014). Genetically modified rice with health benefits as a means to reduce micronutrient malnutrition: global status, consumer preferences and potential health impacts of rice biofortification. In *Wheat and rice in disease prevention and health: benefits, risks and mechanisms of whole grains in health promotion* (pp. 283–299). Elsevier Academic Press.

Deshpande, P., Dapkekar, A., Oak, M. D., Paknikar, K. M., & Rajwade, J. M. (2017). Zinc complexed chitosan/TPP nanoparticles: a promising micronutrient nanocarrier suited for foliar application. Carbohydrate Polymers, 165, 394–401.

Dhaliwal, S. S., Sharma, V., Shukla, A. K., Behera, S. K., Verma, V., Kaur, M., & Hossain, A. (2023). Biofortification of wheat (*Triticum aestivum* L.) genotypes with zinc and manganese lead to improve the grain yield and quality in sandy loam soil. *Frontiers in Sustainable Food Systems*, *7*, 1164011.

Dimkpa, C. O., & Bindraban, P. S. (2016). Fortification of micronutrients for efficient agronomic production: a review. *Agronomy for Sustainable Development*, *36*(1), 7.

Dimkpa, C. O., & Bindraban, P. S. (2018). Nanofertilizers: new products for the industry? *Journal of Agricultural and Food Chemistry*, 66(26), 6462–6473. doi: 10.1021/acs.jafc.7b02150

Dimkpa, C. O., White, J. C., Elmer, W. H., & Gardea-Torresdey, J. (2017). Nanoparticle and ionic Zn promote nutrient loading of sorghum grain under low NPK fertilization. *Journal of agricultural and food chemistry*, *65*(39), 8552–8559.

Dimkpa, C. O., Singh, U., Bindraban, P. S., Elmer, W. H., Gardea-Torresdey, J. L., & White, J. C. (2019). Zinc oxide nanoparticles alleviate drought-induced alterations in sorghum performance, nutrient acquisition, and grain fortification. *Science of the Total Environment*, *688*, 926–934.

Dimkpa, C. O., Andrews, J., Fugice, J., Singh, U., Bindraban, P. S., Elmer, W. H., & White, J. C. (2020). Facile coating of urea with low-dose ZnO nanoparticles promotes wheat performance and enhances Zn uptake under drought stress. *Frontiers in Plant Science*, *11*, 168.

Doolette, C. L., Read, T. L., Howell, N. R., Cresswell, T., & Lombi, E. (2020). Zinc from foliar-applied nanoparticle fertiliser is translocated to wheat grain: a 65Zn radiolabelled translocation study comparing conventional and novel foliar fertilisers. *Science of The Total Environment, 749*, 142369.

Du, W., Yang, J., Peng, Q., Liang, X., & Mao, H. (2019). Comparison study of zinc nanoparticles and zinc sulfate on wheat growth: from toxicity and zinc biofortification. *Chemosphere, 227*, 109–116.

Elbasuney, S., El-Sayyad, G. S., Attia, M. S., & Abdelaziz, A. M. (2022). Ferric oxide colloid: towards green nanofertilizer for tomato plant with enhanced vegetative growth and immune response against fusarium wilt disease. *Journal of Inorganic and Organometallic Polymers and Materials, 32*(11), 4270–4283.

El-Bialy, S. M., El-Mahrouk, M. E., Elesawy, T., Omara, A. E. D., Elbehiry, F., El-Ramady, H., & Solberg, S. Ø. (2023). Biological nanofertilizers to enhance growth potential of strawberry seedlings by boosting photosynthetic pigments, plant enzymatic antioxidants, and nutritional status. *Plants, 12*(2), 302.

Elshayb, O. M., Farroh, K. Y., Amin, H. E., & Atta, A. M. (2021). Green synthesis of zinc oxide nanoparticles: fortification for rice grain yield and nutrients uptake enhancement. *Molecules, 26*(3), 584.

Fakharzadeh, S., Hafizi, M., Baghaei, M. A., Etesami, M., Khayamzadeh, M., Kalanaky, S., & Nazaran, M. H. (2020). Using nanochelating technology for biofortification and yield increase in rice. *Scientific Reports, 10*(1), 4351.

Fathima, S. J., Nallamuthu, I., & Khanum, F. (2017). Vitamins and minerals fortification using nanotechnology: bioavailability and Recommended Daily Allowances. In *Nutrient delivery* (pp. 457–496). Academic Press.

Ferdous, Z., & Nemmar, A. (2020). Health impact of silver nanoparticles: a review of the biodistribution and toxicity following various routes of exposure. *International Journal of Molecular Sciences, 21*(7), 2375.

Ferdousi, M. O. S. (2024). Liquid Nano Urea: Step Forward to Smart Agriculture-A Review. *Journal of Agroforestry and Environment, 17*(1), 26–31.

Feregrino-Perez, A. A., Magaña-López, E., Guzmán, C., & Esquivel, K. (2018). A general overview of the benefits and possible negative effects of the nanotechnology in horticulture. *Scientia Horticulturae, 238*, 126–137.

Fiol, D. F., Terrile, M. C., Frik, J., Mesas, F. A., Álvarez, V. A., & Casalongué, C. A. (2021). Nanotechnology in plants: recent advances and challenges. *Journal of Chemical Technology & Biotechnology, 96*(8), 2095–2108.

Foyer, C. H., Kyndt, T., & Hancock, R. D. (2020). Vitamin C in plants: novel concepts, new perspectives, and outstanding issues. *Antioxidants & Redox Signalling, 32*(7), 463–485.

Frewer, L. J., Fischer, A. R. H., Brennan, M., Bánáti, D., Lion, R., Meertens, R. M., & Vereijken, C. M. (2016). Risk/benefit communication about food—a systematic review of the literature. *Critical Reviews in Food Science and Nutrition, 56*(10), 1728–1745.

Ghosh, T. K., Tompa, N. H., Rahman, M. M., Mohi-Ud-Din, M., Al-Meraj, S. Z., Biswas, M. S., & Mostofa, M. G. (2021). Acclimation of liverwort Marchantia polymorpha to physiological drought reveals important roles of antioxidant enzymes, proline and abscisic acid in land plant adaptation to osmotic stress. *PeerJ, 9*, e12419.

Guo, H., White, J. C., Wang, Z., & Xing, B. (2018). Nano-enabled fertilizers to control the release and use efficiency of nutrients. *Current Opinion in Environmental Science & Health*, *6*, 77–83.

Gupta, G., & Kumar Pal, S. (2025). Applications of AI in precision agriculture. *Discover Agriculture*, *3*, 61.

Gwirtz, J. A., & Garcia-Casal, M. N. (2014). Processing maize flour and corn meal food products. *Annals of the New York Academy of Sciences*, *1312*(1), 66–75.

Haghighi, M., & Teixeira da Silva, J. A. (2014). The effect of N-TiO2 on tomato, onion, and radish seed germination. *Journal of Crop Science and Biotechnology*, *17*(4), 221–227.

Hamed, R., Jodeh, S., & Alkowni, R. (2024). Nano bio fertilizer capsules for sustainable agriculture. *Scientific Reports*, *14*(1), 13646.

Hossain, F., Muthusamy, V., Zunjare, R. U., & Gupta, H. S. (2019). Biofortification of maize for protein quality and provitamin-a content. *Nutritional Quality Improvement in Plants*, 115–136.

Jaberzadeh, A., Moaveni, P., & MOGHADAM, H.R.T., & Zahedi, H. (2013). Influence of bulk and nanoparticles titanium foliar application on some agronomic traits, seed gluten and starch contents of wheat subjected to water deficit stress. *Notulae botanicae horti agrobotanici cluj-napoca*, *41*(1), 201–207.

Jain, A., Ranjan, S., Dasgupta, N., & Ramalingam, C. (2018). Nanomaterials in food and agriculture: an overview on their safety concerns and regulatory issues. *Critical reviews in food science and nutrition*, *58*(2), 297–317.

Kanmani, P., & Rhim, J. W. (2014). Antimicrobial and physical-mechanical properties of agar-based films incorporated with grapefruit seed extract. *Carbohydrate Polymers*, *102*, 708–716.

Kapoor, P., Dhaka, R. K., Sihag, P., Mehla, S., Sagwal, V., Singh, Y., & Kumar, U. (2022). Nanotechnology-enabled biofortification strategies for micronutrients enrichment of food crops: current understanding and future scope. *NanoImpact*, *26*, 100407.

Khan, Z. S., Rizwan, M., Hafeez, M., Ali, S., Javed, M. R., & Adrees, M. (2019). The accumulation of cadmium in wheat (*Triticum aestivum*) as influenced by zinc oxide nanoparticles and soil moisture conditions. *Environmental Science and Pollution Research*, *26*, 19859–19870.

Khan, M. K., Pandey, A., Hamurcu, M., Gezgin, S., Athar, T., Rajput, V. D., Gupta, O. P., & Minkina, T. (2021). Insight into the prospects for nanotechnology in wheat biofortification. *Biology*, *10*(11), 1123.

Khanna, P., Ong, C., Bay, B.H., & Baeg, G.H. (2015). Nanotoxicity: an interplay of oxidative stress, inflammation and cell death. Nanomaterials, 5(3), 1163–1180. https://doi.org/10.3390/nano5031163

Kiran, A., Wakeel, A., Mahmood, K., Mubaraka, R., & Hafsa & Haefele, S.M. (2022). Biofortification of staple crops to alleviate human malnutrition: contributions and potential in developing countries. *Agronomy*, *12*(2), 452.

Klaine, S. J., Alvarez, P. J. J., Batley, G. E., Fernandes, T. F., Handy, R. D., Lyon, D. Y., & Lead, J. R. (2008). Nanomaterials in the environment: behavior, fate, bioavailability, and effects. *Environmental toxicology and chemistry*, *27*(9), 1825–1851.

Kottegoda, N., Sandaruwan, C., Priyadarshana, G., Siriwardhana, A., Rathnayake, U. A., Berugoda Arachchige, D. M., & Amaratunga, G. A. (2017). Urea-hydroxyapatite nanohybrids for slow release of nitrogen. *ACS Nano*, *11*(2), 1214–1221.

Kushwaha, P., Verma, R., Singh, A. K., & Dixit, P. (2025). *Nano-biofortification of millets: a review. PhytoTalks*, *2*(1), 284–291.

Lahiani, M. H., Dervishi, E., Chen, J., Nima, Z., Gaume, A., Biris, A. S., & Khodakovskaya, M. V. (2013). Impact of carbon nanotube exposure to seeds of valuable crops. *ACS applied materials & interfaces*, *5*(16), 7965–7973.

Langridge, P. (2018). Economic and academic importance of barley. In *The barley genome* (pp. 1–10). Cham: Springer International Publishing.

Laskowski, W., Górska-Warsewicz, H., Rejman, K., Czeczotko, M., & Zwolińska, J. (2019). How important are cereals and cereal products in the average polish diet? *Nutrients*, *11*(3), 679.

Lawrencia, D., Wong, S. K., Low, D. Y. S., Goh, B. H., Goh, J. K., Ruktanonchai, U. R., & Tang, S. Y. (2021). Controlled release fertilizers: a review on coating materials and mechanism of release. *Plants*, *10*(2), 238.

León-Silva, S., Fernández-Luqueño, F., & López-Valdez, F. (2018). Nanotechnology and fertilization: controlled release of nutrients for sustainable agriculture. *Agricultural Nanobiotechnology*, *5*(1), 45–67.

Lin, X., Yu, W., Tong, X., Li, C., Duan, N., Wang, Z., & Wu, S. (2024). Application of nanomaterials for coping with mycotoxin contamination in food safety: from detection to control. *Critical Reviews in Analytical Chemistry*, *54*(2), 355–388.

Liu, R., & Lal, R. (2015). Potentials of engineered nanoparticles as fertilizers for increasing agronomic productions. *Science of the Total Environment*, *514*, 131–139.

Liu, Y., Yue, L., Wang, Z., & Xing, B. (2019). Processes and mechanisms of photosynthesis augmented by engineered nanomaterials. *Environmental Chemistry*, *16*(6), 430–445.

Liu, C., Zhou, H., & Zhou, J. (2021). The applications of nanotechnology in crop production. *Molecules*, *26*(23), 7070.

Luyckx, M., Hausman, J. F., Lutts, S., & Guerriero, G. (2017). Silicon and plants: current knowledge and technological perspectives. *Frontiers in plant science*, *8*, 411.

Maharramov, A. M., Hasanova, U. A., Suleymanova, I. A., Osmanova, G. E., & Hajiyeva, N. E. (2019). The engineered nanoparticles in food chain: potential toxicity and effects. *SN Applied Sciences*, *1*(11), 1362.

Majumder, S., Datta, K., & Datta, S. K. (2019). Rice biofortification: high iron, zinc, and vitamin-A to fight against "hidden hunger". *Agronomy*, *9*(12), 803.

Manjunatha, S. B., Biradar, D. P., & Aladakatti, Y. R. (2016). Nanotechnology and its applications in agriculture: a review. *Journal of Farm Sciences*, *29*(1), 1–13.

Mathur, S., Pareek, S., & Shrivastava, D. (2022). Nanofertilizers for development of sustainable agriculture. *Communications in Soil Science and Plant Analysis*, *53*(16), 1999–2016.

Maurice, N. (2023). Impact of nanoparticles on soil microbes for enhancing soil fertility and productivity. In *Nanotechnology for Sustainable Agriculture* (pp. 75–127). Apple Academic Press.

Misra, P., Shukla, P. K., Pramanik, K., Gautam, S., & Kole, C. (2016). Nanotechnology for crop improvement. In *Plant nanotechnology: principles and practices* (pp. 219–256). Cham: Springer International Publishing.

Mittal, S. (2022). Wheat and barley production trends and research priorities: a global perspective. In *New horizons in wheat and barley research: global trends, breeding and quality enhancement* (pp. 3–18). Springer Singapore: Singapore.

Mittal, D., Kaur, G., Singh, P., Yadav, K., & Ali, S. A. (2020). Nanoparticle-based sustainable agriculture and food science: recent advances and future outlook. *Frontiers in Nanotechnology, 2*, 579954.

Moreira, A., Moraes, L. A., de Melo, T. R., Heinrichs, R., & Moretti, L. G. (2022). Management of copper for crop production. *Advances in Agronomy, 173*, 257–298.

Mousavi, S. R., & Rezaei, M. (2011). Nanotechnology in agriculture and food production. *J Appl Environ Biol Sci, 1*(10), 414–419.

Munir, T., Rizwan, M., Kashif, M., Shahzad, A., Ali, S., Amin, N., & Imran, M. (2018). Effect of zinc oxide nanoparticles on the growth and Zn uptake in wheat (*Triticum aestivum* L.) by seed priming method.

Nair, R., Varghese, S. H., Nair, B. G., Maekawa, T., Yoshida, Y., & Kumar, D. S. (2010). Nanoparticulate material delivery to plants. *Plant Science, 179*(3), 154–163.

Nawaz, T., Joshi, N., Rahman, T. U., Khan, M. N. R., Saud, S., & Fahad, S. (2024). Application of nanotechnology in biofortification. In *Revolutionizing agriculture: a comprehensive exploration of agri-nanotechnology* (pp. 67–100). Springer Nature Switzerland: Cham.

Nuthalapati, P., Asaithambi, S., Kumar, M., & Reddy, D. (2022). Nanoparticles: the promising future of advanced diagnosis and treatment of neurological disorders. In *Smart nanodevices for point-of-care applications* (pp. 221–234). CRC Press.

Paramo, L. A., Feregrino-Pérez, A. A., Guevara, R., Mendoza, S., & Esquivel, K. (2020). Nanoparticles in agroindustry: Applications, toxicity, challenges, and trends. *Nanomaterials, 10*(9), 1654.

Parashar, R., Afzal, S., Mishra, M., & Singh, N. K. (2023). Improving biofortification success rates and productivity through zinc nanocomposites in rice (Oryza sativa L.). *Environmental Science and Pollution Research, 30*(15), 44223–44233.

Paris, L., Peghaire, E., Moné, A., Diogon, M., Debroas, D., Delbac, F., & El Alaoui, H. (2020). Honeybee gut microbiota dysbiosis in pesticide/parasite co-exposures is mainly induced by Nosema ceranae. *Journal of Invertebrate Pathology, 172*, 107348.

Parisi, C., Vigani, M., & Rodríguez-Cerezo, E. (2015). Agricultural nanotechnologies: what are the current possibilities? *Nano Today, 10*(2), 124–127.

Patel, D. K., Kim, H. B., Dutta, S. D., Ganguly, K., & Lim, K. T. (2020). Carbon nanotube-based nanomaterials and their agricultural and biotechnological applications. *Materials, 13*(7), 1679.

Pérez-de-Luque, A. (2017). Interaction of nanomaterials with plants: what do we need for real applications in agriculture? *Frontiers in Environmental Science, 5*, 12.

Pradhan, S., Patra, P., Das, S., Chandra, S., Mitra, S., Dey, K. K., Akbar, S., Palit, P., & Goswami, A. (2013). Photochemical modulation of biosafe manganese nanoparticles on Vigna radiata: a detailed molecular, biochemical, and biophysical study. *Environmental Science & Technology, 47*(22), 13122–13131.

Prasad, R., Bhattacharyya, A., & Nguyen, Q.D.(2017). Nanotechnology in sustainable agriculture: recent developments, challenges, and perspectives. *Frontiers in Microbiology*, 8, 1014. https://doi.org/10.3389/fmicb.2017.01014

Rajput, A., Roy, S., Waghmare, M. N., Singh, S., Shukla, V. K., Singh, V., & Singh, P. (2023). Exploring sustainable practices in modern agronomy and their environmental impact-A Review. *International Journal of Environment and Climate Change*, *13*(11), 3146–3161.

Ram, H., Sohu, V. S., Cakmak, I., Singh, K., Buttar, G. S., Sodhi, G. P. S., & Mavi, G. S. (2015). Agronomic fortification of rice and wheat grains with zinc for nutritional security. *Current Science*, 1171–1176.

Reddy, M. K., Asthana, R., Debnath, S., Ray, P., Mandal, N., Arunachalam, A., & Koduru, J. R. (2024). Nanofertilizers for sustainable crop production: A comprehensive review. *BioNanoScience*, *14*(2), 1918–1939.

Rizwan, M., Ali, S., Ali, B., Adrees, M., Arshad, M., Hussain, A., & Waris, A. A. (2019). Zinc and iron oxide nanoparticles improved the plant growth and reduced the oxidative stress and cadmium concentration in wheat. *Chemosphere*, *214*, 269–277.

Rodrigo, S., Santamaría, O., López-Bellido, F. J., & Poblaciones, M. J. (2013). Agronomic selenium biofortification of two-rowed barley under Mediterranean conditions.

Rodrigo, S., Santamaria, O., Chen, Y., McGrath, S. P., & Poblaciones, M. J. (2014). Selenium speciation in malt, wort, and beer made from selenium-biofortified two-rowed barley grain. *Journal of Agricultural and Food Chemistry*, *62*(25), 5948–5953.

Sahoo, J. P., Mohapatra, U., Singh, S. K., Samal, K. C., Yadav, V. K., Moharana, A., & Dash, A. K. (2023). Nanotechnological approaches for biofortification concept and concern in cereal crops. In *Biofortification in cereals: progress and prospects* (pp. 367–384). Springer Nature Singapore: Singapore.

Sakellariou, M., & Mylona, P. V. (2020). New uses for traditional crops: the case of barley biofortification. *Agronomy*, *10*(12), 1964.

Salieri, B., Turner, D. A., Nowack, B., & Hischier, R. (2018). Life cycle assessment of manufactured nanomaterials: Where are we? *NanoImpact*, *10*, 108–120.

Sangeetha, J., Thangadurai, D., Hospet, R., Purushotham, P., Karekalammanavar, G., Mundaragi, A. C., David, M., Shinge, M. R., Thimmappa, S. C., Prasad, R., & Harish, E. R. (2017). Agricultural nanotechnology: concepts, benefits, and risks. In *Nanotechnology: An Agricultural Paradigm* (pp. 1–17). Springer Singapore: Singapore.

Sekhon, B. S. (2014). Nanotechnology in agri-food production: an overview. *Nanotechnology, Science and Applications*, *7*, 31–53.

Sempeho, S. I., Kim, H. T., Mubofu, E., & Hilonga, A. (2014). Meticulous overview on the controlled release fertilizers. *Advances in Chemistry*, *2014*(1), 363071.

Sharma, G., Prajapati, D., Devi, K. A., Pal, A., & Saharan, V. (2022). Chitosan nanomaterials for delivery of micronutrients in plants. In *Role of chitosan and chitosan-based nanomaterials in plant sciences* (pp. 239–253). Academic Press.

Sheoran, P., Grewal, S., Kumari, S., & Goel, S. (2021). Enhancement of growth and yield, leaching reduction in *Triticum aestivum* using biogenic synthesized zinc oxide nanofertilizer. *Biocatalysis and Agricultural Biotechnology*, *32*, 101938.

Siddiqui, M. H., Al-Whaibi, M. H., Faisal, M., & Al Sahli, A. A. (2014). Nano-silicon dioxide mitigates the adverse effects of salt stress on Cucurbita pepo L. *Environmental toxicology and chemistry*, 33(11), 2429–2437.

Siegrist, M., Cousin, M. E., Kastenholz, H., & Wiek, A. (2007). Public acceptance of nanotechnology foods and food packaging: the influence of affect and trust. *Appetite*, *49*(2), 459–466.

Singh, V. (2022). Nanotechnology and its application in agriculture biotechnology. *Knowledgeable Research A Multidisciplinary Journal*, *1*(04), 34–52.

Singh, D., Prasanna, R., Sharma, V., Rajawat, M. V. S., Nishanth, S., & Saxena, A. K. (2020). Prospecting plant–microbe interactions for enhancing nutrient availability and grain biofortification. In *Wheat and barley grain biofortification* (pp. 203–228). Woodhead Publishing.

Singh, D., Jain, D., Rajpurohit, D., Jat, G., Kushwaha, H. S., Singh, A., & Upadhyay, S. K. (2023). Bacteria assisted green synthesis of copper oxide nanoparticles and their potential applications as antimicrobial agents and plant growth stimulants. *Frontiers in Chemistry*, *11*, 1154128.

Singh, S., Angurana, R., Kaur, H., Katoch, V., Dhanjal, D. S., Aguilar-Marcelino, L., & Singh, J. (2024). Nanoencapsulation of fertilizers: delivery and applications. In *Nanofertilizer delivery, effects and application methods* (pp. 97–110). Elsevier.

Sozer, N., & Kokini, J. L. (2020). Nanotechnology and its applications in the food sector. *Trends in Biotechnology*, *38*(1), 82–89.

Stewart, Z. P., Paparozzi, E. T., Djanaguiraman, M., & Shapiro, C. A. (2019). Lipid-based Fe-and Zn-nanoformulation is more effective in alleviating Fe-and Zn-deficiency in maize. *Journal of Plant Nutrition*, *42*(14), 1693–1708.

Sundaria, N., Singh, M., Upreti, P., Chauhan, R. P., Jaiswal, J. P., & Kumar, A. (2019). Seed priming with iron oxide nanoparticles triggers iron acquisition and biofortification in wheat (Triticum aestivum L.) grains. *Journal of Plant Growth Regulation*, *38*(1), 122–131.

Tailor, S., Jain, K., Marwal, A., Meena, M., Anbarasu, K., & Gaur, R. K. (2022). Outlooks of nanotechnology in organic farming management. *Def Life Sci J*, *7*(1), 52–60.

Tang, G., Hu, Y., Yin, S. A., Wang, Y., Dallal, G. E., Grusak, M. A., & Russell, R. M. (2012). Retracted: β-carotene in golden rice is as good as β-carotene in oil at providing vitamin a to children1234. The American Journal of Clinical Nutrition, 96(3), 658–664.

Tarafdar, J. C., Raliya, R., Mahawar, H., & Rathore, I. (2014). Development of zinc nanofertilizer to enhance crop production in pearl millet (Pennisetum americanum). *Agricultural research*, *3*(3), 257–262.

Tarafder, C., Daizy, M., Alam, M. M., Ali, M. R., Islam, M. J., Islam, R., Ahommed, M. S., Aly, M. S. A., & Khan, M. Z. H. (2020). Formulation of a hybrid nanofertilizer for slow and sustainable release of micronutrients. *ACS Omega*, *5*(37), 23960–23966.

Thakur, S., Thakur, S., & Kumar, R. (2018). Bionanotechnology and its role in agriculture and food industry. *Journal of Molecular and Genetic Medicine*, *12*(324), 1747–0862.

Thanh, N. D. (2019). Provitamin A biofortification in maize through genetic engineering and marker-assisted selection. *Journal of Biology/TẠp chí Sinh HỌc*, *41*(4), 37–53.

Tripathi, D. K., Singh, V. P., Prasad, S. M., Chauhan, D. K., & Dubey, N. K. (2015). Silicon nanoparticles (SiNp) alleviate chromium (VI) phytotoxicity in Pisum sativum (L.) seedlings. *Plant physiology and biochemistry*, *96*, 189–198.

Vandermoere, F., Blanchemanche, S., Bieberstein, A., Marette, S., & Roosen, J. (2011). The public understanding of nanotechnology in the food domain: the hidden role of views on science, technology, and nature. *Public Understanding of Science*, *20*(2), 195–206.

Velmurugan, A., & Swarnam, T. P. (2021). Effect of enriched organic manures and traditional practices on soil carbon dynamics and yield of maize-rice cropping system in coastal lowlands. *Journal of Pharmacognosy and Phytochemistry*, *10*(1), 273–278.

Wang, C., Cheng, T., Liu, H., Zhou, F., Zhang, J., Zhang, M., & Cao, T. (2021). Nanoselenium controlled cadmium accumulation and improved photosynthesis in indica rice cultivated in lead and cadmium combined paddy soils. *Journal of Environmental Sciences*, *103*, 336–346.

Watts-Williams, S. J., & Cavagnaro, T. R. (2018). Arbuscular mycorrhizal fungi increase grain zinc concentration and modify the expression of root ZIP transporter genes in a modern barley (*Hordeum vulgare*) cultivar. *Plant Science*, *274*, 163–170.

Wijerathna, P. K. C. B. (2023). *Zinc biofortification of rice using carbon nanodots (Doctoral dissertation)*. Australia: University of Southern Queensland.

Xue, W., Han, Y., Tan, J., Wang, Y., Wang, G., & Wang, H. (2017). Effects of nanochitin on the enhancement of the grain yield and quality of winter wheat. *Journal of Agricultural and Food Chemistry*, *66*(26), 6637–6645.

Xue, Y. F., Li, X. J., Yan, W., Miao, Q., Zhang, C. Y., Huang, M., Sun, J. B., Qi, S. J., Ding, Z. H., & Cui, Z. L. (2023). Biofortification of different maize cultivars with zinc, iron and selenium by foliar fertilizer applications. *Frontiers in Plant Science*, *14*, 1144514.

Yadav, A., Yadav, K., & Abd-Elsalam, K. A. (2023). Nanofertilizers: types, delivery and advantages in agricultural sustainability. *Agrochemicals*, *2*(2), 296–336.

Zaib, M., Raza, I., Zubair, M., Arif, Z., Mumtaz, M. M., Abbas, M. Q., & Uzair, M. (2023). Nanoenabled soil amendments for improved soil structure and water holding capacity: an in-depth review. *International Journal of Educational Technology*, *5*(08), 344–357.

Zaman, W., Ayaz, A., & Park, S. (2025). Nanomaterials in agriculture: a pathway to enhance plant growth and abiotic stress resistance. *Plants*, *14*(5), 716.

Zhang, T., Sun, H., Lv, Z., Cui, L., Mao, H., & Kopittke, P. M. (2017). Using synchrotron-based approaches to examine the foliar application of ZnSO4 and ZnO nanoparticles for field-grown winter wheat. *Journal of agricultural and food chemistry*, *66*(11), 2572–2579.

Zhou, Y., Nie, K., Geng, L., Wang, Y., Li, L., & Cheng, H. (2024). Selenium's role in plant secondary metabolism: regulation and mechanistic insights. *Agronomy*, *15*(1), 54.

Zulfiqar, F., Navarro, M., Ashraf, M., Akram, N. A., & Munné-Bosch, S. (2019). Nanofertilizer use for sustainable agriculture: advantages and limitations. Plant Science, 289, 110270.

5

Role of Nanomaterials in Improving Oil Yield and Quality of Oilseeds Crops

*Zohaib Younas, Ilyas Ahmad, Faiz Ullah, and Zia ur Rehman Mashwani**

Department of Botany, PMAS Arid Agriculture University Rawalpindi, Pakistan

5.1 Introduction of Nanomaterials in Oilseed Crops

Food security means ensuring reliable access to sufficient, safe, nutritious food for all individuals, enabling them to lead active and healthy lives (Kiran et al., 2024). According to Food and Agricultural Organization (FAO), food security is multifaceted and adaptable concept which is built on four key pillars: availability, accessibility, utilization and stability (Coates, 2013). The world ever-increasing population and changing climate are putting heavy pressure on food security. World population is projected to rise, exceeding 9.7 million by 2050 and reaching 11 billion by 2100 reported by Calicioglu et al. (2019). To date, over 800 million people suffer from malnutrition, and the situation is predicted to increase dramatically in the future decades reported by Frona et al. (2019). To prevent this, the United Nations established zero hunger as Sustainable Development Goal 2 (SDG2). According to the FAO, to achieve zero hunger goal, the world needs to produce 15–20% more food than yields predicted from recent trends as reported by Ahmad et al. (2021). To feed this expanding population, worldwide crop production needs to grow from 2.1 to 3 billion tons; however, reports suggest the arable land will not sustain the weight of these demands in coming years (Haris et al., 2023; Voss-Fels et al., 2019). By 2050, globally the requirement of vegetable oil is estimated to be elevated double times compare to current production. This has led crop biotechnologists to shown growing interests in developing innovative approaches to enhance food production (Fanzo et al., 2018).

Agriculture plays a significant role in the development of most emerging economies worldwide, influencing overall economic growth, generating household incomes and ensuring food security reported by Dube et al. (2016). Climatic change adversely affects agriculture and leads to food insecurity. Changing climatic conditions, and various biotic and abiotic stresses, contribute to lower oilseed crop yields (Attia et al., 2021). This, in turn, has increased the number of people suffering from malnutrition. Oilseed crops play a significant role in strengthening the nation's economy and serve as critically important player in the industrialized world (Murphy, 1991). Oilseed products such as meal, oil, and

* Corresponding author.

Nanotechnology Innovations for Food Security and Sustainable Agriculture, First Edition. Edited by Zia Ur Rehman Mashwani, Shah Fahad, Ilyas Ahmad, Shah Saud, Sajid Fiaz, and Taufiq Nawaz.

sunflower hulls serve vital role in human nutrition and industrial applications. Protein is highly nutritious, as the essential amino acids it contains provide significant health benefits (Abiodun, 2017). Furthermore, in addition to their use in food industry, oilseed is utilized in the production of fertilizers, textiles, cosmetics, and biofuels. Such plant-derived oils have enormous economic importance and are widely used in food and other industrial applications, as well as a broad range of industrial products, including pharmaceuticals, surfactants, plasticizers, emulsifiers, detergents, lubricants, adhesives, cosmetics, and oleochemicals as shown in Figure 5.1 (Rahman & de Jimenez, 2016). It is obvious that several species of plants are required to occupy these spaces given the wide range of main and secondary applications of oilseed crops (Jaradat, 2016). Soybeans, sunflowers, canola, cottonseed, and peanuts are among the most prominent oilseed crops that have been widely cultivated and studied in recent years. Furthermore, these crops are valuable dietary sources of essential fatty acids, particularly omega-3 (α-linolenic acid) and omega-6 (linoleic acid), which cannot be synthesized by the human body and must be obtained through food. On average, a healthy adult consumes 20–35% of their daily calories from fats and oils (Mayo Clinic). Numerous studies have shown that unsaturated fats provide greater health benefits than saturated fats (Kris-Etherton & Krauss, 2020). Human body synthesizes omega-9 fatty acids (oleic acid) endogenously (Saini & Keum, 2018). The current body of

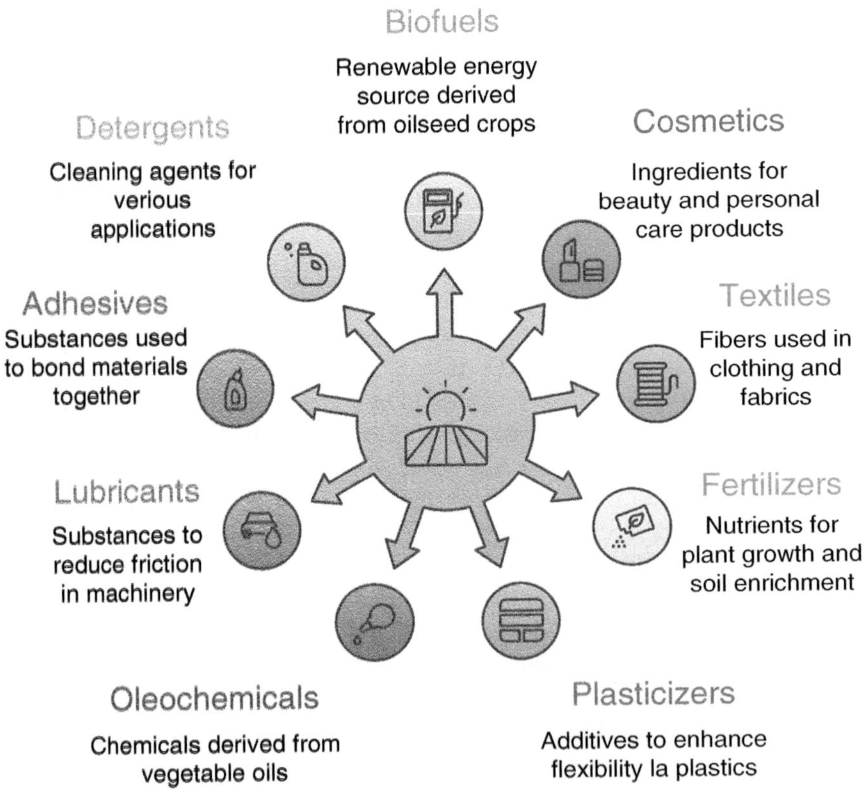

Figure 5.1 Industrial applications of oilseed crops.

literature suggested that a balanced intake of these fatty acids helps prevent cardiovascular diseases, manages cholesterol and blood pressure, decreases inflammation, lowering incidence of certain cancers, and supports maternal and fetal health outcomes (Calder, 2017; Hoshi et al., 2013; Wendel & Heller, 2009).

Nanotechnology and its applications are among the most inventive and promising technologies for revolutionizing the traditional food and oil agribusiness sectors. Nanotechnology has become a groundbreaking platform, offering innovative solution with potential to transform nearly every aspect of agriculture (Scott & Chen, 2012). It has been reported that fertilizers application is crucial for boosting agricultural productivity, excessive use can irreversibly alter the soil chemical composition, reducing the availability of arable land for cultivation (Vermeulen et al., 2012). Sustainable agriculture addresses this challenge by minimizing agrichemical usage, a practice that not only safeguards ecosystems but also help prevent the decline of vulnerable species (Vermeulen et al., 2012). By optimizing the efficiency of agriculture inputs, they enable precise, site-specific delivery of nutrients ensuring minerals use of resources while maximizing productivity. The innovation supports sustainable farming practices by reducing environmental harms and promoting long-term soil health (Shang et al., 2019). It enhances crop protection, tracks plant growth, identifies diseases rapidly, and reduces waste in agriculture to enhance environmental sustainability. A growing number of applications of nanotechnology are being discovered in different fields, including plant pathology. The optimal delivery of nutrients, reducing dependence on agricultural chemicals, and reducing fertilizer loss will also lead to higher yields (Chen et al., 2013; Pirzadah et al., 2019). Thus, plant-based nanomaterials provide biodegradable alternatives while reducing hazardous material emissions from farming operations through their biodegradable nature. Several researchers have proposed that nanomaterials, including nanofertilizers, nanoinsecticides, and non-pesticides, can be used to advance environmentally friendly technologies, such as nanofertilizers, nanoinsecticides, and non-pesticides, enabling safer and more ecologically conscious crop management (Ghidan et al., 2017; Shang et al., 2019). In this chapter, we will discuss how nanomaterials can be used to increase crop health, yields, and oil quality in oilseed plants.

5.2 Mechanisms of Action of Nanomaterials

In the last 20 years, scientists have developed farming practices and advanced crop properties through the application of nanomaterials. The penetration capabilities of nanomaterials (NMs), which include SiO_2, TiO_2, Zn/ZnO, Fe/FeO, CuO, and Ag, which are commonly used in nano-agrochemicals such as nonopesticides and nanofertilizers, vary greatly from plant species to plant species and cannot be accurately predicted (Kah et al. 2018). The researchers had to overcome these obstacles to develop detailed mechanisms to comprehend absorption, translocation, accumulation, barriers, routes, and modes of action of nanomaterials (Rico et al., 2011). Several questions have been raised regarding nanomaterials (NMs) and their entry into plants. Before NMs can access the cytoplasm in plant cells, what kind of physical obstacles must they overcome? Does non-dissolved NM pass through plant systems quickly? Are there any structural characteristics of plants that help them absorb NMs more effectively? In addition, where will the NMs accumulate within the plant? In this chapter, we aim to elucidate the mechanism and pathways through which nanomaterials (NMs) enter plant cells, with an emphasis on identifying the most precise and predictive mechanisms for their uptake. An initial overview of NP entry pathways was provided by Dietz and Herth (2011).

In nature, plants are exposed to both aerosols as well as soil nanomaterials (NMs). Nanomaterials (NMs) are introduced to the plants using various methods such as seed priming, foliar spray, truck feeding, and root application (Su et al., 2019). Among these, foliar and root exposure are particularly practical due to their simplicity and making them suitable for NMs delivery. Understanding NMs penetration mechanism is essential for elucidating the interactions between NMs and plants (Rani et al., 2023; Khodakovskaya et al., 2011). Many new or poorly characterized uptake, transport, and accumulation sites have been considered, including the mycorrhizal community associated with plants, the mucilage of roots, the cuticle, nonendocytic absorption across the cellular membrane, necrotic material, root hair, and trichomes. The exact mechanism through which stomata in the plants and root epidermal cells open up is initiated by nanomaterials (NMs) entering through the cell wall and cell membrane due to a series of complex steps, entering vascular bundles (xylem) and subsequently symplastically propagated to the leaves (Banerjee et al., 2019). A sink is the root, whereas source is leaf for NMs. The most important pathway for nanomaterials (NMs) dispersion and movement is xylem (Aslani et al., 2014). Nanomaterials (NMs) deposition in plant aerial parts can cause extreme tissue damage and phytotoxicity with their concentration surpasses critical thresholds (Shen et al., 2010). Earlier report indicated that NMs might inhibit plant growth and developments by preventing essential processes such as seed germination, fresh and dry plant biomass accumulation, root/shoot elongation, photosynthetic activity, and transpiration rates, and by inducing lipid peroxidation, chromatin condensation, DNA damage, modulation of stress-associated gene expression, and finally apoptosis (Azim et al., 2023; Hossain et al., 2016). These NMs have the potential to modify physio-morphological, anatomical, biochemical, and genetic structure of plants. The nanomaterial deposition in plant tissues undermines crop productivity by inhibiting physiological processes, affecting metabolic processes, and modifying nutrient dynamics (Rico et al., 2015).

NMs entry into plant cells depends on size, shape, stability, chemical structure, surface chemistry, and species of the plant. Wang et al. (2012) studied CuO NMs uptake and translocation routes in plants. Through their research, they proved that CuO NMs under 40 nm are taken up by roots and translocated to shoots through xylem. Upon uptake, they access vascular tissue through apoplastic or symplastic routes as indicated in Figure 5.2. Moreover, they witnessed reverse transport in shoots to roots by the phloem in maize, indicating a two-way movement process (Choudhary et al., 2017). Numerous parameters, including plant physiological characteristics, NM properties (size, shape, and charge), and environmental abiotic factors (humidity, light, and temperature), control the uptake, transport, and accumulation of NMs (Hong et al., 2014). The difference in uptake pattern can be due to uneven distribution of stomata in dicotyledons compared to monocotyledons (Rani et al., 2023). Leaf veins are fewer, petioles are smaller, and leaves are larger in plants that tend to accumulate higher NMs. The developmental stage of the plant and its life cycle also have an effect on the uptake of NMs via the leaves. Characteristics such as waxy coating and minute hairs on leaves slow down NMs entry, and the process is also influenced by

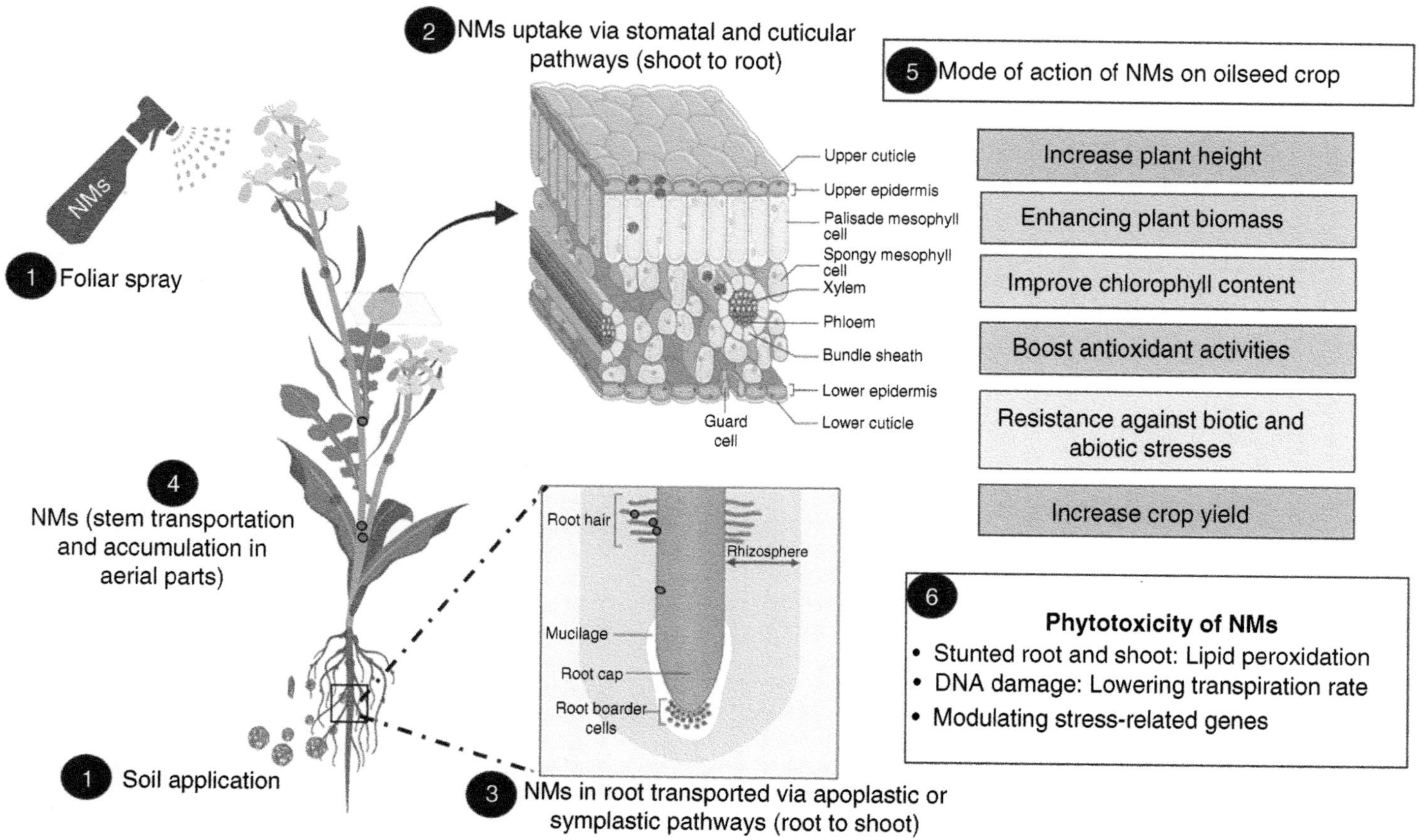

Figure 5.2 Application of nanomaterials on oilseed crop, uptake (pathways), translocation, accumulation, and mode of action (both positive and negative influence on plants).

NMs size, leaf area, outer leaf covering structure, and the development stage of the plant (Perez-de-Luque, 2017). The uptake of NMs is usually thought to be an active-transport system comprising a range of other cell activities, such as recycling, signaling, and control of the plasma membrane. Foliar treatment with NMs provides a more efficient and environmentally friendly methods of plant treatment compared to the root and soil exposure routes reported by Fan et al. (2020). As reported by Tripathi et al. (2012), silicon is the second most prevalent element in the Earth's crust and is important for plant growth and development. Plant roots take up silicon in monosilicic acid form, which is then deposited between and within plant cells to produce phytoliths (Tripathi et al., 2013). These phytoliths contribute mechanical strength to plants and allow them to withstand various biotic stresses (e.g., pathogens and pests) and abiotic stresses (e.g., drought and heavy metal toxicity). Silicon accumulation in plant tissues is therefore an important defense process (Tripathi et al., 2017). With these advantages, agricultural scientists are becoming more interested in the applications of silicon nanoparticles for agriculture, especially how they can be used to maximize plant resilience and productivity (Ma, 2004). Such growing interest highlights the need to learn about nanoparticle interactions in plant systems in order to devise new agricultural solutions (Tripathi et al., 2017).

A study of Khodakovskaya et al. (2009) describes the penetration of nanomaterials (NMs) through the seed coat with a specialize pores on seedcoats known as "aquaporins." Once NMs enter seeds, they can induce physiological changes in both seedling and mature plants. Judy (2013) highlights variations in plant, such as differences in hydraulic conductivity and pore size of the cell wall, which may influence the movement and accumulation of nanomaterials. Another study by Buick et al. (1993) stated that NMs enter the plants through two distinct pathways, i.e., stomatal and cuticular pathways. The plant cuticle, a waxy outermost covering presents on virtually all aerial surfaces, plays a critical role in minimizing water loss and regulate solute exchange. As the primary interface plant and the environments, this hydrophobic structure serves as the first line of defense against both biotic and abiotic stressors, owing to its unique chemical composition and structure resilience (Pollard et al., 2008). The cuticular transport pathway is categorized into two distinct mechanisms based on solute properties: hydrophilic pathway and lipophilic pathway. Furthermore, Zhang et al. (2015) emphasized that the uptake and translocation of NMs depends on the physicochemical properties including shape, size, and composition. Example was coded by Larue et al. (2014) revealing that TiO NPs are able to make a route through cuticle.

5.3 Types of Nanomaterials Used in Oilseed Crops

Nanoscience an emerging field of research focuses on developing cost-effective nanotechnology applications to enhance seed germination, plant growth and development, and environmental resilience (Manjaiah et al., 2019). Seed germination is a critical stage in plant's life cycle; plays a pivotal role in seedling establishments, influencing growth, and survival and population dynamics. However, this process is highly sensitive to numerous factors such as environmental conditions, genetic traits, soil quality, and moisture availability (Joshi et al., 2018). In addressing these challenges, studies have increasing demonstrated that nanomaterials significantly improve germination rates and support overall plant growth and development.

Earlier research has shown that applications of carbon nanotubes have a stimulating effect on the germination characteristics of oilseed plants, cereals, and vegetables (Lahiani et al., 2013; Srivastava & Rao, 2014). Earlier results by Lu et al. (2002) showed increased rates of percentage germination in soybeans treated with applications of silica (Si), titanium (Ti), and zeolite nanomaterials (Laure et al., 2011). Several metal nanomaterials (NMs) were used on oilseed plants, enhancing plant growth as a whole and increasing yield quality. For example, Shojaei et al. (2019) emphasized the effectiveness of zinc, titanium, silicon, iron oxide, copper, and hyroxyfullerenes–metal and carbon nanomaterials in such a context (Figure 5.2). Considering such benefits, Dinh et al. (2018) showed that selenium nanoparticles (Se NPs) are being used in agriculture more and more for crop biofortifications, as well as for resisting biotic and abiotic stresses. In addition, plant-derived Se NPs not only increase crop yield and phytochemical content but also have antifungal activity. Ahmad et al. (2023) showed that they can be used to enhance growth and yield of sesame under biotic stress.

Furthermore, drought stress seriously retards plant physiological processes, which, in most cases, lead to decreased productivity and stunted growth. The severity of these adverse effects on crop developments and yields depends on species specific traits, growth stage, and stress intensity (Ghassemi Golezani et al., 2023). Under severe drought, plants undergo physiological adaptations such as stomatal closure, which lowers transpiration rates and limits photosynthetic activity (Ghassemi Golezani et al., 2023). Increasing the root-to-shoot ratio enhances water uptake efficiency from the soil, allowing plants to mitigate drought impacts by prioritizing leaf hydration and resource allocation (Lu et al., 1998). The higher leaf water contents are the results of higher osmotic regulation of plants under limited water availability. Magnesium (Mg^{2+}) is mostly associated with proteins and chlorophyll molecules in plants, where it functions as enzyme cofactor for the photosynthetic fixation of carbon metabolism (Cai et al., 2018). Plant development, physiology, and metabolism are all adversely impacted by magnesium shortage (Verbruggen & Hermans, 2013). According to Faizan et al. (2022) stated that MgO nanoparticles application on canola enhanced photosynthetic pigments and improved yield performance. Nanomaterials (NMs) have emerged as a promising tool in agriculture, offering innovative solutions to enhanced oil yields in crops. NMs hold the potential to revolutionize agriculture, offering significant advantages while also presenting certain challenges (Choudhary et al., 2017). Previous studies have shown that copper oxide nanoparticles (CuO NPs) improvised plant health by servings is nutrients, therefore strengthening their resistance against fungal diseases. Enormous reports suggested that CuO NPs are a viable alternative to fungicides like Bavistin, carbendazim, and mancozeb due to their unique properties and effectiveness in disease managements (Kanhed et al., 2014). Their advance functionality and efficiency make them a logical and sustainable choice for crop protection. The study of Rui et al. (2016) demonstrated that the application of iron oxide nanoparticles as nanofertilizers in peanut (*Arachis hypogea* L.) significantly enhanced plant growth, physiomorphological parameters, antioxidant enzyme activities, and phytohormones accumulation (Figure 5.3).

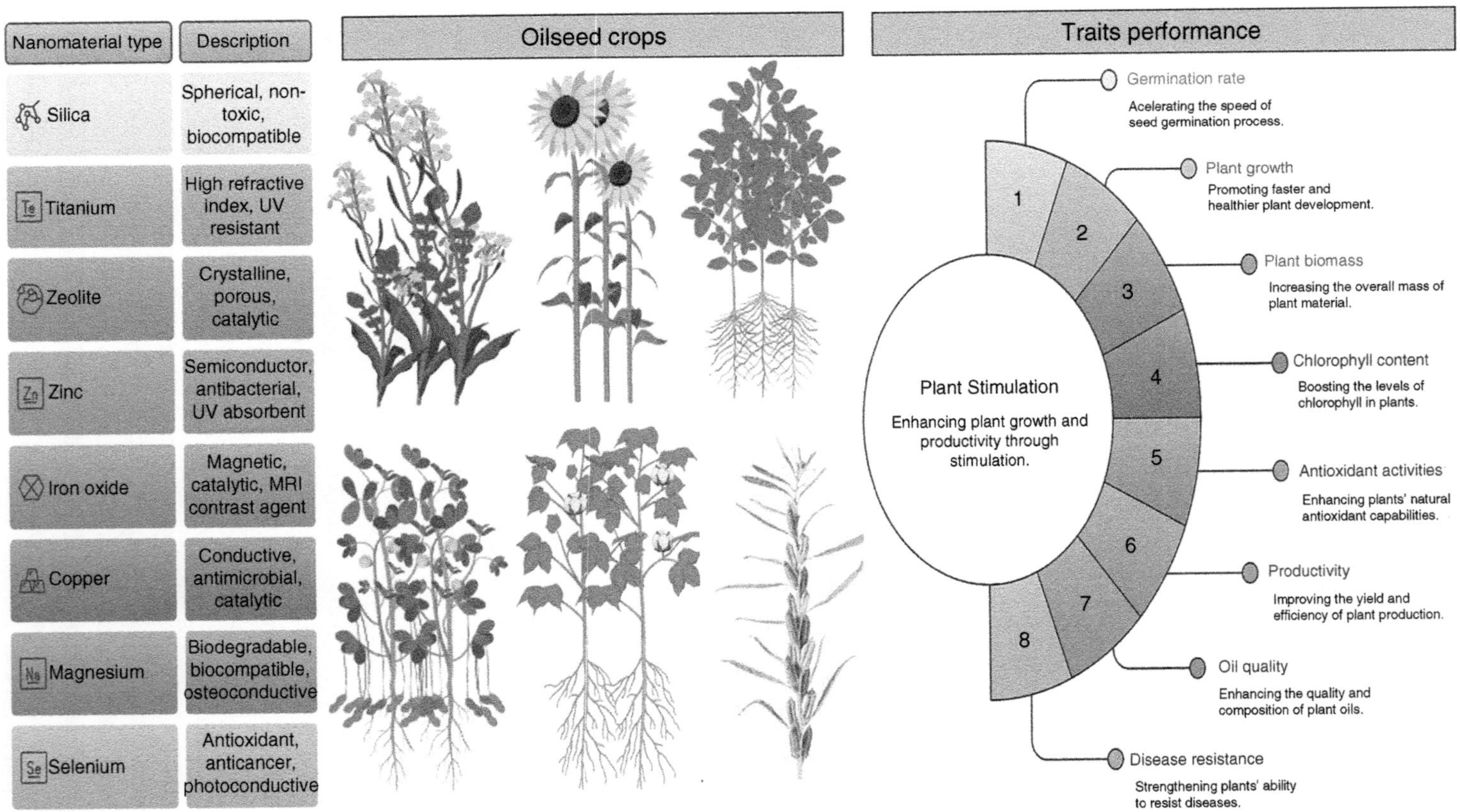

Figure 5.3 Impact of nanomaterials applications on trait performance of oilseed crops. *Source:* Created in BioRender.com.

5.4 Impact of Oil Yield

Despite a sharp increase in production, a significant gap persists between the predicted output and the actual results achieved at the field level. Yield reduction is a major concern in agriculture, frequently caused by a combination of biotic and abiotic stresses. Among biotic stresses, fungal pathogens are responsible for 70% of yield losses in oilseed crops, while abiotic stresses account for the remaining 30% of crop damage (Gaba et al., 2022). The composition of unsaturated fatty acids dynamically adjusts in response to environmental stresses such as drought, cold, and salinity. These modifications regulate cell membrane fluidity, thereby influencing the mobility and functionality of membrane lipids. The growing demand for agricultural products has intensified the need to adopt innovative farming strategies, such as the application of high-yielding chemically synthesized fertilizers, farmyard manure, bio-manures, and other advanced inputs. This movement is spurred by the noted reality that proper fertilizer application greatly increases production and guarantees improved yields and quality of farm produce documented by Fatima et al. (2021). Zhu et al. (2020) reported that overuse and wasteful application of chemical fertilizers can cause land degradation, decreased harvest of crops, and contamination of the environment. Zinc oxide nanoparticles (ZnO NPs) are a good source of micronutrient when used as a nano-fertilizer on oilseed plants such as *Arachis hypogea* (peanut). Zinc uptake is improved through their use, enhancing the crop while increasing the photosynthetic pigments and chlorophyll content by a considerable amount (Prasad et al., 2012). ZnONPs also enhance the antioxidant defense system of the plant, resulting in enhanced growth and an increase in yield (Yusefi-Tanha et al., 2023).

It was discovered that rapeseed oil quality is greatly affected by cultivar types and environmental condition. Enjalbert et al. (2013) proved that raising the plant density greatly lowered oleic acid content throughout rapeseed cultivars and that both plant density and cultivar types had the greatest effect on the proportion of unsaturated fatty acids in rapeseed oil. El-Habbasha and El-Salam (2010) exhibited that zinc oxide nanomaterial application to canola crops enhances yield and oil quality on semi-arid areas with zinc-deficient soil. Fouda and Sofy (2022) unveiled that zinc oxide nanoparticle–treated canola (*Brassica napus*), treated with 100 mg/L of zinc oxide nanoparticles synthesized from *Penicillium chrysogenum*, showed enhanced antioxidant activities and oil content. The research carried out by Ahmad et al. (2024) on sesame varieties using selenium nanoparticles at different concentrations (10–50 mg/L) promotes the enhancement of phenolic content, antioxidant activities, and oil contents considerably. The highest oil content of 60% was noticed for Til-18 variety at 50 mg/L compared to untreated plants. The fundamental nutritional value of soybean can be attributed to its protein and oil content. In this regard, Dola et al. (2022) conducted a study on soybean in which iron oxide nanoparticle applied at 200 ppm substantially improved oil content (10%) and 7% protein content (7%) in soybean.

5.5 Improvement of Oil Quality

Vegetable oils were widely used in machinery before the advent of mineral oil, which later become prevalent due to its superior performance and lower cost (Darminesh et al., 2017). In recent years, biodegradable oils have gained traction in lubrication applications because of price volatility, regulatory challenges, and growing environmental and health concerns (Jayadas & Nair, 2006). Oil such as canola, sunflower, coconut, soyabean, and jojoba have garnered significant attention for their enhanced lubricating properties. The vegetable oils excel in lubrication due to their high lubricity, viscosity index, and low volatility (Zareh-Desari & Davoodi, 2016). Sunflower oil comprises 30% of monounsaturated fat (oleic acid) and 59% of polyunsaturated fats (linolenic acid). Its high oleic acid content contributes to favorable oxidation stability. To improve performance of lubricating oils, additives such as wear inhibitors, viscosity modifiers, film-forming agents, deposit preventers, and high-pressure stabilizers have been incorporated (Batool et al., 2021). Recently, NMs have got significant attention due to their unique properties, which differ markedly from those of their bulk counterparts. Studies on nanomaterial additives have shown that they exhibit lower wear and fractions rate compared to conventional liquid paraffin oil (Peng et al., 2009).

Nanotechnology can be used to assess the quality of oilseeds. A U.S Department of Agriculture (USDA)-funded initiative proposes that food-grade hydrocolloids such as gelatin, chitosan, and starch can be used to develop luminous food sensors based on nanoparticles. The nanoparticles, which measure from 20 to 200 nm, incorporate luminescent probes, food-compatible compounds with specific fluorescence or phosphorescence properties (Subramanian, 2023). In addition to dyes that are approved for use in food (e.g., erythrosine), natural chemicals, such as quinine and flavanols, are also used as probes. In addition to being potential food additives, these nanoparticles have a nanostructure that enables advantages in sensor applications as well. This will also enhance reactivity and precisely integrate into food matrices without compromising taste or texture (Subramanian, 2023).

A study by Younas et al. (2024) demonstrated that selenium-treated sesame oil exhibits enhanced levels of bioactive compounds, as identified through gas chromatography mass spectroscopy (GC-MS) analysis. The selenium-enriched sesame seed oil contained 35 bioactive compounds, indicating improved oil quality following treatments with 50 mg/L selenium nanoparticles (Se NPs). Sunflower is important food source due to the presence of significant nutrient contents. Batool et al. (2021) reported that the application of 40 mg/L silver nanoparticles (Ag NPs) on sunflower significantly increased linolenic acid levels by 59% compared to control and other treatments (Table 5.1). Additionally, a marked reduction in stearic acid and palmitic acid levels was observed in sunflowers treated with varying concentration of silver nanoparticles (Table 5.1). Their finding aligns with the earlier work of Yaseen and Wasan (2015), who documented a decrease in saturated fatty acids and an increase in unsaturated fatty acids following nanomaterials exposure. Unsaturated fatty acids are critical for reducing cholesterol levels and mitigating the risks of cardiovascular diseases.

5.6 Challenges and Limitations

Beyond their role as a source of oil, oilseed crops, which belong to different plant families, serve as the groundwork for several oleochemical businesses. Additionally, they supply key raw materials for electricity generation, functioning as a sustainable energy

Table 5.1 Application of nanomaterials (NMs) in oilseed crop improvements.

NMs type	Synthesis method/plant part used	Characterization techniques	Size of NMs	Concentrations	Oilseed crop	Results	References
ZnONPs	Biological	UV—visible spectroscopy, TEM, zeta potential, XRD	19.4–59.8 nm	100 mg/L	Canola	Increase oil content, minerals, and up-regulating antioxidant activities in *Brassica napus*	Fouda and Sofy (2022)
Ag NPs	Biological	TEM, zeta potential	23.4 nm	32–64 mg/kg	Soybean	Exposure of soybean to Ag NPs significantly decrease seed yield by 39% at 32 mg/kg.	Sharma et al. (2023)
Se NPs	Garlic powder	UV—visible spectroscopy, FTIR, SEM, EDX, XRD	<100 nm	50 mg/L	Sesame	Application of Se NPs significantly enhance antioxidant activities and oil content.	Ahmad et al. (2024)
Cu and Ag nanocolloid	—	SEM, FTIR	13.3–38.3 nm	Varying concentrations	Canola	Nanocolloid improves chlorophyll and carotenoid pigments and nonsignificant variation in oil quality in pressed rapeseed.	Kachel et al. (2023)
Zn–Fe nanocomposites (NCs)	Physical	UV—visible spectroscopy, Transmission electron microscopy (TEM)	13–15 nm	25–60 mg/kg	Soybean	Application of NCs significantly decreased seed phytic acid concentrations by 7.3–59.9%.	Yadav et al. (2024)
ZnO and FeO NPs	Physical	SEM	10–100 nm	300–600 mg/L	Flaxseed	Annual increase in oil percentage in flaxseed by 31% (2019–2020) and 27% for (2020–2021).	Hoseini et al. (2024)
SeNPs	Garlic powder	UV—visible spectroscopy, FTIR, SEM	11.98 nm	40–50 mg/L	Sesame	Application of SeNPs significantly enhancing biochemicals and bioactive compounds in sesame oil.	Younas et al. (2024)

(*Continued*)

Table 5.1 (Continued)

NMs type	Synthesis method/plant part used	Characterization techniques	Size of NMs	Concentrations	Oilseed crop	Results	References
Ag NPs	*Euphorbia helioscopia* leaves	UV–visible spectroscopy, XRD, SEM	< 100 nm	60 mg/L	Sunflower	Ag NPs application of sunflower improved morphological, biochemical metabolites and fatty acid composition.	Batool et al. (2021)
SiO_2 NPs and TiO_2 NPs	Chemical	TEM, EDX, SEM	17–27 nm	—	Cotton	UV-protection and antibacterial activities	Farouk et al. (2013)

source (Ahmad et al., 2024). Despite numerous efforts, Pakistan's domestic production of edible oil and oilseeds has steadily declined, reflecting a significant policy failure. Since the early 1970s, the country has relied heavily on imports to bridge the widening supply demand gap. Currently, only 13–15% of oilseed demand is met through local production, while the remaining shortfall is covered by imports. These imports cost Pakistan approximately 4 billion USD in FY2021, exacerbating trade deficits and straining the nation's balance of payments reported by Rana et al. (2022). The global oilseeds industry faces significant uncertainties, including sudden price volatility, market instability, and preferential import taxes. Given these challenges, import costs are likely to increase, highlighting the urgent need for strategic planning and policy intervention (Ahmad et al., 2024; Rana et al., 2022). Nano-enabled agrochemicals hold significant promise for various applications; they are still in the preliminary phases of research and development (An et al., 2022). Numerous organizations have filed patents covering novel methods for the production and application of nano-based products (An et al., 2022). However, commercially available nanotechnology-based goods are still limited and require further development. According to the Nanotechnology Products Database available at StatNano, out of 9,420 nanotechnology-based products on the market, only 229 are specifically designed for agricultural applications (StatNano, 2021). Moreover, the widespread acceptance and adoption of nanotechnologies in agriculture remain limited growing concern about fate, transport, bioavailability, and toxicity of nanomaterials (NMs), as well as inadequate regulatory frameworks. Additionally, current research trends lack a practical methodology, hindering a comprehensive understanding of risk assessments criteria and the long-term toxicity of NMs to plants, soil, and soil microbiomes within agroecosystems following their environmental release (Mishra et al., 2017).

5.7 Field Applications and Case Studies

Previous field-level studies have highlighted the harmful effects of copper-based fungicides, leading to their ban due to risks to human health (Lamsal et al., 2011). Studies have shown that carbendazim and Bavistin (fungicides) pose serious risks to human health, leading to harmful effects such as cell death, birth defects, embryo damage, liver dysfunction, disruption of hormonal systems, infertility, and genetic mutations (Rama et al., 2014; Yenjerla et al., 2009). Although, the regular application of agrochemicals has failed to effectively control fungal infections, while also adversely impacting soil health by killing beneficial bacteria (Kim et al., 2009). The use of traditional fungicides leads to negative consequences, including imbalanced fertilization, multi-micronutrient deficiencies, and reduced nutrient uptake by plants (Subramanian and Tarafdar, 2011). These problems contribute to widespread environmental contaminations, a frequent issue linked to conventional fertilizers (Yusefi-Tanha et al., 2023). This growing concerns underscore the need for innovative solutions, positioning nano-based fertilizers as a promising alternative. Mahmoodzadeh et al. (2013) reported that a dose concentration of 2000 mg/L of titanium nanoparticles significantly improved the germination rate in canola. Additionally, canola treated with titanium oxide nanoparticle exhibited a greater seed vigor index due to enlarge plumule and radical growth as compared to untreated plants.

5.8 Future Directions and Research Needs

The advancement of nanotechnology has opened a fascinating and innovative option for nearly each industrial application that has significant influence on people's lives (Linkov et al., 2011). This crucial facilitating technology has demonstrated extensive and prominent application in a variety of industries, including electronic, pharmaceuticals, cosmetics, textiles, food science, energy sector, and agriculture (Sastry et al., 2011; Sharon et al., 2010). Nanotechnology has revolutionized agriculture by oilseed crop production and quality through advanced improvements in farming systems. Engineered nanomaterials (NMs) are driving sustainable agriculture innovation, increasing productivity, and the expanding capabilities to meet growing global food demands. With nanoscale precision, this technology paves the way for new a green revolution while reducing agricultural risks. There is a pressing need for new ideas for agricultural research and advancements since the Green Revolution has faced significant challenges related to productivity, consistency, sustainability, and equity reported by Conway and Barbie (1988). The increasing reliance on chemical pesticides and fertilizers during and after Green Revolution had led to major challenge including threats to long-term sustainability, environmental degradation, and health hazards. To access biosafety concerns, innovative strategies involving ecofriendly biofertilizers and biopesticides have emerged as sustainable alternatives to conventional agrochemicals (Mishra & Singh, 2015). A key focus is optimizing resource managements in conservation in sustainable farming, where nanomaterials (NMs) play a vital role in improving efficiency and minimizing environmental pollution. Cutting-edge nanomaterials, nanosensors, and nanoformulations are transforming the agricultural value chain by tackling critical challenges (Cavalcanti et al., 2003). These advancements enhance farm productivity, promote sustainable practices, optimize water and land management, and strengthen food safety and security, paving the way for a more resilient and efficient agricultural system. Applications of nanorobots for agriculture and the quality of water and soil surveillance might have a significant influence on environmental management and reducing the harm that pollution causes to a wide variety of natural species. Nanorobot applications are anticipated to offer incredible opportunities (Adleman, 1995).

5.9 Conclusion

In agriculture, the focus often centers on key SDGs 2 and 13, which address food security, a challenge driven by the rising global population and compounded by climate change, posing significant threat to agricultural productivity. This chapter concludes that the application of nanomaterials (NMs) can address production challenges by enhancing plant resistance to pest attacks, a major cause of crop loss. Furthermore, studies confirmed that NMs application significantly improves the quality of oil and seeds in oilseed crops, offering a pathway to sustainable agricultural advancements. The use of nanorobotics equipped with integrated sensors systems for hydrology presents a promising strategy poised to drive major advancements in agriculture and environmental monitoring. Advanced in nanotechnologies are expected to enable the seamless integration of molecular mechanics with nanoelectronics systems, including nanorobot devices that leverage complementary metal oxide semiconductor (CMOS) technology for precise sensing and real-time communication. The innovation could revolutionize water resource managements, crop monitoring, nano-pesticides delivery, and sustainability efforts in coming decades.

References

Abiodun, O. A. (2017). The role of oilseed crops in human diet and industrial use. *Oilseed Crops: Yield and Adaptations Under Environmental Stress*, 249–263.

Adleman, L. M. (1995). On constructing a molecular computer. *DNA Based Computers, 27*, 1–21.

Ahmad, S., Tang, L., Shahzad, R., Mawia, A. M., Rao, G. S., Jamil, S., & Tang, S. (2021). CRISPR-based crop improvements: a way forward to achieve zero hunger. *Journal of Agricultural and Food Chemistry, 69*(30), 8307–8323.

Ahmad, I., Younas, Z., Mashwani, Z. U. R., Raja, N. I., & Akram, A. (2023). Phytomediated selenium nanoparticles improved physio-morphological, antioxidant, and oil bioactive compounds of sesame under induced biotic stress. *ACS Omega, 8*(3), 3354–3366.

Ahmad, I., Chen, C., Younas, Z., Yousaf, T., & Mashwani, Z. U. R. (2024). Seed and foliar application of nano-selenium improves sesame triacylglycerols and oil yield via photosynthetic pigment and enzymatic and chemical antioxidant enhancement revealed by spectrophotometric, UHPLC-analysis and chemometric modeling. *Frontiers in Plant Science, 15*, 1431877.

Ahmad, I., Yousaf, T., El-Sheikh, M. A., & Ahmad, P. (2024). Comprehensive approaches of phytonanoparticles for stress tolerance, growth performance, and improving oil yield in Sesame (Sesamum indicum): mechanism, applications and future prospects. *Plant Stress, 100*, 498.

An, C., Sun, C., Li, N., Huang, B., Jiang, J., Shen, Y., & Wang, Y. (2022). Nanomaterials and nanotechnology for the delivery of agrochemicals: strategies towards sustainable agriculture. *Journal of Nanobiotechnology, 20*(1), 11.

Aslani, F., Bagheri, S., Muhd Julkapli, N., Juraimi, A. S., Hashemi, F. S. G., & Baghdadi, A. (2014). Effects of engineered nanomaterials on plants growth: an overview. *The Scientific World Journal, 2014*(1), 641759.

Attia, Z., Pogoda, C. S., Reinert, S., Kane, N. C., & Hulke, B. S. (2021). Breeding for sustainable oilseed crop yield and quality in a changing climate. *Theoretical and Applied Genetics, 134*(6), 1817–1827.

Azim, Z., Singh, N. B., Singh, A., Amist, N., Niharika, M., Khare, S., ... Yadav, V. (2023). A review summarizing uptake, translocation and accumulation of nanoparticles within the plants: current status and future prospectus. *Journal of Plant Biochemistry and Biotechnology, 32*(2), 211–224.

Banerjee, K., Pramanik, P., Maity, A., Joshi, D. C., Wani, S. H., & Krishnan, P. (2019). Methods of using nanomaterials to plant systems and their delivery to plants (mode of entry, uptake, translocation, accumulation, biotransformation and barriers). In *Advances in phytonanotechnology* (pp. 123–152). Academic Press. https://doi.org/10.1016/C2017-0-02650-2

Batool, S. U., Javed, B., Sohail, Zehra, S. S., Mashwani, Z. U. R., Raja, N. I., ... Alamri, S. (2021). Exogenous applications of bio-fabricated silver nanoparticles to improve biochemical, antioxidant, fatty acid and secondary metabolite contents of sunflower. *Nanomaterials, 11*(7), 1750.

Buick, R. D., Buchan, G. D., & Field, R. J. (1993). The role of surface tension of spreading droplets in absorption of a herbicide formulation via leaf stomata. *Pesticide Science, 38*(2–3), 227–235.

Cai, L., Chen, J., Liu, Z., Wang, H., Yang, H., & Ding, W. (2018). Magnesium oxide nanoparticles: effective agricultural antibacterial agent against Ralstonia solanacearum. *Frontiers in Microbiology, 9*, 790.

Calder, P. C. (2017). Omega-3 fatty acids and inflammatory processes: from molecules to man. *Biochemical Society Transactions, 45*(5), 1105–1115.

Calicioglu, O., Flammini, A., Bracco, S., Bellù, L., & Sims, R. (2019). The future challenges of food and agriculture: an integrated analysis of trends and solutions. *Sustainability, 11*(1), 222.

Cavalcanti, A., Wood, W. W., Kretly, L. C., & Shirinzadeh, B. (2003). Computational nanorobotics: agricultural and environmental perspectives. *Nanomedicine, 2*(2), 82–87.

Chen, H., Roco, M. C., Son, J., Jiang, S., Larson, C. A., & Gao, Q. (2013). Global nanotechnology development from 1991 to 2012: patents, scientific publications, and effect of NSF funding. *Journal of Nanoparticle Research, 15*, 1–21.

Choudhary, R. C., Kumaraswamy, R. V., Kumari, S., Sharma, S. S., Pal, A., Raliya, R., & Saharan, V. (2017). Cu-chitosan nanoparticle boost defense responses and plant growth in maize (Zea mays L.). *Scientific Reports, 7*(1), 9754.

Chaudhry, N., Dwivedi, S., Chaudhry, V., Singh, A., Saquib, Q., Azam, A., & Musarrat, J. (2018). Bio-inspired nanomaterials in agriculture and food: Current status, foreseen applications and challenges. *Microbial pathogenesis, 123*(1), 196–200.

Coates, J. (2013). Build it back better: deconstructing food security for improved measurement and action. *Global Food Security, 2*(3), 188–194.

Conway, G. R., & Barbie, E. B. (1988). After the green revolution: sustainable and equitable agricultural development. *Futures, 20*(6), 651–670.

Darminesh, S. P., Sidik, N. A. C., Najafi, G., Mamat, R., Ken, T. L., & Asako, Y. (2017). Recent development on biodegradable nanolubricant: a review. *International Communications in Heat and Mass Transfer, 86*, 159–165.

Dietz, K. J., & Herth, S. (2011). Plant nanotoxicology. *Trends in Plant Science, 16*(11), 582–589.

Dinh, Q. T., Cui, Z., Huang, J., Tran, T. A. T., Wang, D., Yang, W., & Liang, D. (2018). Selenium distribution in the Chinese environment and its relationship with human health: a review. *Environment International, 112*, 294–309.

Dola, D. B., Mannan, M. A., Sarker, U., Mamun, M. A. A., Islam, T., Ercisli, S., ... Marc, R. A. (2022). Nano-iron oxide accelerates growth, yield, and quality of Glycine max seed in water deficits. *Frontiers in Plant Science, 13*(992), 535.

Dube, T., Moyo, P., Ncube, M., & Nyathi, D. (2016). The impact of climate change on agro-ecological based livelihoods in Africa: a review. *Journal of Sustainable Development, 9*(1), 256–267.

El-Habbasha, S. F., & Abd El-Salam, M. S. (2010). Response of two canola varieties (Brassica napus L.) to nitrogen fertilizer levels and zinc foliar applications. *Int. J. Acad. Res., 2*(2), 60–66.

Enjalbert, J. N., Zheng, S., Johnson, J. J., Mullen, J. L., Byrne, P. F., & McKay, J. K. (2013). Brassicaceae germplasm diversity for agronomic and seed quality traits under drought stress. *Industrial Crops and Products, 47*, 176–185.

Faizan, M., Bhat, J. A., El-Serehy, H. A., Moustakas, M., & Ahmad, P. (2022). Magnesium oxide nanoparticles (MgO-NPs) alleviate arsenic toxicity in soybean by modulating photosynthetic function, nutrient uptake and antioxidant potential. *Metals, 12*(12), 2030.

Fan, X., Cao, X., Zhou, H., Hao, L., Dong, W., He, C., ... Zheng, Y. (2020). Carbon dioxide fertilization effect on plant growth under soil water stress associates with changes in stomatal traits, leaf photosynthesis, and foliar nitrogen of bell pepper (Capsicum annuum L.). *Environmental and Experimental Botany, 179*, 104203.

Fanzo, J., Davis, C., McLaren, R., & Choufani, J. (2018). The effect of climate change across food systems: implications for nutrition outcomes. *Global Food Security, 18*, 12–19.

Farouk, A., Sharaf, S., & Abd El-Hady, M. M. (2013). Preparation of multifunctional cationized cotton fabric based on TiO_2 nanomaterials. *International Journal of Biological Macromolecules, 61*, 230–237.

Fatima, F., Hashim, A., & Anees, S. (2021). Efficacy of nanoparticles as nanofertilizer production: a review. *Environmental Science and Pollution Research, 28*(2), 1292–1303.

Fouda, H., & Sofy, M. (2022). Effect of biological synthesis of nanoparticles from *Penicillium chrysogenum* as well as traditional salt and chemical nanoparticles of zinc on canola plant oil productivity and metabolic. *Egyptian Journal of Chemistry, 65*(3), 507–516.

Frona, D., Szenderák, J., & Harangi-Rákos, M. (2019). The challenge of feeding the world. *Sustainability, 11*(20), 5816.

Gaba, S., Varma, A., & Goel, A. (2022). Protective and curative activity of biogenic copper oxide nanoparticles against Alternaria blight disease in oilseed crops: a review. *Journal of Plant Diseases and Protection, 129*(2), 215–229.

Ghassemi Golezani, K., Rajabi, M., & Farzi-Aminabad, R. (2023). Improving physiological performance and productivity of oilseed rape under drought stress by foliar application of Zn and Mg nanoparticles. *Journal of Plant Physiology and Breeding, 13*(2), 217–229.

Ghidan, A. Y., Al-Antary, T. M., Awwad, A. M., & Akash, M. W. (2017). Aphidicidal potential of green synthesized magnesium hydroxide nanoparticles using Olea europaea leaves extract. *ARPN Journal of Agricultural and Biological Science, 12*(10), 293–301.

Haris, M., Hussain, T., Mohamed, H. I., Khan, A., Ansari, M. S., Tauseef, A., & Akhtar, N. (2023). Nanotechnology–A new frontier of nano-farming in agricultural and food production and its development. *Science of the Total Environment, 857*(159), 639.

Hong, J., Peralta-Videa, J. R., Rico, C., Sahi, S., Viveros, M. N., Bartonjo, J., ... Gardea-Torresdey, J. L. (2014). Evidence of translocation and physiological impacts of foliar applied CeO_2 nanoparticles on cucumber (Cucumis sativus) plants. *Environmental Science & Technology, 48*(8), 4376–4385.

Hoseini, M. N., Saeidi, M., & Zarei, L. (2024). Comparison of iron and zinc nanoparticles with different sizes and bulk forms on seed yield and its components in flaxseed (Linum usitatissimum L.) under late-season water deficit. *Agrotechniques in Industrial Crops, 5*(1), 10–26.

Hoshi, T., Wissuwa, B., Tian, Y., Tajima, N., Xu, R., Bauer, M., & Hou, S. (2013). Omega-3 fatty acids lower blood pressure by directly activating large-conductance Ca^{2+}-dependent K^{+} channels. *Proceedings of the National Academy of Sciences of the United States of America, 110*(12), 4816–4821.

Hossain, Z., Mustafa, G., Sakata, K., & Komatsu, S. (2016). Insights into the proteomic response of soybean towards Al_2O_3, ZnO, and Ag nanoparticles stress. *Journal of Hazardous Materials, 304*, 291–305.

Jaradat, A. A. (2016). Breeding oilseed crops for climate change. In S. K. Gupta (Ed.), *Breeding oilseed crops for sustainable production* (pp. 421–472). Academic Press.

Jayadas, N. H., & Nair, K. P. (2006). Coconut oil as base oil for industrial lubricants—evaluation and modification of thermal, oxidative and low temperature properties. *Tribology International, 39*(9), 873–878.

Joshi, A., Kaur, S., Dharamvir, K., Nayyar, H., & Verma, G. (2018). Multi-walled carbon nanotubes applied through seed-priming influence early germination, root hair, growth and yield of bread wheat (Triticum aestivum L.). *Journal of the Science of Food and Agriculture, 98*(8), 3148–3160.

Judy, J. D. (2013). Bioavailability of manufactured nanomaterials in terrestrial ecosystems.

Kachel, M., Stryjecka, M., Ślusarczyk, L., Matwijczuk, A., Budziak-Wieczorek, I., & Gładyszewski, G. (2023). Impact of metal nanoparticles on the phytochemical and antioxidative properties of rapeseed oil. *Materials, 16*(2), 694.

Kah, M., Kookana, R. S., Gogos, A., & Bucheli, T. D. (2018). A critical evaluation of nanopesticides and nanofertilizers against their conventional analogues. *Nature Nanotechnology, 13*(8), 677–684.

Kanhed, P., Birla, S., Gaikwad, S., Gade, A., Seabra, A. B., Rubilar, O., & Rai, M. (2014). In vitro antifungal efficacy of copper nanoparticles against selected crop pathogenic fungi. *Materials Letters, 115*, 13–17.

Khodakovskaya, M. V., de Silva, K., Nedosekin, D. A., Dervishi, E., Biris, A. S., Shashkov, E. V., & Zharov, V. P. (2011). Complex genetic, photothermal, and photoacoustic analysis of nanoparticle-plant interactions. *Proceedings of the National Academy of Sciences of the United States of America, 108*(3), 1028–1033.

Khodakovskaya, M., Dervishi, E., Mahmood, M., Xu, Y., Li, Z., Watanabe, F., & Biris, A. S. (2009). Carbon nanotubes are able to penetrate plant seed coat and dramatically affect seed germination and plant growth. *ACS nano, 3*(10), 3221–3227.

Kim, S. W., Kim, K. S., Lamsal, K., Kim, Y. J., Kim, S. B., Jung, M. Y., & Lee, Y. S. (2009). An in vitro study of the antifungal effect of silver nanoparticles on oak wilt pathogen *Raffaelea* sp. *Journal of Microbiology and Biotechnology, 19*(8), 760–764.

Kiran, V., Harini, K., Thirumalai, A., Girigoswami, K., & Girigoswami, A. (2024). Nanotechnology's role in ensuring food safety and security. *Biocatalysis and Agricultural Biotechnology, 103*, 220.

Kris-Etherton, P. M., & Krauss, R. M. (2020). Public health guidelines should recommend reducing saturated fat consumption as much as possible: YES. *The American Journal of Clinical Nutrition, 112*(1), 13–18.

Lahiani, M. H., Dervishi, E., Chen, J., Nima, Z., Gaume, A., Biris, A. S., & Khodakovskaya, M. V. (2013). Impact of carbon nanotube exposure to seeds of valuable crops. *ACS Applied Materials & Interfaces*, *5*(16), 7965–7973.

Lamsal, K., Kim, S. W., Jung, J. H., Kim, Y. S., Kim, K. S., & Lee, Y. S. (2011). Application of silver nanoparticles for the control of *Colletotrichum species* in vitro and pepper anthracnose disease in field. *Mycobiology*, *39*(3), 194–199.

Larue, C., Khodja, H., Herlin-Boime, N., Brisset, F., Flank, A. M., Fayard, B., & Carrière, M. (2011). Investigation of titanium dioxide nanoparticles toxicity and uptake by plants. *Journal of Physics: Conference Series*, *304*(1), 012057.

Larue, C., Castillo-Michel, H., Sobanska, S., Cécillon, L., Bureau, S., Barthès, V., & Sarret, G. (2014). Foliar exposure of the crop Lactuca sativa to silver nanoparticles: Evidence for internalization and changes in Ag speciation. *Journal of Hazardous Materials*, *264*(1), 98–106.

Linkov, I., Bates, M. E., Canis, L. J., Seager, T. P., & Keisler, J. M. (2011). A decision-directed approach for prioritizing research into the impact of nanomaterials on the environment and human health. *Nature Nanotechnology*, *6*(12), 784–787.

Lu, Z., Percy, R. G., Qualset, C. O., & Zeiger, E. (1998). Stomatal conductance predicts yields in irrigated Pima cotton and bread wheat grown at high temperatures. *Journal of Experimental Botany*, *49*, 453–460.

Lu, C. M., Zhang, C. Y., Wen, J. Q., Wu, G. R., & Tao, M. X. (2002). Research of the effect of nanometer materials on germination and growth enhancement of *Glycine max* and its mechanism. *Soybean Science*, *21*, 168–172.

Ma, J. F. (2004). Role of silicon in enhancing the resistance of plants to biotic and abiotic stresses. *Soil Science and Plant Nutrition*, *50*(1), 11–18.

Mahmoodzadeh, H., Nabavi, M., & Kashefi, H. (2013). Effect of nanoscale titanium dioxide particles on the germination and growth of canola (*Brassica napus*). *Journal of Ornamental Plants (Journal of Ornamental and Horticultural Plants)*, *3*(1), 25–32.

Manjaiah, K. M., Mukhopadhyay, R., Paul, R., Datta, S. C., Kumararaja, P., & Sarkar, B. (2019). Clay minerals and zeolites for environmentally sustainable agriculture. In M. Mercurio, B. Sarkar, & A. Langella (Eds.), *Modified clay and zeolite nanocomposite materials* (pp. 309–329). Elsevier.

Mishra, S., & Singh, H. B. (2015). Biosynthesized silver nanoparticles as a nanoweapon against phytopathogens: exploring their scope and potential in agriculture. *Applied Microbiology and Biotechnology*, *99*, 1097–1107.

Mishra, S., Keswani, C., Abhilash, P. C., Fraceto, L. F., & Singh, H. B. (2017). Integrated approach of agri-nanotechnology: challenges and future trends. *Frontiers in Plant Science*, *8*, 471.

Murphy, D. J. (1991). Designer oilseed crops. Genetic engineering of new oilseed crops for edible and non-edible applications. *Agro Food Industry Hi-Tech*, *2*(5), 5–9.

Peng, D. X., Kang, Y., Hwang, R. M., Shyr, S. S., & Chang, Y. P. (2009). Tribological properties of diamond and SiO_2 nanoparticles added in paraffin. *Tribology International*, *42*(6), 911–917.

Perez-de-Luque, A. (2017). Interaction of nanomaterials with plants: what do we need for real applications in agriculture? *Frontiers in Environmental Science*, *5*, 12.

Pirzadah, T. B., Malik, B., Maqbool, T., & Rehman, R. U. (2019). Development of nano-bioformulations of nutrients for sustainable agriculture. *Nanobiotechnology in Bioformulations*, 381–394. https://doi.org/10.1007/978-3-030-17061-5_16

Pollard, M., Beisson, F., Li, Y., & Ohlrogge, J. B. (2008). Building lipid barriers: biosynthesis of cutin and suberin. *Trends in Plant Science, 13*(5), 236–246.

Prasad, T. N. V. K. V., Sudhakar, P., Sreenivasulu, Y., Latha, P., Munaswamy, V., Reddy, K. R., & Pradeep, T. (2012). Effect of nanoscale zinc oxide particles on the germination, growth and yield of peanut. *Journal of Plant Nutrition, 35*(6), 905–927.

Rahman, M., & de Jiménez, M. M. (2016). Designer oil crops. In S. K. Gupta (Ed.), *Breeding oilseed crops for sustainable production* (pp. 361–376). Academic Press.

Rama, E. M., Bortolan, S., Vieira, M. L., Gerardin, D. C. C., & Moreira, E. G. (2014). Reproductive and possible hormonal effects of carbendazim. *Regulatory Toxicology and Pharmacology, 69*(3), 476–486.

Rana, A. W., Gill, S., & Akram, I. (2022). *Promoting oil seed crops in Pakistan: Prospects and constraints.* International Food Policy Research Institute.

Rani, S., Kumari, N., & Sharma, V. (2023). Uptake, translocation, transformation and physiological effects of nanoparticles in plants. *Archives of Agronomy and Soil Science, 69*(9), 1579–1599.

Rico, C. M., Majumdar, S., Duarte-Gardea, M., Peralta-Videa, J. R., & Gardea-Torresdey, J. L. (2011). Interaction of nanoparticles with edible plants and their possible implications in the food chain. *Journal of Agricultural and Food Chemistry, 59*(8), 3485–3498.

Rico, C. M., Barrios, A. C., Tan, W., Rubenecia, R., Lee, S. C., Varela-Ramirez, A., & Gardea-Torresdey, J. L. (2015). Physiological and biochemical response of soil-grown barley (Hordeum vulgare L.) to cerium oxide nanoparticles. *Environmental Science and Pollution Research, 22*(14), 10551–10558.

Rui, M., Ma, C., Hao, Y., Guo, J., Rui, Y., Tang, X., & Zhu, S. (2016). Iron oxide nanoparticles as a potential iron fertilizer for peanut (Arachis hypogaea). *Frontiers in Plant Science, 7*, 815.

Saini, R. K., & Keum, Y. S. (2018). Omega-3 and omega-6 polyunsaturated fatty acids: Dietary sources, metabolism, and significance—A review. *Life Sciences, 203*, 255–267.

Sastry, R. K., Rashmi, H. B., & Rao, N. H. (2011). Nanotechnology for enhancing food security in India. *Food Policy, 36*(3), 391–400.

Scott, N., & Chen, H. (2012). Nanoscale science and engineering for agriculture and food systems. *Industrial Biotechnology, 8*(6), 340–343.

Shang, Y., Hasan, M. K., Ahammed, G. J., Li, M., Yin, H., & Zhou, J. (2019). Applications of nanotechnology in plant growth and crop protection: a review. *Molecules, 24*(14), 2558.

Sharma, S., Singh, G., Wang, Y., White, J. C., Xing, B., & Dhankher, O. P. (2023). Nanoscale sulfur alleviates silver nanoparticle toxicity and improves seed and oil yield in Soybean (*Glycine max*). *Environmental Pollution, 336*, 122423.

Sharon, M., Choudhary, A. K., & Kumar, R. (2010). Nanotechnology in agricultural diseases and food safety. *Journal of Phytology, 2*(4).

Shen, C. X., Zhang, Q. F., Li, J., Bi, F. C., & Yao, N. (2010). Induction of programmed cell death in *Arabidopsis* and rice by single-wall carbon nanotubes. *American Journal of Botany, 97*(10), 1602–1609.

Shojaei, T. R., Salleh, M. A. M., Tabatabaei, M., Mobli, H., Aghbashlo, M., Rashid, S. A., & Tan, T. (2019). Applications of nanotechnology and carbon nanoparticles in agriculture. In Suraya Abdul Rashid, Raja Nor Izawati Raja Othman and Mohd Zobir Hussein *Synthesis, technology and applications of carbon nanomaterials* (pp. 247–277). Elsevier.

Srivastava, A., & Rao, D. P. (2014). Enhancement of seed germination and plant growth of wheat, maize, peanut and garlic using multiwalled carbon nanotubes. *European Chemical Bulletin, 3*(5), 502–504.

StatNano. (2021). Home I Nanotechnology Products Database. https://product.statnano.com/industry/agriculture.

Su, Y., Ashworth, V., Kim, C., Adeleye, A. S., Rolshausen, P., Roper, C., & Jassby, D. (2019). Delivery, uptake, fate, and transport of engineered nanoparticles in plants: a critical review and data analysis. *Environmental Science: Nano, 6*(8), 2311–2331.

Subramanian, K. (2023). Nanotechnologies for vegetable oilseeds production. *Indian Society of Oilseeds Research, 40*(03), 98–104.

Subramanian, K. S., & Tarafdar, J. C. (2011). Prospects of nanotechnology in Indian farming. *Indian Journal of Agricultural Sciences, 81*(10), 887–893.

Tripathi, D. K., Singh, V. P., Kumar, D., & Chauhan, D. K. (2012). Impact of exogenous silicon addition on chromium uptake, growth, mineral elements, oxidative stress, antioxidant capacity, and leaf and root structures in rice seedlings exposed to hexavalent chromium. *Acta Physiologiae Plantarum, 34*, 279–289.

Tripathi, D. K., Mishra, S., Chauhan, D. K., Tiwari, S. P., & Kumar, C. (2013). Typological and frequency based study of opaline silica (phytolith) deposition in two common Indian *Sorghum* L. species. *Proceedings of the National Academy of Sciences, India Section B: Biological Sciences, 83*(1), 97–104.

Tripathi, D. K., Singh, S., Singh, S., Pandey, R., Singh, V. P., Sharma, N. C., & Chauhan, D. K. (2017). An overview on manufactured nanoparticles in plants: uptake, translocation, accumulation and phytotoxicity. *Plant Physiology and Biochemistry, 110*, 2–12.

Verbruggen, N., & Hermans, C. (2013). Physiological and molecular responses to magnesium nutritional imbalance in plants. *Plant and Soil, 368*, 87–99.

Vermeulen, S. J., Aggarwal, P. K., Ainslie, A., Angelone, C., Campbell, B. M., Challinor, A. J., & Wollenberg, E. (2012). Options for support to agriculture and food security under climate change. *Environmental Science & Policy, 15*(1), 136–144.

Voss-Fels, K. P., Stahl, A., & Hickey, L. T. (2019). Q&A: Modern crop breeding for future food security. *BMC Biology, 17*(1), 18.

Wang, Z., Xie, X., Zhao, J., Liu, X., Feng, W., White, J. C., & Xing, B. (2012). Xylem-and phloem-based transport of CuO nanoparticles in maize (*Zea mays* L.). *Environmental Science & Technology, 46*(8), 4434–4441.

Wendel, M., & Heller, A. R. (2009). Anticancer actions of omega-3 fatty acids-current state and future perspectives. *Anti-Cancer Agents in Medicinal Chemistry-Anti-Cancer Agents, 9*(4), 457–470.

Yadav, A., Babu, S., Krishnan, P., Kaur, B., Bana, R. S., Chakraborty, D., ... Lal, S. K. (2024). Zinc oxide and ferric oxide nanoparticles combination increase plant growth, yield, and quality of soybean under semiarid region. *Chemosphere, 352*(141), 432.

Yaseen, A. A. M., & Wasan, M. H. (2015). Response of sunflower (Helianthus annuus L.) to spraying of nano silver, organic fertilizer (Algastar) and salicylic acid and their impact on seeds content of fatty acids and vicine. *American Journal of Experimental Agriculture, 9*(1), 17898.

Yenjerla, M., Cox, C., Wilson, L., & Jordan, M. A. (2009). Carbendazim inhibits cancer cell proliferation by suppressing microtubule dynamics. *The Journal of Pharmacology and Experimental Therapeutics, 328*(2), 390–398.

Younas, Z., Ahmad, I., Yousaf, T., Raja, N. I., & Mashwani, Z. U. R. (2024). Selenium nanoparticle-mediated enhancement of plant biochemistry and metabolite enrichment: GC–MS profiling of selenium-enriched sesame seed oil. *Trends in Phytochemical Research (TPR), 8*(4), 234–247.

Yusefi-Tanha, E., Fallah, S., Pokhrel, L. R., & Rostamnejadi, A. (2023). Addressing global food insecurity: soil-applied zinc oxide nanoparticles promote yield attributes and seed nutrient quality in *Glycine max* L. *Science of the Total Environment, 876*, 162762.

Zareh-Desari, B., & Davoodi, B. (2016). Assessing the lubrication performance of vegetable oil-based nano-lubricants for environmentally conscious metal forming processes. *Journal of Cleaner Production, 135*, 1198–1209.

Zhang, P., Ma, Y., & Zhang, Z. (2015). Interactions between engineered nanomaterials and plants: phytotoxicity, uptake, translocation, and biotransformation. In *Nanotechnology and plant sciences: nanoparticles and their impact on plants* (pp. 77–99). Cham: Springer International Publishing.

Zhu, Q., Liu, X., Hao, T., Zeng, M., Shen, J., Zhang, F., & de Vries, W. (2020). Cropland acidification increases risk of yield losses and food insecurity in China. *Environmental Pollution, 256*(113), 145.

6

Next-Generation Agriculture with Nanopesticides and Fungicides

Md. Parvez Kabir[1], Md. Taharat Al Tauhid[2], Md. Nasir Uddin[3], Nipa Rani Paul[4], Tahmina Iasmin[5], and Akbar Hossain[1]

[1] *Soil Science Division, Bangladesh Wheat and Maize Research Institute, Dinajpur, Bangladesh*
[2] *Plant Pathology Division, Bangladesh Wheat and Maize Research Institute, Dinajpur, Bangladesh*
[3] *Sustainable Agricultural Systems, International Maize and Wheat Improvement Center (CIMMYT), Dhaka, Bangladesh*
[4] *Plant Pathology Division, Bangladesh Institute of Nuclear Agriculture, Mymensingh, Bangladesh*
[5] *Adaptive Research and Extension Division, Bangladesh Institute of Nuclear Agriculture, Mymensingh, Bangladesh*

6.1 Introduction

There is a global increase in demand for food production in the face of challenging environmental and economic conditions. Thus, the agricultural production system must transform to respond to such demand but with minimum environmental impact (Vos & Bellù, 2019). Chemical methods for pest control and disease management (chemical pesticide and fungicide application) have also yielded promising results (El-Baky & Amara, 2021). However, these methods have drawbacks, including environmental side effects, a risk of resistance, and impacts on human health. Owing to these limitations, nanotechnology has become an excellent approach for application in the form of nanopesticides and nanofungicides.

Nanopesticides and nanofungicides are formulated to ensure the release of active ingredients in a controlled, efficient, and targeted manner with far greater effects than equivalent dosages (Vurro et al., 2019). These advances capitalize on the unique properties of nanoparticles, specifically their high surface area, reactivity, and ability to interact directly at the molecular level with biological entities. By exploiting nanotechnology for pest control, these advanced agrochemicals have the potential to revolutionize pest management, minimize environmental pollution, and increase crop productivity (Singh et al., 2021).

This chapter discusses the evolutionary trend of agriculture involving nanopesticides and fungicides. It provides a picture of the various kinds of these nanoformulations, their mode of action and the possible benefits of using these formulations compared with chemical solutions. It also reviews the limitations of their use and discusses their field applications with case studies to illustrate their practical efficacy. This chapter reviews prospects and innovations in this area and highlights the transformational ability of nanopesticides and fungicides, which can increase agricultural sustainability and productivity.

Nanotechnology Innovations for Food Security and Sustainable Agriculture, First Edition. Edited by Zia Ur Rehman Mashwani, Shah Fahad, Ilyas Ahmad, Shah Saud, Sajid Fiaz, and Taufiq Nawaz.

6.2 Overview of Nanopesticides and Nanofungicides

There is an ever-growing need within the agricultural industry to produce more food. Nanopesticides and nanofungicides reflect a new approach in this evolutionary process, uniting nanotechnology advances with adequate and environmentally friendly crop protection requirements.

Nanopesticides refer to pesticide preparations in which the active ingredient is produced with nanoscale materials ranging from 1 to 100 nm (Hajji-Hedfi & Chhipa, 2021). The following figure (Figure 6.1) illustrates the size of nanoparticles compared with everyday objects such as Earth, a marble, and flour with known tiny objects such as bacteria, viruses, proteins, DNA, and hydrogen atoms, emphasizing how small nanopesticides and nanofungicides are related to traditional agricultural solutions. Such formulations utilize nanoparticles or nanostructures, such as silica, chitosan, or polymeric carriers, to deliver insecticides, herbicides, or other chemical agents, facilitating their controlled release more efficiently than conventional approaches do (Jiang et al., 2024; Li et al., 2021). These nano-engineered materials enhance bioavailability and stability and target pests more precisely, as well as, in turn, improve pesticide efficiency, safety, and selectivity, which reduces non-target organisms, thereby minimizing overall pesticide exposure to the environment.

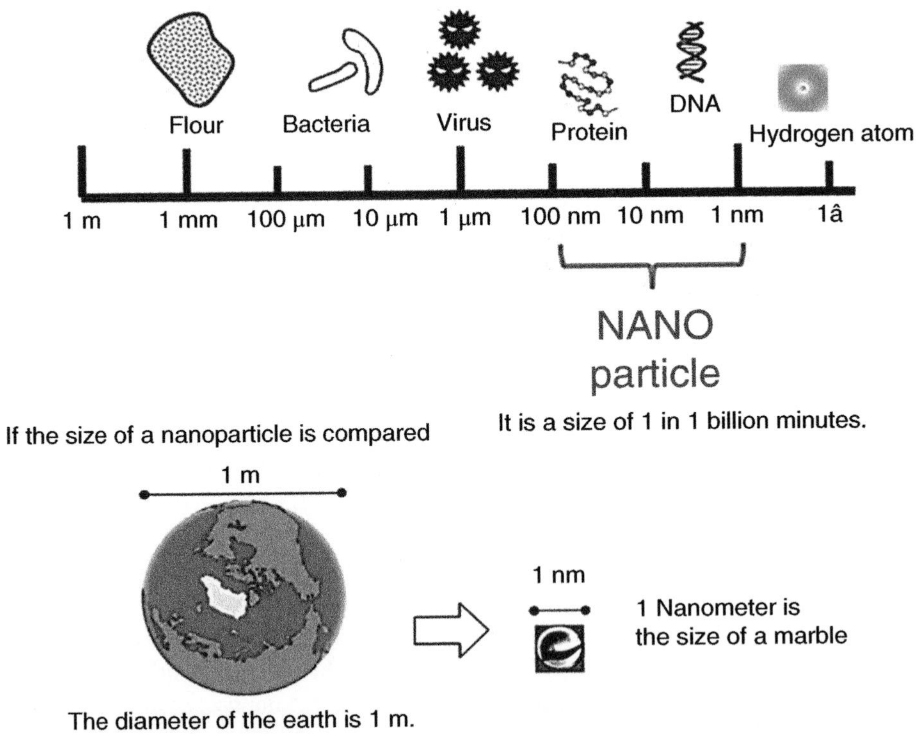

Figure 6.1 The size of the nanoparticles compared with that of known objects. (*Source:* Qureshi et al., 2018/Excellent Publishers).

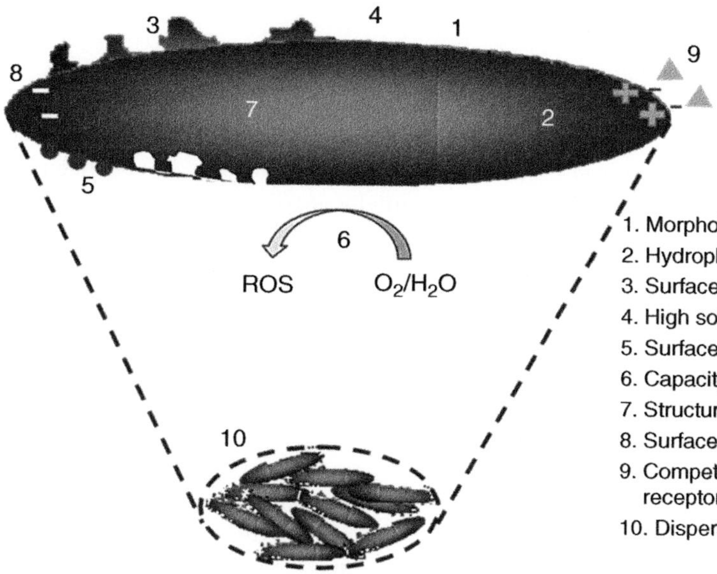

Figure 6.2 Key properties of nanoparticles for targeted delivery. (*Source:* Adapted and modified from Manjunatha et al., 2016).

Nanofungicides are nanosized products that improve the application of antifungal agents to crops. These nanoformulations frequently utilize nanoparticles such as copper oxide, zinc oxide, or titanium dioxide, which have inherent antifungal properties (Vishnu et al., 2024). Nanomaterials can also be used to control the release of fungicides for long-lasting protection with reduced application frequency (Fu et al., 2020). Nanofungicides can also increase the solubility and bioavailability of highly water-insoluble antifungal agents and provide better protection against crop fungal diseases (Kutawa et al., 2021). Some of the properties that make nanopesticides and fungicides highly effective are illustrated in Figure 6.2.

Both nanopesticides and nanofungicides are expected to be part of the next generation of agricultural solutions, which are expected to be more efficient, sustainable, and eco-friendly. These innovative technologies have been developed for improved pest and disease protection and to further decrease the ecological footprint of conventional agrochemicals.

6.3 Types of Nanopesticides and Nanofungicides in Agricultural Applications

Nanopesticides and nanofungicides are a range of formulations employing the unique features of nanomaterials for the improved management of pests and diseases in agriculture. These innovative formulations may be separated according to three main elements: (i) the type of nanomaterial employed, (ii) the pesticide or fungicide release approach, and (iii) the site of action. The following are examples of different forms of nanopesticidal and nanofungicidal formulations:

6.3.1 Nanoemulsion

A nanoemulsion is a colloidal dispersion of oil in water stabilized by a surfactant, the active ingredients contained in the nanoparticles (Adetuyi et al., 2024) (Figure 6.3a). These nano-polymeric systems possess better bioavailability and stability, which favor better penetration into plant tissue. With respect to pesticides, by using nanoemulsions, the effectiveness of the active ingredients can be increased, thus potentially decreasing the amount of the active ingredient used for pest control (Mustafa & Hussein, 2020). This ability can also be exploited by nanofungicides to deposit antifungal active ingredients at the target site, decrease toxicity against nontarget organisms, and provide an extended duration of antifungal efficacy.

6.3.2 Encapsulation of Nanopesticides and Nanofungicides

Insecticide- or fungicide-active compounds are incorporated into nanoparticles (NPs), which are composed of biodegradable polymers, lipids, and silica (Figure 6.3b,g). This encapsulation achieves controlled release of the active ingredients for a prolonged period of time, thereby preventing decomposition due to factors such as UV light or moisture (Mondéjar-López et al., 2024). By regulating release modes for nanoencapsulated products, they are only released to pests or pathogens in more efficient ways, and the application frequency and side effects of pesticide residues in the environment are also lower (Biondo et al., 2022).

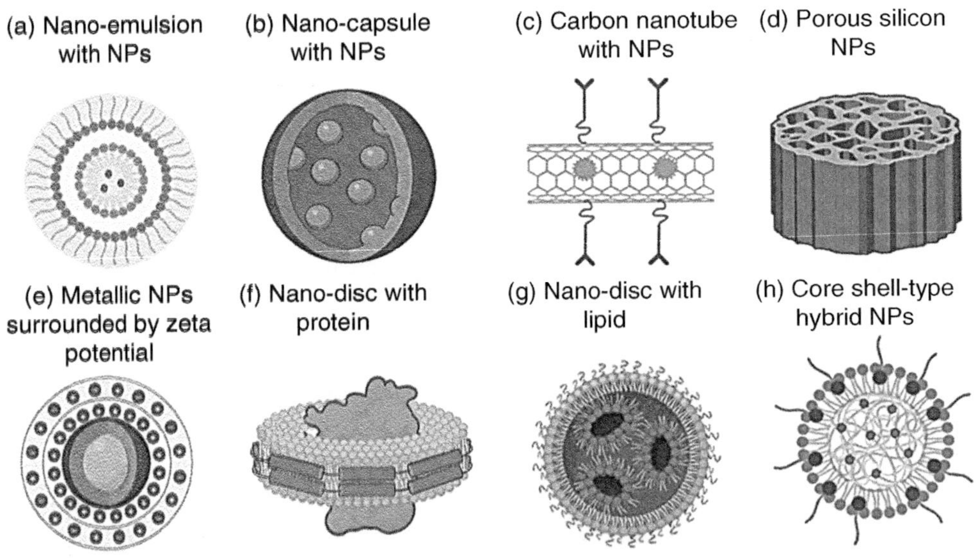

Figure 6.3 Types of nanoparticle-based formulations for nanopesticides and nanofungicides; (a) Nanoemulsion with NPs, (b) Nanocapsule with NPs, (c) Carbon nanotubes with NPs, (d) Porous silicon NPs, (e) Metallic NPs surrounded by zeta potential, (f) Nanodisc with protein, (g) Nanodisc with lipid and (h) Core shell type hybrid NPs. (*Source:* Authors' unpublished materials).

6.3.3 Nanocapsules

Nanocapsules constitute another type of nanoencapsulation, where the pesticide or fungicide is entrapped within the core shell (Kumar et al., 2024) (Figure 6.3b). The first barrier retards external damage and degradation, whereas the second barrier encapsulates active materials, inspiring continuous differential release of the formed NPs (Siddiqui et al., 2023). This specific release of targeted material allows for less waste and the full potential of the chemicals. Nanocapsules may also help solve difficult pest problems, such as the resistance of pests to pesticides, because they are essentially able to deliver treatments in a manner that makes it difficult for pests to become resistant.

6.3.4 Carbon-Based Nanomaterials

The processes of the preparation and application of carbon nanotubes (CNTs), graphene, and fullerene-based materials are used to develop nanopesticides and nanofungicides (Figure 6.3c). Owing to their special structural characteristics, such as high specific surface area, good conductive performance, and high adsorption ability, they have better drug loading and release effects on active ingredients (Madankar et al., 2024). Carbon-based materials can also increase the insecticidal activity of biopesticides such as *Bacillus thuringiensis (Bt)* in the form of nanoformulations (Kumar et al., 2021).

6.3.5 Metal and Metal Oxide Nanoparticles

Metal nanoparticles, such as silver (Ag), copper (Cu), and zinc oxide (ZnO), are promising as pesticide and fungicide agents (Immanuel & Iswareya, 2023; Raza et al., 2024) (Figure 6.3e). These nanoparticles have antimicrobial properties because they can cross the cell walls of pests or pathogens and disrupt their normal biology. For example, silver NPs have been found to inhibit the proliferation of various fungal pathogens (Akpinar et al., 2021; Mansoor et al., 2021; Vera-Reyes et al., 2022), and copper NPs have significant insecticidal activity against insect pests (Badawy et al., 2021; El-Saadony et al., 2020; Rahman et al., 2022). Nanosized ZnO particles have also been used as fungicides (González-Merino et al., 2021; Kamal et al., 2023), in addition to being utilized in the treatment of fungal diseases in crops.

6.3.6 Biodegradable Nanoformulations

Recently, environmentally friendly biodegradable nanopesticides and nanofungicides have emerged because of increasing attention given to environmental concerns. These products utilize biocompatible and biodegradable materials (e.g., chitosan-, starch-, or protein-based nanoparticles) to encapsulate the active ingredient (Singh et al., 2024) (Figure 6.3f,g). Eco-friendly product formulations make it possible to minimize contamination of the environment, thus decreasing the influence on nontarget organisms and ecosystems. Biodegradable nanopesticides and nanofungicides are of special interest for organic farming and sustainable agriculture (Behl et al., 2024).

6.3.7 Hybrid Nanomaterials

Hybrid nanomaterials are combinations of nanoparticles or nanomaterials of different compositions that improve the efficiency of active ingredients in pesticide or fungicide formulations (Abd-Elsalam, 2024; Alghuthaymi et al., 2021) (Figure 6.3h). These hybrids can comprise metal nanoparticles integrated into carbon-containing substances or natural biopolymers combined with artificial nanomaterials. The synergistic effects of hybrid nanomaterials contribute to better pest control and pathogen management via fine-tuning of their physicochemical properties (e.g., stability, dispersion, and biocompatibility) and target-specific delivery of active substances (Martínez-Chávez et al., 2024). Here, we discuss the various types of nanopesticides and nanofungicides currently being examined and employed for agricultural uses. Such advances hold immense potential for improved, more efficient, targeted, and eco-friendly approaches to pest and disease control.

6.4 Mechanisms of Action of Nanopesticides and Nanofungicides

Compared with conventional chemical formulations, nanopesticides and fungicides exhibit novel performance owing to their small size, large surface area, and ability to interact at the molecular level. The mechanisms of action of these nanoformulations are multifold and involve various physical, chemical, and biological processes that increase their efficiency and decrease their environmental impact.

The modes of action of nanoformulations are multifold and involve various physical, chemical, and biological processes that increase their efficiency and decrease their environmental impact, as shown in Figure 6.3. These novel fungal formulations exploit the special properties of nanoparticles, such as increased penetration, controlled release, and physical disruption of fungal cells, to achieve more efficient and sustainable crop protection. Nanofungicides can penetrate the plant cuticle, and fungal membranes enter the site of action, depositing the active ingredient specifically against fungal spores. However, nanoparticles can produce reactive oxygen species (ROS), triggering oxidative stress, which results in damage to fungal DNA and proteins and eventually in cell death. This specific delivery and interaction with fungal cells improve overall fungicide efficiency with minimal environmental pollution and provides a potential eco-friendly alternative to traditional chemical formulations. The basic principles by which nanopesticides and fungicides act are as follows (Figure 6.4):

6.4.1 Enhanced Penetration and Targeted Action

One of the significant advantages of applying nanopesticides and nanofungicides is that nanoparticles penetrate the outer protective layers of the plant (cuticle, cell wall, cell membranes, etc.) more efficiently than traditional formulations do (Maluin & Hussein, 2020; Ramezani et al., 2019). Their small size enables them to pass through protective barriers to bring the active ingredient exactly where it is needed, such as insect pests, fungal spores, or plant cells. This enhances the bioavailability of the active ingredients, and thus, they reach their target more efficiently.

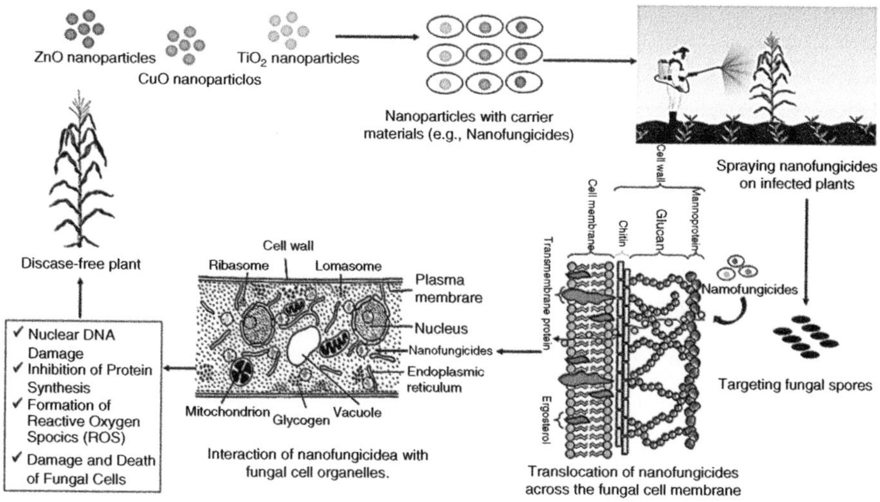

Figure 6.4 Mechanism of action of nanofungicides against fungal pathogens. (*Source:* Authors' unpublished material).

6.4.2 Controlled Release of the Active Ingredient

Nanoencapsulated or nanocapsule formulations ensure the controlled release of active pesticides or fungicides (Kumar et al., 2024). The nanoparticles can also be tuned to release the active ingredient slowly for a more extended period of time to protect against pests and diseases without frequent application. This controlled release minimizes the risk of leaching and breakdown of the active ingredients, leading to better stability and improving overall effectiveness.

6.4.3 Physical Disruption of Pest and Pathogen Cells

The large surface area-to-volume ratio facilitates direct physical contact between the pest and pathogen (Karuvelan et al., 2024). Nanoparticles such as metal and metal oxide nanoparticles can physically damage insect exoskeletons (for insect pests) (Jafir et al., 2023) or fungal cell walls (for fungal pathogens) (Slavin & Bach, 2022). This causes rupture or destabilization of the cell membranes and effectively kills pests or inhibits fungal growth. Metallic nanoparticles, such as silver and copper, can interact with cellular targets and induce oxidative stress and damage vital enzymes and proteins in pests or pathogens (Cao & Wang, 2022; Kale et al., 2021). In addition to the mechanisms described above, various nanoparticles have been tested against different insect pests, which exhibit diverse modes of action, as summarized in Table 6.1.

6.4.4 Generation of Reactive Oxygen Species (ROS)

The generation of ROS is a common mechanism by which metal and metal oxide nanoparticles elicit their pesticidal or fungicidal effects (Ramezani et al., 2019). The high surface reactivity of nanoparticles such as silver (Ag), copper (Cu), and zinc oxide (ZnO) can

Table 6.1 Modes of action of various nanoparticles against insect pests.

Tested nanomaterial	Target insect	Mode of action	References
Ag nanoparticles from *Cassia fistula* extract, $LC_{50} = 3.6$ and 1.7 mg/L	*Aedes albopictus, Culex pipiens pallens*	In the 4th instar larvae, silver nanoparticles (nanoAg) decreased total protein levels and inhibited acetylcholinesterase as well as α- and β-carboxylesterase activities.	Fouad et al. (2018)
Ag nanoparticles from *Punica granatum* peel extract ($LC_{50} = 19.21\,\mu g$/larva)	*Spodoptera litura*	Silver nanoparticles reduced the activities of amylase, protease, lipase, and invertase, along with a decline in gut microflora, extracellular enzyme production, larval weight, gut pH, and total heterotrophic bacterial population.	Bharani and Namasivayam (2017)
ZnO NPs coated with *Bacillus thuringiensis*	*Callosobruchus maculates*	Nanoparticles reduced fecundity, midgut amylase activity, and GST (glutathione S-transferase) activity.	Malaikozhundan et al. (2017)
Ag nanoparticles are synthesized with salicylic acid and 3,5-dinitrosalicylic acid (1–12 ppm)	*A. albopictus*	In the 4th instar larvae, silver nanoparticles reduced total protein content along with esterase, acetylcholinesterase, and phosphatase enzyme activities.	Ga'al et al. (2018)
Mesoporous silica NPs (MSNPs) capped with CS derivatives for pyraclostrobin delivery	*Phomopsis asparagi*	Surface modification with chitosan (CS) derivatives created strong electrostatic interactions, allowing mesoporous silica nanoparticles (MSNPs) to function as effective carriers for plant protectants against *Phomopsis asparagi*.	Cao et al. (2016)
Ag nanoparticles (10–50 μg/mL)	*Drosophila melanogaster*	In fly tissues, silver nanoparticles induced reactive oxygen species (ROS) accumulation, triggering ROS-mediated apoptosis, DNA damage, autophagy, and activation of the Nrf2-dependent antioxidant pathway.	Mao et al. (2018)
Ag NPs	*Bombyx mori*	Silver nanoparticles induced cell necrosis and disrupted signal transduction processes.	Meng et al. (2017)
GO incorporated into solid lipid nanoparticles (SLNPs)	Black cutworm	Graphene oxide (GO) nanoparticles significantly affected larval and pupal development, adult longevity, and female fecundity compared to bulk GO.	Adel et al. (2019)

Table 6.1 (Continued)

Tested nanomaterial	Target insect	Mode of action	References
Nanostructured Al_2O_3 (60–500 ppm)	*Sitophilus oryzae*	Nanoparticles adhere to the beetle cuticle through triboelectric forces, absorbing its wax layer and causing insect dehydration.	Stadler et al. (2017)
Silica NPs	*Callosobruchus maculates*	Nanoparticles inhibited growth and reduced oviposition.	Cáceres et al. (2019)
Carbon-dot-Ag nanohybrid (LC_{50}: 0.30–0.76 ppm)	*Anopheles stephensi, Culex quinquefasciatus*	Silver nanoparticles caused larval body deformation, detected 2.93% Ag accumulation in mosquito tissues, and resulted in damage to the cuticle and cellular organization.	Sultana et al. (2018)
SiO_2 NPs	Different species, with special reference to stored grain pests	Nanoparticles physically sorbed by the insect cuticular lipids caused significant damage, leading to the insect's death.	Athanassiou et al. (2018)
TiO_2 NPs	Egyptian cotton leafworm *Spodoptera littoralis*	A lower lethal dose (LC_{50}) was observed against 2nd and 4th instar larvae.	Shaker et al. (2017)
Ag NPs	*Helicoverpa armigera*	Inhibition of gut protease enzyme activity was observed.	Kantrao et al. (2017)

produce oxidative stress in target pests and pathogens (Jafir et al., 2023). ROS, such as hydroxyl radicals, hydrogen peroxide, or superoxide anions, impair lipids, proteins, and DNA, the most affected cellular constituents. This oxidizing activity ruptures cell membranes, killing pests and pathogens or blocking their ability to proliferate.

6.4.5 Interference with Biological Utilization

The biological utilization of pests and pathogens can interfere with nanoparticles by disrupting their enzymatic systems or metabolic pathways (Kadir et al., 2025; Ramezani et al., 2019). For example, some nanoformulations can interfere with enzymes related to detoxification and the defense systems of pests, increasing their vulnerability to chemical or physical damage (Vishnu et al., 2024). Nanofungicides can also interfere with vital processes in fungal cells, such as cell wall synthesis or DNA replication, which inhibits further growth and reproduction of fungi (El-Beltagi et al., 2022).

6.4.6 Targeted Delivery and Selectivity

One of the benefits that nanotechnology can bring to this field is the development of selective formulations that help the pest or pathogen be the target, with minimal effects on beneficial organisms and the environment. These nanoparticles are designed to bind to specific receptors or molecular markers on the external side of pests or pathogens, which can further increase the specificity of the active ingredients (Ramezani et al., 2019). This type of targeted application has distinct advantages in that it enhances efficacy and reduces nontarget exposure to many organisms, including pollinators, soil microbes, and aquatic organisms.

6.4.7 Synergism with Biological Agents

Biological agents can also synergize with nanopesticides or nanofungicides. For example, nanoparticles can be used to deliver and protect beneficial microorganisms (i.e., *Bacillus thuringiensis* (Bt)) that serve as natural biocontrol agents (Meena & Mishra, 2020). A shell or capsule of nanoparticles protects the biological agent so that it can last longer in the environment and can be applied more efficiently to kill pests (Sukhorukov et al., 2024). This convergence of nanoformulations and biological entities may contribute to reducing the application of chemical insecticides by ensuring appropriate control of pests.

6.4.8 Interference of Nanostructure with Fungal Hyphae

The possible mode of action in the case of fungicides may be physical interference of nanoparticles on the hyphal cells of the fungi. The nanoparticles are able to penetrate the hyphal cells of fungi directly to destroy them, resulting in the inhibition and death of disease-causing fungal pathogens (El-Beltagi et al., 2022). Alterations in the structure of the fungal cell wall by interaction with nanoparticles result in structural limitations, as well as in the loss of integrity and function of fungal cells.

6.5 Challenges and Limitations

Despite the potential of nanopesticides and fungicides for agricultural transformation, several obstacles and difficulties must be overcome before their widespread adoption. These problems arise from technical, regulatory, environmental, and safety issues that must be adequately addressed to facilitate the complete exploitation of the potential of nanotechnology in agriculture. Here are some of the challenges or constraints of nanopesticides and nanofungicides:

6.5.1 Environmental Impact and Ecotoxicity

One of the most significant factors concerning the application of nanomaterials in agriculture is their possible environmental impact. Fine particles are small and highly reactive and can cause undesirable effects through interactions with soil, water, and nontarget living organisms. There are few long-term studies on the effects of nanoparticles on ecosystems (particularly concerning soil health, water quality, and diversity) (Gupta et al., 2024). Undoubtedly, bioaccumulation in the environment or in the food chain represents a significant risk to wildlife and human health. Regulatory authorities also seek to understand how nanopesticides and nanofungicides impact the environment over time.

6.5.2 Regulatory Barriers

The regulatory framework for the approval and use of nanotechnology in agriculture has been slow in many countries. Because nanoformulations are new applications of pesticides and fungicides, they are new "kids on the block" and may not have grown into regulatory definitions (Raj et al., 2021). The lack of established protocols for the testing and approving of nanopesticides and nanofungicides poses a challenge in the commercialization process (Ullah et al., 2024). A clear guideline on safety, efficacy, and environmental impact has to be developed by regulatory authorities so that these nanoproducts can reach the market. More importantly, the absence of harmonized international regulations adds more hurdles to the international trade and application of nanoagrochemicals.

6.5.3 Toxicity and Safety

Even though nanopesticides, as well as nanofungicides are usually promoted as environmentally friendly and more efficient alternatives, concerns related to their toxicity to humans, animals, and nontarget organisms still exist (Dangi & Verma, 2021). NPs can easily pass through biological barriers (cell walls or plasmalemma) and be retained in tissues, which may result in adverse health effects because of the small size of the particles (Prakash et al., 2025). Inhaling or ingesting nanoparticles might also present unknown risks, especially for farm workers, consumers, and wildlife. Hence, extensive studies are needed to investigate the toxicological profiles of these materials and ensure safe exposure levels.

6.5.4 High Production Cost

The production cost of nanopesticides and nanofungicides is usually higher than that of conventional agrochemicals (Abdollahdokht et al., 2022). Specialized devices and cutting-edge technologies with very high costs are inevitable for the synthesis and preparation of nanoparticles. Furthermore, upscaling nanoformulations from the laboratory to large-scale industrial production is challenging in terms of uniformity and economy (Chakraborty et al., 2024). Although cost can be partially offset by less pesticide use and an increase in efficiency in the long run, the cost of producing nanoagrochemicals could be a limiting factor for small-scale farmers and producers in developing countries.

6.5.5 Limited Commercial Availability

There are many studies on nanopesticides and fungicides, but most have been conducted at the level of experimental or laboratory-scale production (Abdollahdokht et al., 2022). Translating laboratory findings to the field with the development of usable products may be slow because of issues related to upscaling production, product stability, and compliance with rules. As a result, nanopesticides and nanofungicides are not currently available very much on the market, and farmers are still not using them. Moreover, the lack of trusted brands and consumer confidence in these new technologies limits market penetration (Singh et al., 2025).

6.5.6 Knowledge Gaps and Limited Field Data

Although laboratory studies and small-scale experiments have indicated that nanopesticides and nanofungicides are promising, insufficient field data are available for these new technologies (Cao & Wang, 2022). The performance of nanoparticles under field conditions is often quite different from that under laboratory conditions because soil, climate, and pest resistance differ. More field studies are needed to establish whether these nanoformulations are promising and effective control approaches for different agronomic systems. In addition, farmers' knowledge or experience with training for the use and application of these technologies can influence their performance in the field.

6.5.7 Nanoparticle Stability and Degradation

The stability of nanoparticles under different environmental conditions is important because it affects the efficacy and persistence of nanopesticides and nanofungicides. NPs can degrade or aggregate depending on their particular circumstances (moisture, ultraviolet (UV) light, and high temperature), diminishing their effectiveness (Andrady et al., 2023; Shrestha et al., 2020). This degradation could lead to the early depletion of active ingredients, thus not only affecting the formulation effect but possibly being responsible for more pesticide applications. Thus, stable nanoparticles, which do not change over time, are necessary for these products to have long-term efficacy.

6.5.8 Consumer Perception and Acceptance

The public's perception of nanotechnology and its application in food production is also a barrier. Many new technologies, especially those related to nanoparticles, are being watched with some war by consumers because of safety and health concerns. The fear of such "unknowns" associated with the chronic effects of nanoformulations may lead to resistance to the acceptance of nanotechnology and its applications, such as in any other new technology that is beneficial or detrimental to us (Prokisch et al., 2024). Establishing confidence and dispelling distrust will also be the further critical mission of public education, which will be realized through open, transparent, and honest dialogue between researchers and the public on the advantages and disadvantages of nanopesticides and nanofungicides.

6.5.9 Intellectual Property and Patent Issues

The advancement and evolution of nanotechnology in agriculture will rely on intellectual property (IP) rights and patent-related issues, which can be considered significant obstacles to the innovation and entry of nanopesticides and nanofungicides. Monopolies or restrictive licensing agreements may restrict access for smaller market participants, and these companies may wish to patent related technologies (Dratler, 2025). This may prevent the application of nanoagrochemicals at the local or global level, particularly in developing countries, where either financial or legislative factors restrict access to these innovative technologies. The advent of these innovative crop protection systems fulfills the promises set by the nano era of agriculture. However, these issues and biotic environments must be compromised to transition these futuristic coatings from the laboratory to land. More research, more straightforward regulatory guidelines, and further consumer education may help address these issues and ensure that the usability of nanotechnology in sustainable agriculture is exploited.

6.6 Field Applications and Case Studies

There is an auspicious point of view for nanopesticides and fungicides since many field trials have shown satisfactory pest or disease control in different agroecosystems. Field trials and case studies are crucial for establishing the practical performance, benefits, and limitations of these nanoformulations across agriculture. This section provides several field practices and case studies, summarized as their applications in terms of how nanopesticides and nanofungicides have been successful in agriculture.

6.6.1 Application of Nanopesticides in Cotton Farming

One of the best examples is cotton pest control via the use of nanoencapsulated pesticides against bollworms. Traditional pesticides are associated with pesticide resistance, pollution, and costliness in field applications (Dhananjayan et al., 2019). Researchers have tested a particular category of nanopesticides, in which a biodegradable polymer surrounds a common insecticide (Mondéjar-López et al., 2024; Sinha et al., 2019). The nanoformulation of the pesticide achieved controlled release of the active agent over time.

According to Chen et al. (2022), graphene-enhanced insecticides have dramatically increased efficacy and reduced pesticide usage by increasing insecticidal activity and permeability. Higher mortality of cotton bollworms was observed in the optimal graphene-insecticide mixtures. In further trials, Said and Abdelaal (2020) reported that the hatchability of pink bollworm (*Pectinophora gossypiella*) eggs, larval and pupal development, and generation time were significantly reduced by Emamectin, ginger aqueous substances, and ginger Ag NPs. These results indicate that these compounds may serve as alternative pest control agents in a nontoxic manner. Ammar and Abd-ElAzeem (2021) established that the biosynthesis of gelatin-copper nanoparticles (G-CuNPs) from fungal filtrates is a promising eco-friendly option for the control of cotton pests. They were more effective than copper nanoparticles in managing spiny bollworms (*Earias insulana*) in terms of larval and pupal mortality, lowered hatchability, and oviposition deterrents.

These experiments demonstrated a markedly lower bollworm population, higher cotton yield, and lower environmental pollution than did conventional pesticide application. This is presented for a case scenario where nanoencapsulated pesticides might be capable of controlling pests effectively with the least overall effect on the ecosystem.

6.6.2 Nano-Silver as Fungicides in Tomato and Strawberry Crops

Another field trial on the application of silver nanoparticles in agriculture with nanofungicides for controlling fungal infection in tomato and strawberry crops. According to an in vitro and in vivo study by Hamouda et al. (2024), *Turbinaria turbinata*-derived biogenic silver nanoparticles (Ag-NPs) can efficiently suppress fungal spoilage by *Penicillium italicum* in tomatoes, showing their fungicidal effectiveness, retardation of spoilage, and prolongation of fruit quality during storage. According to Derbalah et al. (2022), Ag_2O nanoparticles improved strawberry plant resistance to charcoal rot and suppressed *Macrophomina phaseolina* growth by activating an innate defense gene. They significantly enhanced plant growth and crop production, thus representing a new potential disease control approach. However, the plants grew better and left less pesticide residue on the fruits after the nanofungicides treatment. This application demonstrated the promising use of the metal nanoparticle silver as a cost-effective alternative against fungal pathogens for high-value crops.

6.6.3 Nanoparticles for Controlling Fungal Diseases

Potatoes are often susceptible to fungal diseases, such as *Alternaria solani*, which leads to early blight disease. Field trials have been conducted to manage this pathogen via the use of zinc oxide (ZnO) nanoparticles. These nanoformulations facilitated the delivery of the spores to the appropriate target site and achieved better penetration within the plant tissues. Masoud (2018) reported that ZnO NPs are more effective than Kocide as a fungicide for managing early blight disease caused by *Alternaria solani* in potato crops. Compared with those of conventional fungicides, plant growth, tuber yield, the starch content of tubers, and enzyme activity increased, and enzyme activity decreased in response to the use of ZnO NPs. Furthermore, they are the least toxic to nontarget beneficial microorganisms and could be an alternative in integrated pest management (IPM). In support of these findings, various nanoparticles have been successfully applied for the control of different fungal pathogens in plants, as summarized in Table 6.2.

6.6.4 Nanoemulsion for Rice Pest Management

By increasing the surface area and improving dispersibility, nanoemulsions enhance pesticide contact and efficacy. Javanese long pepper *(Piper retrofractum)*-derived essential oils are the most toxic and effective for the treatment of brown planthopper (BPH) (*Nilaparvata lugens*) nymphs (Nuryanti & Budiarti, 2020). The nanoemulsion exhibited more persistent toxicity than maceration extracts and nanoformulations did and could be a potential botanical insecticide for pest management. Similarly, Adak et al. (2020) demonstrated that eucalyptus oil (EO) nanoemulsions prepared with optimum concentrations of EO and Tween 80 were more effective than native EO in controlling stored insect pests such as rice weevil *(Sitophilus oryzae)* and red flour beetle *(Tribolium castaneum)*. Thus, EO nanoemulsions are an environmentally safe alternative to chemical pesticides for controlling pests in stored grains. Additionally, Gao et al. (2020) reported that, compared with the market-available insecticide Benevia, CNAP-HMS-PDAAM nanopesticides, which are hollow mesoporous silica hybrid nanocarriers, had greater sustained effects on rice leaf rollers (*Cnaphalocrocis medinalis)* and striped rice stem borers *(Chilo suppressalis)* under field conditions. Its adhesive nature was potent, and it was released gradually without causing any negative impact on rice growth.

Table 6.2 Applications of various nanoparticles for the control of fungal pathogens in plants.

Nanoparticles	Plant species	Target fungi	Advantages	References
Chitosan and hexaconozole	Palm oil	*Ganoderma boninense*	They serve as an effective nano delivery system.	Maluin and Hussein (2020)
Porous hollow silica nanoparticles (PHSN)	Cucumber	*Botrytis cinerea*	They enable controlled nutrient transfer, enhancing crop protection.	Khoo et al. (2020)
Hexaconozole, Dazomet, and chitosan	Palm oil	*Ganoderma boninense*	They exhibit high biocompatibility and biodegradability of its formulation components.	Maluin et al. (2019)
Chitosan	Maize grains	*Fusarium graminearum*	They reduce volatility and enhances the control of fungal diseases.	Kalagatur et al. (2018)
Polyvinylpyrrolidone (PVP) and PVP copolymer	Southern pine sapwood	*Gloeophyllum trabeum*	They exhibit excellent solubility in solvents of varying polarities, strong binding properties, and a stabilizing effect.	Franco and De Marco (2020)
Bacterial ghosts	Wheat, cucumber, barley	*Leptosphaeria nodorum, Pyrenophora teres, Sphaerotheca fuliginea, Erysiphe graminis*	They provide a multifunctional delivery platform.	Li and Liu (2020)
Calcium carbonate		*Rhizoctonia solani*	They are affordable and acts as a strong antimicrobial agent.	Kutawa et al. (2021)
PVP and PVP copolymer	Southern pine sapwood	*Gloeophyllum trabeum*	They exhibit excellent solubility in solvents of varying polarities, strong binding properties, and a stabilizing effect.	Franco and De Marco (2020)
Eugenol oil, nanoemulsion	Seed cotton	*Fusarium oxysporum, Fusarium oxysporum* f.sp. *vasinfectum*	They exhibited superior antifungal activity compared to its conventional form.	Shao et al. (2018)

Thus, the nanoemulsified insecticide formulations were more effective at reducing pest populations, and fewer pesticides had to be applied compared with conventional formulations. The application frequency was also reduced when nanoemulsions were used, and pest control became more efficient and less labor-intensive. This case study illustrates the beneficial effects of nanoemulsions for pest control in rice cultivation and their potential for sustainable agriculture.

Field applications and case studies highlighted the positive effects of nanopesticides and nanofungicides. With applications ranging from more efficient pest and disease control to minimizing the environmental impact and costs associated with conventional chemical treatments, nanoagrochemicals play an important role in the push towards more sustainable and productive farming around the globe.

6.7 Future Directions and Innovations

Advances in nanotechnology are driving rapid evolution in the fields of nanopesticides and nanopesticides. This field is under rapid development due to materials science, nanotechnology, and farming developments. In this research, the potential of nanomaterials in crop protection has allowed us to discover several innovative findings and future perspectives. These advances can improve pest and disease management, contributing to more sustainable, cost-effective, and environmentally safe agricultural systems. Some of the promising future routes and innovations of nanopesticides and fungicides are discussed below.

6.7.1 Smart and Responsive Nanoformulations

One of the most promising fields of novelty in nanopesticides and nanofungicides is the design of "smart" and responsive formulations (Ullah et al., 2024; Wahab et al., 2024). Such nanoformulations are skillfully engineered and respond to specific environmental factors, including temperature, pH, moisture, and pest conditions. For example, nanoparticles can be designed only to release their active ingredients when pest control conditions are ideal or when they encounter a pest or pathogen (Kannan et al., 2022). This responsive system efficiently delivers agrochemicals by applying less active substances and reducing agrochemical residue in the environment. In addition, such smart formulations could mean that pesticides or fungicides can be applied precisely, at the proper dose, at the right time to ensure that a crop receives the required protection.

6.7.2 Nanobiopesticides and Biocompatible Nanoformulations

An increasing demand for environmentally sustainable agricultural products has driven the development of nanobiopesticides where the efficiency of nanotechnology meets biological control strategies. These formulations often contain natural or biological derivatives (e.g., plant extracts, essential oils or microorganisms) entrapped in nanoparticles (Antunes Filho et al., 2023; Nair et al., 2022). Nanobiopesticides offer targeted, safe and effective pest control. Developments in this field target the stability, solubility, and biological availability of these natural products, which are often unstable or have challenging applications in traditional formulations. Nanobiopesticides may be crucial components of IPM systems, with implications for replacing chemical pesticides to achieve healthy plant growth (de Oliveira, 2021; Hazafa et al., 2021).

6.7.3 Nanosensors for Pest and Disease Diagnostics

Nanosensors might alter how pests and disease onset are detected and handled. Farmers can access real-time information about pest populations, pathogen presence and environmental conditions by deploying nanosensors in their fields (Bharti et al., 2024). These sensors can sense even the most minor environmental changes, for example, detecting certain pest pheromones or specific fungal spores and “switching on” the release of the correct nanopesticide or nanofungicides treatment. Using nanosensors would permit precision agriculture, with applications of treatments being made in the event of need only, limiting overuse and improving the sustainability of crop protection practices.

6.7.4 Nanofertilizer and Pesticide Combinations

Another potential future development is the production of nanofertilizer and pesticide combination formulations. Nanoformulations may be tailored to co-deliver nutrients and plant protective ingredients to plants, promoting growth and protecting against pests and diseases (Sarkar et al., 2021). This two-pronged method could simplify farming and require fewer separate pest control and fertilization applications. Nanofertilizers can also enhance nutrient uptake and efficiency, reducing the environmental footprint of fertilizers and sustaining soil fertility. Combining nutrients and pesticides into one product is a convenient and cost-effective solution for today’s farming (Tang et al., 2023).

6.7.5 Global Market and Regulatory Scenario

The market for nanopesticides and nanofungicides is likely to rise with increasing application, as the advantages of nanopesticides and fungicides have become apparent (Ali et al., 2023; Qazi & Dar, 2020). However, a regulatory framework must be developed for regular use to consider the difficulties and novelties of nanomaterials. International cooperation will be essential in establishing testing standards, safety precautions, and environmental assessments that are far from becoming standard (Isibor, 2024). Consumers increasingly ask for transparent labeling of nanobased products in agriculture and certification systems, which guarantees the safety and effectiveness of products.

6.7.6 Integration with Precision Agriculture

The future challenges associated with nanopesticides and nanofungicides will be integrated with precision agriculture. Another significant task for innovation is the use of nanopesticides and fungicides. By infusing nanoagrochemicals with data-powered farming solutions such as satellite imaging, drones, and soil sensors, farmers can calibrate applications of pesticides and fungicides for maximum benefit and minimal environmental impact (Yadav et al., 2023). Precision agriculture allows farmers to treat only the areas that need it, so they use less and waste less. This synergy between nanotechnology and precision agriculture will assist in promoting more innovative, sustainable farming practices and increasing crop yield.

In summary, the potential of nanopesticides and nanofungicides is significantly brighter for revolutionizing the agricultural landscape. With continual advancements in nanomaterials, innovative formulations, biocompatible ingredients, precise applications, and more than ever before, research on pest and disease control is limited. Owing to progress in sustainability, resistance management and legislation, nanoagrochemicals are at the threshold of having a revolutionary impact on a more efficient, ecologically friendly, and sustainable agricultural system.

6.8 Conclusion

Nanopesticides and nanofungicides are novel pest and disease management solutions in agriculture. While they use the special features of nanomaterials (larger surface area, targeted delivery), such formulations represent more effective and eco-friendly alternatives than traditional chemicals. These compounds promote pest control, environmental compatibility, and yield at reduced dosages. For broad adoption, however, issues such as production costs, legal restrictions, and environmental concerns must be resolved. Their incorporation into agricultural practices requires more field-based research and regulatory frameworks. In the future, developments such as innovative, biodegradable, and nanopesticides formulations and their combination with precision agriculture will increase their sustainability and efficiency. Nanopesticides and nanofungicides have great potential to revolutionize agriculture for food security and environmental sustainability.

References

Abd-Elsalam, K. A. (2024). Chapter 1 - Next-generation fungicides based on nanohybrids: A preliminary review. In: Nanobiotechnology for plant protection, Abd-Elsalam, K.A. (Editor), Nanohybrid Fungicides, (3–23), Elsevier. 10.1016/B978-0-443-23950-2.00017-5

Abdollahdokht, D., Gao, Y., Faramarz, S., Poustforoosh, A., Abbasi, M., Asadikaram, G., & Nematollahi, M. H. (2022). Conventional agrochemicals towards nanobiopesticides: an overview on recent advances. *Chemical and Biological Technologies in Agriculture*, *9*(1), 13.

Adak, T., Barik, N., Patil, N. B., Govindharaj, G. P. P., Gadratagi, B. G., Annamalai, M., ... Rath, P. C. (2020). Nanoemulsion of eucalyptus oil: an alternative to synthetic pesticides against two major storage insects (*Sitophilus oryzae* (L.) and *Tribolium castaneum* (Herbst)) of rice. *Industrial Crops and Products*, *143*, 111849.

Adel, M. M., Salem, N. Y., Abdel-Aziz, N. F., & Ibrahim, S. S. (2019). Application of new nano pesticide Geranium oil loadedsolid lipid nanoparticles for control the black cutworm *Agrotis ipsilon* (Hub.) (Lepi., Noctuidae). *EurAsian Journal of BioSciences*, *13*(2), 1453.

Adetuyi, B. O., Odine, G., Abiodun, P. O., Omowumi, O. S., & Adetunji, C. O. (2024). Nanobioinsecticide and nanoemulsions: recent advances. *Handbook of Agricultural Biotechnology*, *4*, 129–163.

Akpinar, I., Unal, M., & Sar, T. (2021). Potential antifungal effects of silver nanoparticles (AgNPs) of different sizes against phytopathogenic *Fusarium oxysporum* f. sp. radicis-lycopersici (FORL) strains. *SN Applied Sciences, 3*(4), 506.

Alghuthaymi, M. A., Rajkuberan, C., Rajiv, P., Kalia, A., Bhardwaj, K., Bhardwaj, P., Abd-Elsalam, K. A., ... Kuca, K. (2021). Nanohybrid antifungals for control of plant diseases: current status and future perspectives. *Journal of Fungi, 7*(1), 48.

Ali, S., Ahmad, N., Dar, M. A., Manan, S., Rani, A., Alghanem, S. M. S., ... Zhu, D. (2023). Nanoagrochemicals as substitutes for pesticides: prospects and risks. *Plants, 13*(1), 109.

Ammar, H. A., & Abd-ElAzeem, E. M. (2021). Novel treatment of gelatin-copper bionanoparticles as a management method against the spiny bollworm, *Earias insulana*, (Boisd.) (Lepidoptera: Noctuidae) in comparison studies with the uncoated nanoparticles. *Inorganic and Nano-Metal Chemistry, 51*(3), 309–321.

Andrady, A. L., Heikkilä, A. M., Pandey, K. K., Bruckman, L. S., White, C. C., Zhu, M., & Zhu, L. (2023). Effects of UV radiation on natural and synthetic materials. *Photochemical & Photobiological Sciences, 22*(5), 1177–1202.

Antunes Filho, S., Dos Santos, M. S., Dos Santos, O. A. L., Backx, B. P., Soran, M. L., Opriş, O., ... Bououdina, M. (2023). Biosynthesis of nanoparticles using plant extracts and essential oils. *Molecules, 28*(7), 3060.

Athanassiou, C. G., Kavallieratos, N. G., Benelli, G., Losic, D., Usha Rani, P., & Desneux, N. (2018). Nanoparticles for pest control: current status and future perspectives. *Journal of Pest Science, 91*, 1–15.

Badawy, A. A., Abdelfattah, N. A., Salem, S. S., Awad, M. F., & Fouda, A. (2021). Efficacy assessment of biosynthesized copper oxide nanoparticles (CuO-NPs) on stored grain insects and their impact on morphological and physiological traits of wheat (*Triticum aestivum* L.) plant. *Biology, 10*(3), 233.

Behl, K., Jaiswal, P., & Pabbi, S. (2024). Recent advances in microbial and nanoformulations for effective delivery and agriculture sustainability. *Biocatalysis and Agricultural Biotechnology, 50*, 103180.

Bharani, R. A., & Namasivayam, S. K. R. (2017). Biogenic silver nanoparticles mediated stress on developmental period and gut physiology of major lepidopteran pest *Spodoptera litura* (Fab.) (Lepidoptera: Noctuidae)—An eco-friendly approach of insect pest control. *Journal of Environmental Chemical Engineering, 5*(1), 453–467.

Bharti, A., Jain, U., & Chauhan, N. (2024). From lab to field: nanobiosensors for real-time plant nutrient tracking. *Plant Nano Biology, 9*, 100079.

Biondo, F., Baldassarre, F., Vergaro, V., & Ciccarella, G. (2022). Controlled biocide release from smart delivery systems: materials engineering to tune release rate, biointeractions, and responsiveness. In G. M. Balestra & E. Fortunati (Eds.), *Nanotechnology-based sustainable alternatives for the management of plant diseases* (pp. 31–147). Elsevier. https://doi.org/10.1016/C2019-0-05460-X

Cáceres, M., Vassena, C. V., Garcerá, M. D., & Santo-Orihuela, P. L. (2019). Silica nanoparticles for insect pest control. *Current Pharmaceutical Design, 25*(37), 4030–4038.

Cao, X., & Wang, Z. (2022). Application of non-agricultural technology for biotic stress management: mechanisms, optimization, and future perspectives. *Environmental Science: Nano, 9*(12), 4336–4353.

Cao, L., Zhang, H., Cao, C., Zhang, J., Li, F., & Huang, Q. (2016). Quaternized chitosan-capped mesoporous silica nanoparticles as nanocarriers for controlled pesticide release. *Nanomaterials*, *6*(7), 126.

Chakraborty, S., Mal, S., Halder, A., Das, S., Sen, K. K., Mukherjee, A., & Roy, P. (2024). Nano dimensional gold synthesis for biomedical applications: upscaling and challenges. *Particulate Science and Technology*, *42*(1), 145–163.

Chen, Z., Zhao, J., Liu, Z., Bai, X., Li, W., Guan, Z., ... Zhu, H. (2022). Graphene-delivered insecticides against cotton bollworm. *Nanomaterials*, *12*(16), 2731.

Dangi, K., & Verma, A. K. (2021). Efficient & eco-friendly smart nanopesticides: emerging prospects for agriculture. *Materials Today: Proceedings*, *45*, 3819–3824.

Derbalah, A., Essa, T., Kamel, S. M., Omara, R. I., Abdelfatah, M., Elshaer, A., & Elsharkawy, M. M. (2022). Silver oxide nanostructures as a new trend to control strawberry charcoal rot induced by *Macrophomina phaseolina*. *Pest Management Science*, *78*(11), 4638–4648.

Dhananjayan, V., Jayakumar, S., & Ravichandran, B. (2019). Conventional methods of pesticide application in agricultural field and fate of the pesticides in the environment and human health. In Controlled release of pesticides for sustainable agriculture, Rakhimol, K.R., Thomas, S., Volova, T., & Jayachandran, K. (eds.) (pp. 1–39). Cham: Springer International Publishing. https://doi.org/10.1007/978-3-030-23396-9

Dratler, J. (2025). *Licensing of intellectual property*. Law Journal Press.

El-Baky, N. A., & Amara, A. A. A. F. (2021). Recent approaches towards control of fungal diseases in plants: an updated review. *Journal of Fungi*, *7*(11), 900.

El-Beltagi, H. S., Bendary, E. S., Ramadan, K. M., & Mohamed, H. I. (2022). Metallic nanoparticles and nanobased bioactive formulations as nanofungicides for sustainable disease management in cereals. In K. A. Abd-Elsalam & H. I. Mohamed (Eds.), *Cereal diseases: nanobiotechnological approaches for diagnosis and management* (pp. 315–343). Dordrecht, The Netherlands: Springer Singapore. https://doi.org/10.1007/978-981-19-3120-8

El-Saadony, M. T., Abd El-Hack, M. E., Taha, A. E., Fouda, M. M., Ajarem, J. S., Maodaa, N., Maodaa, S., ... Elshaer, N. (2020). Ecofriendly synthesis and insecticidal application of copper nanoparticles against the storage pest *Tribolium castaneum*. *Nanomaterials*, *10*(3), 587.

Fouad, H., Hongjie, L., Hosni, D., Wei, J., Abbas, G., Ga'al, H., & Jianchu, M. (2018). Controlling *Aedes albopictus* and *Culex pipiens* pallens using silver nanoparticles synthesized from aqueous extract of *Cassia fistula* fruit pulp and its mode of action. *Artificial Cells, Nanomedicine, and Biotechnology*, *46*(3), 558–567.

Franco, P., & De Marco, I. (2020). The use of poly (n-vinyl pyrrolidone) in the delivery of drugs: a review. *Polymers*, *12*(5), 1114.

Fu, L., Wang, Z., Dhankher, O. P., & Xing, B. (2020). Nanotechnology as a new sustainable approach for controlling crop diseases and increasing agricultural production. *Journal of Experimental Botany*, *71*(2), 507–519.

Ga'al, H., Fouad, H., Tian, J., Hu, Y., Abbas, G., & Mo, J. (2018). Synthesis, characterization and efficacy of silver nanoparticles against *Aedes albopictus* larvae and pupae. *Pesticide Biochemistry and Physiology*, *144*, 49–56.

Gao, Y., Li, D., Li, D., Xu, P., Mao, K., Zhang, Y., ... He, S. (2020). Efficacy of an adhesive nanopesticide on insect pests of rice in field trials. *Journal of Asia-Pacific Entomology*, *23*(4), 1222–1227.

González-Merino, A. M., Hernández-Juárez, A., Betancourt-Galindo, R., Ochoa-Fuentes, Y. M., Valdez-Aguilar, L. A., & Limón-Corona, M. L. (2021). Antifungal activity of zinc oxide nanoparticles in *Fusarium oxysporum-Solanum lycopersicum* pathosystem under controlled conditions. *Journal of Phytopathology*, *169*(9), 533–544.

Gupta, S., Kumar, D., Aziz, A., AE AbdelRahman, M., Mustafa, A. R. A., Scopa, A., ... Moursy, A. R. (2024). Nanoecology: exploring engineered nanoparticles' impact on soil ecosystem health and biodiversity. *Egyptian Journal of Soil Science*, *64*(4), 1637–1655.

Hajji-Hedfi, L., & Chhipa, H. (2021). Nanobased pesticides: challenges for pest and disease management. *Euro-Mediterranean Journal for Environmental Integration*, *6*(3), 69.

Hamouda, R. A., Almaghrabi, F. Q., Alharbi, O. M., Al-Harbi, A. D., Alsulami, R. M., & Alhumairi, A. M. (2024). Antifungal activities of biogenic silver nanoparticles mediated by marine algae: in vitro and in vivo insights of coating tomato fruit to protect against *Penicillium italicum* blue mold. *Marine Drugs*, *22*(5), 225.

Hazafa, A., Murad, M., Masood, M. U., Bilal, S., Khan, M. N., Farooq, Q., ... & Naeem, M. (2021). Nanobiopesticides as an emerging technology for pest management. In *Insecticides-impact and benefits of its use for humanity*, Ranz, R.E.R. (ed), (pp. 1–31) IntechOpen, UK. 10.5772/intechopen.101285

Immanuel, S. J., & Iswareya, L. V. (2023). Nanoparticles: plant protective agents against pathogenic microbes and pests. In K. Poddar, J. Vijayan, S. Ray, & T. Adak (Eds.), *Nanotechnology for sustainable agriculture* (pp. 129–162). Apple Academic Press. https://doi.org/10.1016/B978-0-12-812160-3.00010-6

Isibor, P. O. (2024). Regulations and policy considerations for nanoparticle safety. In P. O. Isibor, G. Devi, & A. A. Enuneku (Eds.), *Environmental nanotoxicology: combatting the minute contaminants* (pp. 295–316). Cham: Springer Nature Switzerland. https://doi.org/10.1007/978-3-031-54154-4

Jafir, M., Irfan, M., Zia-ur-Rehman, M., Hafeez, F., Ahmad, J. N., Sabir, M. A., ... Moosa, A. (2023). The global trend of nanomaterial usage to control the important agricultural arthropod pests: a comprehensive review. *Plant Stress*, *10*, 100208.

Jiang, X., Yang, F., Jia, W., Jiang, Y., Wu, X., Song, S., ... Shen, J. (2024). Nanomaterials and nanotechnology in agricultural pesticide delivery: a review. *Langmuir*, *40*(36), 18806–18820.

Kadir, M. L., Dageri, A., & Aslan, T. N. (2025). Nanopesticides for managing primary and secondary stored product pests: current status and future directions. *Heliyon*, *11*(4), e42341.

Kalagatur, N. K., Nirmal Ghosh, O. S., Sundararaj, N., & Mudili, V. (2018). Antifungal activity of chitosan nanoparticles encapsulated with *Cymbopogon martinii* essential oil on plant pathogenic fungi *Fusarium graminearum*. *Frontiers in Pharmacology*, *9*, 610.

Kale, S. K., Parishwad, G. V., & Patil, A. S. H. A. S. (2021). Emerging agriculture applications of silver nanoparticles. *ES Food & Agroforestry*, *3*, 17–22.

Kamal, A., Ali, M., Farraj, D. A. A., Al-Zaidi, E. M., Khizar, M., Aljaaidi, R. A., ... Munis, M. F. H. (2023). Diagnosis and control of brown leaf spot of kiwi (*Actinidia deliciosa*) using biochar-zinc oxide nanocomposite (MB-ZnO) as a nontoxic biofungicides. *Crystals*, *13*(1), 98.

Kannan, M., Mohan, M., Elango, K., Govindaraju, K., & Mani, M. (2022). Principles and application of nanotechnology in pest management. In M. Mani (Ed.), *Trends in horticultural entomology* (pp. 49–79). Singapore: Springer. https://doi.org/10.1007/978-981-19-0343-4_2

Kantrao, S., Ravindra, M. A., Akbar, S. M. D., Jayanthi, P. K., & Venkataraman, A. (2017). Effect of biosynthesized silver nanoparticles on growth and development of *Helicoverpa armigera* (Lepidoptera: Noctuidae): interaction with midgut protease. *Journal of Asia-Pacific Entomology*, *20*(2), 583–589.

Karuvelan, M., Kaliyamoorthy, S., Dhanaraj, S., Chelliah, R., & Oh, D. H. (2024). Use of gold nanoparticles as a nano-weapon in plant disease management, Plant response to gold nanoparticles. Smart nanomaterials technology. In A. Husen (Ed.), (pp. 271–310). Singapore: Springer. https://doi.org/10.1007/978-981-97-7174-5_1

Khoo, K. S., Chia, W. Y., Tang, D. Y. Y., Show, P. L., Chew, K. W., & Chen, W. H. (2020). Nanomaterials utilization in biomass for biofuel and bioenergy production. *Energies*, *13*(4), 892.

Kumar, P., Pandhi, S., Mahato, D. K., Kamle, M., & Mishra, A. (2021). Bacillus-based nano bioformulations for phytopathogens and insect–pest management. *Egyptian Journal of Biological Pest Control*, *31*, 1–11.

Kumar, S., Bhuvaneshwari, R., Jain, S., Nirwan, S., Fatima, Z., Kumar, D., ... Poonam. (2024). A systematic review on pesticide-loaded nanocapsules: a sustainable route for pesticide management to enhance crop productivity. *Current Nanoscience*, *20*(3), 280–297.

Kutawa, A. B., Ahmad, K., Ali, A., Hussein, M. Z., Abdul Wahab, M. A., Adamu, A., ... Hossain, M. I. (2021). Trends in nanotechnology and its potentialities to control plant pathogenic fungi: a review. *Biology*, *10*(9), 881.

Li, R., & Liu, Q. (2020). Engineered bacterial outer membrane vesicles as multifunctional delivery platforms. *Frontiers in Materials*, *7*, 202.

Li, N., Sun, C., Jiang, J., Wang, A., Wang, C., Shen, Y., ... Wang, Y. (2021). Advances in controlled-release pesticide formulations with improved efficacy and targetability. *Journal of Agricultural and Food Chemistry*, *69*(43), 12579–12597.

Madankar, R. S., Bhilkar, P. R., Tiple, A. D., Nagmote, M. S., Potbhare, A. K., Sharma, R., ... Chaudhary, R. G. (2024). Role of carbonaceous fullerene materials in crop improvement and targeted pesticide management. In K. Biswas, Y. K. Mohanta, T. K. Mohanta, & M. Saravanan (Eds.), *Carbon-based nanomaterials in biosystems* (pp. 223–242). Academic Press.

Malaikozhundan, B., Vaseeharan, B., Vijayakumar, S., & Thangaraj, M. P. (2017). *Bacillus thuringiensis* coated zinc oxide nanoparticle and its biopesticidal effects on the pulse beetle, *Callosobruchus maculatus*. *Journal of Photochemistry and Photobiology B: Biology*, *174*, 306–314.

Maluin, F. N., & Hussein, M. Z. (2020). Chitosan-based agronanochemicals as a sustainable alternative in crop protection. *Molecules*, *25*(7), 1611.

Maluin, F. N., Hussein, M. Z., Yusof, N. A., Fakurazi, S., Seman, I. A., Hilmi, N. H. Z., & Daim, L. D. J. (2019). Enhanced fungicidal efficacy on *Ganoderma boninense* by simultaneous codelivery of hexaconazole and dazomet from their chitosan nanoparticles. *RSC Advances*, *9*(46), 27083–27095.

Manjunatha, S. B., Biradar, D. P., & Aladakatti, Y. R. (2016). Nanotechnology and its applications in agriculture: a review. *Journal of Farm Sciences*, *29*(1), 1–13.

Mansoor, S., Zahoor, I., Baba, T. R., Padder, S. A., Bhat, Z. A., Koul, A. M., & Jiang, L. (2021). Fabrication of silver nanoparticles against fungal pathogens. *Frontiers in Nanotechnology*, *3*, 679358.

Mao, B. H., Chen, Z. Y., Wang, Y. J., & Yan, S. J. (2018). Silver nanoparticles have lethal and sublethal adverse effects on development and longevity by inducing ROS-mediated stress responses. *Scientific Reports, 8*(1), 2445.

Martínez-Chávez, L. A., Hernández-Ramírez, M. Y., Feregrino-Pérez, A. A., & Esquivel Escalante, K. (2024). Cutting-edge strategies to enhance bioactive compound production in plants: potential value of integration of elicitation, metabolic engineering, and green nanotechnology. *Agronomy, 14*(12), 2822.

Masoud, S. A. (2018). The role of zinc oxide nanoparticles (ZnO NPs) and kocide fungicide on early blight (*Alternaria solani*) of potato and their sides effects on potato plants. *Egyptian Scientific Journal of Pesticides, 4*(4), 1–6.

Meena, R. K., & Mishra, P. (2020). Bio-pesticides for agriculture and environment sustainability. In R. U. E. in Agriculture, S. Kumar, R. S. Meena, & M. K. Jhariya (Eds.), (pp. 85–107). Singapore: Springer. https://doi.org/10.1007/978-981-15-6953-1_3

Meng, X., Abdlli, N., Wang, N., Lü, P., Nie, Z., Dong, X., ... Chen, K. (2017). Effects of Ag nanoparticles on growth and fat body proteins in silkworms (*Bombyx mori*). *Biological Trace Element Research, 180*, 327–337.

Mondéjar-López, M., García-Simarro, M. P., Navarro-Simarro, P., Gómez-Gómez, L., Ahrazem, O., & Niza, E. (2024). A review on the encapsulation of "eco-friendly" compounds in natural polymer-based nanoparticles as next generation nanoagrochemicals for sustainable agriculture and crop management. *International Journal of Biological Macromolecules, 280*, 136030.

Mustafa, I. F., & Hussein, M. Z. (2020). Synthesis and technology of nanoemulsion-based pesticide formulation. *Nanomaterials, 10*(8), 1608.

Nair, A., Mallya, R., Suvarna, V., Khan, T. A., Momin, M., & Omri, A. (2022). Nanoparticles—Attractive carriers of antimicrobial essential oils. *Antibiotics, 11*(1), 108.

Nuryanti, N. S. P., & Budiarti, L. (2020). The toxicity and effectiveness of nanoemulsion formulas from Piper retrofractum essential oil against brown plant hopper (*Nilaparvata lugens* Stål). *Journal of Physics: Conference Series, 1450*(1), 012055.

de Oliveira, J. L. (2021). Nanobiopesticides: present concepts and future perspectives in integrated pest management. In S. Jogaiah, H. B. Singh, L. F. Fraceto, & R. De Lima (Eds.), *Advances in nanofertilizers and nanopesticides in agriculture* (pp. 1–27)). Woodhead Publishing.

Prakash, D. S., Gupta, I., Andanamala, V. R., Nelapati, A. K., Roy, K., & Meena, V. (2025). Nanoparticle uptake, translocation, and biological impact in plants. In R. Pabbati, K. Chepuri, K. V. Reddy, & N. R. Maddela (Eds.), *Plant-based nanoparticle synthesis for sustainable agriculture* (pp. 79–93). CRC Press.

Prokisch, J., Törős, G., Nguyen, D. H., Neji, C., Ferroudj, A., Sári, D., ... El-Ramady, H. (2024). Nanofood farming: toward sustainable applications of proteins, mushrooms, nano nutrients, and nanofibers. *Agronomy, 14*(3), 606.

Qazi, G., & Dar, F. A. (2020). Nanoagrochemicals: economic potential and future trends. In K. R. Hakeem & T. B. Pirzadah (Eds.), *Nanobiotechnology in agriculture: an approach towards sustainability* (pp. 185–193). Cham: Springer International Publishing.

Qureshi, A., Singh, D. K., & Dwivedi, S. (2018). Nanofertilizers: a novel way for enhancing nutrient use efficiency and crop productivity. *International Journal of Current Microbiology and Applied Sciences, 7*(2), 3325–3335.

Rahman, A., Pittarate, S., Perumal, V., Rajula, J., Thungrabeab, M., Mekchay, S., & Krutmuang, P. (2022). Larvicidal and antifeedant effects of copper nanopesticides against *Spodoptera frugiperda* (JE Smith) and its immunological response. *Insects, 13*(11), 1030.

Raj, S. N., Anooj, E. S., Rajendran, K., & Vallinayagam, S. (2021). A comprehensive review on regulatory invention of nano pesticides in agricultural nano formulation and food system. *Journal of Molecular Structure, 1239*, 130517.

Ramezani, M., Ramezani, F., & Gerami, M. (2019). Nanoparticles in pest incidences and plant disease control. In D. G. Panpatte & Y. K. Jhala (Eds.), *Nanotechnology for agriculture: crop production & protection* (pp. 233–272). Singapore: Springer Singapore.

Raza, A., Khandelwal, K., Pandit, S., Singh, M., Kumar, S., Rustagi, S., ... Prasad, R. (2024). Exploring the potential of metallic and metal oxide nanoparticles for reinforced disease management in agricultural systems: a comprehensive review. *Environmental Nanotechnology, Monitoring & Management, 22*, 100998.

Said, S. M., & Abdelaal, A. A. A. (2020). Comparative effect of Ginger extract nanoparticles with some pesticides on Pink bollworm, *Pectinophora gossypiella* (Saunders) (Lepidoptera: Gelechiidae). *International Journal of Scientific Research and Sustainable Development, 3*(3), 1–20.

Sarkar, N., Chaudhary, S., & Kaushik, M. (2021). Nanofertilizers and nanopesticides as promoters of plant growth in agriculture. In P. Singh, R. Singh, P. Verma, R. Bhadouria, A. Kumar, & M. Kaushik (Eds.), *Plant-microbes-engineered nanoparticles (PM-ENPs) nexus in agro-ecosystems: understanding the interaction of plant, microbes and engineered nanoparticles (ENPS)* (pp. 153–163). Cham: Springer International Publishing.

Shaker, A. M., Zaki, A. H., Abdel-Rahim, E. F., & Khedr, M. H. (2017). TiO_2 nanoparticles as an effective nanopesticide for cotton leaf worm. *Agricultural Engineering International: CIGR Journal, Special issue, 68*, 61.

Shao, Y., Wu, C., Wu, T., Li, Y., Chen, S., Yuan, C., & Hu, Y. (2018). Eugenol-chitosan nanoemulsions by ultrasound-mediated emulsification: formulation, characterization and antimicrobial activity. *Carbohydrate Polymers, 193*, 144–152.

Shrestha, S., Wang, B., & Dutta, P. (2020). Nanoparticle processing: understanding and controlling aggregation. *Advances in Colloid and Interface Science, 279*, 102162.

Siddiqui, S. A., Singh, S., Bahmid, N. A., Mehany, T., Shyu, D. J., Assadpour, E., ... Jafari, S. M. (2023). Release of encapsulated bioactive compounds from active packaging/coating materials and its modelling: a systematic review. *Colloids and Interfaces, 7*(2), 25.

Singh, H., Sharma, A., Bhardwaj, S. K., Arya, S. K., Bhardwaj, N., & Khatri, M. (2021). Recent advances in the applications of nanoagrochemicals for sustainable agricultural development. *Environmental Science: Processes & Impacts, 23*(2), 213–239.

Singh, A., Parikh, S., Sethi, N., Patel, S., Modi, N., & Patel, K. (2024). Nanoparticle formulations: a sustainable approach to biodegradable and non-biodegradable products. In V. P. Chavda & V. Apostolopoulos (Eds.), *Nanocarrier vaccines: Biopharmaceutics-based fast track development* (pp. 95–151). John Wiley & Sons.

Singh, A., Ahmed, E., Rather, M. D., Sundararajan, A., Sharma, A., Choudhary, F. S., ... Arya, S. (2025). Marketing strategies in nanomaterials for sensor applications: bridging lab to market. *Global Challenges, 9*(5), 2400294.

Sinha, T., Bhagwatwar, P., Krishnamoorthy, C., & Chidambaram, R. (2019). Polymer based micro and nanoencapsulation of agrochemicals. In T. J. Gutiérrez (Ed.), *Polymers for agri-food applications* (pp. 5–28). Basel, Switzerland: Springer International Publishing.

Slavin, Y. N., & Bach, H. (2022). Mechanisms of antifungal properties of metal nanoparticles. *Nanomaterials*, *12*(24), 4470.

Stadler, T., López García, G. P., Gitto, J. G., & Buteler, M. (2017). Nanostructured alumina: biocidal properties and mechanism of action of a novel insecticide powder. *Bulletin of Insectology*, *70*, 17–25.

Sukhorukov, G. B., Rogach, A. L., Zebli, B., Liedl, T., Skirtach, A. G., Köhler, K., Antipov, A. A., Gaponik, N., Susha, A. S., Winterhalter, M., & Parak, W. J. (2024). Nanoengineered polymer capsules: Tools for detection, controlled delivery, and site-specific manipulation. In W. Parak (Ed.), *Bio-nano interfaces* (pp. 1085–1101). Jenny Stanford Publishing;Sultana, N., Raul, P. K., Goswami, D., Das, B., Gogoi, H. K., & Raju, P. S. (2018). Nanoweapon: control of mosquito breeding using carbon-dot-silver nanohybrid as a biolarvicide. *Environmental Chemistry Letters*, *16*(3), 1017–1023.

Tang, Y., Zhao, W., Zhu, G., Tan, Z., Huang, L., Zhang, P., ... Rui, Y. (2023). Nanopesticides and fertilizers: solutions for global food security. *Nanomaterials*, *14*(1), 90.

Ullah, Z., Iqbal, J., Abbasi, B. A., Ijaz, S., Ahmad, S., Khan, S., ... Mahmood, T. (2024). Nanoformulations for agriculture applicability. In R. Shahzad, S. Fiaz, A. Qayyum, M. U. Islam, & I. J. Lee (Eds.), *Revolutionizing agriculture: a comprehensive exploration of agri-nanotechnology* (pp. 325–367). Cham: Springer Nature Switzerland.

Vera-Reyes, I., Altamirano-Hernández, J., Reyes-De la Cruz, H., Granados-Echegoyen, C. A., Loera-Alvarado, G., López-López, A., ... Loera-Alvarado, E. (2022). Inhibition of phytopathogenic and beneficial fungi applying silver nanoparticles in vitro. *Molecules*, *27*(23), 8147.

Vishnu, M., Malaisamy, K., Soundararajan, R. P., Suganthi, A., Subramanian, A., Senthilkumar, M., ... Kasivelu, G. (2024). Nano-bioformulations: emerging trends and potential applications in next generation crop protection. *Environmental Science: Nano*, *11*(7), 2831–2860.

Vos, R., & Bellù, L. G. (2019). Global trends and challenges to food and agriculture into the 21st century. Sustainable Food and Agriculture, Campanhola, C., Pandey, S., (eds.) (pp. 11–30), Academic Press.

Vurro, M., Miguel-Rojas, C., & Pérez-de-Luque, A. (2019). Safe nanotechnologies for increasing the effectiveness of environmentally friendly natural agrochemicals. *Pest Management Science*, *75*(9), 2403–2412.

Wahab, A., Muhammad, M., Ullah, S., Abdi, G., Shah, G. M., Zaman, W., & Ayaz, A. (2024). Agriculture and environmental management through nanotechnology: eco-friendly nanomaterial synthesis for soil–plant systems, food safety, and sustainability. *Science of the Total Environment*, *926*, 171862.

Yadav, A., Yadav, K., Ahmad, R., & Abd-Elsalam, K. A. (2023). Emerging frontiers in nanotechnology for precision agriculture: advancements, hurdles and prospects. *Agrochemicals*, *2*(2), 220–256.

7

The Role of Nanomaterials in Modern Plant Disease Management

Sajid Hussain, Naveed Iqbal Raja, Zohaib Younas, Chudary Sadam Hussain, and Zia ur Rehman Mashwani

Department of Botany, PMAS Arid Agriculture University, Rawalpindi, Pakistan

7.1 Introduction to Nanomaterials in Plant Disease Management

A range of fungal, bacterial, viral, nematode, and other pathogen diseases pose severe risks to worldwide agricultural productivity while harming crop output and product excellence and food security (Reddy et al., 2024). Research and development activities achieved momentum through economic consequences, which resulted from the activities of these pathogens. Knowledge of complete plant disease impacts, alongside their management methods, helps keep agricultural production healthy and preserves agroecosystems (Cohen et al., 2023).

Plant diseases not only reduce yields but also deteriorate product quality, forcing farmers to adopt costly protective measures that significantly increase agricultural expenses. Laboratory findings on vegetable and fruit pathogens indicate that these organisms can harm complete agricultural production operations, which threatens worldwide food supplies (Islam et al., 2023). The adoption of integrated management strategies now combines chemical methods with biological approaches and cultural interventions for devising an all-encompassing disease control system (Gulzar et al., 2023).

Modern research highlights the significance of disease management systems that combine multiple control measures. Different cultural practices, including crop rotation, along with proper sanitation and resistant cultivars usage, have demonstrated their ability to decrease disease occurrences because they shape environmental conditions less favorable to pathogen growth (Shimada et al., 2021). Through cultural practices, disease impacts can be reduced while avoiding chemical controls, thus establishing a more sustainable farming system (Thapa & Juyal, 2022). Modern biotechnology has led to the creation of disease-resistant plant species, which stands as an environmentally friendly, cost-efficient solution for traditional agricultural practices.

Nanotechnology Innovations for Food Security and Sustainable Agriculture, First Edition. Edited by Zia Ur Rehman Mashwani, Shah Fahad, Ilyas Ahmad, Shah Saud, Sajid Fiaz, and Taufiq Nawaz.

The control of plant diseases through biological approaches stands essential with cultural methods and biotechnological procedures for sustainable plant disease management. The usage of plant growth-promoting rhizobacteria (PGPR) provides dual protection against plant pathogens while triggering systemic immune responses in plant tissues (Gulzar et al., 2023). These biocontrol approaches, coupled with other management systems, cut down pathogen occurrence and intensity substantially, promote environmental well-being, and limit chemical residue accumulation (Shimada et al., 2021). Several physical methods, specifically soil solarization and heat-based approaches, destroy soilborne pathogens to minimize pathogen inoculum in agricultural production environments (Panth et al., 2020; Awasthi et al., 2024).

Modern agricultural practices adopt these integrated methods as part of integrated pest management (IPM) frameworks to demonstrate changes in current pest management approaches. IPM achieves superior management of diseases by combining chemical agents with biological solutions through cultural practices and physical techniques (Carmona et al., 2003). Integrated control methods resulting from these approaches simultaneously benefit crop productivity and maintain ecosystem vitality in addition to securing global food supplies for the future (Manoj et al., 2025). Interdisciplinary teams and ongoing research investigations work collaboratively to enhance these disease management methods for building resilient plants capable of fighting against evolving plant diseases according to Islam et al. (2023).

The distinctive characteristics of nanomaterials prove them to be promising agents that boost agricultural practices through enhanced disease management effectiveness and operational efficiency. Green synthesis represents an environmentally friendly approach to fabricating nanomaterials through plant extracts, which provide sustainable agricultural alternatives to standard agrochemicals (Ali et al., 2020). The synthesis of nanoparticles with antimicrobial properties can be achieved through plant-based reducing agents, including flavonoids and terpenoids, which provide a biological solution for pest and disease control (Bordoloi et al., 2020).

The anti-pathogenic potential of plants is strengthened through the use of zinc oxide (ZnO), silver nanoparticles (AgNPs), and selenium nanoparticles according to current research findings. The disease-suppressive power of Trichoderma fungi against downy mildew in pearl millet receives increased strength through treatment with selenium nanoparticles according to Nandini et al. (2017). The antimicrobial effects of green synthesis silver nanoparticles were confirmed against different plant pathogens that caused bacterial spot disease in tomatoes and powdery mildew disease in cucurbits according to a study performed by Worrall et al. (2018). Because of their versatility, these nanoparticles function both as protective agents and as transport vehicles for active substances that deliver agrochemicals with precision (Pereira et al., 2015).

Nanomaterials demonstrate specific physical and chemical properties as well as extensive surface area and transformed surface charge, which enables their agricultural applications. Nanomaterials serve multiple applications ranging from pesticidal use to nutrient delivery as well as the activation of helpful plant physiological mechanisms to boost crops under stress conditions (Chugh et al., 2021). Nanotechnology provides methods to control the release of nutrients and protective agents, which improve resource use while decreasing environmental damage from normal fertilizers and pesticides (Cao et al., 2021). The implementation of nanomaterials in agroecosystems requires immediate evaluation of their environmental effects and safety aspects. The substantial advantages of nanoparticles require extensive safety evaluations because their potential toxicity makes application potentially dangerous (Kalia et al., 2020; Awasthi & Sharma 2022). The adoption of sustainable practices needs to take precedence because it enables a balance between nanotechnology benefits and environmental safety to protect agricultural health in the long term (Saleem & Zaidi, 2020).

7.2 Mechanisms of Nanomaterials in Disease Management

Plant disease management benefits from nanomaterials through their distinctive physicochemical attributes, which enable simultaneous pathogen diagnosis and disease control. Multiple interactive functions combine to create their overall method, which consists of direct antimicrobial effects as well as targeted substance delivery together with quick pathogen recognition capabilities and plant defense incitement (Padmavathi and Anuradha 2022).

Nanoparticles from silver, copper, zinc oxide, and carbon-based nanostructures possess natural antimicrobial actions at their core. Their high reactivity together with surface characteristics stemming from small size and large surface-to-volume ratio enables direct inhibition of plant pathogens in their bacterial, fungal, and viral forms (Vijayreddy et al., 2023). The scientific community shows that nanoparticles use two mechanisms to protect plants by damaging their cells and interfering with pathogen metabolic pathways, thus lowering overall pathogen counts (Elmer & White, 2018).

Nanomaterials function both as antimicrobial agents and as advanced delivery platforms for agrochemical substances. Engineering techniques enable nanomaterials to adopt either encapsulated or surface-bound forms, which deliver pesticides or genetic materials and bioactive compounds to plant targets precisely. The precisely controlled delivery mechanism of nanomaterials lowers chemical needs while protecting ecosystems from unintended impacts and makes disease management procedures more sustainable (Ray et al., 2023). The protection of active ingredients against breakdown until they reach their target pathogen and plant tissue represents a unique feature demonstrated by nanobioformulations according to Dutta et al. (2023a,b). The implementation of nanomaterials in plant disease management includes modern diagnostic techniques as an additional scope of application. Nanoscale diagnostic devices, which incorporate nanomaterials, use their exceptional selectivity and sensitivity properties to identify pathogens at early stages. The detection capabilities of these sensors help identify fast biochemical plant changes during infection, which allows for on-time therapeutic intervention (Singh et al., 2023). Early detection through these methods helps healthcare providers utilize better disease-management strategies that contain both disease propagation and transmission.

Several nanomaterials demonstrate their capacity to both fuel plant immunity enhancements alongside their other functions. Plant cells express stress-related genes differently due to nanoparticle intervention, while the defense signaling pathways become activated. Through systemic resistance induction, nanoparticles make plants more resistant to various types of pathogens (Dutta et al., 2022). Nanomaterials offer a versatile functionality

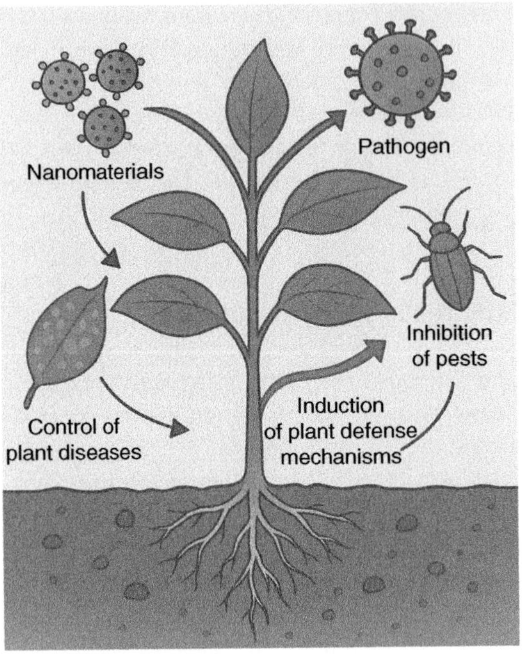

Figure 7.1 Illustration of nanomaterials actions against different plant diseases.

through three distinct capabilities: they function as antimicrobial agents, serve as delivery vehicles, and simultaneously act as stress alleviators to boost plant immune response. Figure 7.1 illustrate the mechanisms of nanomaterials in plant disease management.

7.3 Types of Nanomaterials for Disease Control

Plant disease control through the application of nanomaterials stands as a leading innovative advancement in agricultural science because nanoscale properties make these technologies more effective for disease management tactics. Nanotechnology has produced multiple materials, such as silver nanoparticles (AgNPs), zinc nanoparticles (ZnNPs), and chitosan nanoparticles, which demonstrate powerful properties in fighting plant pathogens. Figure 7.2 demonstrates the types of nanomaterials used to control diseases in plants, while Table 7.1 contains the various types of nanoparticles that are commonly used against the different pathogens that are causing diseases in plants.

7.3.1 Inorganic Nanomaterials

Research indicates the widespread interest in using inorganic nanoparticles, including silver (AgNPs), zinc oxide (ZnO), and magnesium oxide (MgO), because of their critical roles in disease prevention for plants. The antimicrobial properties of AgNPs act effectively to control plant diseases both as nanopesticides and as nanofertilizers according to Anand & Bhagat

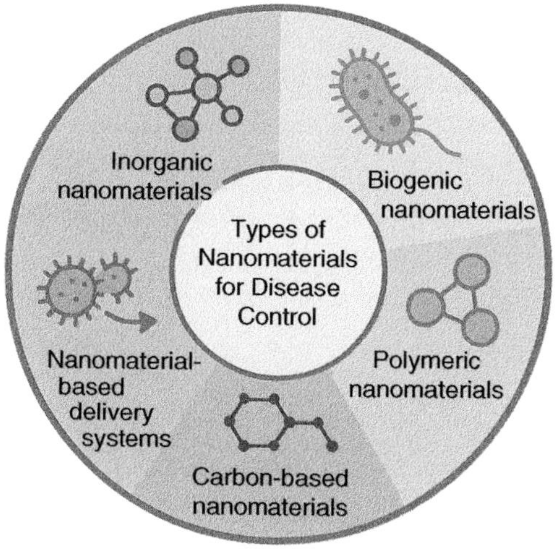

Figure 7.2 Illustration of different types of nanomaterials used to control plants diseases.

(2019). Research shows that zinc-based nanomaterials can achieve both pathogen inhibition combined with plant resistance enhancement as well as enhanced yield production (Kalia et al., 2020; Li et al., 2022). Laboratory studies show that MgO nanoparticles demonstrate remarkable biosafety properties because they successfully decrease bacterial diseases in various plants (Ogunyemi et al., 2020).

7.3.2 Biogenic Nanomaterials

The utilization of biogenic nanomaterials serves as an environmentally friendly pest and disease control method. Scientists produce nanomaterials from natural substances, such as plant extracts along with microorganisms, to create degradable and harmless formulations according to Sharma et al. (2023) and El-Sayed et al. (2023). Research has shown the synthesis of metal nanoparticles, which demonstrate major antimicrobial properties against diverse plant pathogens within organic-derived components (Ali et al., 2020; Cruz-Luna et al. 2021). These biogenic approaches serve as a dual benefit for effective plant protection and sustainable agricultural methods, which support current environmental sustainability objectives (Chavan et al., 2024; Mohanty et al., 2024).

7.3.3 Polymeric Nanomaterials

Scientists examine polyethyleneimine-coated MXene quantum dots as polymeric nanoparticles, which offer disease protection and enhanced plant growth together. Research demonstrates that these nanoparticles preserve reactive oxygen species (ROS) homeostasis essential for pathogen resistance against *Verticillium dahliae* as reported by Qiu et al. (2023). The advanced structure of polymeric nanomaterials enables specific distribution of therapeutic substances along with genetic compounds, which fills a crucial gap for effective plant disease control by boosting natural resistance capabilities (Wu et al., 2024).

Table 7.1 Different types of nanomaterials that are commonly used for plant disease management.

S. No	Nanoparticles	Size	Characterization	Activity	Plant	Citation
1	Silver nanoparticles	5–15 nm	TEM, UV	Rhizosphere antibacterial	Rice	Mirzajani et al. (2013)
2	Green-synthesized silver nanoparticle	~15 nm	UV, TEM, SEM, XRD, FTIR	Antimicrobial activity for plant pathogens	Rice and tomato	Rana et al. (2023)
3	Green-synthesized AgNPs	10–30 nm	UV, TEM, XRD, FTIR	Antibacterial activity	Velvet horn	Asong et al. (2023)
4	Green-synthesized AgNPs	15–25 nm	UV, TEM, SEM, XRD, FTIR	Antimicrobial agents and growth enhancement	Maize	Das et al. (2023)
5	Silver nanoparticles	~20–100 nm	UV, DLS, Zeta Potential, XRD, SEM	Antimicrobial activity	Mentha	Javed & Nadhman (2020)
6	Metallic silver nanoparticles	20–100	UV, DLS, Zeta Potential, XRD, SEM	Biomass and natural antioxidants	Millet	Khan et al. (2019)
7	Silver nanoparticles	~10–50 nm	UV, SEM, XRD, EDX	Biomass and elicitation of secondary metabolites	Wheat	Khan et al. (2023)
8	Silver nanoparticles	~5–15 nm	UV, SEM, XRD, FTIR	Macronutrients and morphology	Tomato	Guzmán-Báez et al. (2021)
9	Silver nanoparticles	~10–50	UV, SEM, XRD, FTIR	Salt tolerance	Bean	Nejatzadeh (2021)
10	Silver nanoparticles	~10–50 nm	UV, SEM, XRD, AFM	Heat stress	Wheat	Iqbal et al. (2019)
11	Green selenium nanoparticles	~5 nm	UV, SEM, XRD, FTIR	Antibacterial activity	Okra	Ghaderi, et al. (2022)
12	Plant-mediated selenium nanoparticles	~33.30–50 nm	UV, DLS, XRD, TEM, FTIR	Antimicrobial activity against pathogenic bacteria	Aloe vera and fungi	Fardsadegh and Jafarizadeh-Malmiri (2019)

13	Green selenium nanoparticles	~16.7 nm	UV, FTIR, SEM, XRD, EDX	Antifungal, physicochemical, and antioxidant activity	Mango	Shahbaz et al. (2023)
14	Bacterial-mediated selenium nanoparticles	~40–45 nm	UV, DLS, XRD, EDX	Antifungal activity	Faba	Hashem et al. (2021)
15	Biosynthesized selenium nanoparticles	~20–40 nm	UV, DLS, TEM, FTIR	Antimicrobial activity	Geranium	Fardsadegh et al. (2019)
16	Green selenium nanoparticles	~60–70 nm	UV, FTIR, XRD, TEM	Antimicrobial and antioxidant activity	Phoenix tree	Matheswari et al. (2022)
17	Biosynthesized slenium nanoparticles	~5–25 nm	UV, TEM, SEM, XRD, FTIR	Drought and heat stress	Wheat	El-Saadony et al. (2021)
18	Selenium nanoparticles	~50–100 nm	UV, SEM, XRD, EDX, FTIR	Environmental stress	Wheat	Iqbal et al. (2022)
19	*Azadirachta indica*-based silver nanoformulation	~5–20 nm	UV, DLS, SEM, FTIR	Biocidal activity against whitefly	Cotton	Shahid et al. (2022)
20	Silicon nanoparticles	~10–20 nm	DLS, EDX, XRD, FTIR	Salt and low temperature stress	Cotton	Liang, et al. (2023)

UV = ultraviolet-visible spectroscopy; SEM = scanning electron microscopy; TEM = transmission electron microscopy; XRD = X-ray Diffraction; EDX = energy dispersive X-ray spectroscopy; FTIR = Fourier transform infrared radiations; DLS = dynamic light scattering, AFM = atomic force microscopy and zeta potential for the effective charge on a nanoparticle's surface.

7.3.4 Carbon-Based Nanomaterials

Agricultural diagnostic practices and disease control receive significant benefits through the use of carbon nanotubes (CNTs) and related carbon-based nanomaterials. Their functionality includes promoting plant growth at early stages and managing seed germination coupled with biosensing detection, which tracks plant wellness and pathogen activity (Sharma et al., 2024b). Nutrient delivery, along with disease resistance, gains optimization through CNT incorporation in agriculture, since these nanomaterials establish molecular bonds with plant cells (Muzammil et al., 2023).

7.3.5 Nanomaterial-Based Delivery Systems

Plant genetic engineering practices have experienced a transformation due to nanomaterials used as delivery vehicles for DNA, RNA, and proteins. Nano-carriers enable better genetic stability and effectiveness of transferred genes, which lead to fundamental changes in disease reactions by promoting plant disease resistance (Alghuthaymi et al., 2015). The integration of modern biotechnology with nanotechnology creates great potential to develop disease-tolerant crops through genetic and nanotechnology methods (Worrall et al., 2018).

7.4 Applications of Nanomaterials in Plant Disease Management

Plant pathology has experienced a substantial transformation due to nanomaterials, which provide environmentally safe and highly effective methods for disease management (Lakshmi et al. 2023). Nanomaterials find multiple uses in disease prevention and control and plant immunity enhancement, which this section addresses in detail.

7.4.1 Nanoparticle-Based Therapeutics

Scientists discovered that the use of silver nanoparticles (AgNPs) demonstrates potential in disease management solutions for plants. Antibacterial as well as antifungal attributes of these nanomaterials allow them to combat numerous plant pathogens effectively. The antifungal effects of silver nanoparticles render them effective against *Fusarium oxysporum* and *Alternaria* species according to research carried out by Shahid et al. (2023) and Kim et al. (2012). The ecological benefits of eco-friendly biosynthesized nanoparticles developed using plant materials and microorganisms prepare them to be viable alternatives against traditional toxic chemical fungicides, thus minimizing plant protection's environmental impact (Mishra et al., 2017).

7.4.2 Enhancing Plant Immunity

Nanomaterials serve as essential compounds that boost the natural defensive mechanisms of plants. Plants employ chitosan nanoparticles as immunomodulators because these nanoparticles release multiple defensive responses in plants. Experiments prove that nanoparticle treatments activate plant cells to enhance antimicrobial compound production and signaling molecule that strengthens disease defense systems (Chandra et al., 2015; Haq et al., 2024). Nanoparticles establish a priming effect that lets plants to protect themselves ahead of pathogen attacks through an enhanced defensive strategy.

7.4.3 Nanotechnology in Disease Diagnosis

Nanomaterials with superior physical and chemical traits enhance their capabilities for fast plant pathogen detection. Nanotechnology has enabled researchers to develop biosensors capable of detecting pathogens throughout their developmental lifecycle so healthcare professionals receive warning signals in time (Singh et al., 2023). Nanotechnology diagnostics help medical personnel detect diseases and establish precise treatment plans, which substantially reduce the requirement for generalized pesticide use.

7.4.4 Delivery Mechanisms for Pesticides and Nutrients

Nanoparticles serve as vectors to transport traditional fungicides, herbicides, and other compounds that strengthen resistance against diseases. This nanocarrier approach enhances compound effectiveness because it directs the agents directly to targets and minimizes unwanted reactions that occur when using traditional methods (Hamid & Saleem, 2022). The encapsulation of biocontrol agents and nutrients inside nanoparticles boosts their stability level and release speed, which results in improved function for plant health management applications.

7.4.5 Sustainable Agricultural Practices

The implementation of nanoparticles for agricultural purposes supports modern sustainable farming methods which focus on precise application. Nanoparticle applications under controlled conditions decrease chemical runoff, which enables environmentally fruitful conservation practices (Vaghasiya et al., 2022). Nanoparticles enhance the stability of soil, which enables plants to take up nutrients better, thus developing stronger disease-resistant plants (Barbinta-Patrascu et al., 2017). Modern farming practices embrace sustainability through this approach because they aim to improve both food security and environmental protection. Figure 7.3 illustrates the applications of nanomaterials in plant disease management.

7.5 Benefits of Using Nanomaterials

The incorporation of nanomaterials in agriculture enables a revolutionary pathway to boost crop yields while ensuring sustainability and defense against global barriers, including climate change and food shortages. The group of substances known as nanomaterials features one minimum dimension within 1–100 nm, which leads to multiple advantageous applications in agricultural operations due to their distinctive physicochemical properties.

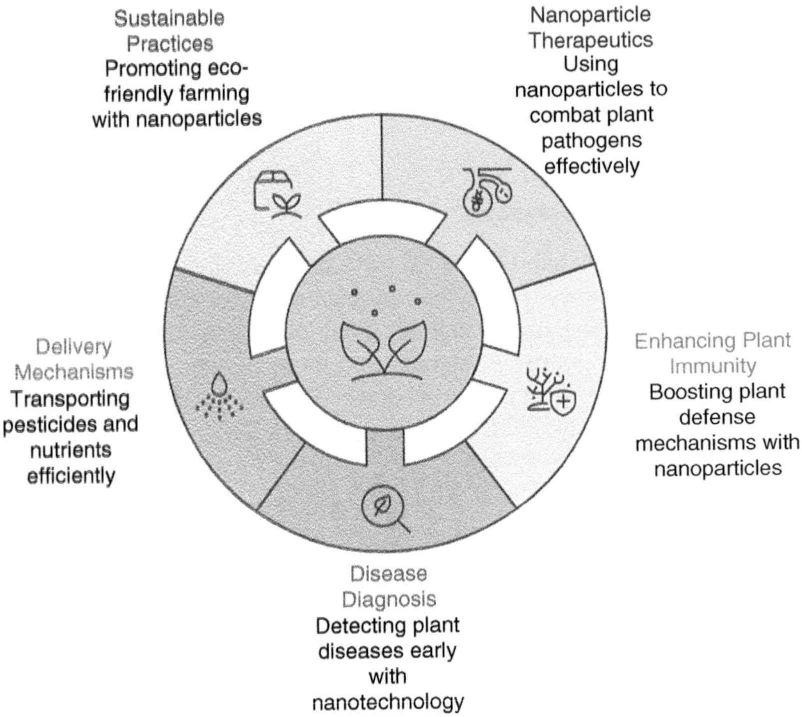

Figure 7.3 Illustration of applications of nanomaterials in plant disease management.

7.5.1 Enhanced Nutrient Delivery

Nanomaterials act as delivery systems to target specific nutrients and agrochemicals so fertilizer and pesticide use becomes more efficient. Unique physical (Meher et al., 2024) and chemical properties of nanomaterials allow for controlled nutrient encapsulation, which enhances plant absorption while minimizing waste according to Meher et al., (2024). The exact delivery capabilities of nanomaterials boost both plant nutrient absorption rates and protect the environment against pollution caused by traditional chemical fertilizers (Dhanapal et al., 2024; Xu, 2022).

7.5.2 Improvement in Crop Yield and Quality

The use of nanomaterials results in substantial improvements in the quantity of harvested crops and their overall quality. Hydroxyapatite nanoparticles function as nanofertilizers in different crops to enhance their nutrient absorption while accelerating plant development according to Alamri et al. (2022) and Ali et al. (2021). The application of nanoparticles in seed priming technology leads to better rates of seed germination and improved crop performance, which becomes vital for fulfilling increasing food requirements of the global human population (Kaler et al., 2024; Bora et al., 2022).

7.5.3 Pest and Disease Management

Nanotechnology enables efficient pest control measures because it leads to the development of nanopesticides. The specific delivery design of these formulations boosts active ingredient delivery precision with reduced unintended side effects based on research by both Sharma et al. (2022) and Shang et al. (2019). Nanomaterials enable early disease pest identification, which allows farmers to prevent crop damage from harmful agents (Solanki et al., 2015; Ahmed et al., 2021).

7.5.4 Environmental Remediation and Sustainability

Nanomaterials demonstrate environmental value by improving soil conditions and serving as effective polluting agents. The application of nanomaterials serves as a tool to recover polluted soils, which boosts safe agricultural conditions while enabling sustainable land usage (Dhanapal et al., 2024; Basak et al., 2023). Through improved water and nutrient use efficiency,0 nanomaterials assist in developing sustainable agricultural methods for addressing plant consequences of drought and high salinity alongside climate change (Fraceto et al., 2016; Rahim et al., 2021).

7.5.5 Innovations in Crop Engineering

Plants that have been genetically modified thanks to nanomaterials obtain improved traits according to Singh and Kumar (2024) and Basak et al. (2023). Through genetic material transport and molecular-level plant physiological control, nanotechnology creates improved crops capable of surviving multiple stresses such as nutritional enrichment and various environmental threats. Proactive research into crop variety development for targeted environments becomes possible through this approach, which benefits food security (Ehsan et al., 2022; Singh & Kumar, 2024).

7.5.6 Real-Time Monitoring and Precision Agriculture

Nanosensors placed in agricultural operations enable ongoing monitoring of soil conditions, combined with observations regarding plant development and monitoring environmental variables. The installed sensors evaluate necessary elements alongside fluid levels and pathogen detection to support strategic mapping of agricultural operations (Parameswari et al., 2024). The advancements enable precision agriculture techniques that let farmers maximize resources and boost operations while keeping farming practices sustainable (Xu, 2022).

7.6 Challenges and Limitations

Nanomaterials applied in plant disease management have created essential barriers and restrictions that need evaluation for proper agricultural utilization. An integration of scholarly articles from different sources establishes the basis of this response.

7.6.1 Toxicity and Safety Concerns

The adoption of nanomaterials for plant disease management faces a main obstacle because they present potential toxicities against plants and unintended organisms such as useful insects and microbes. The antibacterial, antifungal, and anti-parasitic attributes of biogenic nanomaterials established by Sharma et al. (2023) create potential unanticipated ecological problems when such materials are applied. Scientists have proved that excessive soil incorporation of copper nanoparticles can damage microbial soil communities and produce pathogen species resistance (Cao et al., 2021). Elaborate biosafety assessments should be completed as a fundamental step to verify the environmental harmony and ecological protection of these contemporary agents (Kalia et al., 2020).

7.6.2 Regulatory Hurdles and Commercialization

Different regulatory hurdles block the route that leads from laboratory discoveries to commercial market applications. The current regulatory framework creates problems for nanomaterials due to insufficient provisions for their special characteristics, which slows down their agricultural market entry (Worrall et al., 2018). Field experiments, alongside sufficient evidence to confirm nanoparticles' agricultural performance, have proven to be major obstacles for commercial nanotechnology development. Different research success and practical implementation of nanomaterials have prevented their establishment as acceptable solutions for plant disease control systems (Farooq et al., 2021).

7.6.3 Variability in Efficacy

Multiple factors determine nanomaterial performance significantly through a combination of nanomaterial type and dimensions with surface attributes and environmental conditions of soil acidity and wetness (Masood et al., 2024; Zhao et al., 2022). The inconsistent properties of nanomaterials make it difficult to find uniform methods for their application throughout diverse agricultural fields and to develop optimal formulation solutions. The unpredictable effects of engineered nanomaterials when it comes to plant immunity and disease protection make their consistent use for agricultural disease control very difficult across different crops and environmental conditions (Mohanty et al., 2024).

7.6.4 Knowledge Gaps Along the Nanomaterial Lifecycle

Researchers have not sufficiently mapped the complete path of nanomaterial use in agriculture, starting from production, through use, and ending in deterioration. Scientists need to comprehend nanoparticle interactions with soils, plants, and pathogens in order to achieve maximum use with minimal adverse effects (Al-Zaban et al., 2022).

7.6.5 Economic Considerations

Nanotechnology adoption as a practice in agriculture faces major obstacles from its high implementation costs. Crop yield and disease prevention capabilities of nanomaterials come at an expensive cost that could prevent their practical use by some farmers across developing nations (Fu et al., 2019; Sathiyabama & Appu, 2018). Hybrid nanotechnology-based pesticides, which offer competitive performance against traditional chemicals, need to be developed so nanotechnology can successfully implement in current agricultural methods.

7.7 Field Applications and Case Studies

Nanomaterials represent a pioneering solution that scientists use to manage plant diseases throughout agricultural practices. The unique properties of nanomaterials make them suitable for developing innovative methods that diagnose phytopathogens and prevent and treat these pathogens. Nanotechnology-based solutions help farmers achieve dual benefits of improved crop production alongside environmental protection while overcoming conventional methods' negative effects. Nanomaterials demonstrate exceptional diagnostic qualities that enable speedy and exact pathogen identification in plants. The development of advanced biosensing systems using zinc-based nanomaterials enables fast detection of diseases that affect crops through detection strategies. Photocatalytic along with fluorescent characteristics of zinc nanoparticles serve as the foundation for field-based disease assessment, which provides accurate infection results promptly (Kalia et al., 2020). Response time to outbreaks improves substantially through these techniques, therefore enabling appropriate interventions that minimize large-scale crop damage.

An Important study demonstrates that numerous nanomaterials exhibit antifungal capabilities in combatting important plant-infecting pathogens successfully. Active antifungal effects against *Botrytis cinerea* and *Penicillium expansum* have been demonstrated by nano zinc oxide particles. Different plant fungi demonstrate sensitivity to silver nanocomposite treatments (Hosseinzadeh et al., 2017). These antifungal activities mostly depend on creating ROS that destroy pathogen growth and viability; however, traditional pesticides lack this same capability.

The use of nanomaterials as nanopesticides enhances sustainable agriculture through disease resistance in crops and by limiting the presence of chemical residues. This method presents a fundamental approach because it matches the market requirement for sustainable pest control solutions, which preserve ecological well-being. Studies reveal that engineered nanomaterials demonstrate effective biotic stress prevention and management capabilities, which lead to healthier plants producing better yields (Khan et al., 2021; Shafqat et al., 2024). Magnetic nanomaterials with targeted functionalized applications function as delivery systems, which provide precise applications that maximize efficiency and minimize waste (Spanos et al., 2021).

Research on carbon nanomaterials has shown substantial progress because they improve plant immunity systems. Plant defenses become activated by such materials, which work without toxic releases to protect plants from different diseases (Sharma et al., 2024a; Cui et al., 2024). The method proves suitable for strengthening plant stress resistance, which doubles as support for sustainable agricultural development practices.

The beneficial potential of nanomaterials faces major challenges regarding their secure use in environmental conditions and biological systems. Research advancement demands complete assessments of nanomaterial effects on soil health and biodiversity, since these enable sustainable agriculture (Prasad et al., 2017). The benefits of nanotechnology for plant disease management are substantial yet researchers stress that proper safety evaluations of risks together with toxicity assessments are necessary to prevent adverse effects.

7.8 Future Directions and Research Needs

Plant pathology benefits from nanotechnology through its development of innovative detection and preventive methods to control plant diseases. Further comprehensive investigation is essential for the implementation of nanomaterials in plant disease control because their current stage of development needs extensive research for safety and sustainability alongside testing for true effectiveness. Researchers need to enhance their understanding regarding how nanomaterials interact with plant tissues together with their phytopathogen targets. Researchers need to study the mechanisms of pathogen cellular structure breakdown from NPs and the evaluation of both metabolic pathway disruption and immune response triggering in plants. Plant extracts along with microbes and other natural agents attract researchers due to their lowered toxicity and reduced production cost during the green synthesis of nanoparticles. The stability and antimicrobial properties along with constancy of biosynthesized nanomaterials require additional studies for evaluating their performance under diverse environmental conditions. The current research shows an extensive deficiency regarding long-term investigations of nanomaterial behavior in natural environments. Current research on plant protection mainly uses laboratory and greenhouse environments to study plant systems but fails to replicate natural conditions effectively. The next round of research should determine how nanomaterials behave when used in soil–plant systems including studying their degradation compounds and their ability to build up in plant tissue that leads to the food chain. Solid information regarding the environmental dangers posed by nanomaterials continues to be a critical issue. The complete cycle of nanoparticles in agroecological environments demands investigation because it determines their impact on soil microorganisms together with beneficial insects as well as aquatic species. The combination of nanotechnology with contemporary agricultural methods that include smart delivery systems and precision agriculture and omics technologies may demonstrate an advanced strategy (Kumar et al. 2021).

7.9 Conclusion

Nanotechnology has emerged as a revolutionary domain in agriculture, providing novel methods to improve plant development and fortify resilience against abiotic challenges. This study examines the many uses of nanomaterials in agriculture, emphasizing their contribution to enhancing plant growth and increasing resilience to drought, salt, heavy metals, and temperature variations. The technique categorizes nanoparticles often utilized in plant sciences and analyzes their distinct physicochemical features that enable interactions with plants. The chapter asserts that in recent years, nanomaterials have emerged as a potentially useful instrument in the field of plant disease management. The study includes how the nanomaterials help to mitigate the attack of different pathogens. The incorporation of nanomaterials in agriculture enables a revolutionary pathway to boost crop yields while ensuring sustainability and defense against global barriers, including climate changes and food shortages. In the current study, the benefits of using nanomaterials in agriculture are discussed. The study also discusses about the various potential applications of nanomaterials and emphasizes the need to understand the mechanism of action of nanomaterials in pathogen cells. The study shows the concerns about environmental protection and toxicology and highlights the recent advancements that have been made in the field of plant pathology.

References

Ahmed, H., Roy, A., Wahab, M., Ahmed, M., Othman-Qadir, G., Elesawy, B., Islam, M. N., & Emran, T. (2021). Applications of nanomaterials in agrifood and pharmaceutical industry. *Journal of Nanomaterials, 2021*, 1–10. https://doi.org/10.1155/2021/1472096

Alamri, S., Nafady, N., El-Sagheer, A., El-Aal, M., Mostafa, Y., Hashem, M., & Hassan, E. (2022). Current utility of arbuscular mycorrhizal fungi and hydroxyapatite nanoparticles in suppression of tomato root-knot nematode. *Agronomy, 12*(3), 671. https://doi.org/10.3390/agronomy12030671

Alghuthaymi, M., Almoammar, H., Rai, M., Said-Galiev, É., & Abd–Elsalam, K. (2015). Myconanoparticles: Synthesis and their role in phytopathogens management. *Biotechnology & Biotechnological Equipment, 29*(2), 221–236. https://doi.org/10.1080/13102818.2015.1008194

Ali, M., Ahmed, T., Wu, W., Hossain, A., Hafeez, R., Masum, M., Wang, Y., An, Q., Sun, G., & Li, B. (2020). Advancements in plant and microbe-based synthesis of metallic nanoparticles and their antimicrobial activity against plant pathogens. *Nanomaterials, 10*(6), 1146. https://doi.org/10.3390/nano10061146

Ali, S., Mehmood, A., & Khan, N. (2021). Uptake, translocation, and consequences of nanomaterials on plant growth and stress adaptation. *Journal of Nanomaterials, 2021*, 1–17. https://doi.org/10.1155/2021/6677616

Al-Zaban, M., Alhag, S., Dablool, A., Ahmed, A., Alghamdi, S., Ali, B., Al-Saeed, F. A., Saleem, M. H., & Poczai, P. (2022). Manufactured nano-objects confer viral protection against cucurbit chlorotic yellows virus (CCYV) infecting *Nicotiana benthamiana*. *Microorganisms, 10*(9), 1837. https://doi.org/10.3390/microorganisms10091837

Anand, R., & Bhagat, M. (2019). Silver nanoparticles (AgNPs): As nanopesticides and nanofertilizers. *MOJ Biology and Medicine, 4*(1), 19–20. https://doi.org/10.15406/mojbm.2019.04.00107

Asong, J. A., Frimpong, E. K., Seepe, H. A., Katata-Seru, L., Amoo, S. O., & Aremu, A. O. (2023). Green synthesis of characterized silver nanoparticle using *Cullen tomentosum* and assessment of its antibacterial activity. *Antibiotics, 12*(2), 203. https://doi.org/10.3390/antibiotics12020203

Awasthi, S., & Sharma, J. (2022). Toxicity and challenges of nanomaterials and their impact on the environment. *Nature Environment and Pollution Technology, 21*(2), 643–650. https://doi.org/10.46488/nept.2022.v21i02.023

Awasthi, L., Das, S., Lee, R., & Pattanayak, S. (2024). Physical methods of plant disease management. In *Plant Pathology* (pp. 248–258). https://doi.org/10.1201/9781003326915-41

Barbinta-Patrascu, M., Badea, N., Ungureanu, C., Iordache, S., Constantin, M., Purcar, V., Rau, I., & Pîrvu, C. (2017). Ecobiophysical aspects on nanosilver biogenerated from *Citrus reticulatapeels*, as potential biopesticide for controlling pathogens and wetland plants in aquatic media. *Journal of Nanomaterials, 2017*, 1–12. https://doi.org/10.1155/2017/4214017

Basak, A., Amarshettiwar, S., Bordoloi, N., Shahni, Y., Beese, S., Kushwaha, T., & Bhushan, S. (2023). Nanomaterial-induced changes in plant physiology and genetics: Implications for crop improvement strategies. *International Journal of Environment and Climate Change, 13*(11), 2043–2048. https://doi.org/10.9734/ijecc/2023/v13i113364

Bora, K., Hashmi, S., Zulfiqar, F., Abideen, Z., Ali, H., Siddiqui, Z., & Siddique, K. (2022). Recent progress in bio-mediated synthesis and applications of engineered nanomaterials for sustainable agriculture. *Frontiers in Plant Science, 13*. https://doi.org/10.3389/fpls.2022.999505

Bordoloi, M., Sahoo, R., Tamuli, K., Saikia, S., & Dutta, P. (2020). Plant extracts promoted preparation of silver and gold nanoparticles: A systematic review. *Nano, 15*(02), 2030001. https://doi.org/10.1142/s1793292020300017

Cao, X., Wang, C., Luo, X., Yue, L., White, J., Elmer, W., & Xing, B. (2021). Elemental sulfur nanoparticles enhance disease resistance in tomatoes. *Acs Nano, 15*(7), 11817–11827. https://doi.org/10.1021/acsnano.1c02917

Carmona, D., Huarte, M., Arias, G., López, A., Vincini, A., Castillo, H., Manetti, P. L., Capezio, S., Chaves, E., Torres, M., Eyherabide, J. J., Mantecon, J., Chicho, L., & Fernández, D. (2003). Integrated pest management in Argentina. In *Integrated Pest Management in the Global Arena* (pp. 313–326). https://doi.org/10.1079/9780851996523.0313

Chandra, S., Chakraborty, N., Dasgupta, A., Sarkar, J., Panda, K., & Acharya, K. (2015). Chitosan nanoparticles: A positive modulator of innate immune responses in plants. *Scientific Reports, 5*(1). https://doi.org/10.1038/srep15195

Chavan, J., Patil, P., Patil, A., Deshmukh, A., Panari, P., Mohite, A., Lawand, P., Yadav, P., Bodhe, M., Kadam, A., Namdas, D., Pawar, B., Jadhav, A., Shekhawat, M., & Santa-Catarina, C. (2024). *Salacia* spp.: Recent insights on biotechnological interventions and future perspectives. *Applied Microbiology and Biotechnology, 108*(1). https://doi.org/10.1007/s00253-023-12998-z

Chugh, G., Siddique, K., & Solaiman, Z. (2021). Nanobiotechnology for agriculture: Smart technology for combating nutrient deficiencies with nanotoxicity challenges. *Sustainability, 13*(4), 1781. https://doi.org/10.3390/su13041781

Cohen, M., Wohlmuth, H., Williams, C., & Clarke, P. (2023). The impact of plant pathology: Examining diseases and pests in the plant kingdom and strategies for effective control and management. *Advances in Herbal Research, 6*(1), 1–6. https://doi.org/10.25163/ahi.619962

Cruz-Luna, A., Cruz-Martínez, H., Vásquez-López, A., & Medina, D. (2021). Metal nanoparticles as novel antifungal agents for sustainable agriculture: Current advances and future directions. *Journal of Fungi, 7*(12), 1033. https://doi.org/10.3390/jof71210.3

Cui, Y., Wang, K., & Zhang, C. (2024). Carbon nanomaterials for plant priming through mechanostimulation: Emphasizing the role of shape. *ACS Nano, 18*(16), 10829–10839. https://doi.org/10.1021/acsnano.4c00557

Das, I., Sonowal, A., Gogoi, B., Sharma, B., Patowary, K., & Borah, D. (2023). *Citrus maxima* (Burm.) Merr. fruit juice and peel extract mediated synthesis of silver nanoparticles (AgNPs)

and their applications as antimicrobial agents and plant growth enhancers. *Kuwait Journal of Science, 50*(1A). https://doi.org/10.48129/kjs.19889

Dhanapal, A., Thiruvengadam, M., Vairavanathan, J., Venkidasamy, B., Easwaran, M., & Ghorbanpour, M. (2024). Nanotechnology approaches for the remediation of agricultural polluted soils. *Acs Omega, 9*(12), 13522–13533. https://doi.org/10.1021/acsomega.3c09776

Dutta, P., Kumari, A., Mahanta, M., Biswas, K., Dudkiewicz, A., Thakuria, D., Abdelrhim, A. S., Singh, S. B., Muthukrishnan, G., Sabarinathan, K. G., Mandal, M. K., & Mazumdar, N. (2022). Advances in nanotechnology as a potential alternative for plant viral disease management. *Frontiers in Microbiology, 13.* https://doi.org/10.3389/fmicb.2022.935193

Dutta, P., Kumari, A., Mahanta, M., Upamanya, G., Heisnam, P., Borua, S., Kaman, P. K., Mishra, A. K., Mallik, M., Muthukrishnan, G., Sabarinathan, K. G., Puzari, K. R., & Vijayreddy, D. (2023a). Nanotechnological approaches for management of soil-borne plant pathogens. *Frontiers in Plant Science, 14.* https://doi.org/10.3389/fpls.2023.1136233

Dutta, P., Yasin, A., Kumari, A., Mahanta, M., & Sharma, A. (2023b). Nano-bioformulation: A spanking new weapon for plant disease management. *Plant health Archives, 1*(3), 123–129. https://doi.org/10.54083/pha/1.3.2023/1.3-129

Ehsan, M., Raja, N., Mashwani, Z., Zohra, E., Abasi, F., Ikram, M., Mustafa, N., Wattoo, F. H., Proćków, J., & Lastra, J. (2022). Effects of phytogenically synthesized bimetallic Ag/ZnO nanomaterials and nitrogen-based fertilizers on biochemical and yield attributes of two wheat varieties. *Nanomaterials, 12*(17), 2894. https://doi.org/10.3390/nano12172894

Elmer, W., & White, J. (2018). The future of nanotechnology in plant pathology. *Annual Review of Phytopathology, 56*(1), 111–133. https://doi.org/10.1146/annurev-phyto-080417-050108

El-Saadony, M. T., Saad, A. M., Najjar, A. A., Alzahrani, S. O., Alkhatib, F. M., Shafi, M. E., Selem, E., Desoky, E.-S. M., Fouda, S. E. E., El-Tahan, A. M., & Hassan, M. A. (2021). The use of biological selenium nanoparticles to suppress *Triticum aestivum* L. crown and root rot diseases induced by Fusarium species and improve yield under drought and heat stress. *Saudi Journal of Biological Sciences, 28*(8), 4461–4471. https://doi.org/10.1016/j.sj.s.2021.04.043

El-Sayed, E., Mohamed, S., Mousa, S., El-Seoud, M., Elmehlawy, A., & Abdou, D. (2023). Bifunctional role of some biogenic nanoparticles in controlling wilt disease and promoting growth of common bean. *AMB Express, 13*(1). https://doi.org/10.1186/s13568-023-01546-7

Fardsadegh, B., & Jafarizadeh-Malmiri, H. (2019). Aloe vera leaf extract mediated green synthesis of selenium nanoparticles and assessment of their in vitro antimicrobial activity against spoilage fungi and pathogenic bacteria strains. *Green Processing and Synthesis, 8*(1), 399–407. doi.org/10.1515/gps-2019-0007

Fardsadegh, B., Vaghari, H., Mohammad-Jafari, R., Najian, Y., & Jafarizadeh-Malmiri, H. (2019). Biosynthesis, characterization and antimicrobial activities assessment of fabricated selenium nanoparticles using *Pelargonium zonale* leaf extract. *Green Processing and synthesis, 8*(1), 191–198. doi.org/10.1515/gps-2018-0060

Farooq, T., Adeel, M., He, Z., Umar, M., Shakoor, N., Silva, W., Elmer, W., White, J. C., & Rui, Y. (2021). Nanotechnology and plant viruses: An emerging disease management approach for resistant pathogens. *ACS Nano, 15*(4), 6030–6037. https://doi.org/10.1021/acsnano.0c10.10

Fraceto, L., Grillo, R., Medeiros, G., Scognamiglio, V., Rea, G., & Bartolucci, C. (2016). Nanotechnology in agriculture: Which innovation potential does it have? *Frontiers in Environmental Science, 4.* https://doi.org/10.3389/fenvs.2016.00020

Fu, L., Wang, Z., Dhankher, O., & Xing, B. (2019). Nanotechnology as a new sustainable approach for controlling crop diseases and increasing agricultural production. *Journal of Experimental Botany*, *71*(2), 507–519. https://doi.org/10.1093/jxb/erz314

Ghaderi, R. S., Adibian, F., Sabouri, Z., Davoodi, J., Kazemi, M., Amel Jamehdar, S., Meshkat, Z., Soleimanpour, S., & Daroudi, M. (2022). Green synthesis of selenium nanoparticle by *Abelmoschus esculentus* extract and assessment of its antibacterial activity. *Materials Technology*, *37*(10), 1289–1297. https://doi.org/10.1080/10667857.2021.1935602

Gulzar, R., Javed, I., Abbas, M., Fátima, M., Rehman, T., Khan, R., Jahangir, M. S., Nasar, S., Muhammad, R., & Nasar, S. (2023). Emerging trends in plant disease management: A review of sustainable and innovative approaches. *SFS*, 504–509. https://doi.org/10.53555/sfs.v10i3.1855

Guzmán-Báez, G. A., Trejo-Téllez, L. I., Ramírez-Olvera, S. M., Salinas-Ruíz, J., Bello-Bello, J. J., Alcántar-González, G., Hidalgo-Contreras, J. V., & Gómez-Merino, F. C. (2021). Silver nanoparticles increase nitrogen, phosphorus, and potassium concentrations in leaves and stimulate root length and number of roots in tomato seedlings in a hormetic manner. *Dose-Response*, *19*(4), 15593258211044576. https://doi.org/10.1177/155932582110445

Hamid, A., & Saleem, S. (2022). Role of nanoparticles in management of plant pathogens and scope in plant transgenics for imparting disease resistance. *Plant Protection Science*, *58*(3), 173–184. https://doi.org/10.17221/37/2020-pps

Haq, I., Xiang-yun, C., Ali, H., Akhtar, M., Ghafar, M., Hyder, M., & Hou, Y. (2024). Interactions between nanoparticles and tomato plants: Influencing host physiology and the tomato leafminer's molecular response. *Nanomaterials*, *14*(22), 1788. https://doi.org/10.3390/nano14221788

Hashem, A. H., Abdelaziz, A. M., Askar, A. A., Fouda, H. M., Khalil, A. M., Abd-Elsalam, K. A., & Khaleil, M. M. (2021). *Bacillus megaterium*-mediated synthesis of selenium nanoparticles and their antifungal activity against *Rhizoctonia solani* in faba bean plants. *Journal of Fungi*, *7*(3), 195. https://doi.org/10.3390/jof7030195

Hosseinzadeh, A., Davari, M., & Habibi-Yangjeh, A. (2017). Application of nanomaterials in management of fungal plant diseases. *Plant Pathology Science*, *6*(2), 68–77. https://doi.org/10.29252/pps.6.2.68

Iqbal, M., Raja, N. I., Mashwani, Z. U. R., Hussain, M., Ejaz, M., & Yasmeen, F. (2019). Effect of silver nanoparticles on growth of wheat under heat stress. *Iranian Journal of Science and Technology, Transactions A: Science*, *43*, 387–395. https://doi.org/10.1007/s40995-017-0417-4

Iqbal, M., Shafiq, F., Anwar, S., Akram, N. A., Ashraf, M. A., Raza, S. H., Ali, N., & Ashraf, M. (2022). Selenium and nano-selenium-mediated heat stress tolerance in plants. In M. A. Hossain, G. J. Ahammed, Z. Kolbert, H. El-Ramady, T. Islam, & M. Schiavon (Eds.), *Selenium and Nano-Selenium in Environmental Stress Management and Crop Quality Improvement* (pp. 149–171). Cham: Springer International Publishing. https://doi.org/10.1007/978-3-031-07063-1_8

Islam, A., Elango, S., Saini, P., Jan, I., & Sharma, Y. (2023). Advancing plant pathology and the innovative methods for enhancing disease resistance. *JPB*, *2*(2), 14–18. https://doi.org/10.51470/jpb.2023.02.02.14

Javed, B., & Nadhman, A. (2020). Optimization, characterization and antimicrobial activity of silver nanoparticles against plant bacterial pathogens phyto-synthesized by Mentha longifolia. *Materials Research Express*, *7*(8), 085406. https://doi.org/10.1088/2053-1591/abaf19

Kaler, B., Jangir, H., Sarkar, S., Sharma, J., & Das, M. (2024). Nano-pyrite as a plant hormone regulator: Emulating seed hormopriming. *Modern Agriculture*, *2*(1). https://doi.org/10.1002/moda.23

Kalia, A., Abd-Elsalam, K., & Kuča, K. (2020). Zinc-based nanomaterials for diagnosis and management of plant diseases: Ecological safety and future prospects. *Journal of Fungi*, *6*(4), 222. https://doi.org/10.3390/jof6040222

Khan, I., Raza, M. A., Awan, S. A., Khalid, M. H. B., Raja, N. I., Min, S., Zhang, A., Naeem, M., Meraj, T. A., Iqbal, N., Zhang, X., & Huang, L. (2019). In vitro effect of metallic silver nanoparticles (AgNPs): a novel approach toward the feasible production of biomass and natural antioxidants in pearl millet (*Pennisetum glaucum* L.). *Applied Ecology & Environmental Research*, *17*(6). https://doi.org/10.15666/aeer/1706_1287712892

Khan, M., Khan, A., Hasan, M., Yadav, K., Cabral-Pinto, M., Malik, N., Yadav, V. K., Khan, A. H., Islam, S., & Sharma, G. (2021). Agro-nanotechnology as an emerging field: A novel sustainable approach for improving plant growth by reducing biotic stress. *Applied Sciences*, *11*(5), 2282. https://doi.org/10.3390/app11052282

Khan, S., Khan, R. S., Zahoor, M., Islam, N. U., Khan, T., Muhammad, Z., Ullah, R., & Bari, A. (2023). *Alnus nitida* and urea-doped *Alnus nitida*-based silver nanoparticles synthesis, characterization, their effects on the biomass and elicitation of secondary metabolites in wheat seeds under in vitro conditions. *Heliyon*, *9*(3). https://doi.org/10.1016/j.heliyon.2023.e14579

Kim, S., Jung, J., Lamsal, K., Kim, Y., Min, J., & Lee, Y. (2012). Antifungal effects of silver nanoparticles (AgNPs) against various plant pathogenic fungi. *Mycobiology*, *40*(1), 53–58. https://doi.org/10.5941/myco.2012.40.1.053

Kumar, A., Choudhary, A., Kaur, H., Mehta, S., & Husen, A. (2021). Smart nanomaterial and nanocomposite with advanced agrochemical activities. *Nanoscale Research Letters*, *16*(1). https://doi.org/10.1186/s11671-021-03612-0

Lakshmi, K. L., Ayman, R., Shivarudraswamy, D., Bhavana, D., & Prasad, N. (2023). A review on plant pathology in the era of nanotechnology. *Pollution Research*, *42*(03), 367–372. https://doi.org/10.53550/pr.2023.v42i03.010

Li, Y., Zhang, P., Li, M., Shakoor, N., Adeel, M., Zhou, P., Lou, B. Z., & Rui, Y. (2022). Application and mechanisms of metal-based nanoparticles in the control of bacterial and fungal crop diseases. *Pest Management Science*, *79*(1), 21–36. https://doi.org/10.1002/ps.7218

Liang, Y., Liu, H., Fu, Y., Li, P., Li, S., & Gao, Y. (2023). Regulatory effects of silicon nanoparticles on the growth and photosynthesis of cotton seedlings under salt and low-temperature dual stress. *BMC Plant Biology*, *23*(1), 504. https://doi.org/10.1186/s12870-023-04509-z

Manoj, M. S., Revanasiddayya, Gowda, A. K., & Mallikarjun, P. K. (2025). Biotechnological approaches for management of diseases in plants. In *Genetic Horizons: Advancement in Plant Breeding* (pp. 84–102). IIP Series. https://doi.org/10.58532/nbennurghch5

Masood, H., Qi, Y., Zahid, M., Li, Z., Ahmad, S., Lv, J., Shahid, M. S., Ali, H. E., & Qi, X. (2024). Recent advances in nano-enabled immunomodulation for enhancing plant resilience against phytopathogens. *Frontiers in Plant Science*, *15*. https://doi.org/10.3389/fpls.2024.1445786

Meher, R., Thakur, B., Surekha, S., Kiran, K., Pratap, D., Kumar, S., Tripathi, U., & Jayakkannan, L. (2024). Nanomaterial-induced modifications in plant physiology and genetics for optimal crop production strategies. *International Journal of Environment and Climate Change*, *14*(2), 232–240. https://doi.org/10.9734/ijecc/2024/v14i23941

Mirzajani, F., Askari, H., Hamzelou, S., Farzaneh, M., & Ghassempour, A. (2013). Effect of silver nanoparticles on *Oryza sativa* L. and its rhizosphere bacteria. *Ecotoxicology and Environmental Safety*, *88*, 48–54. https://doi.org/10.1016/j.ecoenv.2012.10.018

Mishra, S., Singh, B., Naqvi, A., & Singh, H. (2017). Potential of biosynthesized silver nanoparticles using Stenotrophomonas sp. BHU-s7 (MTCC 5978) for management of soil-borne and foliar phytopathogens. *Scientific Reports*, *7*(1). https://doi.org/10.1038/srep45154

Mohanty, P., Singh, P., Lenka, B., Adhya, T., Verma, S., Ayreen, Z., ... Mishra, S. (2024). Biofabricated nanomaterials in sustainable agriculture: Insights, challenges and prospects. *Biofabrication*, *16*(4), 042003. https://doi.org/10.1088/1758-5090/ad60f7

Muzammil, S., Ashraf, A., Siddique, M., Aslam, B., Rasul, I., Abbas, R., ... Hayat, S. (2023). A review on toxicity of nanomaterials in agriculture: Current scenario and future prospects. *Science Progress*, *106*(4). https://doi.org/10.1177/00368504231221672

Nandini, B., Hariprasad, P., Prakash, H., Shetty, H., & Geetha, N. (2017). Trichogenic-selenium nanoparticles enhance disease suppressive ability of trichoderma against downy mildew disease caused by *Sclerospora graminicola* in pearl millet. *Scientific Reports*, *7*(1). https://doi.org/10.1038/s41598-017-02737-6

Nejatzadeh, F. (2021). Effect of silver nanoparticles on salt tolerance of *Satureja hortensis* l. during in vitro and in vivo germination tests. *Heliyon*, *7*(2). https://doi.org/10.1016/j.heliyon.2021.e05981

Ogunyemi, S., Zhang, M., Abdallah, Y., Ahmed, T., Qiu, W., Ali, M., ... Li, B. (2020). The bio-synthesis of three metal oxide nanoparticles (ZnO, MnO2, and MgO) and their antibacterial activity against the bacterial leaf blight pathogen. *Frontiers in Microbiology*, *11*. https://doi.org/10.3389/fmicb.2020.588326

Padmavathi, S., & Anuradha, C. (2022). Nanotechnology in plant disease management – An overview. *Journal of Advanced Scientific Research*, *13*(10), 01–06. https://doi.org/10.55218/jasr.2022131001

Panth, M., Hassler, S., & Baysal-Gurel, F. (2020). Methods for management of soilborne diseases in crop production. *Agriculture*, *10*(1), 16. https://doi.org/10.3390/agriculture10010016

Parameswari, P., Belagalla, N., Singh, B., Abhishek, G., Rajesh, G., Katiyar, D., Hazarika, B., & Paul, S. (2024). Nanotechnology-based sensors for real-time monitoring and assessment of soil health and quality: A review. *Asian Journal of Soil Science and Plant Nutrition*, *10*(2), 157–173. https://doi.org/10.9734/ajsspn/2024/v10i2272

Pereira, A., Narciso, A., Seabra, A., & Fraceto, L. (2015). Evaluation of the effects of nitric oxide-releasing nanoparticles on plants. *Journal of Physics Conference Series*, *617*, 012025. https://doi.org/10.1088/1742-6596/617/1/012025

Pon Matheswari, P., Jenit Sharmila, G., & Murugan, C. (2022). Green synthesis of selenium nanoparticles using *Delonix regia* and *Nerium oleander* flower extract and evaluation of their antioxidant and antibacterial activities. *Inorganic and Nano-Metal Chemistry*, *1-12*. https://doi.org/10.1080/24701556.2021.2025099

Prasad, R., Bhattacharyya, A., & Nguyen, Q. (2017). Nanotechnology in sustainable agriculture: recent developments, challenges, and perspectives. *Frontiers in Microbiology*, *8*. https://doi.org/10.3389/fmicb.2017.01014

Qiu, P., Li, J., Zhang, L., Chen, K., Shao, J., Zheng, B., Yuan, H., Qi, J., Yue, L., Hu, Q., Ming, Y., Liu, S., & Zhu, L. (2023). Polyethyleneimine-coated MXene quantum dots improve cotton tolerance to *Verticillium dahliae* by maintaining ROS homeostasis. *Nature Communications, 14*(1). https://doi.org/10.1038/s41467-023-43192-4

Rahim, H., Qaswar, M., Uddin, M., Giannini, C., Herrera, M., & Rea, G. (2021). Nano-enable materials promoting sustainability and resilience in modern agriculture. *Nanomaterials, 11*(8), 2068. https://doi.org/10.3390/nano11082068

Rana, A., Kumari, A., Chaudhary, A. K., Srivastava, R., Kamil, D., Vashishtha, P., & Sharma, S. N. (2023). An investigation of antimicrobial activity for plant pathogens by green-synthesized silver nanoparticles using *Azadirachta indica* and *Mangifera indica*. *Physchem, 3*(1), 125–146. https://doi.org/10.3390/physchem3010010

Ray, M., Mishra, A., Mohanta, Y., Mahanta, S., Chakrabartty, I., Kungwani, N., Avula, S. K., Panda, J., & Pudake, R. (2023). Nanotechnology as a promising tool against phytopathogens: A futuristic approach to agriculture. *Agriculture, 13*(9), 1856. https://doi.org/10.3390/agriculture13091856

Reddy, C., Oraon, S., Bharti, S., Yadav, A., & Hazarika, S. (2024). Advancing disease management in agriculture: A review of plant pathology techniques. *PSA, 9*(1), 16–18. https://doi.org/10.51470/psa.2024.9.1.16

Saleem, H., & Zaidi, S. (2020). Recent developments in the application of nanomaterials in agroecosystems. *Nanomaterials, 10*(12), 2411. https://doi.org/10.3390/nano10122411

Sathiyabama, M., & Appu, M. (2018). Application of copper-chitosan nanoparticles stimulate growth and induce resistance in finger millet (eleusine coracana gaertn.) plants against blast disease. *Journal of Agricultural and Food Chemistry, 66*(8), 1784–1790. https://doi.org/10.1021/acs.jafc.7b05921

Shafqat, A., Rasool, A., Fatima, R., Majeed, M., Rahman, K., Narejo, A., Ishtiaq, M., Kousar, S., & Farid, A. (2024). Nanomaterials for sustainable pest and disease management in agriculture. In V. D. Rajput, A. Singh, K. Ghazaryan, A. Alexiou, & A. T. Said (Eds.), *Harnessing NanoOmics and Nanozymes for Sustainable Agriculture* (pp. 100–125). https://doi.org/10.4018/979-8-3693-1890-4.ch006

Shahbaz, M., Akram, A., Raja, N. I., Mukhtar, T., Mehak, A., Fatima, N., Ajmal, M., Ali, K., Mustafa, N., & Abasi, F. (2023). Antifungal activity of green synthesized selenium nanoparticles and their effect on physiological, biochemical, and antioxidant defense system of mango under mango malformation disease. *PLoS One, 18*(2), e0274679. https://doi.org/10.1371/journal.pone.0274679

Shahid, M., Naeem-Ullah, U., Khan, W. S., Saeed, S., & Razzaq, K. (2022). Biocidal activity of green synthesized silver nanoformulation by *Azadirachta indica* extract a biorational approach against notorious cotton pest whitefly, *Bemisia tabaci* (Homoptera; Aleyrodidae). *International Journal of Tropical Insect Science, 42*(3), 2443–2454. https://doi.org/10.1007/s42690-022-00771-0

Shahid, A., Faizan, M., & Raza, M. (2023). Potential role of silver nanoparticles (AgNPs) and zinc nanoparticles (ZnNPs) for plant disease management. *Agrobiological Records, 14*, 59–69. https://doi.org/10.47278/journal.abr/2023.039

Shang, Y., Hasan, M., Ahammed, G., Li, M., Yin, H., & Zhou, J. (2019). Applications of nanotechnology in plant growth and crop protection: A review. *Molecules, 24*(14), 2558. https://doi.org/10.3390/molecules24142558

Sharma, B., Soni, U., Afonso, L., & Cahill, D. (2022). Nanomaterial doping: Chemistry and strategies for agricultural applications. *Acs Agricultural Science & Technology*, *2*(2), 240–257. https://doi.org/10.1021/acsagscitech.1c00273

Sharma, A., Dutta, P., Mahanta, M., Kumari, A., & Yasin, A. (2023). Botanicals as a source of nanomaterial for pest and disease management. *planthealth*, *1*(3), 96–101. https://doi.org/10.54083/pha/1.3.2023/96-101

Sharma, M., Kapoor, D., Sharma, P., & Husen, A. (2024a). Role of carbon nanomaterials in the prevention of plant disease. In A. (J.) Bachheti, R. K. Bachheti, & A. Husen (Eds.), *Carbon-Based Nanomaterials* (pp. 105–121). Springer Nature. https://doi.org/10.1007/978-981-97-0240-4_6

Sharma, S., Yadav, A., Yadav, M., & Verma, S. (2024b). Revolutionizing agriculture with nanotechnology: Advances, applications, and sustainability considerations. *International Journal of Environment and Climate Change*, *14*(7), 1–9. https://doi.org/10.9734/ijecc/2024/v14i74246

Shimada, B., Simon, M., & Cunha, L. (2021). Cultural control: A sustainable method of pest and disease control. *Journal of Experimental Agriculture International*, *19–26*. https://doi.org/10.9734/jeai/2021/v43i630698

Singh, N., & Kumar, A. (2024). Enhancing crop productivity through nanotechnology: a comprehensive review of strategies and results. *Journal of Experimental Agriculture International*, *46*(5), 435–458. https://doi.org/10.9734/jeai/2024/v46i52395

Singh, L., Zeshmarani, S., Hijam, C., Qutub, M., Singh, Y., Prabin, S., & Singh, O. (2023). Nanotechnology and its role in plant pathology. *The Pharma Innovation*, *12*(12), 37–49. https://doi.org/10.22271/tpi.2023.v12.i12a.24452

Solanki, P., Bhargava, A., Chhipa, H., Jain, N., & Panwar, J. (2015). Nano-fertilizers and their smart delivery system. In M. Rai, C. Ribeiro, L. Mattoso, & N. Duran (Eds.), *Nanotechnologies in Food and Agriculture* (pp. 81–101). Springer Cham. https://doi.org/10.1007/978-3-319-14024-7_4

Spanos, A., Athanasiou, K., Ioannou, A., Fotopoulos, V., & Krasia-Christoforou, T. (2021). Functionalized magnetic nanomaterials in agricultural applications. *Nanomaterials*, *11*(11), 3106. https://doi.org/10.3390/nano11113106

Thapa, R., & Juyal, R. (2022). A study on biological control, cultural practices, and chemical control methods in nursery pest and disease management. *Journal of Survey in Fisheries Sciences*. https://doi.org/10.53555/sfs.v8i3.2398

Vaghasiya, T., Kumar, A., & Nakum, K. (2022). A review on wide range application of nanoparticles in agriculture and its implications in plant disease management. *Nanoworld Journal*, *08*(02). https://doi.org/10.17756/nwj.2022-100

Vijayreddy, D., Dutta, P., & Puzari, K. (2023). Nanotechnology in plant disease management. *Research Biotica*, *5*(2), 56–62. https://doi.org/10.54083/resbio/5.2.2023/56-62

Worrall, E., Hamid, A., Mody, K., Mitter, N., & Pappu, H. (2018). Nanotechnology for plant disease management. *Agronomy*, *8*(12), 285. https://doi.org/10.3390/agronomy8120285

Wu, H., Qi, J., Li, Y., Yao, X., Li, G., Xu, W., & Li, Z. (2024). Rational design of ros scavenging and fluorescent gold nanoparticles to deliver sirna to improve plant resistance to *Pseudomonas syringae*. 10.21203/rs.3.rs-3852889/v1

Xu, Z. (2022). Material nanotechnology is sustaining modern agriculture. *ACS Agricultural Science & Technology*, *2*(2), 232–239. https://doi.org/10.1021/acsagscitech.1c00204

Zhao, B., Luo, Z., Zhang, H., & Zhang, H. (2022). Imaging tools for plant nanobiotechnology. *Frontiers in Genome Editing*, *4*. https://doi.org/10.3389/fgeed.2022.1029944

8

Revolutionizing Fruit and Vegetable Farming: Nanotechnology from Soil to Shelf

Muhammad Nauman Khan[1,2], Barkat Ullah[1], Alevcan Kaplan[3], Nasir Assad[4], Marzia Batool Laila[4], Tooba[4], Sana Wahab[5], Amjad Ali[6], Shah Fahad[7], and Majid Iqbal[8,9]

[1] *Department of Botany, Islamia College Peshawar, Peshawar, Pakistan*
[2] *Biology Laboratory, University Public School, University of Peshawar, Peshawar, Pakistan*
[3] *Department of Crop and Animal Production, Sason Vocational School, Batman University, Batman, Turkey*
[4] *Institute of Chemistry, University of Sargodha, Sargodha, Pakistan*
[5] *Department of Plant Sciences, Quaid-i-Azam University, Islamabad, Pakistan*
[6] *Department of Sustainable Crop Production, Università Cattolica Del Sacro Cuore, Piacenza, Italy*
[7] *Department of Agronomy, Abdul Wali Khan University, Pakistan*
[8] *Key Laboratory of Ecosystem Network Observation and Modeling, Institute of Geographic Sciences and Natural Resources Research, Chinese Academy of Sciences, Beijing, China*
[9] *University of Chinese Academy of Sciences, (UCAS), Beijing, China*

8.1 Introduction to Nanotechnology in Agriculture

Farming, also referred to as agriculture, is the process of rearing livestock and cultivating vegetation to produce various products, including food, fodder, and fiber (Neethirajan, 2023). Raw ingredients for food and feed industries come from agriculture, the biggest and most stable sector (Colombo et al., 2023). However, the increasing global population and the scarcity of natural resources pose significant challenges to food security. More than 800 million people are food insecure, and according to the UN, these numbers are growing over the years (de Waal, 2024). In 2020, 811 million people were hungry, and the FAO anticipates 200 million tons of beef production and a population above 9 million by 2050 (Junaid & Gokce, 2024). Food production must increase by 50% to feed 9 million people by 2050 (Van Eenennaam, 2024). Traditional agricultural methods that depend excessively on unsustainable and potentially harmful synthetic materials to meet the increasing demand for food must be upgraded to achieve this goal (Munné-Bosch & Bermejo, 2024). As of now, the worldwide application of pesticides and fertilizers has attained 4.1 million tons and 125 million tons, respectively (Knuth et al., 2024). The overreliance on chemical fertilizers and agricultural products, coupled with a rise in pathogen resistance to pesticides, fungicides, and herbicides, alongside the impacts of climate change, presents a significant challenge to the agriculture and food industry (Zhou et al., 2024). The conventional approach

Nanotechnology Innovations for Food Security and Sustainable Agriculture, First Edition. Edited by Zia Ur Rehman Mashwani, Shah Fahad, Ilyas Ahmad, Shah Saud, Sajid Fiaz, and Taufiq Nawaz.

to applying agrochemicals is often ineffective and requires transitioning to more sustainable, adequate, economical, and efficient methods (Munné-Bosch & Bermejo, 2024).

The quest for enhancing the quality and quantity of crops has been a relentless pursuit of agriculturists and academics alike (Guo et al., 2023). As we stand on the brink of a new era, nanotechnology emerges as the sixth revolutionary technology, following in the footsteps of the Green Revolution of the 1960s and the Industrial Revolution (Malik et al., 2023; Noor et al., 2025). This groundbreaking field offers immense potential to revolutionize agriculture, making it more resilient, sustainable, and environmentally friendly (Mana et al., 2024). The integration of nanotechnology into agriculture has led to the emergence of a novel field focused on improving agricultural productivity through the regulation of plant growth and soil nutrient availability (Javaid et al., 2024). The process involves a comprehensive range of activities, starting from the initial sowing of a seed to the cultivation of crops (Shoukat et al., 2024). Sprouting, processing, managing plant diseases, and every other step of agriculture are all included in it (Javaid et al., 2024). The use of nanoparticles (NPs) in conjunction with the creation of a cutting-edge agricultural system presents a great chance to completely alter conventional farming practices. In addition to improving crop yield and quality, this approach may lessen the environmental impact of contemporary agricultural methods (Wahab et al., 2024). The vast array of uses for nanotechnology has the ability to completely transform the farming industry. New agricultural tools and nanotechnology might be used in many different fields (Figure 8.1) (Donkor & Daliri, 2024). Utilizing nanomaterials in agriculture is to decrease chemical dispersion, lessen soil contamination, and avert nutrient depletion, ultimately enhancing yield through effective management practices (Wahab et al., 2024).

8.2 Nanotechnology for Soil Improvement

Soil is recognized as the most plentiful and economical natural building material utilized in a range of engineering applications. It is characterized by its organic carbon content, encompassing liquid, solid, and gaseous phases that engage in interactions at multiple scales, resulting in a wide array of ecosystem products and services (Bai et al., 2024). The processes of the maintenance of soil structure and the cycling of nutrients are crucial for sustaining soil health (Chauhan et al., 2024). Soil health is defined as the degree to which it sustains biological productivity, maintains environmental quality, and promotes plant and animal health within the framework of an ecosystem and its defined boundaries for land use (Alshameri & Owais, 2022). Biochemical processes and microbial activity are the principal variables that determine these parameters. Soil fertility and health might be greatly improved by using nanomaterials to enhance this activity (Handa & Kalia, 2024). Soil biodiversity is thought to be the most important factor in soil health, and the interplay between soil animals and microbes is vital to soil fertility and production (Shen et al., 2024).

Soil biodiversity is deteriorating at a frightening rate as a result of intensive farming (Cozim-Melges et al., 2024). Nanotechnology, one of the most important inventions of the 21st century, can improve management and conservation practices and reduce agricultural input wastes in various environmental contexts, enabling sustainable development. Nanotechnology combined with agricultural techniques and soil health-improving

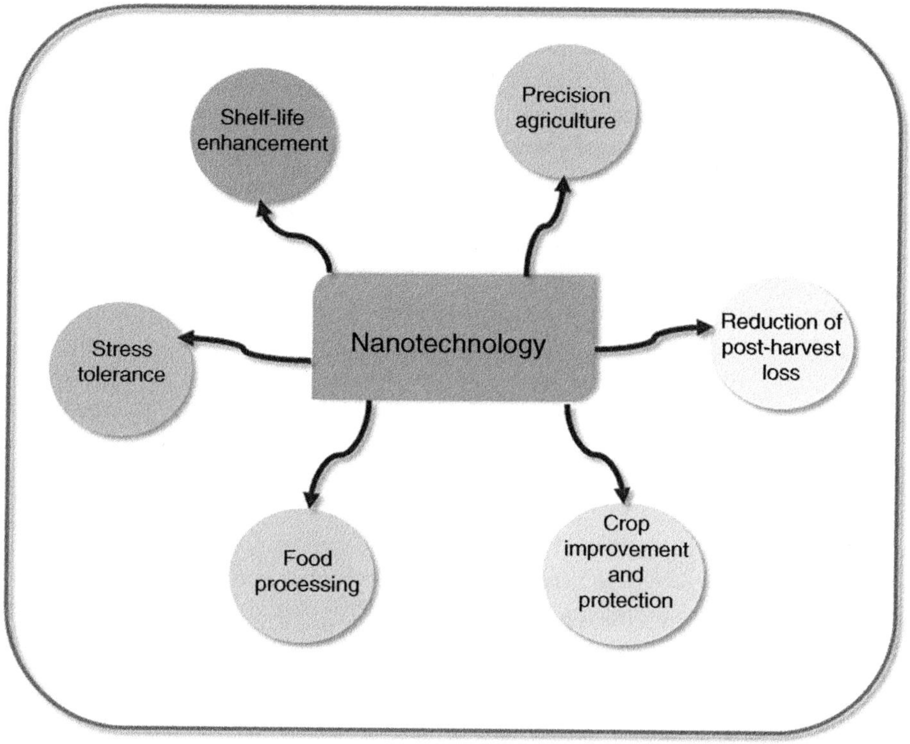

Figure 8.1 Application of nanotechnology in agriculture.

additives might boost soil bacteria and animals. Healthier soil and restored soil will assist to create a climate-resilient agriculture system. It will sustainably increase agricultural revenue (Isibor et al., 2024; Ullah et al., 2023).

As far as plants are concerned, the most essential thing is the soil in which they grow. In this way, plant performance may be negatively impacted by soil stressors (Muhammad et al., 2024). One of the main causes of soil erosion, which harms the surface of the ground and the places surrounding, is soil salinity (Muhammad et al., 2024). Worldwide, salt stress affects around 800 million hectares (Mha), which threatens and decreases agricultural productivity. Therefore, it had a broad impact on plant development and the maturation of developing tissues (Ben Gaied et al., 2024). Several factors influence soil composition over long periods, including changes in average temperatures, water scarcity, cold, alkalinity, and environmental contamination from harmful metals. Soil salinity and urbanization are two problems that need cutting-edge technology to solve (Muhammad et al., 2024). Soil structure was impacted by water deficiency stress or flooding threats, which disrupted the normal equilibrium of water. In response to the challenges faced by agricultural land, several initiatives have been launched, with a focus on nanotechnology. These initiatives have made use of a wide range of nanoparticle combinations (Singh et al., 2024a). There are a variety of NPs that combat soil salinity and have significant effects on soil composition. Silver and silica-based NPs, which also regulated soil degradations, are the most significant for soil applications (Soni et al., 2024). For the present needs of soil fertility, titanium dioxide, being one of the most generated nanomaterials, may be the most released nano-form in soil (Yang et al., 2024). As a result, it mitigates soil salinity, a problem that impacts most farmland. Research on sunflower cultivars has shown that biological treatments of iron sulfate ($FeSO_4$), such as foliar spray, exhibit remarkable resistance to salt stress tolerance (Kaya & Ashraf, 2024).

Soil composition and mineral balance are key factors in plant development, and nano zeolite is one kind of recently changed materials that is effective for the long-term supply of nutrients and minerals (Zeinalipour & Saadati, 2024). In addition to improving drought tolerance in crops, these nanomaterials accelerated plant development and productivity as useful aids in agricultural activities. Sustainable crop production relies on the development of biofortified crops, which are made possible by these agricultural advancements (Wahab et al., 2024). In contrast, nano-Si fertilizers are best used for soil remediation management and are preferable to conventional fertilizers when it comes to lowering heavy metal buildup (Faizan et al., 2024b). Since nanotechnology is so commonly used for transporting NPs to plant tissues, its potential application is more important than that of other technologies. In terms of toxicity, these NPs are far friendlier to the environment (Garg et al., 2024).

Nanomaterial delivery is effective because it allows for the adsorption of the majority of substances sprayed on. Agricultural goods are processed more quickly due to the noticeable change in soil properties as crops develop. Soil and agricultural development are best served by a nano-based targeted delivery strategy that employs gene transfer. Additionally, it promotes the updating nutrients, which enhances the soil's nutrition equilibrium (Sharma & Kumar, 2024). The current environment's metal absorption also lowers the effectiveness of updated nutrients, which disrupts soil nutrition equilibrium. In addition to protecting crops, nanopesticides improve soil quality, increase plant stress tolerance, and other similar benefits. The effectiveness of mineral uptake is crucial for plant development since it determines the course of most stages of growth (Singh et al., 2024b). Boosting output with little impact on water and soil pollution while providing defense against a wide range of pest insects and microbiological illnesses is the goal of most products and nutrient levels (Ahmad et al., 2024). Plants' development is stunted when they don't get enough minerals because their absorption power drops (Welch, 1999). Appropriately placed NPs improved soil growth and soil characteristics. Shihua Wang et al. used nano-Si to extract harmful metals from soil is one prime example (Wang et al., 2016). Because different plant tissues may become very poisonous when hazardous metals accumulate. Therefore, in an atmosphere with reactive oxygen species (ROS), plant development is stunted. Nano-Si that controls Cd buildup and alleviates cadmium toxicity in rice plants (Wang et al., 2016). Nano-Si that alleviates a wide range of additional plant stressors, including chromium and nickel. In one case, hazardous metals collected in plant tissues retarded development, whereas in another, silica-based NPs deposited in plant tissues stimulated growth. Thus, nanotechnology has been useful in soil practices, particularly in the areas of soil structure and prevention (Thiruvengadam et al., 2024a).

Soil remediation techniques vary, but the majority of them have major downsides. Chemical and physical methods include applying chemicals to the soil *via* sprays and solutions. The soil's characteristics are impacted by the chemical concentrations that are applied. The absorption effectiveness is not substantial, therefore a low dosage is effective for soil and plant health. Since there is no discernible absorption effectiveness, using a high dose is counterproductive and harmful to soil health. The uptake effectiveness is very crucial under environmental challenges, and the optimal concentration is effective for soil and plant health (Ajmal et al., 2023; Azam et al., 2022).

Optimal levels also boost plant defense mechanisms, which help plants fend off various insect attacks. Nanofertilizers, such as magnetic fertilizers or nanocomposite fertilizers, employ a nano-device to deliver a variety of macro- and micronutrients at acceptable concentrations; they are also excellent for controlling the loss of fertilizers and reducing soil damage (Haydar et al., 2024). Several nano-remediation methods, nanofertilizers, nano-biosensors, and nanopesticides have significantly improved soil applications of nanotechnology by eliminating substances containing heavy metals. When compared to more conventional techniques of chemical delivery, nanofertilizers and biochemical uses of nanomaterials are much more efficient. Because of the significant environmental risks associated with conventional soil application techniques, they are no longer allowed. For soil nutritional products, these technologies have been superseded by nanotechnology-based ones (Wahab et al., 2024).

Examining the interparticle diffusion distance, redox reactions, adsorption, ion exchange, and electrostatic interaction as they pertain to soil applications of nanomaterials reveals how these materials work as catalysts to remove the problematic soil contaminants. These NPs are useful for increasing agricultural yields as a whole because they increase contact throughout development. If you apply NPs to soil, they will destroy harmful metals that are present in plant tissues. This will reduce the likelihood of environmental hazards. Second, they promote the development of plant tissues while inhibiting the growth of attacking insects, forming intact combinations. Therefore, they serve a dual purpose in preserving the soil structure (Faizan et al., 2024a). The use of nanomaterials is shown in Table 8.1.

To amplify the delivery issues with conventional techniques, a wide range of NP combinations including organic, metallic, inorganic, and bioinspired materials have been developed (Chaudhary et al., 2024). The surface of titanium oxide may be connected with humic acids by a multi-walled carbon nanotube (CNT), and these newly generated NPs are natural short-ordered aluminosilicate (Zhao et al., 2024). Optimal concentrations of a broad variety of macro- and micronutrients were maintained by them as well. The interaction between soil and

Table 8.1 Application of nanomaterials in agriculture and soil improvement.

NPs type	Usage	Application	Soil applications
Nano zeolite	Agricultural practices also enhance resistance in drought	For long-term delivery of nutrients and minerals	Soil composition also improved
Iron sulfate	Excellent resistance to salinity stress	Non-flower cultivars	Soil layering and composition
Silica based	Chlorophyll content with proline accumulation	Seed germination	Soil composition also improved
Nanofertilizers	Heavy metals that can easily be eliminated	Effective for the loss of soil damage	Soil composition also improved
Nano-Si	Regulating Cd accumulation	Remediation	Soil remediation

bioorganic metallic NPs not only strengthens the root system, but also allows for the bioabsorption of nutrients and vital minerals. Contrarily, inorganic NPs are useful for soil fertility, and when used in combination, they help keep soil structure intact as well (Semenova et al., 2024).

Nanotechnology has the added benefit of making the soil more resistant to crop tolerance, which in turn lowers the likelihood of bioaccumulation of different anions that cause roots system injury. Ions such as SO_4^{2}– and Cl^-, together with cations like Na^+, Mg^{2+}, and Ca^{2+}, make up these substances (Chen et al., 2024). Pollution and other environmental problems might result from an overabundance of these ions. Soil contamination concerns trigger several genes that may suppress plant tissue functioning. Hence, it is critical to reduce environmental pollution to boost the total yield, since this will allow newly produced plant kinds to grow more rapidly in soil. Soil temperature, color, structure, porosity, density, consistency, aggregate stability, and texture have all been improved by the use of NPs. Biological processes, nitrogen cycling, infiltration, and erosion are all impacted by the features (Asiminicesei et al., 2024).

The activation of antioxidant enzyme activity was another critical function of NPs. These enzymes are activated as nano-enzymes capable of scavenging oxidative stress. In addition to their roles in soil remediation, oxidative stress reduction, and plant activity maintenance in the face of nutrient shortage, salt, water, drought, and heat, these nano-enzymes play an important role in these areas. Plants also produce antioxidant enzymes such as glutathione peroxidase (GSH), ascorbate peroxidase (APX), catalase (CAT), and superoxide dismutase (SOD). The use of cutting-edge technology has the potential to solve the worldwide issue of soil contamination (Singh et al., 2024b). When soil is degraded and plant activity decrease under stressful circumstances, many processes come into play. Oxidative stress in plants is caused by ROS, which are activated by the harmful soil. Increased lipid synthesis, which in turn damages various plant sections, may occur when certain plant organelles are unable to carry out their activities effectively. The total survival rate of plants is reduced by these unsavory circumstances. The soil's pH also affects the nutrients that plants may use for growth, reducing the amount of magnesium, calcium, and phosphorus that are accessible to plants (Thiruvengadam et al., 2024b).

To manage cleanup and other soil-damaging concerns, many agricultural improvements with integrated tactics are also useful. As an excellent example, consider cucumber plants, when sprayed with foliar-delivered Mn_3O_4 NPs, they trigger a cascade of reactions that protect plants from oxidative stress. These reactions include the activation of enzymes that neutralize free radicals, such as GSH, APX, CAT, SOD, and CAT. One major mechanism under nano-enabled plant salt tolerance is enhancing the capacity to maintain ROS homeostasis. Hence, there are several possible advantages of using nanotechnology monitoring in agriculture, including better food quality and safety, less agricultural inputs, and improved soil (Shah et al., 2024).

8.3 Precision Agriculture and Nanotechnology

Since the beginning of time, agriculture has been the foundation upon which our civilizations have been built. When one takes into account the significance of agriculture, it is impossible to ignore the reality that the agri-food industry is in a pitiful position. Exacerbating the problem are factors such as the depletion of soil, the lack of available food, the prevalence of poverty, and the reduction of biodiversity (Harris, 2024). Nanotechnology has garnered unquestionable attention when it comes to circumstances of such crucial importance. Crop protection, crop promotion, pathological analysis, customized farming, and stress tolerance are just a few of the many applications of nanomaterials, sometimes called the particles of the century (Jiang et al., 2024).

A modern farming method known as precision agriculture (PA) makes use of data-driven decision-making tools and state-of-the-art technology to maximize crop yields while minimizing waste of resources. Its goals are to boost efficiency, cut down on waste, and boost profits. In contrast, nanotechnology is an area of study that focuses on structures and materials at the nanoscale. Precision agriculture stands to benefit greatly from nanotechnology, which might provide farmers and producers with cutting-edge methods for increasing harvest yields. There are certain shared features between nanotechnology and precision agriculture, despite their divergent applications. Advances in nanotechnology have the potential to revolutionize precision agriculture by producing innovative materials and equipment. To better manage crops, nanosensors can monitor soil and plant health in real time. Mini sensors engineered to detect certain substances or environmental factors are known as nanosensors. Nutrients and insecticides may be more efficiently delivered to plants with the help of NPs, which minimizes waste (Babar & Akan, 2024).

Nanotechnology enhances agricultural yields, decreases waste, and lessens environmental consequences in precision agriculture. Nanoscale material modification allows for distinct properties and advantages over traditional agricultural practices. To better manage their crops, farmers may benefit from using nanosensors to track soil and plant health in real-time. This would provide them with more precise information on when and how much water and fertilizer to apply. One way to lessen the environmental effect of pesticides and fertilizers is to employ NPs as delivery vehicles. Plant fibers and seeds are examples of agricultural materials that may have their qualities improved by nanotechnology, which can increase their resistance to pests and weathering (Zain et al., 2024). But further study is required to determine the hazards and their effects on water, soil, and human health. The use of nanotechnology has the potential to dramatically alter agricultural practices by making them more efficient, environmentally friendly, and less wasteful.

Innovative agricultural practices that might boost crop yields while reducing environmental consequences are urgently needed to meet the ever-increasing worldwide need for more food. It has been shown that using nanoscale materials, especially nanofertilizers (NFs), may increase the efficiency with which crops use nutrients while decreasing the negative impacts of over-fertilization (Salem & Mekky, 2024). By encasing nutrients in NPs and enabling targeted and controlled release, nanotechnology may greatly improve the efficiency of nutrient delivery to crops (Semenova et al., 2024). The usual method for creating nanofertilizers involves encasing nutrients in nanoscale carriers, including polymeric nanocapsules or metal oxide NPs. These carriers enable the nutrients to be released at a regulated rate, reducing the amount of nutrients lost to leaching or volatilization (Semenova et al., 2024). Research has shown that nanofertilizers have the potential to greatly enhance plants' ability to absorb nutrients. One research found that urea treated with zinc oxide nanoparticles (ZnO NPs) enhanced wheat grain production by 51% relative to a control group (Dimkpa et al., 2020). ZnO NPs were shown to enhance plant growth and zinc absorption efficiency in rice plants, leading to greater grain zinc content (Francis et al., 2024). In addition to improving the efficiency of nutrient use, precision agriculture based on nanotechnology may help create more targeted and long-term nutrition management strategies. As an example, farmers may now adjust the rates and timings of fertilizer applications based on real-time data collected from soil nutrient levels monitored using nanosensors. This method lessens the negative effects of agriculture on the environment, including eutrophication and emissions of greenhouse gases, while simultaneously improving crop nutrient usage (Zain et al., 2024).

By tackling issues including safety, environmental effect, and efficacy, precision agriculture based on nanotechnology provides a potential substitute for conventional pest management approaches. Targeted and effective administration of insecticides and other pest control agents is made possible by the development of nanoscale materials and systems. For example, nanopesticides have the potential to enhance the stability and solubility of active components, which in turn enables controlled release and targeted delivery. This, in turn, reduces the quantity of pesticide that has to be applied, while also limiting nontarget effects and environmental contamination. By maximizing the administration of pesticides in response to real-time monitoring of insect populations and environmental factors, the integration of nanopesticides with precision agriculture improves pest control. The efficient delivery of pesticides and the creation of novel kinds of pesticides that are effective at lower dosages may be achieved, for instance, by engineering NPs to target particular pests or plant structures (Sun et al., 2024a). Because NPs enable more targeted administration of pesticides, precision agriculture based on nanotechnology also minimizes pesticide release into the environment, thus reducing environmental impact. A few examples are the targeted delivery of RNA interference (RNAi) molecules using nanocapsules for improved effectiveness (Yadav et al., 2023b) and less environmental release and the use of NPs for extremely efficient and focused pest management (Zain et al., 2024). Additionally, precision agriculture based on nanotechnology enhances consumer and farmer safety by reducing pesticide exposure *via* the development of less hazardous pesticides. The use of nanobiosensors for early disease and pest detection enhances farmer and consumer safety while decreasing the requirement for widespread pesticide applications (Getahun et al., 2024).

In addition to its many other benefits, precision agriculture based on nanotechnology promotes sustainable farming techniques and increases food yields. By using NPs for the precise distribution of pesticides and control agents, we can ensure that these approaches will be effective in the long run and help make agriculture more sustainable as a whole. As a result, there will be less chemical spraying and less chance of pesticide-resistant bugs developing. Innovative pest-control strategies may also be made possible by precision agriculture based on nanotechnology. For better pest control, scientists employ nanoformulations that combine chemical pesticides with biopesticides and other pest control agents. Better pest control with less use of chemical pesticides is possible with this method. In addition, by using sophisticated sensing and diagnostic procedures, nanotechnology may assist in the monitoring and management of insect populations. To help farmers decide when and where to apply pesticides most effectively, nano-biosensors combined with remote sensing technology may provide real-time data on insect populations, crop health, and environmental factors (Getahun et al., 2024).

When it comes to precision agriculture and high-tech environmental monitoring, nanotechnology is a game-changer. The main use case for it is in conjunction with nanosensors that can continuously measure soil, water, and plant factors. In order to improve agricultural management techniques, data gathered from monitoring using nanosensors is essential. One example of how these nanosensors may be used is with ZnO NPs, which are very good at detecting soil phosphorus levels (Morchid et al., 2024). Other soil characteristics that these nanosensors can detect include moisture, nutrient levels, and pH (Zain et al., 2024). By precisely detecting when water and fertilizers are needed, these nanosensors allow for more efficient use of both, leading to better crop development by avoiding over- or under-fertilization. Nanosensors like these, when combined with precision agriculture technology, greatly improve the reliability and effectiveness of decision-making, leading to more sustainable and productive harvests (Huo et al., 2024). Water conservation and nutrient runoff reduction are two additional benefits of integrating nanosensors with precision agriculture technology in this way.

An integral part of precision agriculture is variable rate technology (VRT), which differs depending on soil type and crop health and enables the targeted application of inputs including fertilizers, herbicides, and insecticides. By directing inputs precisely where they are required, farmers may optimize crop growth, increase yields, and minimize waste. By acting as transporters for these inputs, NPs may further improve VRT's efficacy. In scenarios where there are variations in soil type or crop health, VRT-based NPs can be utilized as carriers for pesticides, herbicides, and fertilizers. This enables the precise delivery of these inputs to specific field areas, which is especially beneficial (Zain et al., 2024). Also, depending on the kind of field, VRT may adjust the seeding rate accordingly. Nanocarriers provide several benefits when used in VRT. One advantage of nanocarriers is their ability to transmit inputs precisely, which cuts down on waste and protects the environment. Furthermore, nanocarriers have the ability to shield inputs from deterioration, which in turn increases their efficacy and decreases the frequency of reapplication. By selectively providing nutrients to plants, nanocarriers may maximize yields (Zain et al., 2024).

More efficient and precise agricultural activities are made possible by precision agriculture's essential automated gear. The accuracy and productivity of farming may be significantly improved by combining nanosensors with automated technology. When it comes to automated equipment, nanosensors have a lot of potential for efficiency and precision gains. Automated irrigation systems, for instance, may adapt the water distribution rates in real-time based on soil moisture levels monitored by nanosensors. Overwatering or underwatering crops may stunt their development, but nanosensors can help farmers avoid these problems by providing them real-time data on soil moisture levels (Zain et al., 2024).

When we talk about planting, fertilizing, and harvesting crops, data analytics plays a crucial role. Data collection may be made more accurate and precise with the use of NPs, which in turn leads to more trustworthy analytics. By analyzing soil or water samples for the presence of certain pollutants or illnesses, NPs might help farmers get a better understanding of their crops' health. As sensors, NPs may detect pollutants in soil and water. As an example, gold NPs (AuNPs) may be used to identify heavy metals in soil samples. Researchers were able to create an electrochemical sensor that could detect cadmium ions in soil and food samples with good sensitivity and selectivity. The sensor was built on graphene oxide (GO)/AuNPs/tetraphenyl porphyrin nanoconjugates (Yadav et al., 2023b). In a similar vein, magnetic NPs may effectively eliminate viruses and bacteria from water samples (Abdel-Aziz et al., 2020). A highly sensitive and selective biosensor based on magnetic NPs was created for the purpose of detecting *Escherichia coli* in water samples (Bu et al., 2019).

The promise of nanomaterials to improve crop yields and plant health has piqued researchers' interests in their possible use in agriculture. Nanocarbon, nanocellulose, and nanolignocellulose are three kinds of nanomaterials that show promise for enhancing plant development (Ren et al., 2020). There is promising evidence that nanocarbons, such as graphene, fullerenes, and CNTs, may enhance plant development. As an example, one research found that tomato plants whose growth was stimulated by CNTs grew at a much faster rate (Khodakovskaya et al., 2013). The use of GO has a similar effect on plant development. Research indicated that wheat seedlings treated with GO exhibited an increase in dry weight, root length, and plant height. Scientific evidence suggests that nanocellulose, namely cellulose nanofibrils (CNFs) and cellulose nanocrystals (CNCs), may promote plant growth and development. Using CNFs improves plant water and nutrient uptake and speeds up root development (Xu et al., 2013). Furthermore, nanofibers derived from cellulose anionic hydrogels promote the germination of sesame seeds (Qin et al., 2022). There have been reports of nanolignocellulose, a mixture of lignin and cellulose NPs, having a favorable effect on plant development as well. It is well known that nanolignocellulose encourages the growth of root hairs and improves plants' ability to absorb water and nutrients. According to research by Saito et al. (2021), soybean leaves may become more resistant to the obligate biotrophic fungal disease *Phakopsora pachyrhizi* after being treated with cellulose nanofibers.

Nanotechnology and precision agriculture together have the power to revolutionize sustainable agriculture by making it more efficient and environmentally friendly. Potentially fruitful areas of integration include data analytics, automated equipment, nutrient delivery, pest management, environmental monitoring, and variable rate technologies; as well as data-driven decision-making tools and nanoscale materials and devices. Through more precise input delivery, real-time monitoring of soil and plant traits, targeted and effective pesticide application, and optimized nutrient utilization, this integration may enhance agricultural yields, decrease waste, and minimize environmental repercussions in the long term.

8.4 Innovations in Pest and Disease Management

Modern disease and pest management practices have made extensive use of nano-agrochemicals, including nano-herbicides, nanofungicides, and nanoemulsions, which hold tremendous potential for addressing the drawbacks of conventional pesticides. Sustainable agriculture is made possible by nanoherbicides, which are target-specific (Ali et al., 2023). The target Brassica species had favorable responses to the herbicide atrazine when encased in poly(ε-caprolactone) (Joye & McClements, 2014). Using carboxymethyl cellulose nanocapsules filled with Clodinafop-propargyl herbicide instead of standard herbicides was shown to boost crop production (Pramanik et al., 2020). In contrast to using the herbicide 2,4-dichlorophenoxyacetate in its synthetic form, using it in hollow porous silica nanocapsules allowed for its regulated and slow release (Singh et al., 2021).

Nanofungicides, which are nanoformulations with fungicidal action, are very useful for reducing plant illnesses caused by fungi and increasing crop yields. Bouisol, a commercial fungicide made of nano copper solution, increased agricultural yields by protecting grapes and other fruit-bearing crops from fungal diseases (Varshney et al., 2024). Polyethylene terephthalate punnets sealed with chitosan nanocrystals (NCs) were used to combat *Monilinia laxa*, the plant pathogen responsible for brown rot on peach trees. The sickness was found to be less severe. Sulfur nanoparticles (SNPs) fungicide formulation was efficient in treating *Fusarium solani* early blight and wilt disease and *Venturia inaequalis* apple scab disease (Rao & Paria, 2013). *Macrophomina phaseolina* is a fungal disease that affects oilseed crops and pulses; nevertheless, ZnO NPs and TiO_2 NPs shown fungicidal action against this disease (Saritha et al., 2022).

The use of traditional chemical-based fungicides such as isoprothiolane and azoxystrobin was reduced because a silver NP (Ag NP)- based fungicide was able to combat *Magnaporthe grisea*, the causal agent of rice blast disease. SNPs were shown to inhibit the growth of two fungal diseases: wilt disease (*Fusarium solani*) and apple scab (*Venturia inaequalis*) (Tao et al., 2023). When it comes to *Curcuma longa* (turmeric), the fungus *Pythium aphanidermatum*, which causes rhizome rot, the plant may protect itself against fungal development by producing antifungal hydrolases and chitosanase enzymes when treated with chitosan NPs (Gangwar et al., 2023). By limiting spore germination and lowering mycelial development, Ag NPs applied at 100 mg/kg might overcome powdery mildew in cucurbits and pumpkins (Srivastava et al., 2024). Mycelium is the vegetative structure of fungus; nanofungicides based on molybdenum (Mo) may kill it and also limit the formation of conidia and conidiophore in plants. Solid lipid polymeric nanocapsules containing active ingredients such as carbendazim and tebuconazole were used to create a fungicide formulation (Salama et al., 2021). Nanocapsules' delayed and steady release of active chemicals helped counteract fungal impacts and decreased fungicide dosage in comparison to synthetic fungicides (Ding et al., 2023).

Additionally, pathological research have made extensive use of nanoemulsions. These colloidal structures, manufactured by nanotechnology, are optically isotropic and complex, with diameters ranging from 20 to 200 nm. They are kinetically stable and nonequilibrium. The functional and active substances are suspended in water and enclosed in droplets. They have spherical forms and are surfactant-assisted, making them homogeneous suspensions (Srivastava et al., 2024). Nanoemulsions' composition and manufacturing process determine their functionalities and qualities. A dry nanoemulsion is created by extracting water from an aqueous solution because of its physicochemical fragility when stored (Malik et al., 2023). Extreme circumstances, such as high pressure, temperatures, and salinity, are no match for nanoemulsions. This is due to their remarkable morphological stability, which remains unaffected by variations in oil volume ratio. As an example, a nanoemulsion of the water-insoluble pesticide β-cypermethrin shown improved stability when administered to plants. Similarly, a nanoemulsion derived from neem oil had a stronger larvicidal action, even when the droplet size was lowered, compared to the original (Saritha et al., 2022).

In the field of diagnosis and analysis of diseases, specific nano-biosensor and nano-barcode system are yet another emerging technology that has garnered significant praise all over the globe. Biosensors, nanofabricated instruments, identify dangerous substances and infectious particles. Biosensors use a transducer to identify chemical molecules, microbes, and other biological components in analytes, such as phenols, benzene, and pesticides. The ability to transform a physiological reaction into an electrical signal is a remarkable and exciting property of these instruments. In order to provide information on the physicochemical process, it detects, analyzes, and transmits signals by electric current or heat. The distinctive properties of NPs make them an attractive analytical tool, and nanosensors are gaining popularity as a result (Valenzuela-Amaro et al., 2023). Nanobiosensors have made use of a variety of NPs, including AuNPs and CNTs. Here are a few prominent examples: bioconjugated dye-doped SiNPs for *E. coli* detection, amperometric detection of *Salmonella typhi* using AuNPs, and electrochemical detection of pesticides with CNTs (Narayanan et al., 2013). Using single-walled carbon nanotubes (SWCNTs) allowed researchers to detect nitric oxide in *Arabidopsis thaliana*. With the use of polystyrene NPs and fluorescence resonance energy transfer (FRET), scientists were able to create a nanosensor that could detect and quantify phytoalexins found in various plants (Narayanan et al., 2013). Insecticides such as dichlorodiphenyltrichloroethane (DDT) and methyl parathion were detected and their levels were monitored using CNT- and AuNP-based sensors, respectively (Valenzuela-Amaro et al., 2023).

NPs known as quantum dots (QDs) are semiconducting and have a diameter of less than 10 nm. A quantum-based nanosensor, for instance, was able to detect as little as five phytoplasma cells per μL of *Candidatus Phytoplasma aurantifolia*, the causative agent of witches' broom disease in lime plants (Saritha et al., 2022). AuNPs and QDs play a key role in bioassays for biomolecule detection. The size range of AuNPs is typically 5–200 nm, and they have a rod form (Sadeghi et al., 2017). One example is the use of AuNPs coupled with certain oligonucleotides to detect complementary deoxyribonucleic acid (cDNA) in biological tests, which manifests as a color shift. Due to their unique electrochemical and optical characteristics, they are also used in pathogen detection (Vega-Vásquez et al., 2020). One example is the Surface Plasmon Resonance (SPR) (Jabbar et al., 2023) method, which uses electromagnetic waves to probe biomolecular interactions. Its nanogold based immunosensor was used to detect Karnal bunt disease in wheat plants, which is caused by the fungal pathogen *Tilletia indica* (Alghuthaymi et al., 2021). As a pesticide delivery method, AuNPs are also found in tobacco and rice plants. Colorimetric assays using AuNPs have also been used to identify various pesticides in food samples (Elnahal et al., 2022). Using AuNPs, researchers were able to track the concentrations of the pesticide DDT in several foods, including mango juice, milk, cauliflower, and grapes. Kitazine is a fungicide; AuNPs were employed to detect it in cucumbers, tomatoes, oranges, and grapes (Wu et al., 2014).

The most prudent approach to environmentally friendly farming would be to first determine the disease and then treat the affected areas with pesticides or fungicides. To diagnose plant diseases, NPs may detect pathogens (bacteria, fungi, viruses) or chemicals that suggest their presence (Joudeh & Linke, 2022). The fungus species *Magnaporthe oryzae* causes blight disease in several Poaceae plants, the most common of which are rice, wheat, and millets. The disease manifests mostly as grey to dark patches on the leaves of affected plants. In order to boost production and avoid losses, early and precise infection identification is now crucial. Nanotechnology, along with real-time polymerase chain reaction (RT-PCR), fluorescence microscopy, the loop-mediated isothermal amplification (LAMP) test, next-generation sequencing, and a few others, is among the most prominent approaches now used to diagnose blast infections (Kumar et al., 2021). In order to detect the

Magnaporthe oryzae-specific chitinase enzyme at concentrations as low as 17 pg/mL during the early stages of infection. Yang et al. (2014) used nanotechnology principles to develop an electrochemical device. The catalyst in this device is palladium NPs (PdNPs). *Xanthomonas axonopodis* is a bacterium that causes bacterial spot disease in Solanaceae plants, such as pepper and tomato. To identify it, researchers utilized antibodies and fluorescent silica NPs (Alghuthaymi et al., 2021). Additionally, different NPs are utilized for the diagnosis of various plant pathogens. For example, copper NPs (Cu NPs) can detect the fungal pathogen *Sclerotinia sclerotiorum*, which infects oil seeds (Wang et al., 2010), AgNPs can detect *Phytophthora* species (Schwenkbier et al., 2015), copper oxide NPs (CuO NPs) can detect the *Aspergillus niger* fungus, and nanowires can detect the Papaya ringspot virus and Cauliflower mosaic virus (Alghuthaymi et al., 2021). There was no direct detection of *S. sclerotiorum* in oilseed plants. One of the plant hormones that responds to disease conditions was found to be salicylic acid (Wang et al., 2010). In response to a variety of stressors, plants secrete hormones that control their self-defense systems. When certain infections are present, the levels of certain hormones reveal it. An electrochemical sensor that detects salicylic acid levels was developed using CuNPs and a modified gold electrode (Carrillo-Lopez et al., 2024).

An additional new technology that has received considerable worldwide acclaim for disease diagnosis and analysis is a specialized nano-barcode system, which complements nano-biosensors. Examples of biological experiments that use these nanosized tags include histopathology and multiplexed gene expression analysis. Agrifood items and other dairy products may also be tracked using these tags, as can generic encodings (Haris et al., 2023). For the purpose of multiplex pathogen detection, Um et al. (2006) created fluorescent microbarcodes based on dendrimer-like DNA (DL-DNA). A Y-shaped DNA molecule was first created, with one stick end and two ends labeled with molecular probes or fluorescent dyes. To create a DL-DNA nano-barcode system, the adhesive ends were covalently joined. Using ultraviolet light or flow cytometry, this nanofabricated technology may detect infections (Um et al., 2006). In a similar vein, Eastman et al. (2006) developed a device for analyzing multiplexed gene expression using QDs, nano-barcodes, and microbeads. Nanotechnology has supposedly made plant disease diagnosis better, according to these research.

8.5 Enhancing Crop Yield and Quality

The NPs encapsulate, dissolve, entrap, or absorb active components like drugs or bioactive substances. Nanofertilizers are efficient plant nutrition inputs that boost agricultural yields without negatively impacting the environment or the bottom line. Nanofertilizers are referred to as "smart fertilizers" because they function as intelligent material delivery systems. Nanofertilizer is a kind of nutrition delivery system that uses encapsulated nanotubes, nanoemulsions, or nanomaterials to transfer nutrients (Ijaz et al., 2023). They may be modified from conventional fertilizers or manufactured industrially using physical, chemical, or biological processes. Traditional farming methods rely on chemical fertilizers, which plants do not completely absorb. Production costs rise and environmental damage occurs when 50–70% of the material is lost to rainfall, agricultural runoff, soil erosion, or leaching. According to Kanjana's estimations, synthetic fertilizers release 40–70% nitrogen, 50–70% potassium, 80–90% phosphorus, and around 95% micronutrients including zinc, iron, copper, and molybdenum into the environment. As a result, nanofertilizers are preferred because they provide regulated nutrient delivery, enhance micronutrient accessibility, and target-specific function. There are three methods in which nanofertilizers provide nutrients to agricultural plants, first by encasing the nutrient in nanoporous materials or nanotubes, second by encasing it in a thin protective layer of polymer, and third by providing it as an emulsion or particles with tiny measurements. Nanofertilizers are distributed to plants in a controlled, targeted, and efficient manner. For instance, urea ($CO(NH_2)_2$), calcium phosphate ($Ca_3(PO_4)_2$), and potassium chloride (KCl)—sources of potassium, phosphorus, and methacrylic acid (MAA), nitrogen, phosphorus, and potassium (NPK)—were encapsulated with chitosan and MAA NPs. Slow and regulated release of NPK fertilizers was shown by this smart fertilizer based on nanotechnology. For soils lacking in sulfur, urea and other phosphorus-based fertilizers that were coated with sulfur NPs also worked as slow-release fertilizers (Shanmugavel et al., 2023).

Compared to conventional fertilizers, nanofertilizers have several advantages, such as better nutritional value and health, higher yields, lower production costs, and more efficiency in terms of both time and energy. Tomatoes grown with nanofertilizer based on silicon dioxide had a faster germination rate, whereas plants grown with nano-composites based on nitrogen, potassium, and sulfur showed improved absorption, translocation, and accumulation of nutrients. Research on *Nicotiana tabacum* (tobacco) has shown that fertilizers based on carbon NPs (CNPs) are more effective than traditional fertilizers. Improved plant development with CNPs was shown by increased nitrogen and potassium content in several plant organs at maturity. The application of CNPs resulted in the early blooming of *A. thaliana*. The photoperiod-dependent pathways and phytochrome B were influenced by CNPs to cause flowering. Titanium dioxide NPs (TiO_2 NPs), when applied to spinach seeds or sprayed over spinach leaves, increased enzymatic activity and nitrate absorption, leading to improved plant development. ZnO NPs applied to the rice plant's leaves helped alleviate zinc deficit and boost dehydrogenase enzyme activity. Compared to traditional fertilizers, phosphate-based nanofertilizers enhanced soybean seed production by 20% and growth rate by 32% (Yadav et al., 2023a). Table 8.2 provides information on several types of nanobased fertilizers, including their properties and potential uses.

In a similar vein, increasing the concentration of biomolecules like chlorophyll, proteins, and phenols in maize grains by applying silicon dioxide NPs (SiO_2 NPs) increases plant biomass. With a modest concentration of CNTs, hexaploidy wheat roots are improved, mustard seeds and seedlings are grown better, black gram is grown better, rice is grown better, and tobacco cell growth is increased by 16%. Nanomaterials made of titanium dioxide and silicon dioxide improve nitrate reductase activity, which seems to speed up the germination and development of soybean seeds. The use of nanomaterials not only aids field crops in their development and growth but also horticulture crops (Wang et al., 2024).

Nanofertilizers are employed to improve floral fertility, facilitate more effective plant pollination, and enhance the productivity and quality of fruit trees. In conditions of salt stress, blueberries treated with external nano-Ca exhibit enhanced leaf chlorophyll content and improved vegetative growth. The utilization of nano-boron sprays on mango tree leaves has demonstrated an enhancement in fruit yield and chemical characteristics. The likely

Table 8.2 Nanobased fertilizers and their properties.

Method	Characteristics	Application	Types	Limitations	References
Carbon nanotubes	• High tensile strength • Good biocompatibility • High aspect ratio	• Delivery of plasmid DNA and small interfering RNA (siRNA) for gene silencing • Protection of mammalian DNA from degradation	• Single-walled carbon nanotubes (SWCNT) • Multiwalled carbon nanotubes (MWCNT)	• Penetrance issues because of strong cell wall • Low water dispersible • Nontoxicity of CNTs is not clearly explored	Lv et al. (2020)
Clay nanosheets	• Nontoxic • Can be easily washed off from leaves • Degradable	• Delivery of dsRNA • Protecting the plant by silencing the target		• No clear data available about the mechanism and process	Lv et al. (2020)
Peptide NPs	• Cell wall penetrating capacity • Extracellular stability	• DNA delivery in tomato root cells and BY-2 cells of tobacco		• The mechanism of DNA–protein complex formation, DNA association and dissociation are not yet clearly studied	Lv et al. (2020)
DNA nanostructures	• High biostability • High biocompatibility • Easy and flexible	• Gene delivery • Drug delivery • Delivery of siRNA for gene silencing in *Nicotiana benthamiana* (benth or benthi), *Nasturtium officinale* (watercress), *Nicotiana tabacm* (tobacco), *Eruca sativa* (arugula)	• Nanostrings • Nanostars • Nanocubes • 3D geometric figures	• Penetrance issues due to strong cell wall	Lv et al. (2020)

explanation for this phenomenon is the elevated concentrations of chlorophyll and vital nutrients such as nitrogen, phosphorus, potassium, magnesium, boron, zinc, and iron present in the leaves. A study found that mango trees treated with nano-zinc exhibited enhancements in fruit yield, fruit weight, leaf chlorophyll levels, carotene concentration, and the levels of various nutrients such as nitrogen, potassium, and zinc. The application of nano-zinc and nano-boron fertilizers demonstrates comparable impacts on pomegranates, improving fruit quality while also boosting fruit yield, total sugar and phenol content, maturity index, and total soluble sugar ratio (Wang et al., 2024).

While micronutrients are essential for plants, they are only needed in trace levels for plants to grow and develop normally. For optimal health and balance, it is crucial to consume the recommended amounts of macro- and micronutrients. Therefore, in these cases, nanofertilizers based on metals are considered. Copper (Cu) and iron (Fe) nanofertilizers shown fewer side effects and improved performance when compared to conventional fertilizers. Due to their high solubility, nanosize, and reduced environmental effect, nanofertilizers have shown promise even when used in smaller quantities compared to traditional fertilizers, which are sometimes utilized in excessive amounts. After being treated with iron nanofertilizer, plants showed improvements in many areas, including photosynthetic activity, chlorophyll content, fruit mass, and leaves. Micronutrients derived from nanotechnology have applications in both plant and human systems. One example is the substitution of iron NPs (Fe NPs) for traditional tablets in micronutrient supplements. Nanotechnology-based target delivery systems also facilitate the absorption of other micronutrients, including minerals, folic acid, and vitamins (Welch, 1999).

8.6 Harvesting and Postharvest Processing

The development of several fruit varieties is made possible by the diverse agroclimatic zones and soil types found around the globe. It is crucial to take good care of fruits both before and after harvesting. The use of nanotechnology is revolutionizing harvesting methods and is quickly becoming a potent tool in agriculture. Precision real-time monitoring of fruit ripeness, moisture content, and other quality indicators is possible with the integration of nanosensors into harvesting equipment. Reduced postharvest losses are a result of the precise scheduling of harvests made possible by this technology. The effectiveness and longevity of harvesting implements may both be improved by nanotechnology. To improve hygiene and save downtime, equipment may be coated with NPs that resist dirt and debris. More effective, accurate, and environmentally friendly harvesting is on the horizon thanks to these nanotech breakthroughs (Yadav et al., 2023c).

Fruit crops need special handling after harvesting until they are ready to be sold. Collection, transportation, and packing all contribute to physical harm to fruits, shortening their shelf life. Furthermore, illnesses caused by microbes, including fungi, viruses, and bacteria, are a significant source of crop losses. Because of their delicate nature, fruits are ripe targets for microbial diseases (Shakeel et al., 2022). Decomposition by *Rhizopus* and *Aspergillus*, as well as decay by *Penicillium* and *Fusarium*, are among the microbial illnesses that may affect citrus fruits (Alegbeleye et al., 2022). After being picked, grapefruits are prone to fungal diseases. Anthracnose, crown rot, and blossom end rot are among the most frequent diseases that may infect bananas (Zakaria, 2021). Various fungal diseases may affect apples, causing significant crop failures across the globe. Speck rot, Sphaeropsis rot, Mucor rot, powdery mildew, bull's eye rot, blue mold (*Penicillium expansum*), and gray mold (*Botrytis cinerea*) are some of the diseases that may harm apple harvests (Głos et al., 2022). A number of diseases may infect mango fruits. These include *Rhizopus* rot, *Aspergillus niger* rot, *Botrytis theobromae*, stalk end rot, and anthracnose (Oztekin et al., 2023). Numerous fruits, such as guavas, papayas, avocados, dragonfruits, and apricots, are susceptible to these illnesses, which result in significant financial losses for the fruit sector globally (Pedro et al., 2024). The acceptable storage life of the majority of fruits and vegetables is exploited when kept in a typical refrigerator because of their perishable nature. While there are several time-honored methods for preserving food, they are all prohibitively costly, inefficient, or leave behind unwanted residues. Consequently, it is necessary to organize the postharvest processing of fruits methodically. Exploitation of shelf-life extension strategies based on nanotechnology has the potential to mitigate the downsides of conventional approaches due to the various regulatory properties of nanomaterials (Yadav and Sinha, 2022).

Nanofilms serve as protective barriers that carry metal NPs (Hashmi et al., 2024; Jabbar et al., 2023; Khan et al., 2023; Noor et al., 2025), which are antibacterial agents. Crops may have their shelf life extended by using these innovative materials, which reduce respiration rate, regulate breakdown and color change, stabilize storage nature, and more. An effective strategy for agricultural goods' storage and shelf-life balance is the use of edible nanofilms (Gupta et al., 2023). Glycerol, *aloe vera* gel, and ZnO NPs solutions were recently combined to form an edible film by Dubey et al. (2019). The ZnO NPs edible film has the potential to extend the mango fruit's shelf life and quality by up to nine days at room temperature. This is achieved through many mechanisms, including a decrease in mass loss, the maintenance of lower soluble solids, an increase in the fruit's titratable acidity and ascorbic acid concentration, the maintenance of pH, an increase in thickness, and an improvement in transmittance (Dubey et al., 2019).

To ensure the quality and freshness of food during long-distance travel, it is crucial to package it safely. Traditional approaches do have their limits. It would seem that nanotechnology offers a viable substitute for traditional food packaging. The material used to package food is a big drawback as it reduces the food's quality. Common polymers used for food packaging include polyethylene, polystyrene, and polyvinyl chloride, these materials are impermeable to water and ambient gases and prevent the formation of microorganisms (del Rosario Herrera-Rivera et al., 2024). Biodegradability, safety, nutrition, spoilage, and contamination are some issues with conventional food packaging. There is now widespread worldwide acceptance of the idea that combining conventional approaches with nanotechnology might provide superior outcomes (Malik et al., 2023).

Polymer nanocomposites (PNCs) are a new kind of food packaging material made by mixing a polymer matrix with nanomaterials. These PNCs may take several shapes, including nanotubes, nanosheets, nanoplatelets, nanorods, and nano-whiskers. There are a variety of NPs used as organic or inorganic fillers in the packaging, including clay and silica NPs, CNTs, TiO_2, and ZnO (del Rosario Herrera-Rivera et al., 2024). Better food packaging, active packaging, smart packaging, and bio-based packaging are just a few examples of the many approaches to food packaging that rely on functional nanomaterials. Various packaging techniques track food conditions, which enhances flavor, prolongs shelf life, and keeps food fresher for longer (Malik et al., 2023). Sensors that scan the characteristics and interior conditions of packaged goods and provide signals allow for the tracking of spoiled or contaminated food. For instance, smart packaging makes use of perylene nanofibrils, AuNPs, ZnO, and TiO_2 nanocomposites to identify gaseous amines released by spoiling meat, microbial growth, and organic volatile chemicals, respectively (Malik et al., 2023). The Aflatoxin B1 toxin in milk may be tracked using AuNPs, and fizzy beverage packaging uses nanocomposites as gas barriers (Malik et al., 2023). PNCs are safe for the environment since they break down quickly in the body. An example of a PNC would be an organic one, made with polyvinyl acetate (PVA), poly 3-hydroxybutyrate-*co*-3-hydroxyvalerate (PHBV), or polylactic acid (PLA). On the other hand, PNCs of polysaccharides would include polysaccharides such as cellulose, chitosan, starch, or thermoplastic starch (TPS) (Fahmy et al., 2020). Polymer preparation from solution, melt mixing process, in situ polymerization, and in situ synthesis of NPs are some of the several ways utilized to produce PNCs. Polymer preparation involves creating a solution with many layers of silicate, as the name implies. In order to create intercalated nanocomposites, the prepolymers are soaked in this mixture. This technique is known as exfoliation adsorption. It is possible to combine NPs with polymer using the melt mixing technique, which requires controlled temperatures and pressures. When heat or radiation is used to start the process, the result is a polymer solution. This process is known as in situ polymerization.

Similar steps, with the exception of polymerization, are involved in the in situ synthesis of nanocomposites. Nowadays, active food packaging using biopolymer matrices based on NPs is widely recognized as a sustainable and environmentally favorable option. By integrating Ag NPs into biopolymer matrices, this cutting-edge technique now provides edible coatings and packaging materials. Antimicrobial activity against many species, including *Salmonella, E. coli, Pseudomonas aeruginosa,* and *Bacillus cereus*, has been shown by these nanocomposite films. Cellulose, chitosan, agar/banana, and gelatin are some of the most common biopolymer matrices that include Ag NPs. Biopolymer nanocomposites made with nanofillers such as TiO_2 NPs, ZnO_2 NPs, SiO_2 NPs, and Ag NPs have antibacterial properties, oxygen scavenging activity, and oxidation prevention.

Due to its exceptional properties, such as a high volume-to-surface ratio, target-specific action, and great efficiency, nanotechnology has expanded its reach beyond agriculture and into food processing and preservation. Rethink Food Waste through Economics and Data (ReFED) found that 13% of all agricultural goods, 26% of all dairy products, 42% of all fruits and vegetables, and 19% of all grains are lost during harvesting, processing, and transportation. Therefore, nanotechnology is used for food preservation in addition to traditional methods such as heat treatment, freezing, ultrasound, ozone, and pulsed electric field therapy (Sridhar et al., 2021). The food sector makes use of both organic and inorganic NPs. Combining selenium NPs (Se NPs) with Cu NPs boosted tomato output and chlorophyll content.

Among the many uses for silver NPs, one is in packaging, and another is in storage containers where they inhibit the proliferation of harmful microbes (Sridhar et al., 2021). There are many food additives that include nanotechnology, including nanosilica, nanocharcoal, nano-Se, and nano-Ag, all of which act as anti-odorants (Gelaye, 2024). When cucumber plants were treated with calcium oxide NPs (CaO NPs), their storage life was significantly increased. During the postharvest storage phase, Cid-López et al. studied the effects of CaO NPs applied to poly(vinyl acetate-*co*-vinyl alcohol) [P(VAc-*co*-VA)] latex coatings on *Cucumis sativus* (cucumber), a plant species in the Cucurbitaceae family. According to their findings, cucumbers had a longer shelf life after being treated with P (VAc-co-VA) combined with CaO NPs (Zhang et al., 2024).

Nanosensors made of NPs are ideal for use in the food business because of their unique optical and electrochemical characteristics. One application of graphene nanoribbons is the detection of antioxidants in mixed fruit juices. Chili sauce and chili powder adulterants may be detected with the use of Fe_3O_4 NPs (Yin HuanShun et al., 2011). Melamine in milk may be detected using CNPs (Cuong et al., 2021). The quality control of packaged meals makes use of recycled QDs. Bacterial staining and fluorescence labeling are two applications for these (Saritha et al., 2022). Foods that seem fine on the exterior but may have gone bad on the inside may be detected by certain bacteria that function as biosensors. When this occurs, QDs step in to detect food spoilage by changing the color of the food (Vijayalakshmi et al., 2017).

A lot of effort is going into making and selling nano-meals. Processing and preservation methods including nanoencapsulated components are becoming more commonplace in the food industry. Examples of bioactive substances that may be used as food additives or preservatives include coloring agents, vitamins, fatty acids, carotenoids, and flavonoids, all of which are nanoencapsulated. The German firm "Aqua nova" produces "Nova sol" (active chemicals encased in NPs) for sale in stores and online (Neme et al., 2021). The encapsulations used for delivery and projection sheets should be of food-grade material (Sarkar et al., 2017). This kind of food, which contains NPs, is known as an interactive meal since it can be customized to suit individual preferences and dietary restrictions (Saritha et al., 2022). Improved absorption efficiency is achieved by using nanodroplets of vitamins and nanocapsules of zinc and iron as nanosupplements. The encapsulated foods' functional properties are preserved in these (Malik et al., 2023). Bioactive substances with low water solubility may be encapsulated in nanocochleates, and molecules such as phospholipids, polysaccharides, and proteins can be delivered in nanoemulsions, where they can be confined within droplets and protected from chemical destruction. You can see this in ice creams and other similar products, where nanoemulsions keep the texture. This makes the meal more appealing to consumers (Neme et al., 2021). Nanoemulsions have many advantages over their larger counterparts, such as being thermally stable, having a large surface area, being tiny, protecting bioactive chemicals from enzymatic degradation, and eliminating contaminants. The improved flavor, taste, nutritional content, and extended shelf life are all benefits of these nanogoods. They effectively transport to the target location, inhibit bioactive ingredient degradation in high acidic environments, and keep food products fresh with edible coatings. Probiotics, like other bioactive compounds or nutraceuticals, are encapsulated to protect them from the stomach's acidity and the harmful effects they have on the human body when taken in sufficient doses (Saritha et al., 2022).

To get the soil ready to hold seeds or seedlings until harvest, farmers engage in a variety of tasks. There are a number of farming practices that contribute significantly to agricultural pollution. There are now environmental problems due to the buildup of improperly handled agro-wastes, which is particularly true in emerging nations. Some potential advantages of this approach include the following: increasing crop productivity and soil fertility, decreasing reliance on mineral fertilizers, decreasing reliance on fossil fuels, creating protein-based feedstock for animal feeding, creating bioactive compounds, creating nanomaterials and NPs, creating pharmaceutical compounds, using it as a feedstock in fermentation industries, making it a binder for asphalt or a natural aggregate in concrete, and making it a nano-adsorbent. However, improper handling of these agro-wastes may lead to soil, water, and air pollution, among other environmental issues. There are potential health hazards and economic costs associated with hypoxia, hazardous algal blooms, and eutrophication in aquatic bodies caused by this pollution (Padhiary and Kumar, 2024).

Deterioration of the environment will result from the interplay between the energy crisis and the handling of ever-increasing solid waste loads caused by a rapidly expanding human population. Certain solutions for turning agro-waste into energy, including processing based on nanotechnology, should be considered within the circular economy idea in order to satisfy sustainable development objectives. As a result, a lot of recent research has been on finding ways to turn agricultural waste into renewable energy as part of a circular economy. One environmentally friendly approach is to turn agricultural waste into nano-enabled energy applications, which can be found in literature such as (Preethi et al., 2024; Sun et al., 2024b). Many another researcher produce nanomaterials or NPs from agricultural trash, as shown in Table 8.3. Hazards to human and environmental health may be

Table 8.3 Nanomaterials produce from agro-wastes.

Method of synthesis	Types of agro-wastes	Main findings of the review	Type of NPs	References
Biogenic synthesis	Groundnut shell, coconut husk, banana peel, orange peel, and walnut shell	Coconut produced a higher content of biogenic silica compared to the other agro-wastes.	Nano silica	Peerzada and Chidambaram (2021)
Green method	Extracts from *Persea americana* seed, *Zea mays* silk, and *Eucalyptus globulus* leaves	*Eucalyptus globulus* leaves were recommended to produce NiO-NPs as compared to other studied agro-wastes. The NiO-NPs could be used as a bacterial inhibitor.	Nickel oxide NPs (NiO NPs)	Razanamahandry et al. (2021)
Biological methods	Rice straw and others	Nanosilica has many remarkable applications in environmental clean-up, water treatment, and other types of nanoremediation.	Silica NPs	Yadav et al. (2022)
Green method with azolla help	Coconut husk and sugarcane bagasse	Biodiesel generation was effectively achieved using graphene oxide nanocatalyst derived from agro-waste by transesterification using azolla.	Nano graphene oxide	Sathish et al. (2022)
Acid and enzymatic hydrolysis	Agro-industrial wastes from *Salicornia ramosissima*	*Salicornia ramosissima* wastes were considered a renewable source of nanofiber production for circular economy concepts.	Cellulose nanofibers	Thakur et al. (2024)
Chemical method/acid hydrolysis	Cut garlic stalk, corncob, and giant cane wastes	Produced nanocellulose for application as new materials for food packaging, purification, and multiple industrial sectors.	Cellulose nanocrystals	Rovera et al. (2023)

Figure 8.2 (a) Banana peel, (b) orange peel, (c) pomegranate peel and (d) mango peel as agro-wastes materials for synthesis of NPs.

posed by agri-food wastes. Worldwide, the food business produces around one-third of the world's food, which results in about 1.3 billion tons of food waste per year, bringing the total to approximately 2 billion tons. Different agro-waste materials were used for the synthesis of NPs as shown in Figure 8.2 (Peerzada & Chidambaram, 2021).

8.7 Conclusions

Nanotechnology is modernizing agriculture by introducing new methods and materials that further enhance productivity, sustainability, and environmental safety in view of the whole food production chain. From soil improvement to the management of pests, precision farming, and preservation after harvest, nanotechnology has something to offer for almost all the critical challenges being faced by modern agriculture. By enabling efficient nutrient delivery, targeted pesticide application, real-time monitoring of plant health, and environmentally friendly packaging, nanotechnology reduces waste and minimizes ecological footprint in farming.

Application fields include nanofertilizers, nanosensors, and nanobased remediation systems—very promising to improve crop yields, soil health, and plant resilience for abiotic stresses and biotic stresses under different environmental conditions. These developments consequently contribute toward sustainable agriculture with the quest to ensure food safety in the process of climate and resource depletion. For example, longevity, quality, reduction of losses from agri-produce—five major concerns for nanotechnology in food packaging and postharvest management—include global food supply chains. More significantly, the use of agro-waste for the synthesis of nanomaterials closes a loop in a circular economy, thereby reducing environmental pollution. All these developments are in tune with global goals on sustainability and open up the possibilities of making agriculture greener and more resource-use efficient. Though challenges with safety, scalability, and regulatory frameworks exist, continuous research and development have been unlocking immense potential through the use of nanotechnology to transform agriculture from soil to shelf. By integrating these innovations, the agricultural industry is poised to meet the growing demand for food in a sustainable and environmentally conscious manner.

References

Abdel-Aziz, M. M., Emam, T. M., & Elsherbiny, E. A. (2020). Bioactivity of magnesium oxide nanoparticles synthesized from cell filtrate of endobacterium *Burkholderia rinojensis* against *Fusarium oxysporum*. *Materials Science and Engineering: C*, *109*, 110617.

Ahmad, M. F., Ahmad, F. A., Alsayegh, A. A., Zeyaullah, M., Alshahrani, A. M., Muzammil, K., Saati, A. A., Wahab, S., Elbendary, E. Y., & Kambal, N. (2024). Pesticides impacts on human health and the environment with their mechanisms of action and possible countermeasures. *Heliyon*, *10*(7), e29128.

Ajmal, M., Ullah, R., Muhammad, Z., Khan, M. N., Kakar, H. A., Kaplan, A., Okla, M. K., Saleh, I. A., Kamal, A., & Abdullah, A. (2023). Kinetin capped zinc oxide nanoparticles improve plant growth and ameliorate resistivity to polyethylene glycol (PEG)-induced drought stress in *Vigna radiata* (L.) R. Wilczek (Mung Bean). *28*, Molecules, 5059.

Alegbeleye, O., Odeyemi, O. A., Strateva, M., & Stratev, D. (2022). Microbial spoilage of vegetables, fruits and cereals. *Applied Food Research*, *2*, 100122.

Alghuthaymi, M. A., Rajkuberan, C., Rajiv, P., Kalia, A., Bhardwaj, K., Bhardwaj, P., Abd-Elsalam, K. A., Valis, M., & Kuca, K. (2021). Nanohybrid antifungals for control of plant diseases: current status and future perspectives. *Journal of Fungi*, *7*, 48.

Ali, S., Ahmad, N., Dar, M. A., Manan, S., Rani, A., Alghanem, S. M. S., Khan, K. A., Sethupathy, S., Elboughdiri, N., & Mostafa, Y. S. (2023). Nano-agrochemicals as substitutes for pesticides: prospects and risks. *Plants*, *13*, 109.

Alshameri, A. W., & Owais, M. (2022). Antibacterial and cytotoxic potency of the plant-mediated synthesis of metallic nanoparticles Ag NPs and ZnO NPs: a review. *OpenNano*, *8*, 100077.

Asiminicesei, D.-M., Fertu, D. I., & Gavrilescu, M. (2024). Impact of heavy metal pollution in the environment on the metabolic profile of medicinal plants and their therapeutic potential. *Plants, 13*, 913.

Azam, N., Khan, M. W., Sardar, S., Yousaf, I., Zahid, A., Ismail, A., Summiya, S., Mushtaq, R., & Javed, M. S. (2022). New insights for improving agriculture soils through nanotechnology. *Haya: The Saudi Journal of Life Sciences, 7*, 244–248.

Babar, A. Z. & Akan, O. B. 2024. Sustainable and precision agriculture with the internet of everything (IoE). *arXiv preprint arXiv:2404.06341*.

Bai, Z., Zhang, D., Wang, Z., Harrison, M. T., Liu, K., Song, Z., Chen, F., & Yin, X. (2024). Challenges and strategies in estimating soil organic carbon for multi-cropping systems: a review. *Carbon Footprints, 3*. Article 19. https://doi.org/10.20517/cf.2024.15

Ben Gaied, R., Brígido, C., Sbissi, I., & Tarhouni, M. (2024). Sustainable strategy to boost legumes growth under salinity and drought stress in semi-arid and arid regions. *Soil Systems, 8*, 84.

Bu, T., Jia, P., Liu, J., Liu, Y., Sun, X., Zhang, M., Tian, Y., Zhang, D., Wang, J., & Wang, L. (2019). Diversely positive-charged gold nanoparticles based biosensor: a label-free and sensitive tool for foodborne pathogen detection. *Food Chemistry: X, 3*, 100052.

Carrillo-Lopez, L. M., Villanueva-Verduzco, C., Villanueva-Sánchez, E., Fajardo-Franco, M. L., Aguilar-Tlatelpa, M., Ventura-Aguilar, R. I., & Soto-Hernández, R. M. (2024). Nanomaterials for plant disease diagnosis and treatment: a review. *Plants, 13*, 2634.

Chaudhary, R. G., Potbhare, A. K., Aziz, S. T., Ayyub, M. M., Kahate, A., Madankar, R. S., Wankar, S., Dutta, A., Abdala, A., & Adhikari, R. (2024). Bioinspired graphene-based metal oxide nanocomposites for photocatalytic and electrochemical performances: an updated review. *Nanoscale Advances, 6*(10), 2539–2568.

Chauhan, R., Awasthi, S., Tiwari, P., Upadhyay, M. K., Srivastava, S., Dwivedi, S., Dhankher, O. P., & Tripathi, R. D. (2024). Biotechnological strategies for remediation of arsenic-contaminated soils to improve soil health and sustainable agriculture. *Soil & Environmental Health, 2*, 100061.

Chen, Q., Xing, G., Cao, X., Liang, T., Chen, L., Dai, L., Ci, L., & Yan, M. (2024). Functional carbon nanodots enhance tomato tolerance to zinc deficient soils: mechanisms and structure-function relationships. *Science of the Total Environment, 953*, 176113.

Colombo, S. M., Roy, K., Mraz, J., Wan, A. H., Davies, S. J., Tibbetts, S. M., Øverland, M., Francis, D. S., Rocker, M. M., & Gasco, L. (2023). Towards achieving circularity and sustainability in feeds for farmed blue foods. *Reviews in Aquaculture, 15*, 1115–1141.

Cozim-Melges, F., Ripoll-Bosch, R., Veen, G., Oggiano, P., Bianchi, F. J., Van Der Putten, W. H., & Van Zanten, H. H. (2024). Farming practices to enhance biodiversity across biomes: a systematic review. *npj Biodiversity, 3*, 1.

Cuong, N. M., Cao, D. T., Thu, V. T., & Ngan, L. T.-Q. (2021). Direct detection of melamine in liquid milk and infant formula using surface-enhanced Raman scattering combined with silver nanodendrites. *Optik, 243*, 167504.

Dimkpa, C. O., Andrews, J., Fugice, J., Singh, U., Bindraban, P. S., Elmer, W. H., Gardea-Torresdey, J. L., & White, J. C. (2020). Facile coating of urea with low-dose ZnO nanoparticles promotes wheat performance and enhances Zn uptake under drought stress. *Frontiers in Plant Science, 11*, 168.

Ding, Y.-Y., Zhou, H., Zhang, B.-Q., Zhang, Z.-J., Wang, G.-H., Zhang, S.-Y., Wu, Z.-R., Wang, Y.-R., & Liu, Y.-Q. (2023). Antimicrobial activity of natural and semi-synthetic carbazole alkaloids. *European Journal of Medicinal Chemistry, 259*, 115627.

Donkor, L., & Daliri, E. B.-M. (2024). The role of nanotechnology in agriculture and food security. In S. C. George & B. Tawiah (Eds.), *Nanotechnology in societal development* (pp. 149–190). Springer.

Dubey, P. K., Shukla, R. N., Srivastava, G., Mishra, A. A., & Pandey, A. (2019). Study on quality parameters and storage stability of mango coated with developed nanocomposite edible film. *International Journal of Current Microbiology and Applied Sciences, 8*, 2899–2935.

Eastman, P. S., Ruan, W., Doctolero, M., Nuttall, R., De Feo, G., Park, J. S., Chu, J. S., Cooke, P., Gray, J. W., & Li, S. (2006). Qdot nanobarcodes for multiplexed gene expression analysis. *Nano Letters, 6*, 1059–1064.

Elnahal, A. S., El-Saadony, M. T., Saad, A. M., Desoky, E.-S. M., El-Tahan, A. M., Rady, M. M., Abuqamar, S. F., & El-Tarabily, K. A. (2022). The use of microbial inoculants for biological control, plant growth promotion, and sustainable agriculture: a review. *European Journal of Plant Pathology, 162*, 759–792.

Fahmy, H. M., Eldin, R. E. S., Serea, E. S. A., Gomaa, N. M., Aboelmagd, G. M., Salem, S. A., Elsayed, Z. A., Edrees, A., Shams-Eldin, E., & Shalan, A. E. (2020). Advances in nanotechnology and antibacterial properties of biodegradable food packaging materials. *RSC Advances, 10*, 20467–20484.

Faizan, M., Ahmed, S. M., Rajput, P., Rajput, V. D., Shende, S. S., Minkina, T., Sushkova, S., Mandzhieva, S., Hayat, S., & Hussain, A. (2024a). Nanotechnology in the restoration of polluted/degraded soil. In J. K. Biswas & M. Ra (Eds.), *Nanotechnology for environmental management* (pp. 23–34). CRC Press.

Faizan, M., Alam, P., Kumari, A., Suresh, G., Sharma, P., Karabulut, F., Soysal, S., Djalovic, I., Trivan, G., & Adil, M. F. (2024b). Unraveling the nano-biochar mediated regulation of heavy metal stress tolerance for sustaining plant health. *Plant Stress, 14*, 100615.

Francis, D. V., Abdalla, A. K., Mahakham, W., Sarmah, A. K., & Ahmed, Z. F. (2024). Interaction of plants and metal nanoparticles: exploring its molecular mechanisms for sustainable agriculture and crop improvement. *Environment International, 14*, 108859.

Gangwar, J., Kadanthottu Sebastian, J., Puthukulangara Jaison, J., & Kurian, J. T. (2023). Nano-technological interventions in crop production—a review. *Physiology and Molecular Biology of Plants, 29*, 93–107.

Garg, S., Rumjit, N. P., & Roy, S. (2024). Smart agriculture and nanotechnology: technology, challenges, and new perspective. *Advanced Agrochem, 3*, 115–125.

Gelaye, Y. (2024). Application of nanotechnology in animal nutrition: bibliographic review. *Cogent Food & Agriculture, 10*, 2290308.

Getahun, S., Kefale, H., & Gelaye, Y. (2024). Application of precision agriculture technologies for sustainable crop production and environmental sustainability: a systematic review. *The Scientific World Journal, 2024*, 2126734.

Głos, H., Bryk, H., Michalecka, M., & Puławska, J. (2022). The recent occurrence of biotic postharvest diseases of apples in Poland. *Agronomy, 12*, 399.

Guo, Y., Zhao, G., Gao, X., Zhang, L., Zhang, Y., Cai, X., Yuan, X., & Guo, X. (2023). CRISPR/Cas9 gene editing technology: a precise and efficient tool for crop quality improvement. *Planta, 258*, 36.

Gupta, R. K., Abd El Gawad, F., Ali, E. A., Karunanithi, S., Yugiani, P., & Srivastav, P. P. (2023). Nanotechnology: current applications and future scope in food packaging systems. *Measurement: Food*, *13*, 100131.

Handa, M., & Kalia, A. (2024). Exploring the dynamics of nanoparticle-plant-microbe interactions to realize the goal of improved crop productivity and food security. *Rhizosphere*, *30*, 100884.

Haris, M., Hussain, T., Mohamed, H. I., Khan, A., Ansari, M. S., Tauseef, A., Khan, A. A., & Akhtar, N. (2023). Nanotechnology–a new frontier of nano-farming in agricultural and food production and its development. *Science of the Total Environment*, *857*, 159639.

Harris, D. R. (2024). Introduction: themes and concepts in the study of early agriculture. In D. R. Harris (Ed.), *The origins and spread of agriculture and pastoralism in Eurasia* (pp. 1–9). Routledge.

Hashmi, S. S., Ibrahim, M., Adnan, M., Ullah, A., Khan, M. N., Kamal, A., Iqbal, M., Kaplan, A., Assad, N., & Elshikh, M. S. (2024). Green synthesis of silver nanoparticles from *Olea europaea* L. extracted polysaccharides, characterization, and its assessment as an antimicrobial agent against multiple pathogenic microbes. *Open Chemistry*, *22*, 20240016.

Haydar, M. S., Ghosh, D., & Roy, S. (2024). Slow and controlled release nanofertilizers as an efficient tool for sustainable agriculture: recent understanding and concerns. *Plant Nano Biology*, *7*, 100058.

Huo, D., Malik, A. W., Ravana, S. D., Rahman, A. U., & Ahmedy, I. (2024). Mapping smart farming: addressing agricultural challenges in data-driven era. *Renewable and Sustainable Energy Reviews*, *189*, 113858.

Ijaz, M., Khan, F., Ahmed, T., Noman, M., Zulfiqar, F., Rizwan, M., Chen, J., Siddique, K. H., & Li, B. (2023). Nanobiotechnology to advance stress resilience in plants: current opportunities and challenges. *Materials Today Bio*, *22*, 100759.

Isibor, P. O., Kayode-Edwards, I. I., Oyewole, O. A., Agbontaen, D. O., Onwaeze, O. O., Paul, E., Oyegbade, S. A., Ilyasu, U. S., & Chimbekujwo, K. I. (2024). Risk assessment and management in nanotoxicology. In G. Devi & A. A. Enuneku (Eds.), *Environmental nanotoxicology: combatting the minute contaminants* (pp. 267–293). Springer.

Jabbar, A., Abbas, A., Assad, N., Naeem-ul-Hassan, M., Alhazmi, H. A., Najmi, A., Zoghebi, K., Al Bratty, M., Hanbashi, A., & Amin, H. M. (2023). A highly selective Hg^{2+} colorimetric sensor and antimicrobial agent based on green synthesized silver nanoparticles using *Equisetum diffusum* extract. *RSC Advances*, *13*, 28666–28675.

Javaid, A., Hameed, S., Li, L., Zhang, Z., Zhang, B., & Rahman, M.-U. (2024). Can nanotechnology and genomics innovations trigger agricultural revolution and sustainable development? *Functional & Integrative Genomics*, *24*, 216.

Jiang, X., Yang, F., Jia, W., Jiang, Y., Wu, X., Song, S., Shen, H., & Shen, J. (2024). Nanomaterials and nanotechnology in agricultural pesticide delivery: a review. *Langmuir*, *40*, 18806–18820.

Joudeh, N., & Linke, D. (2022). Nanoparticle classification, physicochemical properties, characterization, and applications: a comprehensive review for biologists. *Journal of Nanobiotechnology*, *20*, 262.

Joye, I. J., & McClements, D. J. (2014). Biopolymer-based nanoparticles and microparticles: fabrication, characterization, and application. *Current Opinion in Colloid & Interface Science*, *19*, 417–427.

Junaid, M., & Gokce, A. (2024). Global agricultural losses and their causes. *Bulletin of Biological and Allied Sciences Research, 2024*, 66–66.

Kaya, C., & Ashraf, M. (2024). Foliar fertilization: a potential strategy for improving plant salt tolerance. *Critical Reviews in Plant Sciences, 43*, 94–115.

Khan, Z. U. R., Assad, N., Naeem-ul-Hassan, M., Sher, M., Alatawi, F. S., Alatawi, M. S., Omran, A. M., Jame, R. M., Adnan, M., & Khan, M. N. (2023). *Aconitum lycoctonum* L. (Ranunculaceae) mediated biogenic synthesis of silver nanoparticles as potential antioxidant, anti-inflammatory, antimicrobial and antidiabetic agents. *BMC Chemistry, 17*, 128.

Khodakovskaya, M. V., Kim, B. S., Kim, J. N., Alimohammadi, M., Dervishi, E., Mustafa, T., & Cernigla, C. E. (2013). Carbon nanotubes as plant growth regulators: effects on tomato growth, reproductive system, and soil microbial community. *Small, 9*, 115–123.

Knuth, D., Gai, L., Silva, V., Harkes, P., Hofman, J., Sudoma, M., Bílková, Z., Alaoui, A., Mandrioli, D., & Paskovic, I. (2024). Pesticide residues in organic and conventional agricultural soils across Europe: measured and predicted concentrations. *Environmental Science & Technology, 58*, 6744–6752.

Kumar, S., Kashyap, P. L., Mahapatra, S., Jasrotia, P., & Singh, G. P. (2021). New and emerging technologies for detecting *Magnaporthe oryzae* causing blast disease in crop plants. *Crop Protection, 143*, 105473.

Lv, Z., Jiang, R., Chen, J., & Chen, W. (2020). Nanoparticle-mediated gene transformation strategies for plant genetic engineering. *The Plant Journal, 104*, 880–891.

Malik, S., Muhammad, K., & Waheed, Y. (2023). Nanotechnology: a revolution in modern industry. *Molecules, 28*, 661.

Mana, A., Allouhi, A., Hamrani, A., Rehman, S., El Jamaoui, I., & Jayachandran, K. (2024). Sustainable AI-based production agriculture: exploring AI applications and implications in agricultural practices. *Smart Agricultural Technology, 7*, 100416. *Go to original source.*

Morchid, A., El Alami, R., Raezah, A. A., & Sabbar, Y. (2024). Applications of Internet of Things (IoT) and sensors technology to increase food security and agricultural sustainability: benefits and challenges. *Ain Shams Engineering Journal, 15*, 102509.

Muhammad, M., Waheed, A., Wahab, A., Majeed, M., Nazim, M., Liu, Y.-H., Li, L., & Li, W.-J. (2024). Soil salinity and drought tolerance: an evaluation of plant growth, productivity, microbial diversity, and amelioration strategies. *Plant Stress, 11*, 100319.

Munné-Bosch, S., & Bermejo, N. F. (2024). Fruit quality in organic and conventional farming: advantages and limitations. *Trends in Plant Science, 29*(8), 878–894.

Narayanan, A., Sharma, P., & Moudgil, B. M. (2013). Applications of engineered particulate systems in agriculture and food industry. *Kona Powder and Particle Journal, 30*, 221–235.

Neethirajan, S. (2023). Artificial intelligence and sensor technologies in dairy livestock export: charting a digital transformation. *Sensors, 23*, 7045.

Neme, K., Nafady, A., Uddin, S., & Tola, Y. B. (2021). Application of nanotechnology in agriculture, postharvest loss reduction and food processing: food security implication and challenges. *Heliyon, 7*(12), e08539.

Noor, A., Ahmad, N., Ali, A., Ali, M., Iqbal, M., Khan, M. N., Laila, M. B., Shah, S. N., Kaplan, A., & Ercişli, S. (2025). Green synthesis of silver-doped ZnO nanoparticles from *Adiantum venustum* D. Don (Pteridaceae): antimicrobial and antioxidant evaluation. *Journal of Basic Microbiology, 65*(4), e2400543.

Oztekin, S., Dikmetas, D. N., Devecioglu, D., Acar, E. G., & Karbancioglu-Guler, F. (2023). Recent insights into the use of antagonistic yeasts for sustainable biomanagement of postharvest pathogenic and mycotoxigenic fungi in fruits with their prevention strategies against mycotoxins. *Journal of Agricultural and Food Chemistry*, *71*, 9923–9950.

Padhiary, M., & Kumar, R. (2024). Assessing the environmental impacts of agriculture, industrial operations, and mining on agro-ecosystems. In M. Azrour, J. Mabrouki, A. Alabdulatif, A. Guezzaz, & F. Amounas (Eds.), *Smart Internet of Things for environment and healthcare* (pp. 107–126). Springer.

Pedro, A. C., Maciel, G. M., Lima, N. P., Lima, N. F., Ribeiro, I. S., Pinheiro, D. F., & Haminiuk, C. W. I. (2024). Valorization of bioactive compounds from juice industry waste: applications, challenges, and future prospects. *Trends in Food Science & Technology*, *152*, 104693.

Peerzada, J. G., & Chidambaram, R. (2021). A statistical approach for biogenic synthesis of nano-silica from different agro-wastes. *Silicon*, *13*, 2089–2101.

Pramanik, P., Krishnan, P., Maity, A., Mridha, N., Mukherjee, A., & Rai, V. (2020). Application of nanotechnology in agriculture. *Environmental Nanotechnology*, *4*, 317–348.

Preethi, B., Karmegam, N., Manikandan, S., Vickram, S., Subbaiya, R., Rajeshkumar, S., Gomadurai, C., & Govarthanan, M. (2024). Nanotechnology-powered innovations for agricultural and food waste valorization: a critical appraisal in the context of circular economy implementation in developing nations. *Process Safety and Environmental Protection*, *184*, 477–491.

Qin, C.-C., Abdalkarim, S. Y. H., Zhou, Y., Yu, H.-Y., & He, X. (2022). Ultrahigh water-retention cellulose hydrogels as soil amendments for early seed germination under harsh conditions. *Journal of Cleaner Production*, *370*, 133602.

Rao, K. J., & Paria, S. (2013). Use of sulfur nanoparticles as a green pesticide on *Fusarium solani* and *Venturia inaequalis* phytopathogens. *RSC Advances*, *3*, 10471–10478.

Razanamahandry, L., Nwanya, A., Bashir, A., Furqan, C., Sackey, J., Ameh, A., Kavitha, G., Manikandan, E., Ntwampe, S., & Fosso-Kankeu, E. (2021). Thermal valorisation extracts of selected agro-waste for human pathogen antibacterial NiO nanoparticles synthesis. *Materials Today: Proceedings*, *36*, 559–565.

Ren, W., Chang, H., Li, L., & Teng, Y. (2020). Effect of graphene oxide on growth of wheat seedlings: insights from oxidative stress and physiological flux. *Bulletin of Environmental Contamination and Toxicology*, *105*, 139–145.

del Rosario Herrera-Rivera, M., Torres-Arellanes, S. P., Cortés-Martínez, C. I., Navarro-Ibarra, D. C., Hernández-Sánchez, L., Solis-Pomar, F., Pérez-Tijerina, E., & Román-Doval, R. (2024). Nanotechnology in food packaging materials: role and application of nanoparticles. *RSC Advances*, *14*, 21832–21858.

Rovera, C., Carullo, D., Bellesia, T., Büyüktaş, D., Ghaani, M., Caneva, E., & Farris, S. (2023). Extraction of high-quality grade cellulose and cellulose nanocrystals from different lignocellulosic agri-food wastes. *Frontiers in Sustainable Food Systems*, *6*, 1087867.

Sadeghi, R., Rodriguez, R. J., Yao, Y., & Kokini, J. L. (2017). Advances in nanotechnology as they pertain to food and agriculture: benefits and risks. *Annual Review of Food Science and Technology*, *8*, 467–492.

Saito, H., Yamashita, Y., Sakata, N., Ishiga, T., Shiraishi, N., Usuki, G., Nguyen, V. T., Yamamura, E., & Ishiga, Y. (2021). Covering soybean leaves with cellulose nanofiber changes leaf surface hydrophobicity and confers resistance against *Phakopsora pachyrhizi*. *Frontiers in Plant Science, 12*, 726565.

Salama, D. M., Abd El-Aziz, M., Rizk, F. A., & Abd Elwahed, M. (2021). Applications of nanotechnology on vegetable crops. *Chemosphere, 266*, 129026.

Salem, S. S., & Mekky, A. E. (2024). Biogenic nanomaterials: synthesis, characterization, and applications. In M. P. Shah, N. Bharadvaja, & L. Kumar (Eds.), *Biogenic nanomaterials for environmental sustainability: principles, practices, and opportunities* (pp. 13–43). Springer. https://doi.org/10.1007/978-3-031-45956-6_2

Saritha, G. N. G., Anju, T., & Kumar, A. (2022). Nanotechnology-big impact: how nanotechnology is changing the future of agriculture? *Journal of Agriculture and Food Research, 10*, 100457.

Sarkar, P., Choudhary, R., Panigrahi, S., Syed, I., Sivapratha, S., & Dhumal, C. V. (2017). Nano-inspired systems in food technology and packaging. *Environmental Chemistry Letters, 15*, 607–622.

Sathish, S., Supriya, S., Andal, P., Prabu, D., Rajasimman, M., Ansar, S., & Rezania, S. (2022). Effective utilization of azolla filiculoides for biodiesel generation using graphene oxide nano catalyst derived from agro-waste. *Fuel, 329*, 125412.

Schwenkbier, L., Pollok, S., König, S., Urban, M., Werres, S., Cialla-May, D., Weber, K., & Popp, J. (2015). Towards on-site testing of *Phytophthora species*. *Analytical Methods, 7*, 211–217.

Semenova, N. A., Burmistrov, D. E., Shumeyko, S. A., & Gudkov, S. V. (2024). Fertilizers based on nanoparticles as sources of macro-and microelements for plant crop growth: a review. *Agronomy, 14*, 1646.

Shah, S. T., Ahmad, N., Basit, A., Sajid, M., Jamal, A., Saeed, M. F., Iqbal, W., Seleiman, M. F., Radicetti, E., & Mancinelli, R. (2024). Optimizing biochemical and phytochemical attributes in peaches through foliar applications of silicon and zinc. *Horticulturae, 10*, 1031.

Shakeel, Q., Shaheen, M. R., Ali, S., Ahmad, A., Raheel, M., & Bajwa, R. T. (2022). Postharvest management of fruits and vegetables. In Inamuddin & C. O. Adetunji (Eds.), *Applications of biosurfactant in agriculture* (pp. 1–16). Elsevier. https://doi.org/10.1016/B978-0-12-822921-7.00001-5

Shanmugavel, D., Rusyn, I., Solorza-Feria, O., & Kamaraj, S.-K. (2023). Sustainable SMART fertilizers in agriculture systems: a review on fundamentals to in-field applications. *Science of the Total Environment, 904*, 166729.

Sharma, R., & Kumar, V. (2024). Nano enabled agriculture for sustainable soil. *Waste Management Bulletin, 2*(1), 152–161.

Shen, X., Bai, X., Zhao, C., Tan, Q., Luo, G., Li, C., Ran, C., Zhang, S., Xiong, L., & Liao, J. (2024). Global response of soil biodiversity to climate and land use changes. *Journal of Cleaner Production, 471*, 143381.

Shoukat, A., Pitann, B., Zafar, M. M., Farooq, M. A., Haroon, M., Nawaz, A., Wahab, S. W., & Saqib, Z. A. (2024). Nanotechnology for climate change mitigation: enhancing plant resilience under stress environments. *Journal of Plant Nutrition and Soil Science, 187*, 604–620.

Singh, R. P., Handa, R., & Manchanda, G. (2021). Nanoparticles in sustainable agriculture: an emerging opportunity. *Journal of Controlled Release, 329*, 1234–1248.

Singh, A., Rajput, V. D., Lalotra, S., Agrawal, S., Ghazaryan, K., Singh, J., Minkina, T., Rajput, P., Mandzhieva, S., & Alexiou, A. (2024a). Zinc oxide nanoparticles influence on plant tolerance to salinity stress: insights into physiological, biochemical, and molecular responses. *Environmental Geochemistry and Health, 46*, 148.

Singh, A., Sharma, A., Singh, O., Rajput, V. D., Movsesyan, H. S., Minkina, T., Alexiou, A., Papadakis, M., Singh, R. K., & Singh, S. (2024b). In-depth exploration of nanoparticles for enhanced nutrient use efficiency and abiotic stresses management: present insights and future horizons. *Plant Stress, 14*, 100576.

Soni, S., Jha, A. B., Dubey, R. S., & Sharma, P. (2024). Nanowonders in agriculture: unveiling the potential of nanoparticles to boost crop resilience to salinity stress. *Science of the Total Environment, 925*, 171433.

Sridhar, A., Ponnuchamy, M., Kumar, P. S., & Kapoor, A. (2021). Food preservation techniques and nanotechnology for increased shelf life of fruits, vegetables, beverages and spices: a review. *Environmental Chemistry Letters, 19*, 1715–1735.

Srivastava, A., Joshi, M., & Rengan, A. K. (2024). Feeding the future: the role of nanotechnology in tailored nutrition. *The Nucleus, 68*(1), 179–193.

Sun, X.-D., Ma, J.-Y., Feng, L.-J., Duan, J.-L., & Yuan, X.-Z. (2024a). Precise tracking of nanoparticles in plant roots. *Nature Protocols, 20*, 248–271. https://doi.org/10.1038/s41596-024-01044-5

Sun, Y., Jiang, Y., Wei, H., Zhang, Z., Irshad, S., Liu, X., Xie, Y., Rui, Y., & Zhang, P. (2024b). Nano-enabled strategies for greenhouse gases emission mitigation: a comprehensive review. *Nano Today, 57*, 102378.

Tao, R., You, C., Qu, Q., Zhang, X., Deng, Y., Ma, W., & Huang, C. (2023). Recent advances in the design of controlled-and sustained-release micro/nanocarriers of pesticide. *Environmental Science: Nano, 10*, 351–371.

Thakur, A., Kumar, A., & Somya, A. (2024). Forestry and agricultural residues-based wastes: Fundamentals, classification, properties, and applications. In C. Verma & S. Dubey (Eds.), *Biomass wastes for sustainable industrial applications: a waste-to-wealth approach* (pp. 95–139). Boca Raton: CRC Press. https://doi.org/10.1201/9781003466833-6

Thiruvengadam, M., Chi, H. Y., & Kim, S.-H. (2024a). Impact of nanopollution on plant growth, photosynthesis, toxicity, and metabolism in the agricultural sector: an updated review. *Plant Physiology and Biochemistry, 207*, 108370.

Thiruvengadam, R., Venkidasamy, B., Easwaran, M., Chi, H. Y., Thiruvengadam, M., & Kim, S.-H. (2024b). Dynamic interplay of reactive oxygen and nitrogen species (ROS and RNS) in plant resilience: unveiling the signaling pathways and metabolic responses to biotic and abiotic stresses. *Plant Cell Reports, 43*, 198.

Ullah, R., Bibi, S., Khan, M. N., Al Mohaimeed, A. M., Naz, Q., & Kamal, A. (2023). Application of bio-inspired gold nanoparticles as advanced nanomaterial in halt nociceptive pathway and hepatotoxicity via triggering antioxidation system. *Catalysts, 13*, 786.

Um, S. H., Lee, J. B., Kwon, S. Y., Li, Y., & Luo, D. (2006). Dendrimer-like DNA-based fluorescence nanobarcodes. *Nature Protocols, 1*, 995–1000.

Valenzuela-Amaro, H. M., Aguayo-Acosta, A., Meléndez-Sánchez, E. R., de la Rosa, O., Vázquez-Ortega, P. G., Oyervides-Muñoz, M. A., Sosa-Hernández, J. E., & Parra-Saldívar, R. (2023). Emerging applications of nanobiosensors in pathogen detection in water and food. *Biosensors, 13*, 922.

Van Eenennaam, A. L. (2024). Addressing the 2050 demand for terrestrial animal source food. *Proceedings of the National Academy of Sciences of the United States of America, 121*, e2319001121.

Varshney, V., Singh, J., Dutta, S., Mishra, V., & Jha, S. (2024). Understanding the problem at hand: nanotechnological implications that cannot be foreseen. In H. Tombuloglu, G. Tombuloglu, & K. R. Hakeem (Eds.), *Molecular impacts of nanoparticles on plants and algae* (pp. 397–424). Elsevier.

Vega-Vásquez, P., Mosier, N. S., & Irudayaraj, J. (2020). Nanoscale drug delivery systems: from medicine to agriculture. *Frontiers in Bioengineering and Biotechnology*, *8*, 79.

Vijayalakshmi, S., Sachin, C., & Kirtan, T. (2017). Nanotechnology: a growing need for agriculture and food sectors. *Integrated Ferroelectrics*, *185*, 73–81.

de Waal, A. 2024. The history and future of famine. Oxford Research Encyclopedia of Food Studies.

Wahab, A., Muhammad, M., Ullah, S., Abdi, G., Shah, G. M., Zaman, W., & Ayaz, A. (2024). Agriculture and environmental management through nanotechnology: eco-friendly nanomaterial synthesis for soil-plant systems, food safety, and sustainability. *Science of the Total Environment*, *926*, 171862.

Wang, Z., Wei, F., Liu, S.-Y., Xu, Q., Huang, J.-Y., Dong, X.-Y., Yu, J.-H., Yang, Q., Zhao, Y.-D., & Chen, H. (2010). Electrocatalytic oxidation of phytohormone salicylic acid at copper nanoparticles-modified gold electrode and its detection in oilseed rape infected with fungal pathogen *Sclerotinia sclerotiorum*. *Talanta*, *80*, 1277–1281.

Wang, S., Wang, F., Gao, S., & Wang, X. (2016). Heavy metal accumulation in different rice cultivars as influenced by foliar application of nano-silicon. *Water, Air, & Soil Pollution*, *227*, 1–13.

Wang, M., Sun, G., Li, G., Hu, G., Fu, L., Hu, S., Yang, J., Wang, Z., & Gu, W. (2024). Effects of multi walled carbon nanotubes and nano-SiO_2 on key enzymes for seed germination and endogenous hormone level in maize seedling. *Agronomy*, *14*, 2908.

Welch, R. M. (1999). Importance of seed mineral nutrient reserves in crop growth and development. In Z. Rengel (Ed.), *Mineral nutrition of crops: Fundamental mechanisms and implications* (pp. 205–226). New York, NY: CRC Press. https://doi.org/10.1201/9781003578468-8

Wu, P., Zhao, T., Wang, S., & Hou, X. (2014). Semicondutor quantum dots-based metal ion probes. *Nanoscale*, *6*, 43–64.

Xu, X., Liu, F., Jiang, L., Zhu, J., Haagenson, D., & Wiesenborn, D. P. (2013). Cellulose nanocrystals vs. cellulose nanofibrils: a comparative study on their microstructures and effects as polymer reinforcing agents. *ACS Applied Materials & Interfaces*, *5*, 2999–3009.

Yadav, K. C. & Sinha, K. (2022). Storage behavior of fresh fruits and vegetables: a review.

Yadav, R., Singh, S., & Singh, A. N. (2022). Cost-benefit analysis act as a tool for evaluation of agricultural waste to the economy: a synthesis. In P. Singh & P. Verma (Eds.), *Waste management and resource recycling in the developing world* (pp. 647–663). Amsterdam, The Netherlands: Elsevier. https://doi.org/10.1016/B978-0-323-90463-6.00009-9

Yadav, A., Yadav, K., & Abd-Elsalam, K. A. (2023a). Exploring the potential of nanofertilizers for a sustainable agriculture. *Plant Nano Biology*, *5*, 100044.

Yadav, A., Yadav, K., Ahmad, R., & Abd-Elsalam, K. A. (2023b). Emerging frontiers in nanotechnology for precision agriculture: advancements, hurdles and prospects. *Agrochemicals*, *2*, 220–256.

Yadav, N., Garg, V. K., Chhillar, A. K., & Rana, J. S. (2023c). Recent advances in nanotechnology for the improvement of conventional agricultural systems: a review. *Plant Nano Biology, 4*, 100032.

Yang, W., Zhang, H., Li, M., Wang, Z., Zhou, J., Wang, S., Lu, G., & Fu, F. (2014). Early diagnosis of blast fungus, *Magnaporthe oryzae*, in rice plant by using an ultra-sensitive electrically magnetic-controllable electrochemical biosensor. *Analytica Chimica Acta, 850*, 85–91.

Yang, X., Li, Q., Lu, Y., Zhang, L., & Bian, X. (2024). Insight into the short-term effects of TiO_2 nanoparticles on the cultivation of medicinal plants: comprehensive analysis of *Panax ginseng* physiological indicators, soil physicochemical properties and microbiome. *Science of the Total Environment, 951*, 175581.

Yin Huanshun, Y. H., Zhou Yunlei, Z. Y., Meng Xiaomeng, M. X., Tang Tiantian, T. T., Ai Shiyun, A. S., & Zhu Lusheng, Z. L. (2011). Electrochemical behaviour of Sudan I at Fe_3O_4 nanoparticles modified glassy carbon electrode and its determination in food samples. *Food Chemistry, 127*(3), 1348–1353.

Zain, M., Ma, H., Rahman, S. U., Nuruzzaman, M., Chaudhary, S., Azeem, I., Mehmood, F., Duan, A., & Sun, C. (2024). Nanotechnology in precision agriculture: advancing towards sustainable crop production. *Plant Physiology and Biochemistry, 206*, 108244.

Zakaria, L. (2021). Diversity of *Colletotrichum species* associated with anthracnose disease in tropical fruit crops—a review. *Agriculture, 11*, 297.

Zeinalipour, N., & Saadati, S. (2024). Physiological and biochemical response of strawberry cv. diamond to nano zeolite soil application and cinnamic acid foliar application. *Scientific Reports, 14*, 28908.

Zhang, Y., Kong, Q., Niu, B., Liu, R., Chen, H., Xiao, S., Wu, W., Zhang, W. & Gao, H. 2024. The dual function of calcium ion in fruit edible coating: regulating polymer internal crosslinking state and improving fruit postharvest quality. *Food Chemistry*, 447, 138952.

Zhao, J., Zeng, D., Wang, Q., Lin, Z., Vogel, F., Li, W., & Zhang, P. (2024). Effects of a dual functional filler, polyethersulfone-g-carboxymethyl chitosan@MWCNT, for enhanced antifouling and penetration performance of PES composite membranes. *Journal of Environmental Management, 365*, 121611.

Zhou, W., Li, M., & Achal, V. (2024). A comprehensive review on environmental and human health impacts of chemical pesticide usage. *Emerging Contaminants, 11*(1), 100410.

9

Role of Nanotechnology in Increasing the Shelf Life of Fruits

Sanam Ashraf[1], *Nirma Mubeen*[2], *Javed Iqbal*[3], *Banzeer Ahsan Abbasi*[4], *Muhammad Anas*[1,5], *and Shah Fahad*[6]

[1] *Department of Plant Sciences, Faculty of Biological Sciences, Quaid-i-Azam University, Islamabad, Pakistan*
[2] *Department of Botany, Faculty of Plant Science, University of Narowal, Pakistan*
[3] *Department of Botany, Bacha Khan University, Charsadda, Khyber Pakhtunkhwa, Pakistan*
[4] *Department of Botany, Rawalpindi Women University, Rawalpindi, Pakistan*
[5] *Department of Soil Science & Plant Nutrition, University of Selcuk. Campus, Konya, Turkiye*
[6] *Department of Agronomy, Abdul Wali Khan University Mardan, Khyber Pakh-tunkhwa, Pakistan*

9.1 Introduction to Nanotechnology in Food Preservation

Many experts consider nanotechnology to be a novel, cutting-edge, and sustainable approach for this century. The application of nanotechnology might overcome the constraints of classical methods of storing and preserving fruits and vegetables. A nanomaterial is said to be one that has at least one dimension at the nanoscale level and is created using a "bottom-up" approach (formation of a complex structure through the rearrangement of molecules one at a time) or a "top-down" approach (cutting down of a bulk material into smaller pieces using a physical or chemical process). The use of nanotechnology in agriculture and food processing is increasing rapidly. Higher barrier property, broad surface-to-volume ratio, and higher antimicrobial efficacy are some of the features of nanomaterials, which overcome the setbacks of traditional methods of shelf-life extension (Sagar et al., 2022). Nanotechnology is also targeted toward the overhauling of the agricultural sector productivity, thus enabling better security in terms of food abundance in the future. It is well known that in horticulture, the combination of retaining quality and controlling the deterioration of harvested fruits and vegetables is a serious problem. Different types of nanomaterials have been shown to have increased productivity, improved shelf-life, reduced postharvest damage, and enhanced quality of horticultural crops (Curtis et al., 2006).

Nanotechnology is a novel method for creating new fit for edible coating components for the postharvest management of fresh produce by reducing the size of the edible coating particles to a nanometric scale ranging from 1 to 100 nm (Gul et al., 2025; Ullah et al., 2025). The effectiveness of nanotechnology has tested to be one of the high-quality strategies for extending the

Nanotechnology Innovations for Food Security and Sustainable Agriculture, First Edition. Edited by Zia Ur Rehman Mashwani, Shah Fahad, Ilyas Ahmad, Shah Saud, Sajid Fiaz, and Taufiq Nawaz.

shelf life of fruits, and nanoparticles used for edible coatings additionally possess particular properties, which include antibacterial and antifungal qualities (Odetayo et al., 2022). Recently, there were studies on the effects of the addition of nanoparticles to edible coatings for human consumption and these studies have shown that the addition of nanoparticles is safe. Edible coatings have opened new possibilities no longer only for boosting better properties of antimicrobial, antifungal, and antiviral, but also for improving the cost-effectiveness of edible coatings (Sorrentino et al., 2007, Ullah et al., 2024, Ahmad et al., 2022, Haris et al., 2023). Furthermore, the use of nanomaterials improves food processing, packaging, food protection, plant nutrition, the effectiveness of pesticides and fertilizers, the decrease of environmental pollutants, and the production of nutraceuticals (Rana et al., 2021). The role of nanotechnology in productivity and quality of shelf life of fruits is illustrated in Figure 9.1.

Consumers seek premium fruits and vegetables containing nutritious compounds, and they choose fresh produce since they offer maximum nutritional value. The attainment of health benefits requires that these foods maintain their favorable physicochemical and sensory qualities, together with food safety, and should remain free from contaminants and pathogenic microorganisms (Meng et al., 2014). The expanding market has led to a necessity for the food and horticulture sectors to create suitable preservation methods. Scientific experts along with food processing industries recognize the need to develop multiple methods for improving food preservation quality because natural edible and biodegradable polymers provide promising solutions (Mahalik & Nambiar, 2010).

The human need for fruit keeps rising because fruits address essential dietary requirements. Fresh produce consumption rates have grown since people started paying attention to their health and wellness. Natural components of fruits such as vitamins and minerals alongside dietary fibers maintain excellent health for individuals. Products that support microbial multiplication consistently make them vulnerable to degradation (Heaton & Jones, 2008). They show both nutritional value to consumers as well as serving as dietary material for microbes which attack fruits. Ripe fruits in compost quickly spoil because their nutrient-rich cellular structures are highly perishable. The nutritional elements in fruits including carbohydrates along with minerals and vitamins and proteins create a suitable environment that allows parasitic and saprophytic forms of fungi to thrive successfully (Thiyam & Sharma, 2013). After product consumption, traditional packaging made from petroleum-based plastic materials,

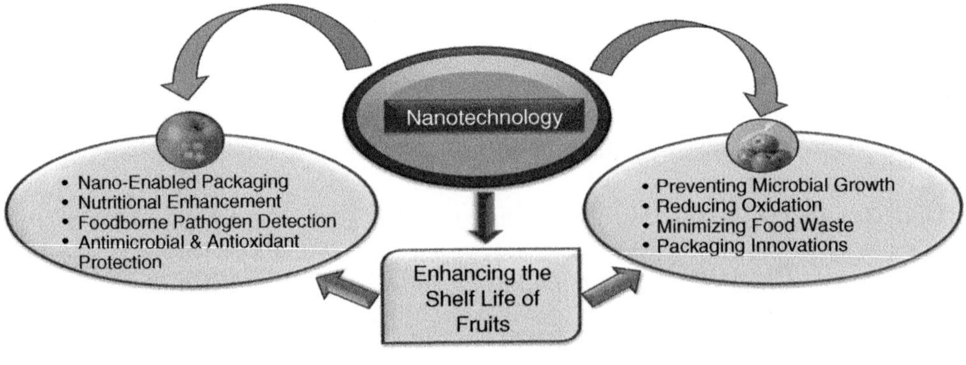

Figure 9.1 Role of nanotechnology in productivity and quality of shelf life of fruits.

including polypropylene, polyethylene, and polystyrene, becomes a substantial environmental concern because of their persistence in the environment (Ncube et al., 2020).

Research focuses heavily on biodegradable packaging production because non-biodegradable materials generate major environmental issues that require scientists to investigate natural polymers from renewable sources for food packaging use. Expert researchers have examined edible coatings as well as films since they represent sustainable food packaging options. Edible packaging serves as an upcoming food protection system, which extends shelf life by providing microbial resistance, moisture control, and gas product barriers (Kondle et al., 2022). The worldwide production of fruits and vegetables during a year reaches 3.7 billion tons (Barrett and Lloyd 2012). Markets for edible food packages reached $697 million in 2016 before analysts predicted it would grow to $1097 million during 2017 to 2023 with international market expansion rates of 6.81% (Kondle et al., 2022). The worldwide fruit and vegetable deterioration rate reaches between 20% and 40% due to insufficient shelf-life extension techniques, thus making the preservation of perishable foods a critical matter in food science (El-Ramady et al. 2015).

Nanotechnology stands as a leading-edge technology of the twenty-first century according to modern classification. The use of nanotechnology in extending product shelf life for fruits and vegetables aims to address the weaknesses found in conventional preservation approaches (Ma et al., 2017). Three main types exist for engineered nanomaterials used in fruit and vegetable shelf-life extension: these materials include inorganic nanomaterials together with organic nanomaterials and mixed nanomaterials. Silver oxide, titanium oxide, zinc oxide, and silver nanoparticles serve the primary purpose of food packaging by acting as detectors and disease preventers while enhancing shelf life (Khot et al., 2012).

Research has revealed that edible coatings (ECs) succeed at extending fresh food shelf life while supporting eco-friendly practices during the last few years. Food surface protection occurs through edible coatings, which are thin food-layer applications that possess filmogenic properties (Kocira et al., 2021). Among the intriguing features of coating development are compounds with antibacterial properties incorporated into the polymeric matrix. Anti-fungal agents in edible coatings keep fungal growth at bay because fungal decay leads to the major discarding of perishable foods (Iñiguez-Moreno et al., 2021). Nanotechnology enhances food safety alongside quality because it delivers antimicrobial and antioxidant and antifungal properties to maintain fruit nutritional elements while preserving texture and flavor. The chapter shows how these technological methods provide nature-friendly and eco-efficient preservation alternatives to conventional methods including chemical preservatives alongside refrigeration, thus enabling better sustainable outcomes for the fruit industry. This study examines both advancing nanosensor applications for continuous monitoring and how nanotechnology can enhance food supply sustainability by improving fruit preservation methods.

9.2 Mechanisms of Nanotechnology for Shelf-Life Extension

The field of nanotechnology enables effective and innovative advances for food preservation especially related to food packaging methods. Extended packaging efficiencies are possible through nanotechnology which introduces nanomaterials to systems to produce

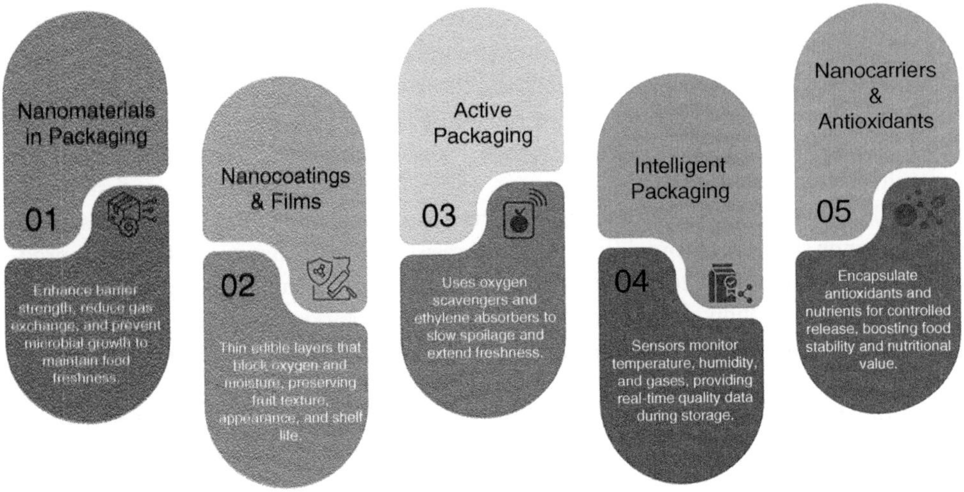

Figure 9.2 Mechanisms of nanotechnology in extending fruit shelf life.

barriers that decrease mold formation while providing antimicrobial protection and real-time tracking. These functional capabilities both expand food durability and support quality retention and minimize food waste amounts. Nanotechnology enables three main food packaging applications through nanocoatings while also including films and active intelligent packaging technologies as shown in Figure 9.2.

9.3 Nanomaterials for Food Packaging

Food packaging manufacturers integrate nanomaterials because these materials exhibit exceptional characteristics, which include large surface areas and enhanced strength and controlled gas, water, and light transport properties. The materials improve packaging functionality, since they protect food products better and preserve them more effectively (El Sheikha et al., 2018). The implementation of strong nanomaterials within packaging barriers maintains the shipping stability of food products while enhancing storage quality for fruits and vegetables and protecting meat products against microbial infection. Due to their superior barrier function as well as mechanical strength and heat-resistant traits they offer an innovative substitute for additives that enhance polymeric properties in food packaging materials (Singh et al., 2017). The employing of nanomaterials in food packaging can reduce the growth of various microorganisms, such as *Lactobacillus plantarum* (*Escherichia coli*, *Staphylococcus aureus*, *Salmonella*, *Alicyclobacillus acidoterrestris*, and aerobic psychrotrophic, molds and yeasts, besides the further improvements in the shelf life of food. The application of antimicrobial nanocomposite, such as Ag and ZnO, increased the shelf life of fresh orange juice or chitosan-based composite nanolayers improved the shelf life of meat (Emamhadi et al., 2020). Nano-food packaging represents a prime chance to adopt lightweight, high-performance nanomaterials as non-degradable plastic packaging materials' replacements (Bahrami et al., 2020).

9.3.1 Nanocoatings and Films

Nanotechnology methods are applied throughout different fields for nanocoating material production, which also supports edible food coatings in the industry (Yemmireddy et al., 2015). The application of nanomaterial thin layers on food packaging surfaces through nanocoatings and nanofilms functions to enhance packaging performance (Jeevahan & Chandrasekaran 2019). The functions of these coatings, include preventing transmission of oxygen, moisture, and light, since these elements play a major role in food spoilage (Rovera et al., 2020). Nanocoatings consist of different materials with content based on silver nanoparticles alongside clay nanoparticles, chitosan nanoparticles, and cellulose nanoparticles (Cano et al., 2017). Silver nanoparticles show antimicrobial behavior that stops fungi and bacteria growth over the packaging surface, which decreases contamination risk (Xu et al., 2017). The mechanical strength and flexibility of packaging increases through nanofilm applications, thus making the packaging more durable and less prone to physical damage. The coatings serve to protect food texture and appearance because they minimize moisture and nutrient loss (Zhao et al., 2025). Consolidated research indicates that food loss prevention requires high-quality packaging materials to stop moisture and oxygen from passing through products to their surrounding environment. Oxygen and water changes play a significant role in causing food to deteriorate. Companies take two main steps to lengthen product shelf stability through improved packaging methods by creating air-tight packaging and selecting more effective barrier materials. Packaging methods use desiccants and oxygen absorbers or scavengers, together with newer techniques that monitor gas transmission, to maintain the shelf life of food items, particularly food materials with a risk of rancidification (Pereda et al., 2017; Wang et al., 2018). Current research demonstrates that nanocomposite films together with edible coatings significantly improve fresh fruit and vegetable preservation similarly to modified atmosphere methods that have proven successful in quality preservation of both fruits and vegetables. The use of clay-based packaging stands as a sustainable nanotechnology application for food packaging, since polymers get reinforced by the addition of organic and synthetic nano clay materials. Nanoclays form a fundamental component in packaging films because they enhance the barrier properties to moisture and other elements (Miteluţ et al., 2021). Table 9.1 presents different nanomaterials that serve as moisture regulators for fruit packaging.

9.3.2 Active and Intelligent Packaging

Latest developments in active and intelligent packaging systems occur because of modern society's growing challenges. Food packaging technologies transform continuously because modern society faces escalating problems. Fast-moving consumer goods packaging faces present-day as well as future challenges because of legislation together with global markets along with requirements for longer shelf stability and convenience and safety and health benefits for food and environmental sustainability and authenticity and minimizing food waste (Kerry, 2014). Multiple entities across the world consider food safety to be a primary concern, while the current food legislation prioritizes this matter as its key objective. The microbial dangers present in food products continue to be among the leading causes of food-related illnesses during the present day. Active

Table 9.1 Different types of nanomaterials, which function as moisture regulators for fruit packaging.

Nanomaterials	Function in Packaging
Nanoclays	The materials boost their protective functions by decreasing water permeation without restricting flow movement.
Silica nanoparticles	These materials should absorb moisture before slowly releasing it, which stabilizes the packaging environment's humidity levels.
Silver nanoparticles	The protection of food quality and safety depends on films that stop microbial growth due to excessive moisture.
Starch nanocrystals	The films function as barriers, which both soak up moisture from the air and hold it inside to avoid fruit dehydration.
Cellulose nanofibers (CNFs)	The material functions as a strong and waterproof compound, which creates a solid protection against vapor.
Polymer nanocomposites	The films gain improved mechanical strength as well as increased barrier resistance against moisture penetration.
Zinc oxide nanoparticles	The protection from spoiling due to excess moisture comes with built-in antimicrobial qualities.
Titanium dioxide nanoparticles	The structure forms an effective water protection layer, which safeguards produce from both mold growth and UV radiation to maintain fresh state.
Graphene oxide	This active agent strengthens the films and functions to boost their moisture blocking capabilities while giving them better flexibility.
Activated carbon nanoparticles	The material acts as a moisture regulator through its ability to absorb waste fluids in addition to ethylene emissions that preserve inner environmental conditions.

packaging demonstrates an innovation by advancing beyond typical packaging functions since it links food items with their containers and surrounding environment to enhance product safety and latent attributes without compromising food quality (Ahvenainen, 2003). Packaging technology delivers environmental and food condition data to both food processors as well as retailers and end consumers (Kerry et al. 2006). Active packaging provides reactive components that extract or accept substances including gases and moisture in order to improve storage time. Nanotechnology in film production enables the incorporation of oxygen scavengers and ethylene absorbents which extends food shelf life by reducing spoilage through oxidation. Packaging, which incorporates nanoparticles, serves to release antioxidants and antimicrobials that protect food from spoilage-causing microorganisms (Realini & Marcos, 2014; Taghipour & Assar, 2025). Intelligent packaging implements sensors both inside and outside the package, which continuously tracks food condition during storage and transportation period. Sensors embedded in packaging systems monitor temperature variations together with changes in humidity and gas composition to send immediate quality assessments about the food products. Nanotechnology-based sensors monitor fruit ripening status and spoilage conditions since they provide vital information for consumer and producer food management. Nanotechnology merges properties of nanocoatings with films and active/intelligent packaging to produce superior packaging methods, which tailor specific preservation solutions to different fruits (Sohail et al., 2018; Khedr & Khayri, 2024).

9.3.3 Silver Nanoparticles and Antimicrobial Action

Research activities focusing on inorganic nanoparticles have gained significant attention throughout the past few decades because of their potential usages in biological applications and pharmaceutical solutions along with food packaging (De Azeredo, 2009). Research demonstrates that silver ions as well as silver-based compounds, display first-rate antibacterial properties by impeding bacteria growth and eliminating numerous bacterial species, including *Escherichia coli* (Yang et al., 2009). The study of silver nanoparticles (AgNPs) as nanomaterials ranks among the highest because of their strong antimicrobial capabilities. Food preservation benefits greatly from silver nanoparticles because they show antimicrobial activity against bacteria, fungi, and viruses. Several mechanisms activate the antimicrobial properties of silver nanoparticles. Through their mechanism of action, AgNPs break down microbial cell membranes, then interrupt cellular processes while setting free toxic ion particles that hurt microbes (Costa et al., 2011). The incorporation of Ag NPs in polymeric materials as active ingredients enables controlled ion release through which the negative impact of $Ag+$ ions on organisms can be minimized. Food receives silver through two different mechanisms, which include the detachment of Ag NPs from composites or the oxidative release of silver ions. The most silver release from food simulants occurred through acidic food medium based on research by Echegoyen & Nerín (2013). Microwave heating and food heating processes trigger $Ag/Ag+$ migration from packaging films, which results in toxic substances found in food stuff according to Yu et al. (2019). Silver nanoparticles are used in food packaging through insertion into films, coatings, and packaging materials to provide persistent antimicrobial effects (Hoseinnejad et al., 2018). Utilizing nanoparticles in food packaging protects the surfaces of fruits from harmful microorganisms, which could trigger spoilage. The use of silver nanoparticles improves fruit storage duration by fighting against spoilage while eliminating the requirement for toxic additives in food preservation. Silver nanoparticles receive preference due to their non-toxicity in regulated quantities as well as their permission to use in food applications pursuant to established regulatory requirements (Omerovićet al., 2021).

9.3.4 Nanocarriers for Antioxidants and Nutrient Retention

Nanocarriers are new delivery systems used to encapsulate bioactive compounds, i.e., antioxidants, vitamins, and nutrients, which can protect and release them in a controllable manner. Active molecule encapsulation techniques are useful tools for preserving and optimizing bioactive compound efficacy, ensuring high quality of food, medicines, and other materials that include them, both for human and veterinary use (Pegg & Shahidi, 2007). Therefore, in commercially available products, encapsulation allows for the improvement of foods and beverages with high active molecules that are used only commercially due to their sensory properties, poor stability, bioaccessibility, and bioavailability (Ahmed et al., 2022). In essence, encapsulation is the covering of active compounds via a matrix. The compounds are further protected under this shield, and this layer helps make the compounds more stable. Moreover, encapsulation is effective to enhance pharmacokinetics of bioactive compounds, particularly bioaccessibility and bioavailability. It may be used as a food component in functional food and nutraceutical development to create impact on health. The encapsulated bioactive compounds can be used (Lengyel et al., 2019). Encapsulation is a technology that consists in enclosing solid, liquid, or gaseous compounds into small sealed receptacles called capsules. Structures in the nano- ($<1.0\,\mu m$) or the micrometric (1.0–$1000\,\mu m$) size can be produced using different materials for the formation of the capsule's membrane, which among the vast range of materials currently used are the biobased materials, i.e., polysaccharides, gums, proteins, lipids, or blends of these (Guía-García et al., 2022).

9.3.5 Types of Nanomaterials Used in Fruit Preservation

Nanomaterials exist in four main classes, including carbon-based, inorganic, organic, and composite materials. Food processing applications might benefit from nanomaterials based on composites because of their excellent stability properties and gas barrier characteristics as well as their size, shape, and matching capabilities. The utilization of nanomaterials for food packaging includes three basic purposes related to intelligent, enhanced, and active food packaging (Makwana et al., 2020). Multiple food preservation applications use current nanomaterials, which include SiO_2, Ag, nano-ZnO, CNT, CuO, and TiO_2 NPs. Nanomaterials of these types were applied to construct barrier systems, which combined thermal stability with light-strength and gas blocking properties for food preservation purposes. Each functional nanomaterial packaging application, including active, preservation, and nano-sensing depends on the unique chemical structures and functional properties exhibited by particular nanomaterials (Manikandan & Min, 2023). Scientific research has recently focused on nanoparticles incorporating clay and TiO_2 and Ag and Zn in addition to other materials because these substances demonstrate better antimicrobial features and physicochemical properties than traditional polymers (De Azeredo, 2009). The growth of clay nanoparticle-based nanocomposite films has experienced explosive expansion in the market. Nanocomposite thermomechanical properties are optimized with ideal permeability according to macro-level investigations (Zanetti et al., 2001). The effects of nanotechnology on food packaging containers have been thoroughly researched by scientists. Food manufacturers enhance food packaging quality and develop new delivery systems and nanoencapsulation methods while extending food freshness due to nanotechnology applications (Ahari & Naeimabadi, 2021). The packaging industry, which involves a multibillion-dollar market, spends efforts to resolve spoilage issues that affect products before their stated storage duration expires. Through the implementation of nanoparticles, the innovative food packaging technologies introduced by nanotechnology enhance food quality (de Sousa et al., 2023).

Food safety techniques upgraded by functional nanoparticles decrease spoilage risks while they preserve moisture stability, mechanical strength, and durability to maintain product safety and quality (PrimoŽič et al., 2021). Secure packaging systems require development, which can be efficiently achieved through nanoparticles. Among all promising nanoparticles, silver nanoparticles (AgNPs) stand out as top choices because of their advanced functional properties for the food industry. The antimicrobial strength of silver nanoparticles made from different biological materials, including plants and microorganisms, has led to their broad adoption against multiple pathogens. Scientists find biological silver nanoparticle production using microorganisms better than chemical nanoparticle synthesis because biological methods produce environmentally friendly nanoparticles that are minimally toxic and affordable. Active packaging materials with such applications show great potential to extend the shelf life of food products. The elimination of harmful microbes through toxic spoilage enables more secure or protective packaging results than typical preservation methods (Beyene et al., 2017) (Gudikandula & Charya Maringanti, 2016). One of the nutrients that human bodies require is zinc. In human bodies, it serves a variety of physiological purposes. Zinc has crucial structural, functional, and regulatory roles through its interactions with several proteins and enzymes. Because zinc shortage will result in numerous physiological problems, it is crucial to consume adequate amount of zinc. Natural sources of zinc include animal-based foods, including beef, chicken, and pork. Since zinc is typically not well absorbed by the human body, plant-based diets are not a good supply of it (Cakmak and Kutman, 2018; Mwangi et al., 2018). The most prevalent carbohydrate polymer on the planet, cellulose, has notable structures and characteristics. Particularly for use in meals and in environmentally friendly and biocompatible products, this renewable natural biopolymer and the polymers derived from it have garnered a lot of interest (Stern & Jedrzejas, 2008). A linear homopolysaccharide made up of glucose monomers joined by β-1-4-glycosidic bonds, cellulose has a unique molecular structure that gives it special qualities like hydrophilicity, chirality, degradability, and wide chemical variability brought on by the high donor reactivity of the OH groups. Furthermore, cellulose's techno-functional qualities could be further promoted thanks to its isolation and modification, particularly with the use of sophisticated nanotechnology instruments (Chaudhary et al., 2022). The nanoscale physicochemical properties of cellulose, such as its high specific surface area and aspect ratio, high crystallinity, purity, superior mechanical qualities, and low density and thermal expansion, create new opportunities for the use of nanocellulose in a variety of industries, including energy, biological medicine, and the environment (Tayeb et al., 2018). Various types of nanomaterials with examples and applications utilized in fruit preservation are shown in Table 9.2.

9.3.6 Application of Nanotechnology in Fruit Storage

Nanotechnology has emerged as an effective tool over the last two decades to advance postharvest systems, particularly for fruit storage longevity improvements. Postharvest losses occur massively because fruits tend to perish easily due to their unstable nature. Recent nanotechnology applications offer fresh and cutting-edge solutions to reduce spoiled food quantities while maintaining high fruit quality throughout storage and shipment stages.

9.3.7 Nano-Edible Coatings for Fruits

The effective use of nanotechnology to reduce fruit and vegetable losses depends heavily on sustaining food quality attributes as well as ruling out any unfavorable modifications of functional properties and nutritional value. Edible coatings represent an efficient technology to safeguard product quality since these thin edible surface layers on fruits and vegetables produce extended shelf life (de Oliveira Filho et al., 2022). The primary edible coating operating principle consists of carbon dioxide retention by edible films, which leads to decreased oxygen consumption during respiration (Figure 9.3). The respiration process requires lesser amounts

Table 9.2 Types of nanomaterials with examples and applications utilized in fruit preservation.

Types of nanomaterials	Examples	Applications in fruit preservation
Metal and metal oxide nanoparticles	Silver (AgNPs), zinc oxide (ZnO NPs), titanium dioxide (TiO_2 NPs), copper oxide (CuO NPs), gold (AuNPs)	The substances exhibit antimicrobial effects while blocking UV rays, reducing oxidation levels, stopping microbial growth, and decomposing ethylene.
Biopolymeric nanoparticles	Chitosan, alginate, starch-based NPs	Edible coatings, antimicrobial activity, moisture control, prevention of microbial contamination.
Carbon-based nanomaterials	Carbon nanotubes (CNTs), graphene oxide NPs	Smart packaging solutions and sensors that track gases combine with fruit ripening control functions as well as antimicrobial qualities.
Lipid-based nanomaterials	Nanoemulsions, liposomes	Controlled release of bioactive compounds, antimicrobial protection, moisture control.
Silicon-based nanomaterials	Silica nanoparticles	The application of protective coatings produces two main advantages: it preserves freshness while extending storage time.
Nanoclays	Montmorillonite, bentonite	Packaging materials require enhanced gas barrier properties since this helps decrease oxygen and moisture permeability.
Smart and active nanomaterials	Nanosensors, nanoencapsulation	The technique allows both freshness monitoring in real time and delivers antimicrobial agents and antioxidants through controlled systems.

of O_2 due to which it initiates various metabolic processes including reduced ethylene formation and decreased moisture loss as well as improved product firmness, which results in extended shelf life (McHugh, & Senesi, 2000).

The action mechanism requires edible coatings with particular properties including water and mechanical resistance, as well as the ability to provide proper product coverage, maintain internal oxygen levels, and enhance appearance without melting at high temperatures, but allowing light passage while ensuring economic viability (de Oliveira Filho et al., 2022). About any type of edible coating can be applied through different methods on fruits and vegetables' surfaces, and traditional application techniques involve dipping, spraying, and manual brushing. Food and pharmaceutical industries have applied fluidized bed processing and panning as newer methods for edible coating technology during the past few decades. Any edible coating composition benefits from various surface application methods for fruits and vegetables, including dipping, spraying, and brushing. Advanced coating application methods such as fluidized bed processing and panning now serve the industrial food sector because they have been adopted from their previous usage in pharmaceuticals and foods. Surface coating distribution for nanostructured coatings becomes possible through micro-sprays and electro-sprays, which subject the deposition solution to high-electric potential capillary nozzle processing to generate droplets from tens of nanometers to hundreds of micrometers. A method known as

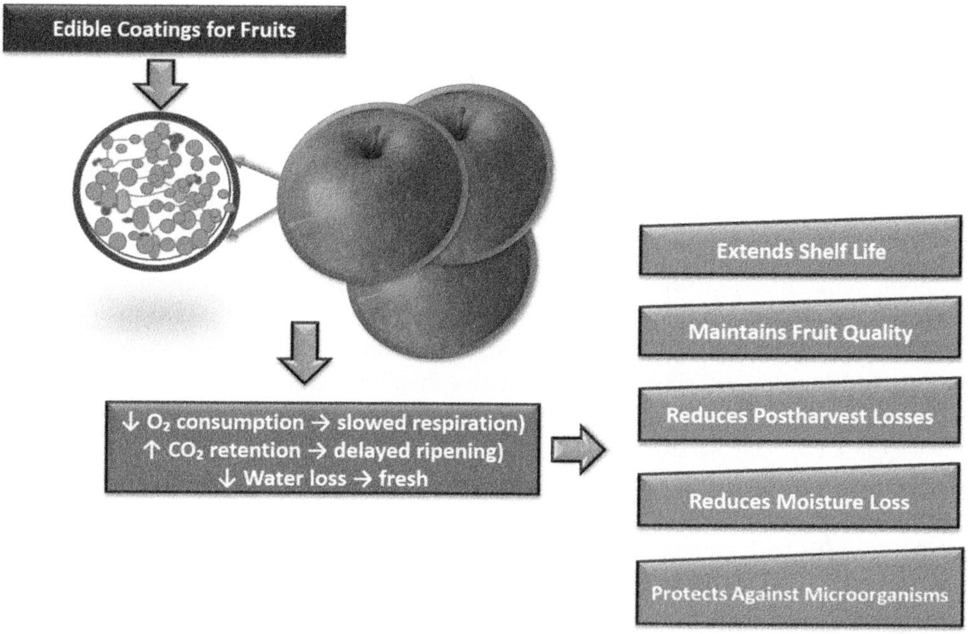

Figure 9.3 The primary way that edible coatings work.

atomic layer deposition (ALD) allows users to build ultra-thin films through sequential processes, which helps users control the formed film thickness (Dewettinck & Huyghebaert, 1999; Makwana et al., 2014).

9.3.8 Activing Packaging Technology Preserve the Quality of Fruits

The packaging system works as an effective shield to protect food against contaminants of biological nature as well as physical and chemical agents, while keeping sensory features unchanged. The technology serves basic functions that preserve food safety standards, while it extends preservation times and minimizes food waste (Khaneghah et al., 2018). Active food packaging technologies improve product quality and safety along with shelf-life extension by embedding antimicrobial agents into polymeric structures. The main molecular bond between antimicrobial agents and polymeric chains establishes van der Waals and electrostatic attractions that enable additive-free transfer through the polymeric medium until it reaches package surfaces to touch microbial cell membranes (Galli et al., 2020).

9.3.9 Nanosensors for Monitoring Fruits and Vegetables Quality

Real-time quality tracking of fruits and vegetables depends on nanosensors, which operate as an up-and-coming feature in intelligent packaging systems. Nanosensors provide opportunities for the current market by implementing them within packaging to measure food quality throughout delivery stages, thus maintaining product quality for end users.

Nanosensors act as sensing devices that examine pathogenic bacteria alongside food-contaminating toxins and adulterants, within vitamin, dye, fertilizer, pesticide, flavor, and odor parameters (Kumar et al., 2017). The sensors identify and detect the signs of ethylene gas together with volatile organic compounds (VOCs) and moisture and temperature changes, which indicate the state of product firmness and ripeness. Through their integration smart packaging and RFID (Radio Frequency Identification) systems provide retail customers and consumers with specific visual and digital updates about the freshness state and aging condition of their products. Greater supply chain management with this technology enables better supply chain control, which contributes to reduced postharvest waste and lengthened shelf life and allows delivery of produce in optimized ripeness levels to buyers. The application of nanosensors enhances food packaging technology through real-time monitoring along with active labeling while improving both product safety and quality, thus making them an innovative solution for food preservation (Kuswandi 2020). During storage, along with manufacturing processes, the antioxidant components together with vitamins in food items display significant sensitivity along with degradation susceptibility. Food item vitamin detection relies on effective nanoparticle methods, which both guarantee nutritive value quality and safeguard food safety standards. Scientific studies have created MWCNT and SWCNT–ionic liquid nanocomposites that detect folic acid in wheat flour, fruit juices, and milk to prevent vitamin deficiencies that cause anemia, carcinogenesis, and heart disease (Kumar et al., 2017). The sensing ability of modified MWCNTs functionalized with *N*-(3,4-dihydroxyphenethyl)-3,5-dinitrobenzamide stretches to both ascorbic acid detection as well as the fundamental amino acid tryptophan. Separately, nickel oxide nanoparticles serve for detecting ascorbic acid (vitamin C). The nanosensors act as monitoring tools, which simultaneously protect food nutrients while maintaining product quality throughout transport (Ensafi et al., 2012; Karimi-Maleh et al., 2014).

9.3.10 Case Study: Nanotechnology in the Preservation of Fruits

Improving survival and growth happens when humans consume food either as raw or processed materials. Food wastage has turned into a global crisis that impacts worldwide nations during the present time. Food waste develops throughout all stages of both food making and distribution and consumption. A worldwide shortage of food continuously affects populations because supply networks fail to deliver adequate supplies simultaneously with environmental changes alongside increasing human numbers (Leisner 2020). According to the study mentioned (Gomez Zavaglia et al. 2020) food reduction from cooking processes together with supply chain limitations, strict consumer requirements and manufacturing defects and climate changes and government constraints and soil erosion represent the main sources of food waste. The growing number of people worldwide leads customers to search for fresh nutritious food that is also healthy. Daily production of food in the world exceeds demand yet not all people who need benefits from this supply can access them. Current scientific research focuses on Nanotechnology since it develops applications for food processing alongside agricultural and packaging systems (Sridhar et al., 2021).

9.3.11 Nanotechnology in the Preservation of Citrus Fruits

Non-climacteric citrus fruits comprise a major worldwide fruit production (more than 100 countries) along with China, Brazil, and America as significant growing areas (Ladaniya, 2010). The area devoted to citrus cultivation reaches up to 11.42 million hectares, which yields 179 million tons of citrus fruits. The Food and Agriculture Organization reported 73.3 million tons of oranges, 33.4 million tons of mandarins, and 17.2 million tons of lemons and grapefruits amounted to 9.5 million tons and others totaled 13.5 million tons for the year 2017. The world produces its largest quantity of citrus fruits in China, amounting to 82.7 million tons per year, while Brazil ranks second with 18.14 million tons and India follows with 1.053 million tons (Kumar et al., 2024; Bora et al., 2020). The main purpose behind growing citrus fruits is to provide both fresh produce for human consumption and raw materials for juice processing operations. The global market uses orange juice for 76% of its supply as reported by Neves et al. (2020), which creates both economic strength and employment opportunities. The consumer population values citrus fruits for their taste and aroma as well as health-promoting effects and enjoyable flavor. The nutritional value of this food item includes vital components such as nutrients in combination with Vitamin C, phenolic compounds, anthocyanins, and other bioactive compounds that lead to better antioxidant function, antimicrobial efficacy, cardiovascular disease prevention, and cancer risk reduction (Patil et al., 2017). The antioxidant along with anti-inflammatory and anticancer effects and neuroprotective and cardiovascular protection exhibited in citrus fruits arise from their bioactive compounds, which include phenolic acids, flavonoids, ascorbic acid, ferulic acid, naringin, hydrocinnamic acid, carotenoids, essential oils, and others. Growing species within the citrus genus includes oranges as well as desert lime, ginger lime, hyuganatsu, kabosu, kawachi bankan, koji orange, mangshanyegan, round lime, satsuma, sudachi, and multiple hybrid types (Richa et al., 2023; Zhong and Nicolosi, 2020). Nanotechnology stands as a cost-effective method to increase fresh food longevity in the present market. The increased surface area of small particles against bulkier forms creates opportunities for finding more durable biologically active nanoformulations, which increase fresh food shelf life. Nanotechnology stands as an advanced solution for creating nanoformulations, which lengthen the storage time of fruits and vegetables while providing active components like colorants, flavoring agents, antimicrobial agents, preservatives, and nutraceuticals (Liu et al., 2020). Nanomaterials, together with nanoformulations, display a superior capability to sustain fruit and vegetable shelf life because of their nanoscale dimensions, which impact their mechanical composition, thermal properties, barrier properties, and antimicrobial functionality. Synthesis technologies in nanotechnology demonstrate a good interplay between biopolymers and active agents that influences how packaging materials function while preventing postharvest quality degradation in fruits and vegetables. The shelf-life extension of various foods takes advantage of silver nitrate (Ag), gold (Au), zinc (Zn), and copper (Cu), utilized in nanoformulations (Paidari et al., 2024). Recent research indicates that nanocoatings derived from biopolymers like chitosan, lipid-based nanoparticles, and cellulose nanocrystals can effectively act as a barrier to respiration and moisture loss, while eliminating dried pulp and preserving the fruit's form and freshness (Poonia & Mishra, 2022). Silver and zinc oxide nanoparticles have received considerable attention in regard to their antimicrobial activities, which retard decay-causing fungi and bacteria, that destroy citrus fruits (Sharma, 2020). Also, nanoencapsulation with natural antioxidants and essential oils like thymol and limonene offers a way to slow down oxidative and microbial activity, thus enhancing and maintaining the nutritional and sensory attributes of the citrus fruits (Oprea et al., 2022). Novel thin-film edible nanoparticles, consisting of bioactive materials, also support the functionality of intelligent nanotechnology-based packages with nanodevice sensors. These devices detect freshness indicators such as ethylene gas or microbial activity and can signal when alarms to avoid spoiling need to be activated (Chelliah et al., 2022). Furthermore, the resistant coatings for the surfaces of the fruits created by nanochemistry provide enhanced resistance to environmental stress, which allows to increase the postharvest life of the citrus fruits by up to 30–50% using these methods rather than standard techniques (Ishaque et al., 2019).

9.3.12 Benefits of Nanotechnology for Fruits Shelf Life

Nanotechnology provides strong prospects for enhancing fruit shelf life through its ability to solve spoilage and microbiological contamination problems as well as postharvest quality deterioration during storage and handling processes. The perishable nature of fruits caused by high water content and metabolisms leads to rapid decomposition before numerous points along distribution chains result in significant waste (Sridhar et al., 2021; Onwude et al., 2020). Current fruit preservation methods, employing cold storage and waxing combined with chemical treatments, can extend shelf life slightly yet generate suspected health risks for people consuming preserved food (Ogwu & Ogunsola, 2024). Fruits benefit from nanotechnology, which creates nano-coatings, nano-packaging materials, nanoencapsulation systems, and nanosensors to enhance preservation as well as improve quality and safety (Singh et al., 2023). Scientists have discovered nanocomposite edible coatings with film functionality as a leading nanotechnology breakthrough that creates barriers, which protect against gas transmission and moisture loss while blocking microorganisms to maintain fresh fruit quality (Basumatary et al., 2022).

Research indicates that antimicrobial properties in chitosan-based nanocomposite coatings become more significant through the addition of AgNPs, ZnO NPs, and TiO_2 NPs. The research demonstrated that chitosan-ZnO nanocomposite-based papaya coatings function both as antifungal protection and texture-maintaining barriers, resulting in twice the normal seven-day shelf-life extension to fourteen days. Studies have confirmed that banana storage life increases when the product receives silver nanoparticle (AgNP)-incorporated coatings that inhibit microbial proliferation and delay natural aging processes. Nano-coatings establish protective permeable membranes that let fruits breathe as they decrease ethane formation and decelerate food aging (Motelica et al., 2020; Chitena, 2022; Nayab & Akhtar, 2023). Nanotechnology has improved packaging materials that provide strong protection to fruits. Active packaging systems that contain antimicrobial nanoparticles such as silver and copper nanoparticles along with clay nanoparticles operate as physical barriers to protect against pathogens and also work actively to prevent microbial growth on fruit surfaces (Omerović et al., 2021). The use of montmorillonite among other clay nanoparticles serves as packaging film additives that enhance material strength and reduce oxygen penetration to guard against apple and mango fruit oxidation. The nano-packaging materials demonstrate strong thermal properties suitable for transportation and extended storage safety of products (El Mouzahim et al., 2023; Poongavanam et al., 2023). The antimicrobial functionality and antifungal properties of essential oil extracts come with unstable volatile elements. Bioactive components maintained within nanocontainers experience both environmental protection and controlled substance distribution allowing them to sustain fruit protection against microbial threats (Pateiro et al., 2021; Reyes-Jurado et al., 2015). The application of thyme and cinnamon essential oils encapsulated in nanocontainers by scientists successfully prevents fungal growth in citrus fruits alongside berries, which reduces spoilage rates. The nanoencapsulation of ascorbic acid or vitamin C serves as an enzymatic browning prevention system for fresh-cut apples that preserves nutritional content and taste quality (Das et al., 2021, Vhangani, 2022). Nanosensors developed through nanotechnology serve two functions: they track how fruits age in real time and they identify ripening progress. The capability of intelligent systems serves to detect gas levels along with temperature measurements, pathogen contents, and environmental humidity conditions for immediate assessment of packaged food quality (Khedr et al., 2024). These ethylene detection nanosensors hold exceptional market value because ethylene acts as the main hormone responsible for regulating fruit maturation.

The sensors monitor ethylene concentrations to develop improved storage conditions and estimate the product expiration. The integration of CNT sensors into packaging allows users to monitor ethylene gas release from bananas through which they can develop better storage procedures and distribution methods. The detection of Salmonella surface pathogens on fruits for food safety enforcement and product quality security relies on gold nanoparticle nanodevices according to Kathirvelan and Vijayaraghavan (2020) and Sreeram and Kathirvelan (2024). The technologies of nanotechnology support fruit nutrients while protecting their taste characteristics with procedures which conventional storage methods degrade. Nanotechnology systems use microbial and reactive substance control systems to safeguard fruit nutritional value combined with taste attributes and touch qualities thus extending their usefulness length. When nanotech methods enter fruit preservation operations they produce multiple enhanced benefits that combine longer storage stability with safer products and diminished waste and enhanced consumer confidence. Research and development shows that nanotechnology will form a vital base for sustainable food systems through its ability to minimize waste, maintain the nutritional value, and provide users with fresh high-quality fruits from production areas (Kalia & Parshad, 2015; Liu et al., 2020; Neme et al., 2021).

9.3.13 Challenges and Limitations

Food preservation benefits from nanotechnology, which develops modern packaging solutions along with coatings for extending the storage life of fruits. Research teams that work with nanomaterials have discovered preservation solutions, which deal with the preservation issues that often affect fruits like microbial growth, moisture evaporation, and chemical oxidation breakdown. The application of nanotechnology to fruit preservation requires further analysis because several obstacles and constraints need attention (Kumar, 2024; Kalia & Parshad, 2015).

Nanomaterials present major risks to human health since their adoption is extensive in our modern world. The entry of nanoparticles through packaging materials and coatings into the fruit constitutes potential health risks because they present an ingestion hazard that may accumulate inside human bodies. Scientific evidence shows that silver and zinc oxide nanoparticles demonstrate preservation-enhancing antimicrobial actions yet they create potential harm by being toxic to cells and damaging genes when humans eat them. Complete toxicological assessments help scientists determine the permanent health effects of eating nanomaterials present in foods (Ahari & Lahijani, 2021; Ślosarczyk et al., 2023). The quick progress of nanotechnology applications in food preservation surpasses the development rate of regulatory system standards for nanomaterial use in food preservation methods. Industry players encounter approval hurdles while launching nanotechnology-based products because of undefined regulatory standards, which lead to market acceptance doubts and consumer safety issues. The preservation of fruit through nanotechnology requires juridical alignment of safety assessments and regulatory standards to achieve responsible market deployment (Singh & Kumar, 2023; Porter et al., 2012).

Environmental consequences related to using nanomaterials in fruit preservation should be taken seriously as an immediate consideration. The functional lifetime of nanoparticles ends once they enter waste disposal systems, where they pose risks to soil and water quality. The presence of nondegradable nanoparticles in agricultural lands creates two problems: they change the composition of microbial communities, which leads to soil fertility problems. The creation of nanomaterials that degrade naturally while harming the environment is necessary to reduce ecological threats (Isibor, 2024; Sridhar et al., 2021). The implementation of nanotechnology in fruit preservation faces economic obstacles from manufacturing nanomaterials and integrating them into both packaging and coating systems. Technical expenses related to nanotechnology-based solutions pose difficulties for small- and medium-sized food enterprises to implement widespread solutions. Research investments focusing on developing economical production procedures must combine with scalable technologies to extend nanotechnology access to new markets (Bellanthudawa et al., 2023; Handford et al., 2014). Food preservation market success depends strongly on how people perceive nanotechnology in the food industry. Some consumers refrain from using nanoenhanced food products because they have doubts regarding product safety and ethical concerns in addition to their priority for natural foods that underwent minimal processing. Educational programs and clear communication about nanotechnology's potential advantages and dangers will construct public confidence, thus leading to greater market reception (Cobb & Macoubrie, 2004).

Technical obstacles exist during the process of preserving fruit through nanotechnology applications. Product quality demands attention to how nanoparticles impact the sensory qualities of fruits through alterations in taste attributes, texture aspects, and visual appearance features. New advances in nanomaterial engineering combined with modern application methods need to be developed in order to resolve existing technical limitations. To surmount these challenges, future research should focus on conducting rigorous studies to evaluate the toxicological and environmental impacts of nanomaterials used in fruit preservation, establishing comprehensive and harmonized regulations that ensure the safe use of nanotechnology in the food industry, designing eco-friendly, biodegradable nanomaterials that minimize the environmental footprint, developing efficient manufacturing processes to lower production costs and enhance accessibility for SMEs, and implementing educational programs to inform the public about the benefits and safety measures associated with nanotechnology in food preservation. The successful realization of nanotechnology for extending fruit shelf life requires proper attention to safety aspects together with sustainable practices and profitable business models as displayed in Figure 9.4.

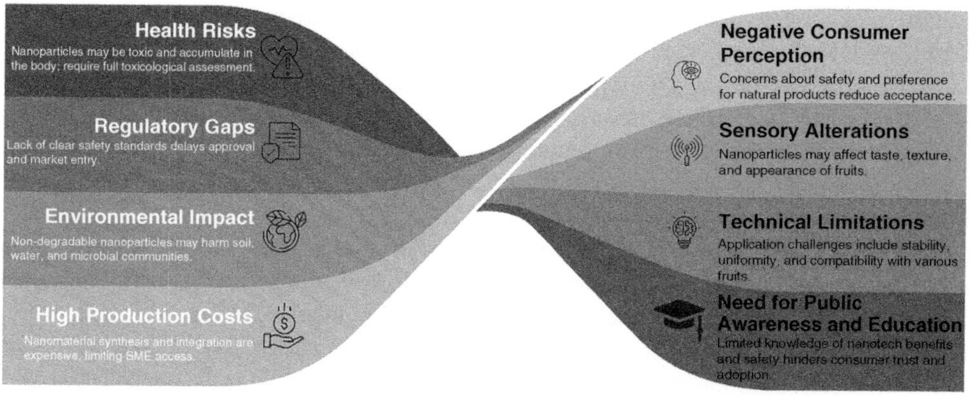

Figure 9.4 Key challenges and limitations in the application of nanotechnology for fruit preservation.

9.3.14 Conclusion and Future Prospect

The emergence of nanotechnology provides a novel solution for fruit shelf-life preservation through the management of microbial, oxidative, and physical degradation factors. Research integration of nanomaterials into food packaging systems and storage promotes better environmental defense, thus boosting preservation qualities of fruit products. The application of nanocarriers as antioxidants and nutrients has achieved enhanced protection of fresh fruit quality through inhibition of enzymatic degradation and oxidative stress. The innovative developments improve food protection standards while eliminating the requirement for chemical preservatives thereby allowing fruit storage to become safer and more sustainable and health-focused. Technological implementation at an industrial scale faces multiple barriers even though it provides many benefits for fruit preservation methods. The main obstacle to the use of nanoparticles in food products arises from their potential toxic properties as well as their ability to cross into edible materials. Scientists need to conduct additional studies in order to properly gauge how nanomaterials affect human health together with the environment throughout extended periods of time.

The global implementation of nanotechnology regulatory rules caused food industry delays and reduced commercialization activities throughout worldwide agencies. The expensive processes involved in nanoparticle manufacturing along with packaging integration create obstacles that prevent small-scale farmers and food producers from employing them. The necessary solutions for resolving these hurdles need thorough interdisciplinary studies alongside rising financial backing of nanotechnology-based food methods and uniform safety regulations for consumers. The advancement of nanotechnology applications in food preservation relies on sustainable practices that protect the environment throughout the upcoming period. Biodegradable nanomaterials present a promising solution to environmental priorities because they maintain effectiveness in fruit preservation strategies. Consumer and retail customers gain visual access to fresh condition and potential contamination levels of fruits through integrated nanobiosensors built into smart package systems. The new technologies will provide enhanced protection for food safety while minimizing losses from food waste for supply chain positions. The combination of artificial intelligence (AI) along with machine learning analysis of data through nanotechnology optimizes storage conditions in real-time. AI-based predictive models assist in selecting the optimal preservation approaches depending on different produce varieties thus they extend shelf life sustainably. The investigation of plant-derived nanoparticles together with naturally occurring nanomaterials offers safe alternatives to synthetic nanomaterials thus making nanotechnology use in food preservation more suitable for regulators and consumers.

Scientific progress in this field depends on the combined efforts between researchers along with legislators and business partners who need to address existing obstacles for innovation. Large-scale trials together with international standards development efforts will create trust between consumers and regulators toward the adoption of nanotechnology for fruit preservation. Widespread education efforts about nanotechnology benefits will facilitate food industry businesses to use these technologies. Although challenges persist in its use the potential benefits of nanotechnology for fruit preservation remain immense. The positive transformation of the food industry becomes possible through smart monitoring systems and safety features together with novel discovery materials. The deployment of nanotechnology-based solutions will accomplish two essential goals: it will lengthen fruit storage duration, support worldwide food supplies, and reduce discard amounts along with creating eco-friendly preservation strategies. Research and innovation continued in nanotechnology will lead to fundamental shifts in fruit storage procedures and transportation processes to guarantee better health outcomes for the growing human population.

References

Ahari, H., & Lahijani, L. K. (2021). Migration of silver and copper nanoparticles from food coating. *Coatings, 11*(4), 380.

Ahari, H., & Naeimabadi, M. (2021). Employing nanoemulsions in food packaging: Shelf life enhancement. *Food Engineering Reviews, 13*(4), 858–883.

Ahmad, M., Ali, A., Ullah, Z., Sher, H., Dai, D. Q., Ali, M., & Ali, I. (2022). Biosynthesized silver nanoparticles using *Polygonatum geminiflorum* efficiently control fusarium wilt disease of tomato. *Frontiers in Bioengineering and Biotechnology, 10*, 988607.

Ahmed, M. W., Haque, M. A., Mohibbullah, M., Khan, M. S. I., Islam, M. A., Mondal, M. H. T., & Ahmmed, R. (2022). A review on active packaging for quality and safety of foods: Current trends, applications, prospects and challenges. *Food Packaging and Shelf Life, 33*, 100913.

Ahvenainen, R. (2003). Active and intelligent packaging: An introduction. In *Novel food packaging techniques* (pp. 5–21). Woodhead Publishing.

Bahrami, A., Delshadi, R., Assadpour, E., Jafari, S. M., & Williams, L. (2020). Antimicrobial-loaded nanocarriers for food packaging applications. *Advances in Colloid and Interface Science, 278*, 102140.

Barrett, D. M., & Lloyd, B. (2012). Advanced preservation methods and nutrient retention in fruits and vegetables. *Journal of the Science of Food and Agriculture, 92*(1), 7–22.

Basumatary, I. B., Mukherjee, A., Katiyar, V., & Kumar, S. (2022). Biopolymer-based nanocomposite films and coatings: Recent advances in shelf-life improvement of fruits and vegetables. *Critical Reviews in Food Science and Nutrition, 62*(7), 1912–1935.

Bellanthudawa, B. K. A., Nawalage, N. M. S. K., Handapangoda, H. M. A. K., Suvendran, S., Wijayasenarathne, K. A. S. H., Rathnasuriya, M. L. D., ... Perera, I. J. J. U. N. (2023). A perspective on biodegradable and non-biodegradable nanoparticles in industrial sectors: Applications, challenges, and future prospects. *Nanotechnology For Environmental Engineering*, *8*(4), 975–1013.

Beyene, H. D., Werkneh, A. A., Bezabh, H. K., & Ambaye, T. G. (2017). Synthesis paradigm and applications of silver nanoparticles (AgNPs), a review. *Sustainable Materials and Technologies*, *13*, 18–23.

Bora, H., Kamle, M., Mahato, D. K., Tiwari, P., & Kumar, P. (2020). Citrus essential oils (CEOs) and their applications in food: An overview. *Plants*, *9*(3), 357.

Cakmak, I., & Kutman, U. Á. (2018). Agronomic biofortification of cereals with zinc: A review. *European Journal of Soil Science*, *69*(1), 172–180.

Cano, L., Pollet, E., Avérous, L., & Tercjak, A. (2017). Effect of TiO_2 nanoparticles on the properties of thermoplastic chitosan-based nano-biocomposites obtained by mechanical kneading. *Composites Part A: Applied Science and Manufacturing*, *93*, 33–40.

Chaudhary, S., Jain, V. P., & Jaiswar, G. (2022). The composition of polysaccharides: Monosaccharides and binding, group decorating, polysaccharides chains. In *Innovation in nano-polysaccharides for eco-sustainability* (pp. 83–118). Elsevier.

Chelliah, R., Khan, I., Wei, S., Madar, I. H., Sultan, G., Daliri, E. B. M., ... Oh, D. H. (2022). Intelligent packaging systems: Food quality and intelligent medicine box based on nano-sensors. In *Smart nanomaterials in biomedical applications* (pp. 555–587). Cham: Springer International Publishing.

Chitena, L. (2022). *Investigation of the efficacy of a ZnO-starch nanocomposite packaging to prolong the shelf life of Fragaria ananassa (strawberries)* (Master's thesis, Botswana International University of Science and Technology (Botswana)).

Cobb, M. D., & Macoubrie, J. (2004). Public perceptions about nanotechnology: Risks, benefits and trust. *Journal of Nanoparticle Research*, *6*, 395–405.

Costa, C. O. N. T. E., Conte, A., Buonocore, G. G., & Del Nobile, M. A. (2011). Antimicrobial silver-montmorillonite nanoparticles to prolong the shelf life of fresh fruit salad. *International Journal of Food Microbiology*, *148*(3), 164–167.

Curtis, J., Greenberg, M., Kester, J., Phillips, S., & Krieger, G. (2006). Nanotechnology and nanotoxicology: A primer for clinicians. *Toxicological Reviews*, *25*, 245–260.

Das, S., Ghosh, A., & Mukherjee, A. (2021). Nanoencapsulation-based edible coating of essential oils as a novel green strategy against fungal spoilage, mycotoxin contamination, and quality deterioration of stored fruits: An overview. *Frontiers in Microbiology*, *12*, 768414.

De Azeredo, H. M. (2009). Nanocomposites for food packaging applications. *Food Research International*, *42*(9), 1240–1253.

de Oliveira Filho, J. G., Bertolo, M. R. V., da Costa Brito, S., Malafatti, J. O. D., Bertazzo, G. B., Colacique, M. N., ... Ferreira, M. D. (2022). Recent advances in the application of nanotechnology to reduce fruit and vegetable losses during post-harvest. *Brazilian Journal of Physics*, *52*(4), 126.

de Sousa, M. S., Schlogl, A. E., Estanislau, F. R., Souza, V. G. L., dos Reis Coimbra, J. S., & Santos, I. J. B. (2023). Nanotechnology in packaging for food industry: Past, present, and future. *Coatings*, *13*(8), 1411.

Dewettinck, K., & Huyghebaert, A. (1999). Fluidized bed coating in food technology. *Trends in Food Science & Technology*, *10*(4-5), 163–168.

Echegoyen, Y., & Nerín, C. (2013). Nanoparticle release from nano-silver antimicrobial food containers. *Food and Chemical Toxicology*, *62*, 16–22.

El Mouzahim, M., Eddarai, E. M., Eladaoui, S., Guenbour, A., Bellaouchou, A., Zarrouk, A., & Boussen, R. (2023). Food packaging composite film based on chitosan, natural kaolinite clay, and *Ficus carica* leaves extract for fresh-cut apple slices preservation. *International Journal of Biological Macromolecules*, *233*, 123430.

El Sheikha, A. F., Levin, R., & Xu, J. (2018). *Molecular techniques in food biology*. New York, NY, USA: John Wiley & Sons Ltd.

El-Ramady, H. R., Domokos-Szabolcsy, É., Abdalla, N. A., Taha, H. S., & Fári, M. (2015). Postharvest management of fruits and vegetables storage. *Sustainable Agriculture Reviews*, *15*, 65–152.

Emamhadi, M. A., Sarafraz, M., Akbari, M., Fakhri, Y., Linh, N. T. T., & Khaneghah, A. M. (2020). Nanomaterials for food packaging applications: A systematic review. *Food and Chemical Toxicology*, *146*, 111825.

Ensafi, A. A., Karimi-Maleh, H., & Mallakpour, S. (2012). Simultaneous determination of ascorbic acid, acetaminophen, and tryptophan by square wave voltammetry using N-(3,4-dihydroxyphenethyl)-3,5-Dinitrobenzamide-modified carbon nanotubes paste electrode. *Electroanalysis*, *24*(3), 666–675.

Galli, R., Hall, M. C., Breitenbach, E. R., Colpani, G. L., Zanetti, M., de Mello, J. M. M., ... Fiori, M. A. (2020). Antibacterial polyethylene-ethylene vinyl acetate polymeric blend by incorporation of zinc oxide nanoparticles. *Polymer Testing*, *89*, 106554.

Gomez-Zavaglia, A., Mejuto, J. C., & Simal-Gandara, J. (2020). Mitigation of emerging implications of climate change on food production systems. *Food Research International*, *134*, 109256.

Gudikandula, K., & Charya Maringanti, S. (2016). Synthesis of silver nanoparticles by chemical and biological methods and their antimicrobial properties. *Journal of Experimental Nanoscience*, *11*(9), 714–721.

Guía-García, J. L., Charles-Rodríguez, A. V., Reyes-Valdés, M. H., Ramírez-Godina, F., Robledo-Olivo, A., García-Osuna, H. T., ... Flores-López, M. L. (2022). Micro and nanoencapsulation of bioactive compounds for agri-food applications: A review. *Industrial Crops and Products*, *186*, 115198.

Gul, F., Ullah, Z., Iqbal, J., Abbasi, B. A., Ali, S., Kanwal, S., & Mahmood, T. (2025). Ecofriendly synthesis characterization and biological activities of *Eruca sativa* mediated silver oxide nanoparticles. *Scientific Reports*, *15*(1), 13466.

Handford, C. E., Dean, M., Spence, M., Henchion, M., Elliott, C. T., & Campbell, K. (2014). Nanotechnology in the agri-food industry on the island of Ireland: Applications, opportunities and challenges. In *Institute for global food security at Queen's university, Belfast, and the Teagasc Ashtown food research centre, Dublin*.

Haris, M., Fatima, N., Iqbal, J., Chalgham, W., Mumtaz, A. S., El-Sheikh, M. A., & Tavafoghi, M. (2023). *Oscillatoria limnetica* mediated green synthesis of iron oxide (Fe_2O_3) nanoparticles and their diverse in vitro bioactivities. *Molecules*, *28*(5), 2091.

Heaton, J. C., & Jones, K. (2008). Microbial contamination of fruit and vegetables and the behaviour of enteropathogens in the phyllosphere: A review. *Journal of Applied Microbiology*, *104*(3), 613–626.

Hoseinnejad, M., Jafari, S. M., & Katouzian, I. (2018). Inorganic and metal nanoparticles and their antimicrobial activity in food packaging applications. *Critical Reviews in Microbiology*, *44*(2), 161–181.

Iñiguez-Moreno, M., Ragazzo-Sánchez, J. A., & Calderón-Santoyo, M. (2021). An extensive review of natural polymers used as coatings for postharvest shelf-life extension: Trends and challenges. *Polymers*, *13*(19), 3271.

Ishaque, F., Hossain, M. A., Sarker, M. A. R., Mia, M. Y., Dhrubo, A. S., Uddin, G. T., & Rahman, M. H. (2019). A study on low cost post-harvest storage techniques to extend the shelf life of citrus fruits and vegetables. *Journal of Engineering Research and Reports*, *9*(1), 1–17.

Isibor, P. O. (2024). Regulations and policy considerations for nanoparticle safety. In *Environmental nanotoxicology: Combatting the minute contaminants* (pp. 295–316). Springer Nature Switzerland: Cham.

Jeevahan, J., & Chandrasekaran, M. (2019). Nanoedible films for food packaging: A review. *Journal of Materials Science*, *54*(19), 12290–12318.

Kalia, A., & Parshad, V. R. (2015). Novel trends to revolutionize preservation and packaging of fruits/fruit products: Microbiological and nanotechnological perspectives. *Critical Reviews in Food Science and Nutrition*, *55*(2), 159–182.

Karimi-Maleh, H., Moazampour, M., Yoosefian, M., Sanati, A. L., Tahernejad-Javazmi, F., & Mahani, M. (2014). An electrochemical nanosensor for simultaneous voltammetric determination of ascorbic acid and Sudan I in food samples. *Food Analytical Methods*, *7*, 2169–2176.

Kathirvelan, J., & Vijayaraghavan, R. (2020). Review on sensitive and selective ethylene detection methods for fruit ripening application. *Sensor Review*, *40*(4), 421–435.

Kerry, J. P. (2014). New packaging technologies, materials and formats for fast-moving consumer products. In *Innovations in food packaging* (pp. 549–584). Academic Press.

Kerry, J. P., O'grady, M. N., & Hogan, S. A. (2006). Past, current and potential utilisation of active and intelligent packaging systems for meat and muscle-based products: A review. *Meat Science*, *74*(1), 113–130.

Khaneghah, A. M., Hashemi, S. M. B., & Limbo, S. (2018). Antimicrobial agents and packaging systems in antimicrobial active food packaging: An overview of approaches and interactions. *Food and Bioproducts Processing*, *111*, 1–19.

Khedr, E. H., & Al-Khayri, J. M. (2024). Biosensors and nanosensors for determination of harvest quality parameters and fruit handling processes monitoring. In *Innovative methods in horticultural crop improvement: Biosensors and nanosensors* (pp. 143–190). Springer Nature Switzerland: Cham.

Khedr, E. H., Al-Qahtani, W. H., & Khedr, N. (2024). Biosensors and nanosensors for determination of fruit safety. In *Innovative methods in horticultural crop improvement: Biosensors and nanosensors* (pp. 191–234). Springer Nature Switzerland: Cham.

Khot, L. R., Sankaran, S., Maja, J. M., Ehsani, R., & Schuster, E. W. (2012). Applications of nanomaterials in agricultural production and crop protection: A review. *Crop Protection*, *35*, 64–70.

Kocira, A., Kozłowicz, K., Panasiewicz, K., Staniak, M., Szpunar-Krok, E., & Hortyńska, P. (2021). Polysaccharides as edible films and coatings: Characteristics and influence on fruit and vegetable quality—A review. *Agronomy*, *11*(5), 813.

Kondle, R., Sharma, K., Singh, G., & Kotiyal, A. (2022). Using nanotechnology for enhancing the shelf life of fruits. In *Food Processing and Packaging Technologies-Recent Advances*. IntechOpen.

Kumar, N., Upadhyay, A., Shukla, S., Bajpai, V. K., Kieliszek, M., Yadav, A., & Kumaravel, V. (2024). Next generation edible nanoformulations for improving post-harvest shelf-life of citrus fruits. *Journal of Food Measurement and Characterization, 18*(3), 1825–1856.

Kumar, S. (2024). Prospects and challenges of nanomaterials in sustainable food preservation and packaging: A review. *Discover Nano, 19*(1), 1–18.

Kumar, V., Guleria, P., & Mehta, S. K. (2017). Nanosensors for food quality and safety assessment. *Environmental Chemistry Letters, 15*, 165–177.

Kuswandi, B. (2020). Real-time quality assessment of fruits and vegetables: Sensing approaches. In *Sensor-based quality assessment systems for fruits and vegetables* (pp. 1–30). Apple Academic Press.

Ladaniya, M. (2010). *Citrus fruit: Biology, technology and evaluation*. Academic press.

Leisner, C. P. (2020). Review: Climate change impacts on food securityfocus on perennial cropping systems and nutritional value. *Plant Science, 293*, 110412.

Lengyel, M., Kállai-Szabó, N., Antal, V., Laki, A. J., & Antal, I. (2019). Microparticles, Microspheres, and Microcapsules for Advanced Drug Delivery. *Scientia Pharmaceutica, 87*, 20.

Liu, W., Zhang, M., & Bhandari, B. (2020). Nanotechnology–A shelf life extension strategy for fruits and vegetables. *Critical Reviews in Food Science and Nutrition, 60*(10), 1706–1721.

Ma, L., Zhang, M., Bhandari, B., & Gao, Z. (2017). Recent developments in novel shelf life extension technologies of fresh-cut fruits and vegetables. *Trends in Food Science & Technology, 64*, 23–38.

Mahalik, N. P., & Nambiar, A. N. (2010). Trends in food packaging and manufacturing systems and technology. *Trends in Food Science & Technology, 21*(3), 117–128.

Makwana, D., Castano, J., Somani, R. S., & Bajaj, H. C. (2020). Characterization of Agar-CMC/Ag-MMT nanocomposite and evaluation of antibacterial and mechanical properties for packaging applications. *Arabian Journal of Chemistry, 13*(1), 3092–3099.

Makwana, S., Choudhary, R., Dogra, N., Kohli, P., & Haddock, J. (2014). Nanoencapsulation and immobilization of cinnamaldehyde for developing antimicrobial food packaging material. *LWT- Food Science and Technology, 57*(2), 470–476.

Manikandan, V., & Min, S. C. (2023). Roles of polysaccharides-based nanomaterials in food preservation and extension of shelf-life of food products: A review. *International Journal of Biological Macromolecules, 252*, 126381.

McHugh, T. H., & Senesi, E. (2000). Apple wraps: A novel method to improve the quality and extend the shelf life of fresh-cut apples. *Journal of Food Science, 65*(3), 480–485.

Meng, X., Zhang, M., Zhan, Z., & Adhikari, B. (2014). Changes in quality characteristics of fresh-cut cucumbers as affected by pressurized argon treatment. *Food and Bioprocess Technology, 7*, 693–701.

Miteluț, A. C., Popa, E. E., Drăghici, M. C., Popescu, P. A., Popa, V. I., Bujor, O. C., ... Popa, M. E. (2021). Latest developments in edible coatings on minimally processed fruits and vegetables: A review. *Food, 10*(11), 2821.

Motelica, L., Ficai, D., Ficai, A., Trușcă, R. D., Ilie, C. I., Oprea, O. C., & Andronescu, E. (2020). Innovative antimicrobial chitosan/ZnO/Ag NPs/citronella essential oil nanocomposite—Potential coating for grapes. *Food, 9*(12), 1801.

Mwangi, M. N., Oonincx, D. G., Stouten, T., Veenenbos, M., Melse-Boonstra, A., Dicke, M., & Van Loon, J. J. (2018). Insects as sources of iron and zinc in human nutrition. *Nutrition Research Reviews, 31*(2), 248–255.

Nayab, D. E., & Akhtar, S. (2023). Green synthesized silver nanoparticles from eucalyptus leaves can enhance shelf life of banana without penetrating in pulp. *PLoS One*, *18*(3), e0281675.

Ncube, L. K., Ude, A. U., Ogunmuyiwa, E. N., Zulkifli, R., & Beas, I. N. (2020). Environmental impact of food packaging materials: A review of contemporary development from conventional plastics to polylactic acid based materials. *Materials*, *13*(21), 4994.

Neme, K., Nafady, A., Uddin, S., & Tola, Y. B. (2021). Application of nanotechnology in agriculture, postharvest loss reduction and food processing: Food security implication and challenges. *Heliyon*, *7*(12).

Neves, M. F., Trombin, V. G., Marques, V. N., & Martinez, L. F. (2020). Global orange juice market: A 16-year summary and opportunities for creating value. *Tropical Plant Pathology*, *45*, 166–174.

Odetayo, T., Tesfay, S., & Ngobese, N. Z. (2022). Nanotechnology-enhanced edible coating application on climacteric fruits. *Food Science & Nutrition*, *10*(7), 2149–2167.

Ogwu, M. C., & Ogunsola, O. A. (2024). Physicochemical methods of food preservation to ensure food safety and quality. In *Food Safety and Quality in the Global South* (pp. 263–298). Springer Nature Singapore: Singapore.

Omerović, N., Djisalov, M., Živojević, K., Mladenović, M., Vunduk, J., Milenković, I., ... Vidić, J. (2021). Antimicrobial nanoparticles and biodegradable polymer composites for active food packaging applications. *Comprehensive Reviews in Food Science and Food Safety*, *20*(3), 2428–2454.

Onwude, D. I., Chen, G., Eke-Emezie, N., Kabutey, A., Khaled, A. Y., & Sturm, B. (2020). Recent advances in reducing food losses in the supply chain of fresh agricultural produce. *PRO*, *8*(11), 1431.

Oprea, I., Fărcaş, A. C., Leopold, L. F., Diaconeasa, Z., Coman, C., & Socaci, S. A. (2022). Nano-encapsulation of citrus essential oils: Methods and applications of interest for the food sector. *Polymers*, *14*(21), 4505.

Paidari, S., Esmaeili, Y., Ibrahim, S. A., Vahedi, S., Al-Hilifi, S. A., & Zamindar, N. (2024). Application of nanoparticles to enhance the microbial quality and shelf life of food products. In *Microbial biotechnology in the food industry: Advances, challenges, and potential solutions* (pp. 75–102). Cham: Springer International Publishing.

Pateiro, M., Gómez, B., Munekata, P. E., Barba, F. J., Putnik, P., Kovačević, D. B., & Lorenzo, J. M. (2021). Nanoencapsulation of promising bioactive compounds to improve their absorption, stability, functionality and the appearance of the final food products. *Molecules*, *26*(6), 1547.

Patil, B. S., Jayaprakasha, G. K., & Murthy, K. N. C. (2017). Beyond vitamin C: The diverse, complex health-promoting properties of citrus fruits. *Citrus Research & Technology*, *38*(1), 107–121.

Pegg, R. B., & Shahidi, F. (2007). Encapsulation, stabilization, and controlled release of food ingredients and bioactives. In *Handbook of food preservation* (pp. 527–586). CRC Press.

Pereda, M., Marcovich, N. E., & Ansorena, M. R. (2017). Nanotechnology in food packaging applications: Barrier materials, antimicrobial agents, sensors, and safety assessment.

Poongavanam, S. S., Subramaniyan, V., Sellamuthu, P. S., Jarugala, J., & Sadiku, E. R. (2023). Fabrication of bio-nanocomposite packaging films with PVA, MMt clay nanoparticles, CNCs, and essential oils for the postharvest preservation of sapota fruits. *Polymers*, *15*(17), 3589.

Poonia, A., & Mishra, A. (2022). Edible nanocoatings: Potential food applications, challenges and safety regulations. *Nutrition & Food Science, 52*(3), 497–514.

Porter, R. D., Breggin, L., Falkner, R., Pendergrass, J., & Jaspers, N. (2012). Regulatory responses to nanotechnology uncertainties. In *The nanotechnology challenge*. New York: *Cambridge University Press.*

Primožič, M., Knez, Ž., & Leitgeb, M. (2021). (Bio) Nanotechnology in food science—food packaging. *Nanomaterials, 11*(2), 292.

Rana, R. A., Siddiqui, M. N., Skalicky, M., Brestic, M., Hossain, A., Kayesh, E., ... Islam, T. (2021). Prospects of nanotechnology in improving the productivity and quality of horticultural crops. *Horticulturae, 7*(10), 332.

Realini, C. E., & Marcos, B. (2014). Active and intelligent packaging systems for a modern society. *Meat Science, 98*(3), 404–419.

Reyes-Jurado, F., Franco-Vega, A., Ramírez-Corona, N., Palou, E., & López-Malo, A. (2015). Essential oils: Antimicrobial activities, extraction methods, and their modeling. *Food Engineering Reviews, 7*, 275–297.

Richa, R., Kohli, D., Vishwakarma, D., Mishra, A., Kabdal, B., Kothakota, A., ... Naik, B. (2023). Citrus fruit: Classification, value addition, nutritional and medicinal values, and relation with pandemic and hidden hunger. *Journal of Agriculture and Food Research, 14*, 100718.

Rovera, C., Ghaani, M., & Farris, S. (2020). Nano-inspired oxygen barrier coatings for food packaging applications: An overview. *Trends in Food Science & Technology, 97*, 210–220.

Sagar, N. A., Kumar, N., Choudhary, R., Bajpai, V. K., Cao, H., Shukla, S., & Pareek, S. (2022). Prospecting the role of nanotechnology in extending the shelf-life of fresh produce and in developing advanced packaging. *Food Packaging and Shelf Life, 34*, 100955.

Sharma, N. (2020). Managing postharvest diseases: Classical versus new technologies. In *Bio-management of postharvest diseases and mycotoxigenic fungi* (pp. 7–56). CRC Press.

Singh, R., & Kumar, S. (2023). Regulatory and safety concerns regarding the use of active nanomaterials in food industry. In *Nanotechnology advancement in agro-food industry* (pp. 269–306). Springer Nature Singapore: Singapore.

Singh, R., Dutt, S., Sharma, P., Sundramoorthy, A. K., Dubey, A., Singh, A., & Arya, S. (2023). Future of nanotechnology in food industry: Challenges in processing, packaging, and food safety. *Global Challenges, 7*(4), 2200209.

Singh, T., Shukla, S., Kumar, P., Wahla, V., Bajpai, V. K., & Rather, I. A. (2017). Application of nanotechnology in food science: Perception and overview. *Frontiers in Microbiology, 8*, 1501.

Ślosarczyk, A., Klapiszewska, I., Skowrońska, D., Janczarek, M., Jesionowski, T., & Klapiszewski, Ł. (2023). A comprehensive review of building materials modified with metal and metal oxide nanoparticles against microbial multiplication and growth. *Chemical Engineering Journal, 466*, 143276.

Sohail, M., Sun, D. W., & Zhu, Z. (2018). Recent developments in intelligent packaging for enhancing food quality and safety. *Critical Reviews in Food Science and Nutrition, 58*(15), 2650–2662.

Sorrentino, A., Gorrasi, G., & Vittoria, V. (2007). Potential perspectives of bio-nanocomposites for food packaging applications. *Trends in Food Science & Technology, 18*(2), 84–95.

Sreeram, A., & Kathirvelan, J. (2024). Design and development of ethylene gas sensor for non-destructive analysis of food quality: An updated review. *Sensor Review, 44*(3), 267–283.

Sridhar, A., Ponnuchamy, M., Kumar, P. S., & Kapoor, A. (2021). Food preservation techniques and nanotechnology for increased shelf life of fruits, vegetables, beverages and spices: A review. *Environmental Chemistry Letters, 19*, 1715–1735.

Stern, R., & Jedrzejas, M. J. (2008). Carbohydrate polymers at the center of life's origins: The importance of molecular processivity. *Chemical Reviews, 108*(12), 5061–5085.

Taghipour, L., & Assar, P. (2025). Nano-packaging technology for conserving the quality of fruits and vegetables. In *Sustainable Postharvest Technologies for Fruits and Vegetables* (pp. 227–243). CRC Press.

Tayeb, A. H., Amini, E., Ghasemi, S., & Tajvidi, M. (2018). Cellulose nanomaterials—Binding properties and applications: A review. *Molecules, 23*(10), 2684.

Thiyam, B., & Sharma, G. D. (2013). Isolation and identification of fungi associated with local fruits of Barak Valley, Assam. *Current World Environment, 8*(2), 319–322.

Ullah, R., Iqbal, J., Abbasi, B. A., Ijaz, S., Zarshan, F., Yaseen, T., & Iqbal, R. (2025). Biosynthesis of phytochemical-adorned green iron oxide nanoparticles via *Astragalus anisacanthus* extract and evaluation of biological potentials. *Journal of Inorganic and Organometallic Polymers and Materials*, 1–21.

Ullah, Z., Iqbal, J., Gul, F., Abbasi, B. A., Kanwal, S., Elsadek, M. F., & Mahmood, T. (2024). Biogenic synthesis, characterization, and in vitro biological investigation of silver oxide nanoparticles (AgONPs) using *Rhynchosia capitata*. *Scientific Reports, 14*(1), 10484.

Vhangani, L. N. (2022). *The use of nano-encapsulated plant extracts in inhibiting non-enzymatic browning in fruit canned in juice* (Doctoral dissertation, Cape Peninsula University of Technology).

Wang, J., Gardner, D. J., Stark, N. M., Bousfield, D. W., Tajvidi, M., & Cai, Z. (2018). Moisture and oxygen barrier properties of cellulose nanomaterial-based films. *ACS Sustainable Chemistry & Engineering, 6*(1), 49–70.

Xu, W., Xie, W., Huang, X., Chen, X., Huang, N., Wang, X., & Liu, J. (2017). The graphene oxide and chitosan biopolymer loads TiO_2 for antibacterial and preservative research. *Food Chemistry, 221*, 267–277.

Yang, W., Shen, C., Ji, Q., An, H., Wang, J., Liu, Q., & Zhang, Z. (2009). Food storage material silver nanoparticles interfere with DNA replication fidelity and bindwith DNA. *Nanotechnology, 20*(8), 085102.

Yemmireddy, V. K., Farrell, G. D., & Hung, Y. C. (2015). Development of titanium dioxide (TiO_2) nanocoatings on food contact surfaces and method to evaluate their durability and photocatalytic bactericidal property. *Journal of Food Science, 80*(8), N1903–N1911.

Yu, Z., Wang, W., Kong, F., Lin, M., & Mustapha, A. (2019). Cellulose nanofibril/silver nanoparticle composite as an active food packaging system and its toxicity to human colon cells. *International Journal of Biological Macromolecules, 129*, 887–894.

Zanetti, M., Camino, G., Thomann, R., & Mülhaupt, R. (2001). Synthesis and thermal behaviour of layered silicate–EVA nanocomposites. *Polymer, 42*(10), 4501–4507.

Zhao, Z., Huai, X., Zhang, B., Huang, N., Sun, Y., Miao, J., ... Zhou, X. (2025). Types of nanomaterials commonly used in food packaging, film formation techniques, and recent advances in their applications. *International Journal of Food Science and Technology, vvae036*.

Zhong, G., & Nicolosi, E. (2020). Citrus origin, diffusion, and economic importance. *The Citrus Genome*, 5–21.

10

Harnessing Nanoherbicides and Insecticides for Eco-Friendly Solutions: Advancing Green Agriculture

Syeda Anber Zahra[1], Javed Iqbal[2], Banzeer Ahsan Abbasi[3], Aimen Fatima[4], Muhammad Anas[5,6], Tabassum Yaseen[2], Akhtar Munir[7], Tariq Mahmood[8], and Shah Fahad[9]

[1] *Institute of Biomaterials and biomedical, School of Food and Pharmacy, Shanghai Zhongqiao Vocational and Technical University, China*
[2] *Department of Botany, Bacha Khan University, Charsadda, Khyber Pakhtunkhwa, Pakistan*
[3] *Department of Botany, Rawalpindi Women University, Rawalpindi, Pakistan*
[4] *Department of Microbiology, Faculty of biological sciences Quaid-i-Azam University, Islamabad*
[5] *Department of Plant Sciences, Faculty of Biological Sciences, Quaid-I-Azam University, Islamabad, Pakistan*
[6] *Department of Soil Science & Plant Nutrition, University of Selcuk. Campus, Konya, Turkiye*
[7] *Department of Chemistry, Quaid-i-Azam University, Islamabad, Pakistan*
[8] *Department of Plant Sciences, Faculty of Biological Sciences, Quaid-i-Azam University, Islamabad, Pakistan*
[9] *Department of Agronomy, Abdul Wali Khan University Mardan, Khyber Pakh-tunkhwa, Pakistan*

10.1 Introduction

Climate change and the decline in soil fertility present considerable challenges to the advancement of sustainable agricultural practices (Forini et al., 2022). According to the United Nations, the global population is projected to reach 9.7 billion by the end of 2050. In order to support this growing population, various agrochemicals are being used to improve agricultural production. But the improper use of these agrochemicals causes harm to the environment and also to human health (Okeke et al., 2022). With the decline of farming land and water resources, the agricultural sector must enhance resource utilization and protect the agroecological system by effectively implementing modern technology (Pandey & Mishra, 2024).

Furthermore, swift urbanization and industrialization worldwide have resulted in a substantial reduction in the entire expanse of arable land, presenting challenges to agricultural output. FAO statistics reveal that annually, up to 40% of crops are compromised by pests (Lord, 2024), resulting in direct harm to plants, with the extent of damage occasionally amplified by associated pests and bacteria. Moreover, it has been proven that climate change, particularly extreme weather phenomena, has precipitated a substantial proliferation of herbivorous insects, hence exacerbating the detrimental impact on the yield and quality of agricultural produce (Lord, 2024).

Nanotechnology Innovations for Food Security and Sustainable Agriculture, First Edition. Edited by Zia Ur Rehman Mashwani, Shah Fahad, Ilyas Ahmad, Shah Saud, Sajid Fiaz, and Taufiq Nawaz.

In order to improve food production during that time, the use of chemical fertilizers, insecticides, and herbicides developed into cutting-edge technologies. Weeds significantly contribute to agricultural damage, resulting in a reduction of around 10–25% in crop yield in underdeveloped countries (Gharde and Singh, 2018; Horvath et al., 2023). The reliance on physical labor for hand-weeding has become increasingly challenging due to manpower shortages caused by urbanization and unsatisfactory results stemming from climatic irregularities. This scenario facilitated the use of herbicides that provide immediate protection against weeds and may be applied by a single worker for spraying (Das et al., 2024).

The careful use of herbicides is a complex procedure that requires accuracy and strategic planning (Triantafyllidis et al., 2023). Success depends on the accuracy of timing, source, location, and dose; the harmonic integration of these elements is essential for effective weed control with little environmental impact. Several studies consistently state that extended pesticide application results in lasting and permanent environmental consequences (Kaur et al., 2024).

Recent decades have yielded minimal study findings on nano-herbicides. In order to control and manage pests, insecticides are commonly used in agriculture. Insecticides have become an essential tool for enhancing crop yield and safeguarding plants with the advancement of agriculture. A diverse array of insecticides is essential for effective pest control to mitigate pest infestations and enhance agricultural productivity, as such infestations contribute to the loss of 45% annually in food production (Schmitz et al., 2022). It is frequently employed to avert vector-borne diseases, encompassing the protection of crops, food preservation, and significant functions in commercial and food-related industrial activities such as aquaculture, agriculture, food processing, and preservation (Sharma et al., 2019).

Crop damage from insects, such as defoliation and reduced agricultural yields, can be substantial. The effects differ based on the type of bug, the stage of the crop's development, and the particular crop being grown (Gardarin et al., 2022). The most common pathogens that inflict diseases on plants are bacteria, viruses, and fungi. Diseases affecting plants can have a negative impact on farming and ecosystems in general, leading to lower crop yields, financial losses, and ecological disturbances (Kulabhusan et al., 2022). A wide variety of chemicals are used as insecticides to reduce or eradicate bug populations.

The use of these pesticides may negatively affect the environment that is not the intended aim, such as soil, water, and wildlife (Ren et al., 2023). Recently, there has been a dramatic increase in the use of pesticides. The U.S. Environmental Protection Agency (U.S. EPA) oversees regulations for the registration, usage, importation, and exporting of insecticides. The U.S. EPA initiated the gradual ban of chlorpyrifos, a type of herbicide that has been associated with a decline in children's brain health and neurocognitive abilities, starting in 2021 (Raj et al., 2023). Chlorpyrifos was last evaluated for prohibition in 2021 because of the substantial hazards it posed, attracting public scrutiny (Raj et al., 2023). China and the United States are not the only countries that use pesticides; Argentina, Brazil, Russia, Canada, Australia, Colombia, France, and India are also major players. South Korea, South Africa, Canada, Turkey, India, and Tanzania have all reported child fatalities from acute pesticide exposure (Islam et al., 2022).

Agriculture has historically depended on chemical pesticides and herbicides to safeguard crops and optimize output. Global pesticide consumption is estimated at roughly 2 million metric tons per year, with herbicides representing about 40% of the total pesticide market (Brunel, et al., 2013). However, these pesticides are necessary for controlling pests, diseases, and weeds as they are responsible for up to 40% of worldwide agricultural losses (Junaid & Gokce, 2024). Though they offer benefits by increasing agricultural output, the use of traditional pesticides and herbicides has led to environmental and health problems on a massive scale.

This increase in problems has forced us to seek out increasingly efficient and eco-friendly options. Among various alternatives to traditional pesticides and herbicides (Yadav et al., 2023), the use of nanotechnology in the manipulation of materials at the atomic or molecular levels might be applicable to overcome the associated issues. By engineering nanoparticles to deliver active compounds with higher efficiency and precision, less total amount of chemicals is needed, while they can still achieve desired efficacy. Nanoencapsulation of pesticides keeps active chemicals away from the harsh environmental conditions such as sunlight and high temperatures, releasing in a controlled and prolonged manner.

The use of this method will significantly reduce the frequency of application of pesticides, thereby decreasing the environmental burden. In addition, the small size and large surface area of nanoparticles allow them to penetrate biological barriers more readily. This enhances the availability of active chemicals to pests, leading to less active chemical dose than traditional formulations for the same or even better pest control. Targeted delivery systems are deployed using nanotechnology where systems of nanoparticles are created that selectively bind to pest or weed of concern to avoid the attack on nontarget species (Punniyakottai et al., 2024; Ullah et al., 2023).

It adds to the uniqueness of the property, reducing the risk of environmental contamination and safeguarding critically needed creatures for the purpose of ecosystem balance. It is not merely about ensuring more potent insecticides and herbicides, although such an achievement would be a significant boost to nanotechnology in agriculture. Additionally, it provides new solutions to the early detection and monitoring of pest and disease . In the field, nanosensors can send real-time data on insect populations and field conditions, providing farmers with time and field conditions at which decisions on when and where to treat are made (Yadav et al., 2023). Finally, through this precision method of agriculture, pesticide and herbicide use is further optimized while reducing overspray and thereby enhancing environmental impact.

10.2 Nanoherbicides and Insecticides for Eco-Friendly Solutions: Advancing Green Agriculture

Agricultural pollution is a phenomenon that has been set in motion by modern farming methods. Due to the current agricultural by-products, this process can degrade ecosystems, land, and the environment. Chemical fertilizers, herbicides, and the use of polluted water for irrigation all contribute to the already devastating effects on farming. Accordingly, the food and agricultural industries' present state is no longer viable (Singh et al., 2021). Concerningly, synthetically produced chemicals have far-reaching consequences that are contributing to a myriad of problems in the fields of ecology and medicine. In addition to being resistant to biodegradability, many of these chemicals have recently been found to be carcinogenic, which poses serious health risks. Also, contaminants like dichlorodiphenyltrichloroethane (DDT), benzene hexachloride (BHC), and others can end up in the food chain (Kulkarni, 2023).

Planning for food security would be quite difficult with the predicted world population rise from 7.35 billion to 8.60 billion by 2030. According to Saha and Bharadwaj (2023), the current demand for crops can only be satisfied if production is boosted by as much as 70%. However, the best way to ensure a bountiful harvest is to use fertilizer according to the manufacturer's instructions, as well as to choose enhanced kinds that are resistant to pests and diseases. To feed a world that is expanding at an alarming rate, conventional farmers have turned to synthetic pesticides and other chemical fertilizers in an effort to rein out pests (Sharma et al., 2019). The biosphere's biotic and abiotic components are negatively affected by these compounds' persistence and the hazardous intermediates they generate during degradation (Abhiram, 2023). All four ecosystems (air, soil, water, and vegetation) have been negatively impacted by the chemical components in question; they have bioaccumulated in food webs and endangered nontarget flora and fauna. Harvests can be reduced by an estimated 20–90% due to biotic stress and weeds, which are significant issues in agriculture.

Classical herbicides are widely used by farmers and are extremely poisonous to people as well as other animals and plants (Malik et al., 2022). At a compound annual growth rate (CAGR) of 8.90%, the worldwide pesticide market is anticipated to expand from USD 78.16 billion in 2021 to USD 85.11 billion in 2022. In 2022, the worldwide herbicide market was valued at a substantial USD 31.50 billion (Shehbaz, 2023). With limited resources, there needs to be a concerted effort to find healthy and environmentally sustainable ways to increase food production despite these major health and environmental concerns. New and creative methods need to be implemented in agriculture in order to meet the European Union's Farm to Fork Strategy's requirement of a 20% reduction in chemical fertilization (Heyl et al., 2023).

Nanotechnology has introduced new and resourceful advancements to the agricultural sector by enhancing conventional agricultural methods and practices. Agri-nanotechnology is likely to significantly influence agricultural sustainability and crop growth. The agricultural industry stands to benefit greatly from nanotechnology, according to recent studies, which might increase food production and security by solving agricultural challenges and making agricultural inputs more efficient (Figure 10.1). Conventional farming methods have been revolutionized by the potential application of nanoscale agrochemicals such as nanofertilizers, nanopesticides, nanosensors, and nanoformulations, which have increased their efficiency and sustainability (Ahmad et al., 2022; Haris et al., 2023;).

Weeds significantly contribute to agricultural damage, resulting in a reduction of around 10–25% in crop yield in underdeveloped countries (Gharde and Singh, 2018; Horvath et al., 2023). A novel approach to weed management in the context of climate change is exemplified by nanoherbicides (NHs). These novel nanotechnologies are regarded as a potential discovery in the realm of agrochemicals. Nanoherbicides ensure minimal toxicological effects and limited residues in soil and the environment; due to their targeted nature, they necessitate smaller quantities, hence reducing costs (Amna et al., 2019). Primarily, NHs are utilized as a foliar spray. They penetrate the stomata and subsequently utilize xylem or apoplastic transport to arrive at the membrane (Akintelu et al., 2023).

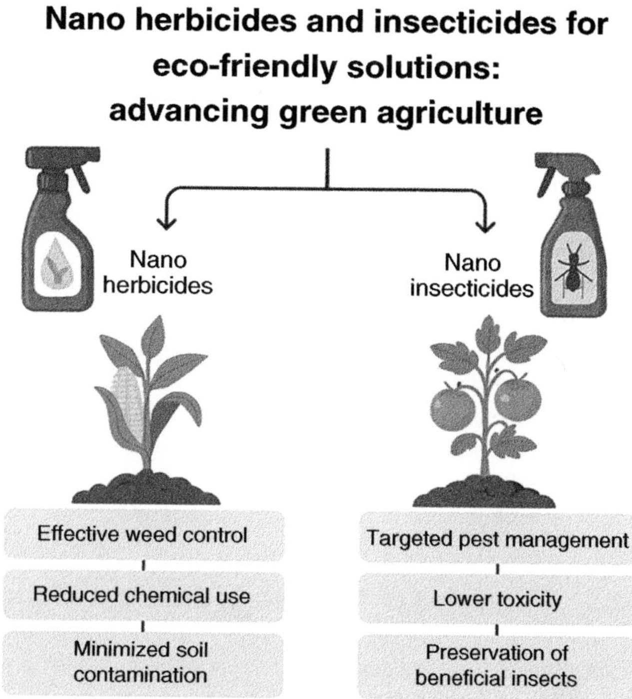

Figure 10.1 Eco-friendly benefits of nanoherbicides and nanoinsecticides.

Insect pests pose a significant threat to agricultural and post-harvest grain crops. The quality, quantity, and commercial worth of the products are significantly influenced by these insect pests. To mitigate the significant losses inflicted by insects on agriculture, many pesticides have been employed to exterminate them or suppress their reproduction and feeding behaviors (Vinutha et al. 2013). Approximately 2 million tons of pesticides are utilized globally each year, with Europe accounting for 45%, the USA for 25%, and the remainder for the rest of the globe. The excessive dependence on chemicals has resulted in numerous issues: heightened occurrences of pesticide poisoning among farmers, diminished soil biodiversity, mammalian toxicity, the emergence of insecticide resistance, bioaccumulation of pesticides, the presence of insecticidal residues on commercially available vegetables and fruits, threats to nontarget organisms within the ecosystem, and a decline in pollinator populations. Due to the detrimental effects of pesticides, significant efforts have been dedicated to discovering new pest control tactics through the implementation of revolutionary pest-resistant concepts and improved pest management technology (Kapinder & Verma, 2021).

Nanopesticides are characterized as formulations or products that incorporate manufactured nanoparticles possessing biocidal properties. The incorporation of nanotechnology into pesticide applications offers novel possibilities for traditional agricultural methods. Integrating nanotechnology into pesticide formulation has the potential to markedly improve pesticide efficacy while simultaneously reducing negative environmental effects (Adisa et al., 2019).

Nanopesticides exist in several forms, including lipids, polymers, and metal-organic frameworks. These materials demonstrate distinct processes for pesticide delivery and release. Nanopesticides made of polymers have the potential to achieve microstructural tunability, allowing for precise control over pesticide attachment and release on crop surfaces (Sun et al., 2020). However, because lipids are soluble in water, nanopesticides based on these layers may improve persistence and stability by releasing the pesticides gradually into the water column. In addition, due to their highly organized pore structure, metal-organic frameworks may be able to achieve efficient loading and progressive release of pesticides. Compared to their commercial counterparts, certain nanopesticide compounds have shown to be far more effective (Tang et al., 2024).

In a nutshell, nanoherbicides and nanoinsecticides offer targeted, efficient, and controlled release of active ingredients (AIs), minimizing environmental contamination and reducing harm to nontarget organisms. By enhancing pesticide stability and bioavailability, these nanoformulations help optimize agricultural productivity while addressing concerns related to chemical overuse. As agriculture moves toward sustainability, nanobased solutions provide a promising alternative to conventional agrochemicals, ensuring better crop protection with minimal ecological impact.

10.3 Principles of Nanoherbicides and Nanoinsecticides

Nanotechnology utilizes devices and nanomaterials at the nanoscale to develop advanced pest management systems. The specific delivery approach of nanoparticles includes directing pesticides to pest locations along with specific areas to minimize environmental contamination. The technique of controlled release lets active components distribute over time, thus extending effectiveness and requiring less frequent applications. Real-time detection of insect demographics and environmental elements relies on nanotechnology due to its employment of miniature sensors. The development of bug-killing compounds and plant-deterrent compounds through nanoparticles and nanomaterials has led to better results while generating diminished environmental effects (Figure 10.2). Nanoformulations execute three key functions: enhancing active substance solubility, along with stability, and then offer to control substance release behavior, which enables precise pest targeting in addition to minimizing harm toward both nontarget animals and ecosystems. Additionally, biopesticides—which incorporate natural pest control agents at the nanoscale for increased efficacy and environmental safety—are made possible by nanotechnology. Engineered nanoparticles are presently employed or hold promise as advanced carriers for the delivery of pesticides. Many different types of nanoformulations have been suggested, including emulsions, encapsulations, vesicles, gels, and nanofibers, which can either improve the environmental safety profiles or increase the effectiveness of currently used AIs, or both (Kookana et al., 2014).

10.3.1 Nanoemulsion

A biphasic dispersion system known as a nanoemulsion is formed by integrating surfactants and AIs that are distributed within the water and oil phases (Figure 10.3). This system demonstrates kinetic stability and exhibits a transparent or translucent appearance attributed to the presence of nanometer-sized droplets (approximately 20–200 nm). A

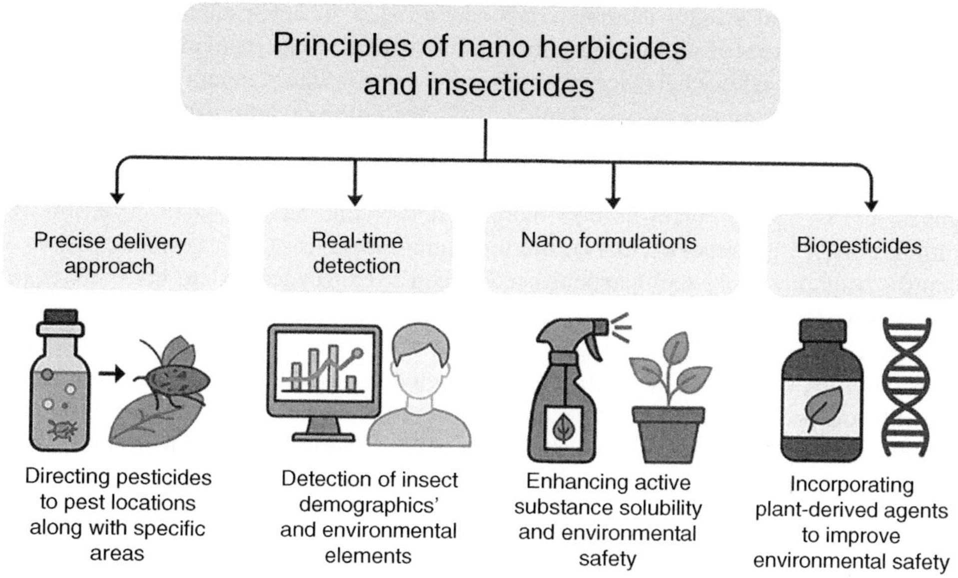

Figure 10.2 Principles of nanoherbicides and nanoinsecticides.

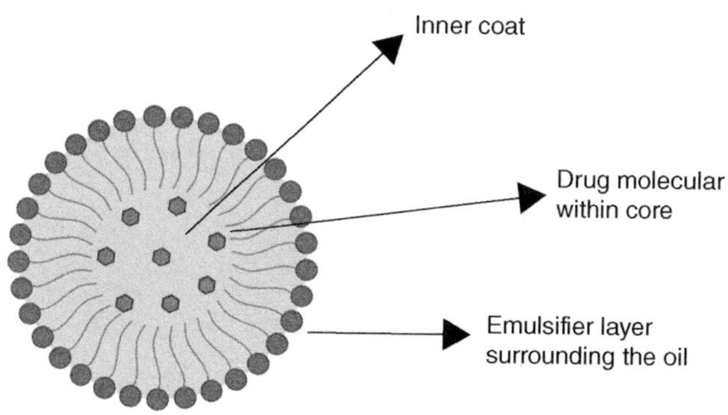

Figure 10.3 Structure of nanoemulsion.

dispersion of oil in water (O/W) known as a nanoemulsion of pesticides has the potential to dissolve the least water-soluble pesticides into tiny oil droplets, greatly increasing their bioavailability and effectiveness (Feng et al., 2018; Wang et al., 2019).

10.3.2 Nanoencapsulation

Nanoencapsulation is actually a delivery technique wherein AIs are encapsulated within diverse nanomaterials and subsequently disseminated in a controlled manner (Nuruzzaman et al., 2016). Nanomaterial-based encapsulation can prevent premature degradation of AIs (e.g., photolysis, hydrolysis, biodegradation) and reduce losses from leaching and volatilization, outperforming traditional pesticide formulations in real-world applications (Zhao et al., 2017).

Following rational design, nanomaterials can provide an encapsulating system that exhibits characteristics of stimulus-responsive release for accurate regulation or that demonstrates sustained release behavior to lengthen the control time (Camara et al., 2019).

Different types of nanomaterials (Figure 10.4), including organic substances such as synthetic and natural polymers (Arasoglu et al., 2016; Liang et al., 2018; Pang et al., 2019; Shakiba et al., 2020; Tong et al., 2017; Yan et al., 2019), lipids (de Oliveira et al., 2015; Nguyen et al., 2012), plant-derived nanoparticles (Cao et al., 2015; Chariou & Steinmetz 2017; Chariou et al., 2019), and inorganic substances (such as silica-, carbon-, calcium-, and clay-based nanoparticles) have been used to encapsulate insecticides in nanoparticles (Chen et al., 2018). By integration, electrostatic complexation, or covalent bonding, AIs are commonly encased within nanocarriers, improving absorption, dispersibility, mobility, adhesion, and controlled or targeted release, thereby augmenting bioavailability and prolonging the effective lifespan of AIs. Lipid-based nanoencapsulation of pesticides facilitates the effective encapsulation of both hydrophobic and hydrophilic AIs without the necessity of organic solvents, while penetration and absorption in insect epidermis also improved, hence augmenting insecticidal efficacy. The use of plant viral nanoparticles for pesticide nanoencapsulation, specifically tobacco mild green mosaic virus and red clover necrotic mosaic virus, has been demonstrated in recent studies to

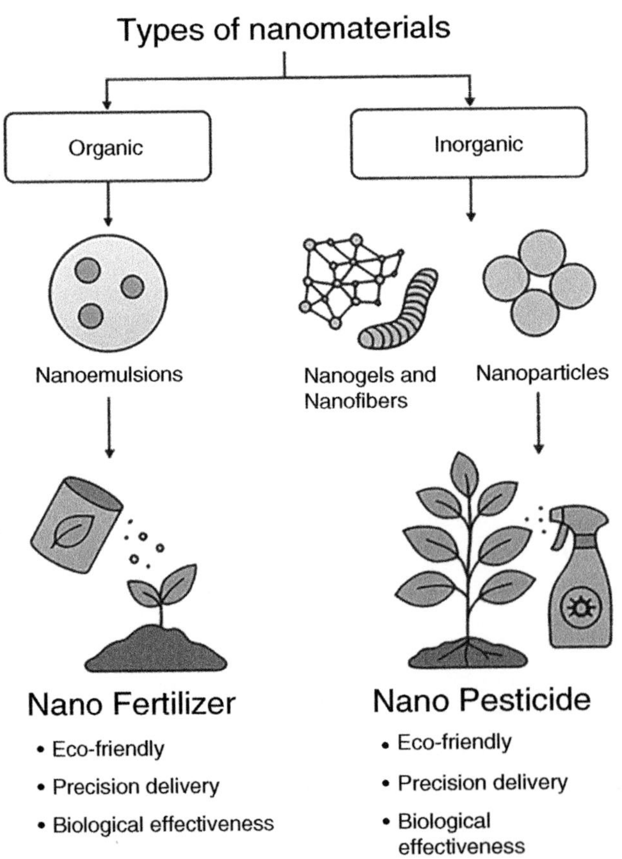

Figure 10.4 Types of nanomaterials (organic and inorganic).

improve crop protection against plant-parasitic nematodes by increasing the mobility and distribution of the nematicide *Abamectin* (Abn) soil in comparison to its unencapsulated form (Chariou et al., 2019).

Silica and calcium carbonate, two examples of inorganic materials with pores, are perfect pesticide transporters due to their extensive surface area, modifiable pore size, substantial loading capacity, and excellent biocompatibility (Abdollahdokht et al., 2022). The encapsulation of AIs in these materials considerably enhances their controlled release capabilities and anti-photolysis capability. Certain articles indicate that by manipulating the pore structures, the release dynamics of encapsulated AIs may be modulated (e.g., 2D hexagonal channels, 3D open network configurations, single large pores, etc.), thereby enhancing practical efficacy (Nuruzzaman et al., 2016). Moreover, many inorganic materials displaying distinct forms or architectures have been recorded for the modulation of encapsulated AI loss (Xiang et al., 2013).

There are two types of release behaviors exhibited by nanopesticides, one is sustained release, and the other is stimulus-responsive release. By slowly releasing nanoencapsulated insecticides over time, AI concentrations can be kept within the desired range for a long time, hence reducing the necessity for repeated pesticide applications. To attain maximal control efficacy, it is essential to produce stimuli-responsive nanoencapsulated pesticides capable of precisely delivering AIs to the target. Numerous stimulus-responsive nanopesticides have been documented to facilitate the intelligent release of AIs in reaction to biotic (plant pathogens, insects, and weeds) or abiotic (temperature, sunlight, drought, soil texture, flooding, and salinity) stimuli, as elaborated in the reviews by Camara et al., (2019), Gao et al., (2018), Gao et al., (2020), Huang et al., (2018), Liang et al., (2020), Lowry et al., (2019), and Tang et al., (2024). The controlled release characteristics suggest the possibility of nanoencapsulated insecticides to enhance pesticide-use efficiency.

10.3.3 Nanoparticles

The nanoparticle-based pesticides addressed in this discussion primarily consist of two categories: nanoparticles utilized directly as pesticides and pesticide-loaded nanoparticles. Pesticides containing organic and inorganic nanoparticles have been the subject of substantial research for their potential to manage plant diseases caused by bacteria and fungi and to ward off insect pests (Pestovsky & Martínez-Antonio, 2017). For instance, weevils (*Sitophilus oryzae*), cattle ticks (*Rhipicephalus microplus*, *Hyalomma anatolicum anatolicum*), mosquitoes (*Anopheles subpictus*, *Culex quinquefasciatus*, and *Culex gelidus*), and *Haemaphysalis bispinosa* have all been shown to be susceptible to the insecticidal effects of silica, nickel, and titanium dioxide nanoparticles (Debnath et al., 2011). Because of their antifungal or antibacterial qualities, a variety of metalloids (such as gold, silver, copper, and platinum) and metal oxides (such as silver oxide, zinc oxide, copper oxide, and titanium dioxide) as well as nonmetal nanoparticles (such as sulfur, reduced graphene oxide, and chitosan) can be used directly as nanopesticides (Roy Choudhury et al., 2012; Saharan et al., 2013).

Recent advancements in bottom-up and top-down nanotechnology for the formulation of pesticide AIs, such as nanocrystalline particles, have demonstrated considerable potential in enhancing the solubility and dispersion of pesticides with low water solubility (Zhao et al., 2017). According to Cui et al. (2019), pesticides' solubility, permeability, dispersion, penetration, and bioactivity are all much improved by nanosizing them. Nanosuspensions have demonstrated greater retention on leaf surfaces, superior anti-photolysis, and increased bioavailability relative to commercial water-based emulsions. Wang and colleagues used the self-emulsifying technique to create a pyraclostrobin solid nanodispersion with a zeta potential of −29.3 mV and an average particle size of 20 nm, exhibiting improved wettability, retention, and storage stability (Wang et al., 2019). The solid nanodispersion exhibited about a fourfold enhancement in fungicidal efficacy relative to the traditional water-dispersible granule. Because they are easily produced on a wide scale and can be used for almost all pesticides that are poorly soluble in water, pesticide nanoparticles are quickly becoming an important new way to reduce pesticide consumption.

10.3.4 Nanogels

Research have shown interest in the nanogels used as nanoscopic drug carriers, which are mainly used for the delivery of bioactive mediators in a controlled time- or site-specific manner. Cross-linking hydrophilic polymers using van der Waals forces or covalent bonds enables them to absorb large volumes of water. Essential oils and pheromones delivered by nanogels have been the subject of extensive research (Abreu et al., 2012). This demonstrates the potential of nanogel systems for the regulated, targeted, and continuous delivery of AIs.

Biological control methods that use pheromones must be slow-released, protected against environmental degradation, and specific to the target population. Evidence has shown that evaporation of AIs in nanogel carriers is much reduced than when pure AIs are applied. This achieves efficacy of up to 33 weeks beyond their efficacy when pure AIs can only be effective for 3 weeks. One case in point is pure chitosan nanogels improve the efficacy of copper-based antifungal therapies (Brunel et al., 2013).

10.3.5 Electrospun Nanofibers

Electrospinning is emerging as a valuable nanotechnology used for the fabrication of nanofiber materials and is available for several biomedical fields. Electrospinning techniques are capable of scaling up to fulfill the demands of manufacturing industries. Recently, scientists have investigated electrospun nanofibers for use in plant protection. Despite being in the nascent phase of agricultural implementation, electrospun nanofibers possess the advantage of circumventing release surges, thereby enhancing the field application of essential oils and pheromones (Sun et al., 2020).

According to Hellman et al., cellulose acetate or polyamide nanofibers can be successfully modified to include pheromones (about 30% of weight), producing a release that is almost linear over a period of weeks (Hellmann et al., 2011).

Similarly, nanocrystals of poly(lactic acid) and cellulose were used to create a comparable nanofiber network by Xiang et al. The thiamethoxam-impregnated fibers showed effectiveness against whitefly for 9 days in a greenhouse testing at 50% of the recommended dosage for the active components (Xiang et al., 2013). Pesticides were administered locally when seeds were coated with electrospun cellulose diacetate (CDA) nanofibers infused with fluopyram or Abm, demonstrating that nanofibrous seed coatings could serve as a viable alternative for managing plant pathogens, including fungi and nematodes (Farias et al., 2019). In a separate work, the authors synthesized the thiram/hydroxypropyl-β-cyclodextrin inclusion complex nanofiber (thiram/HPβCD-IC-NF) via electrospinning for the effective delivery of pesticides (Gao et al., 2020).

10.3.6 Silica

There is a lot of evidence that shows how silicon can make plants more resistant to various stress conditions. Thus, one possible approach for better pest management in agriculture is the use of silica nanoparticles (Kah & Hofmann, 2014), which have subsequently been evaluated for their pesticidal efficiency. The porous structure, high surface reactivity, and strong adsorption properties of silicon dioxide nanoparticles (SiO_2 NPs) make them very appropriate for a variety of applications, including the development of nanoinsecticides. The porosity improves the UV-shielding properties of the siliceous frameworks, which interact with the pesticides to increase their efficacy (Zhang et al., 2023) and enable them to serve as superior nanocarriers for diverse agrochemicals. In agriculture, they perform a dual purpose: first, as pesticides that farmers apply directly to the ground to kill insects; second, as herbicides and insecticide carriers, since they extend the life and efficiency of other commercial pesticides (Rastogi et al., 2019).

But the efficacy of silica nanoparticles as an insecticide may vary on the basis of their composition and origin. For example, the efficacy of a chemical sourced from crystalline silica samples was shown to be less successful in preventing the potato tuber moth than that obtained from amorphous silica powders (Idris et al., 2023).

10.4 Types of Nanomaterials Used in Nanoherbicides and Nanopesticides

10.4.1 Nanoemulsions

Nanoemulsions function as colloidal systems of nanodroplets that enhance pesticide delivery systems because of their key advantages in delivery systems. The abuse resistance of nanoemulsions improves hydrophobic AI solubility, thus increasing pest insect response to these substances. Prolonged shelf life of pesticides results from nanoemulsion stability because it minimizes degradation and enhances the uniformity of pest control applications (Mossa et al., 2018). Multiple research studies indicate that nanoemulsions enhance pesticide penetration into insect bodies while requiring reduced dosages, thus reducing adverse environmental effects (Rao and Geckeler, 2011).

10.4.2 Nanoparticles

According to Hegde et al. (2022), silver and gold along with zinc oxide metallic nanoparticles exhibit strong insecticidal effects. Nanoparticles possess the capability to enter insect cuticles while they simultaneously create oxidative stress that damages cell components, resulting in insect death (Raj et al., 2023). Silver nanoparticles exhibit strong insecticidal behavior against various insect pests even among populations that demonstrate resistance to traditional pesticides (Pathak et al., 2022). The diminutive size of nanoparticles accommodates cellular and molecular interactions through their novel approach, thus decreasing the possibility of resistance advancement (Iravani, 2011).

10.4.3 Nanogels and Nanofibers

AIs are sustained-released using nanogels and nanofibers so that there is prolonged pest control and reduced frequency of application (Usman et al., 2020). Cross-linked polymer networks, known as nanogels, can absorb great amounts of water and encapsulate pesticides that are released slowly over time, protecting against insect pests continuously (Servin & White, 2016). While mats and films made with nanofibers can be created that gradually release insecticides and offer a physical coverage of chemical control (Usman et al., 2020), nanofibers are also used to just release chemical control. They help with the efficacy of pest management as well as reduce the environmental footprint from the release of pesticides into the environment at any particular moment (Chithrani et al., 2006).

10.4.4 Nanocarriers

Delivery of pesticides is enhanced using nanocarriers like liposomes and polymeric nanoparticles, which also serve to provide controlled release and targeted delivery of the pesticides to insect pests (Chandraker et al., 2021). By placing the AIs in these carriers, these carriers protect the AIs from the degradation of the environment, and they prevent the release of the AIs until they have reached the target site, reducing nontarget effects and minimizing environmental contamination (Chithrani et al., 2006). Particularly, liposome-based nanocarriers have been extensively studied for the delivery of hydrophilic and hydrophobic pesticides and therefore provide versatility in pest management strategies (Kumar et al., 2023). In addition, polymeric nanoparticles can be built to respond to particular environmental types for instance, pH or temperature adjustments, which makes pesticide application even more specific (Kerry et al., 2019).

10.5 Benefits of Nanoherbicides and Insecticides

In agricultural fields, it encourages farmers to use more pesticides, which raises application costs, causes herbicide resistance, and accumulates toxins in the environment. Nanoenabled herbicide delivery alleviates environmental problems and reduces labor expenses by utilizing a smaller quantity of herbicide (An et al., 2022). They are effective against variety of plant diseases caused by different organisms such as fish, microbes, oomycetes, and other pathogens (Paul et al., 2019).

Because AI molecules are released over an extended period of time without experiencing external losses, nanoenabled herbicides lower the total amount of herbicide that is needed (de Oliveira et al., 2015; Zargar et al., 2023). The results of applying nanoencapsulated atrazine, clodinofop propargyl, and fenoxaprop-*p*-ethyl against weeds at 10 times the concentration of commercial formulations are similar (Khan et al., 2022a, 2023). Nanoherbicides that are encapsulated possess an extensive surface area, hence necessitating significantly lower quantities of active chemicals during their initial practical use (Kremer, 2019). Using encapsulated

Table 10.1 Examples of nanoinsecticides formulations.

Name	Nanoformulation	Active ingredients (AIs)
Neonicotinoids	Nanoencapsulation	Imidacloprid
Pyrethroid	Chitosan encapsulation	Etofenprox
Rotenoids	Chitosan encapsulation	Rotenone
Pyrethroid	Suspension	Beta-cypermethrin
Pyrethroid	Encapsulation	Cyfluthrin
Organophosphate	Emulsion	Triazofos

herbicide formulations improved by nanotechnology enhances the weed-control efficacy and reduces environmental hazards (Kah and Hofmann, 2014). A high surface area-to-volume ratio of nanoparticles guarantees their efficiency as a vector for bioactive chemicals (Xin et al., 2020) as well as a transporter. Insecticides formulated with encapsulation are the most common controlled-release formulation (Frandsen et al. 2010), although some insecticides have appeared recently with a new approach that mixes different formulations such as encapsulation of nanoemulsions or liposome coating, which are more effective against some insects (Table 10.1) (Liu et al. 2008).

Nanoherbicides are meticulously designed to surmount many limitations, including evaporation, volatilization, leaching, drainage, erosion, photochemical degradation, and both microbiological and chemical degradation (An et al., 2022; Paul et al., 2019). Nanoencapsulated herbicides enhance the wettability and dispersion of herbicide formulations because of their large surface area and small amounts of herbicidal AIs (Forini et al., 2022). The enhancement of nanoherbicides' herbicidal performance relative to conventional herbicides largely focuses on improving their physicochemical characteristics, such as size, shape, surface chemistry, and concentration. The way that nanomaterials interact with weed leaves, roots, and seeds is determined by their chemical makeup (Forini et al., 2022).

The encapsulation method of biodegradable polymers provides numerous advantages, such as improved efficacy, sustained release of biological compounds, protection from environmental factors (Sahoo et al., 2022), and extended material shelf life (An et al., 2022; Campos et al., 2023). In contrast to conventional herbicides, which exhibit reduced efficacy in arid settings (Sahoo et al., 2022) because the weed seed bank resides in the inert subsurface soil, nanoencapsulated herbicide molecules contained in the soil can restrict weed seeds to germinate if the seeds receive suitable water content for germination.

Nanohydrogels are a special form of agrochemicals that have higher service efficiency and greater environmental sustainability. These formulations have great stability and low solubility, ensuring the prolonged release of AIs while mitigating pest issues without affecting other organisms (Artusio et al., 2021; Ghaderpoori et al., 2020; Kashyap et al., 2015).

According to Khot et al. (2012), the use of nanopesticides in agricultural formulations improves wettability and dispersion, which reduces the need for organic solvents and prevents pesticide displacement by accident. The development of controlled-release formulations, the improvement of nutrient use, and the reduction of environmental problems all depend on the integration of nanoparticles into nanoscale materials. Silica nanoparticles serve as versatile components in this regard, acting as efficient agents for nanopesticides, nanoherbicides, and nanofertilizers, with considerable promise in many agricultural operations (Awad-Allah, 2023).

Nanopesticides provide an increased surface area for greater insect contact and can be tailored for selective application via surface changes. Permethrin, a pyrethroid insecticide, was transformed into a hydro-dispersible nanopowder via the rapid solvent evaporation of a pesticide-laden oil-in-water microemulsion. When compared to bulk and control treatments, nanopermethrin was much more harmful to mosquito larvae, damaging epithelial cells and midgut contents while reducing quantities of cellular macromolecules and biomarker enzyme activity (Mishra et al., 2017).

10.6 Challenges to Nanoherbicides and Insecticides

At present, nanopesticides are garnering significant attention, and with the advancement of novel formulations, they are expected to attract considerable focus in the future (Kah et al., 2013; Kah and Hofmann, 2014). Conventional pesticides present multiple issues, including soil and water contamination due to the slow breakdown of chemicals, disruption of ecosystems, and hazardous effects on non-target creatures (Patel et al., 2014; Stamatis et al., 2010). Conversely, nanopesticides specifically target plant pests while leaving nontarget species unharmed, rendering them environmentally benign and safer alternatives. Nonetheless, the enduring impacts of these nanoparticles on both people and the environment remain ambiguous and even unforeseen.

The efficacy of nanopesticides is examined for their potential nontarget impacts on other creatures within the ecosystem. Nanotechnology has led to new insecticides, but research on the molecular principles behind their activity in insects, microorganisms, plants, soil, and the environment remains limited. Nanoscale pesticides have complicated and unclear consequences due to the indeterminate degradation mechanisms of these nanochemicals in soil and the environment (Kah et al., 2016; Scheringer et al., 2014). Only a few studies have investigated the bioaccumulation, transportation, and ultimate fate of nanoscale materials. Despite their microscopic size, certain organic groups and AIs of nanosized pesticides persist and accumulate in the environment because of their complex nature, posing both short-term and long-term dangers (Ramadass & Thiagarajan, 2017).

Although many studies have shown that silver nanoparticles formulated fungicides showed a great activity against fungal plant infections, these formulations are also linked with soil toxicity and soil interaction (Rafique et al., 2017; Worrall et al., 2018). Residues from nanoformulations in food and the toxicity of nanoparticles to human health upon direct exposure are significant concerns that hinder the regulation and legislation governing the marketing of novel nanoformulation insecticides. Human cells, tissues, organs, and DNA can all suffer serious damage from the long-term effects of nanomaterials, especially metal nanoparticles (He et al., 2019; McClements & Xiao, 2017). Chemical and physical methods used to create nanosized active compounds have the potential to increase chemical toxicity, which can lead to genotoxicity and behavioral abnormalities among organisms. Under the present pesticide regulatory framework, neither the field behavior of nanopesticides nor their evaluation has been adequately investigated. The growing interest in employing nanopesticides for agricultural protection requires a comprehensive assessment of the environmental dangers linked to these nanomaterials for particular regulatory purposes (Kookana et al., 2014). Currently, our understanding of the ecotoxicological implications of nanopesticides is poor. To better understand the nano-bio-eco interactions between nanomaterials and both biotic and abiotic components within an interconnected ecosystem, additional toxicological research and risk assessment methodologies are necessary. To definitively evaluate nanotoxicity's effects on people and the planet, a case-by-case approach is needed (Grillo et al., 2021; He et al., 2019; Iavicoli et al., 2017).

10.7 Conclusion and Future Perspective

Contemporary agricultural practices are focused on sustainability. Non-toxic "green herbicides" are growing in demand in modern agriculture. To ensure global food security, the scientific community fervently supports the use of nanoenabled biodegradable pesticides. The global pesticide market is projected to reach 82.9 billion dollars by 2027 (Ali et al., 2023), necessitating the development of natural, eco-friendly, and sustainable alternatives due to humans and environmental concerns associated with their use. Most ecologically sustainable pesticides are biopesticides, derived from natural sources rather than chemically manufactured. Nonetheless, their effectiveness and ecological consequences require additional research to optimize dosage, species specificity, physical ambient conditions, and administration methods. Nano-agrochemicals have emerged as a possible alternative owing to their distinctive characteristics, including accessibility, cost-effectiveness, mode of action, environmental sustainability, enhanced efficacy at reduced dosages, and appealing nanoscale dimensions. They have the potential to transform farming, increase harvests, and lessen the toll on the environment. For example, metal nanoparticles, for instance, can kill germs by releasing reactive oxygen species, rupturing cell membranes, and interfering with cellular processes; these and other nanopesticides and nanoherbicides can be modified for use in agriculture. They possess tremendous potential; yet their environmental dangers and conditions require additional examination and comprehensive optimization. Comprehending and mitigating these hazards necessitates perpetual research and constant surveillance.

Research on nanotechnology in agriculture is in its nascent phases but is advancing rapidly. The use of nanomaterials in agrochemicals for protection and growth of plants requires investigation due to their unique physicochemical properties (Raliya et al., 2017). These nanoparticles exhibit distinct nano–bio interactions, transport mechanisms, and environmental fate, enabling prolonged release and resistance for effective plant protection management (Fraceto et al., 2020). The bulk of recently funded studies and forthcoming research initiatives appear to focus on the development of safer, eco-friendly nanomaterials for effective applications (Raliya et al., 2017). Future studies should assess how nano-agrochemicals interact with other contaminants and use natural sources rather than traditional media. Shortly, the application of nanostructured catalysts will be feasible, hence improving the efficacy of pesticides and herbicides and enabling the utilization of lower dosages.

Experts must research how nanoparticles affect the environment and human health and improve the techniques to evaluate and reduce any possible threats they may pose. Scientists must also identify renewable techniques for the fabrication of nanoscale materials intended for agricultural applications. Furthermore, for improved technological validation, broad approaches should be developed in top universities or nations to assess laboratory-scale nanoproducts in actual agricultural settings. Aligning users and specialists in satisfying prospective customers' desires should be the ultimate goal of educational and developmental activities carried out through research initiatives.

References

Abdollahdokht, D., Gao, Y., Faramarz, S., Poustforoosh, A., Abbasi, M., Asadikaram, G., & Nematollahi, M. H. (2022). Conventional agrochemicals towards nano-biopesticides: An overview on recent advances. *Chemical and Biological Technologies in Agriculture*, *9*(1), 13.

Abhiram, G. (2023). Contributions of nano-nitrogen fertilizers to sustainable development goals: A comprehensive review. *Nitrogen*, *4*(4), 397–415.

Abreu, F. O., Oliveira, E. F., Paula, H. C., & de Paula, R. C. (2012). Chitosan/cashew gum nanogels for essential oil encapsulation. *Carbohydrate Polymers*, *89*(4), 1277–1282.

Adisa, I. O., Pullagurala, V. L. R., Peralta-Videa, J. R., Dimkpa, C. O., Elmer, W. H., Gardea-Torresdey, J. L., & White, J. C. (2019). Recent advances in nano-enabled fertilizers and pesticides: A critical review of mechanisms of action. *Environmental Science: Nano*, *6*(7), 2002–2030.

Ahmad, M., Ali, A., Ullah, Z., Sher, H., Dai, D. Q., Ali, M., & Ali, I. (2022). Biosynthesized silver nanoparticles using *Polygonatum geminiflorum* efficiently control fusarium wilt disease of tomato. *Frontiers in Bioengineering and Biotechnology*, *10*, 988607.

Akintelu, S. A., Olabemiwo, O. M., Ibrahim, A. O., Oyebamiji, J. O., Oyebamiji, A. K., & Olugbeko, S. C. (2023). Biosynthesized nanoparticles as a rescue aid for agricultural sustainability and development. *International Nano Letters*, 13(1), 15–40.

Ali, S., Ahmad, N., Dar, M. A., Manan, S., Rani, A., Alghanem, S. M. S., Khan, K. A., Sethupathy, S., Elboughdiri, N., Mostafa, Y. S., & Alamri, S. A. (2023). Nano-agrochemicals as substitutes for pesticides: prospects and risks. *Plants*, *13*(1), 109.

Amna, A.,. H. F., Hakeem, K. R., & Qureshi, M. I. (2019). Weed control through herbicide-loaded nanoparticles. In *Nanomaterials and plant potential* (pp. 507–527). Cham: Springer International Publishing.

An, C., Sun, C., Li, N., Huang, B., Jiang, J., Shen, Y., Wang, C., Zhao, X., Cui, B., Wang, C., & Li, X. (2022). Nanomaterials and nanotechnology for the delivery of agrochemicals: strategies towards sustainable agriculture. *Journal of Nanobiotechnology*, *20*(1), 11.

Arasoglu, T., Mansuroglu, B., Derman, S., Gumus, B., Kocyigit, B., Acar, T., & Kocacaliskan, I. (2016). Enhancement of antifungal activity of juglone (5-hydroxy-1, 4-naphthoquinone) using a poly (d, l-lactic-co-glycolic acid)(PLGA) nanoparticle System. *Journal of Agricultural and Food Chemistry*, *64*(38), 7087–7094.

Artusio, F., Casà, D., Granetto, M., Tosco, T., & Pisano, R. (2021). Alginate nanohydrogels as a biocompatible platform for the controlled release of a hydrophilic herbicide. *PRO*, *9*(9), 1641.

Awad-Allah, E. F. (2023). Effectiveness of silica nanoparticle application as plant nano-nutrition: A review. *Journal of Plant Nutrition*, *46*(11), 2763–2776.

Brunel, F., El Gueddari, N. E., & Moerschbacher, B. M. (2013). Complexation of copper (II) with chitosan nanogels: Toward control of microbial growth. *Carbohydrate Polymers, 92*(2), 1348–1356.

Camara, M. C., Campos, E. V. R., Monteiro, R. A., do Espirito Santo Pereira, A., de Freitas Proença, P. L., & Fraceto, L. F. (2019). Development of stimuli-responsive nano-based pesticides: emerging opportunities for agriculture. *Journal of Nanobiotechnology, 17*, 1–19.

Campos, E. V., Ratko, J., Bidyarani, N., Takeshita, V., & Fraceto, L. F. (2023). Nature-based herbicides and micro-/nanotechnology fostering sustainable agriculture. *ACS Sustainable Chemistry & Engineering, 11*(27), 9900–9917.

Cao, J., Guenther, R. H., Sit, T. L., Lommel, S. A., Opperman, C. H., & Willoughby, J. A. (2015). Development of abamectin loaded plant virus nanoparticles for efficacious plant parasitic nematode control. *ACS Applied Materials & Interfaces, 7*(18), 9546–9553.

Chandraker, S. K., Ghosh, M. K., Lal, M., & Shukla, R. (2021). A review on plant-mediated synthesis of silver nanoparticles, their characterization and applications. *Nano Express, 2*(2), 022008.

Chariou, P. L., & Steinmetz, N. F. (2017). Delivery of pesticides to plant parasitic nematodes using tobacco mild green mosaic virus as a nanocarrier. *ACS Nano, 11*(5), 4719–4730.

Chariou, P. L., Dogan, A. B., Welsh, A. G., Saidel, G. M., Baskaran, H., & Steinmetz, N. F. (2019). Soil mobility of synthetic and virus-based model nanopesticides. *Nature Nanotechnology, 14*(7), 712–718.

Chen, C., Zhang, G., Dai, Z., Xiang, Y., Liu, B., Bian, P., Zheng, K., Wu, Z., & Cai, D. (2018). Fabrication of light-responsively controlled-release herbicide using a nanocomposite. *Chemical Engineering Journal, 349*, 101–110.

Chithrani, B. D., Ghazani, A. A., & Chan, W. C. (2006). Determining the size and shape dependence of gold nanoparticle uptake into mammalian cells. *Nano Letters, 6*(4), 662–668.

Cui, B., Lv, Y., Gao, F., Wang, C., Zeng, Z., Wang, Y., Sun, C., Zhao, X., Shen, Y., Liu, G., & Cui, H. (2019). Improving abamectin bioavailability via nanosuspension constructed by wet milling technique. *Pest Management Science, 75*(10), 2756–2764.

Das, T. K., Behera, B., Nath, C. P., Ghosh, S., Sen, S., Raj, R., Ghosh, S., Sharma, A. R., Yaduraju, N. T., Nalia, A., & Dutta, A. (2024). Herbicides use in crop production: An analysis of cost-benefit, non-target toxicities and environmental risks. *Crop Protection*, 106691.

Debnath, N., Das, S., Seth, D., Chandra, R., Bhattacharya, S. C., & Goswami, A. (2011). Entomotoxic effect of silica nanoparticles against *Sitophilus oryzae* (L.). *Journal of Pest Science, 84*, 99–105.

Farias, B. V., Pirzada, T., Mathew, R., Sit, T. L., Opperman, C., & Khan, S. A. (2019). Electrospun polymer nanofibers as seed coatings for crop protection. *ACS Sustainable Chemistry & Engineering, 7*(24), 19848–19856.

Feng, J., Zhang, Q. I., Liu, Q. I., Zhu, Z., McClements, D. J., & Jafari, S. M. (2018). Application of nanoemulsions in formulation of pesticides. In *Nanoemulsions* (pp. 379–413). Academic Press.

Forini, M. M., Pontes, M. S., Antunes, D. R., de Lima, P. H., Santos, J. S., Santiago, E. F., & Grillo, R. (2022). Nano-enabled weed management in agriculture: From strategic design to enhanced herbicidal activity. *Plant Nano Biology, 1*, 100008.

Fraceto, L. F., De Castro, V. L. S., Grillo, R., Ávila, D., Oliveira, H. C., & Lima, R. (2020). *Nanopesticides*. Springer International Publishing.

Frandsen, S., Lehmann, H., Hekker, S., Southworth, J., Debosscher, J., Beck, P., ... & Elsworth, Y. (2013). KIC 8410637: a 408-day period eclipsing binary containing a pulsating giant star. Astronomy & Astrophysics, 556, A138.

Gao, C., Huang, Q., Lan, Q., Feng, Y., Tang, F., Hoi, M. P., Zhang, J., Lee, S. M., & Wang, R. (2018). A user-friendly herbicide derived from photo-responsive supramolecular vesicles. *Nature Communications, 9*(1), 2967.

Gao, Y., Xiao, Y., Mao, K., Qin, X., Zhang, Y., Li, D., Zhang, Y., Li, J., Wan, H., & He, S. (2020). Thermoresponsive polymer-encapsulated hollow mesoporous silica nanoparticles and their application in insecticide delivery. *Chemical Engineering Journal, 383*, 123169.

Gardarin, A., Celette, F., Naudin, C., Piva, G., Valantin-Morison, M., Vrignon-Brenas, S., Verret, V., & Mediene, S. (2022). Intercropping with service crops provides multiple services in temperate arable systems: A review. *Agronomy for Sustainable Development, 42*(3), 39.

Ghaderpoori, M., Jafari, A., Nazari, E., Rashidipour, M., Nazari, A., Chehelcheraghi, F., Kamarehie, B., & Rezaee, R. (2020). Preparation and characterization of loaded paraquat-polymeric chitosan/xantan/tripolyphosphate nanocapsules and evaluation for controlled release. *Journal of Environmental Health Science and Engineering, 18*, 1057–1066.

Gharde, Y., & Singh, P. K. (2018). Assessment of yield and economic losses in agriculture due to weeds in India. *Crop Protection, 107*, 12–18.

Grillo, R., Fraceto, L. F., Amorim, M. J., Scott-Fordsmand, J. J., Schoonjans, R., & Chaudhry, Q. (2021). Ecotoxicological and regulatory aspects of environmental sustainability of nanopesticides. *Journal of Hazardous Materials, 404*, 124148.

Haris, M., Fatima, N., Iqbal, J., Chalgham, W., Mumtaz, A. S., El-Sheikh, M. A., & Tavafoghi, M. (2023). *Oscillatoria limnetica* mediated green synthesis of iron oxide (Fe_2O_3) nanoparticles and their diverse in vitro bioactivities. *Molecules, 28*(5), 2091.

He, X., Deng, H., & Hwang, H. M. (2019). The current application of nanotechnology in food and agriculture. *Journal of Food and Drug Analysis, 27*(1), 1–21.

Hegde, M., Pai, P., Shetty, M. G., & Babitha, K. S. (2022). Gold nanoparticle based biosensors for rapid pathogen detection: A review. *Environmental Nanotechnology, Monitoring & Management, 18*, 100756.

Hellmann, C., Greiner, A., & Wendorff, J. H. (2011). Design of pheromone releasing nanofibers for plant protection. *Polymers for Advanced Technologies, 22*(4), 407–413.

Heyl, T. P., Panieri, G., Fornari, D. J., Mattingsdal, R., Sauer, S., Yao, H., ... Shank, T. M. (2023). Implications of transient methane flux on associated biological communities in high-arctic seep habitats, Storbanken, Norwegian Barents Sea. *Deep Sea Research Part I: Oceanographic Research Papers, 201*, 104156.

Horvath, D. P., Clay, S. A., Swanton, C. J., Anderson, J. V., & Chao, W. S. (2023). Weed-induced crop yield loss: A new paradigm and new challenges. *Trends in Plant Science, 28*(5), 567–582.

Huang, B., Chen, F., Shen, Y., Qian, K., Wang, Y., Sun, C., Zhao, X., Cui, B., Gao, F., Zeng, Z., & Cui, H. (2018). Advances in targeted pesticides with environmentally responsive controlled release by nanotechnology. *Nanomaterials, 8*(2), 102.

Iavicoli, I., Leso, V., Beezhold, D. H., & Shvedova, A. A. (2017). Nanotechnology in agriculture: Opportunities, toxicological implications, and occupational risks. *Toxicology and Applied Pharmacology, 329*, 96–111.

Idris, I., Naddaf, M., Harmalani, H., Alshater, R., & Alsafadi, R. (2023). Efficacy of olive stones and corncobs crystalline silica nanoparticles (SiO_2, NPs) treatments on potato tuber moths (*Phthorimaea operculella*). *Silicon, 15*(8), 3591–3598.

Iravani, S. (2011). Green synthesis of metal nanoparticles using plants. *Green Chemistry, 13*(10), 2638–2650.

Islam, M. A., Amin, S. N., Rahman, M. A., Juraimi, A. S., Uddin, M. K., Brown, C. L., & Arshad, A. (2022). Chronic effects of organic pesticides on the aquatic environment and human health: A review. *Environmental Nanotechnology, Monitoring & Management, 18*, 100740.

Junaid, M. D., & Gokce, A. F. (2024). Global agricultural losses and their causes. *Bulletin of Biological and Allied Sciences Research, 2024*(1), 66–66.

Kah, M., & Hofmann, T. (2014). Nanopesticide research: Current trends and future priorities. *Environment International, 63*, 224–235.

Kah, M., Beulke, S., Tiede, K., & Hofmann, T. (2013). Nanopesticides: state of knowledge, environmental fate, and exposure modeling. *Critical Reviews in Environmental Science and Technology, 43*(16), 1823–1867.

Kah, M., Weniger, A. K., & Hofmann, T. (2016). Impacts of (nano) formulations on the fate of an insecticide in soil and consequences for environmental exposure assessment. *Environmental Science & Technology, 50*(20), 10960–10967.

Kapinder, K. B., & Verma, A. K. (2021). Corona outbreak: Is it nature's wake-up call. *International Journal of Applied Biology and Pharmaceutical Technology, 12*, 301–321.

Kashyap, P. L., Xiang, X., & Heiden, P. (2015). Chitosan nanoparticle based delivery systems for sustainable agriculture. *International Journal of Biological Macromolecules, 77*, 36–51.

Kaur, R., Choudhary, D., Bali, S., Bandral, S. S., Singh, V., Ahmad, M. A., Rani, N., Singh, T. G., & Chandrasekaran, B. (2024). Pesticides: An alarming detrimental to health and environment. *Science of the Total Environment, 915*, 170113.

Kerry, R. G., Rout, J. R., Das, G., Fraceto, L. F., Paramithiotis, S., & Patra, J. K. (2019). Applications of nanotechnology in food and agriculture. In *Food molecular microbiology* (pp. 223–249). CRC Press.

Khan, B. A., Nadeem, M. A., Javaid, M. M., Maqbool, R., Ikram, M., & Oraby, H. (2022a). Chemical synthesis, characterization, and dose optimization of chitosan-based nanoparticles of clodinofop propargyl and fenoxaprop-p-ethyl for management of *Phalaris minor* (little seed canary grass): First report. *Green Processing and Synthesis, 11*(1), 1118–1127.

Khan, F., Pandey, P., & Upadhyay, T. K. (2022b). Applications of nanotechnology-based agrochemicals in food security and sustainable agriculture: An overview. *Agriculture, 12*(10), 1672.

Khan, B. A., Nadeem, M. A., Iqbal, M., Yaqoob, N., Javaid, M. M., Maqbool, R., Elnaggar, N., & Oraby, H. (2023). Chitosan nanoparticles loaded with mesosulfuron methyl and mesosulfuron methyl+ florasulam+ MCPA isooctyl to manage weeds of wheat (*Triticum aestivum* L.). *Green Processing and Synthesis, 12*(1), 20228152.

Khot, L. R., Sankaran, S., Maja, J. M., Ehsani, R., & Schuster, E. W. (2012). Applications of nanomaterials in agricultural production and crop protection: A review. *Crop Protection, 35*, 64–70.

Kookana, R. S., Boxall, A. B., Reeves, P. T., Ashauer, R., Beulke, S., Chaudhry, Q., Cornelis, G., Fernandes, T. F., Gan, J., Kah, M., & Lynch, I. (2014). Nanopesticides: guiding principles for regulatory evaluation of environmental risks. *Journal of Agricultural and Food Chemistry, 62*(19), 4227–4240.

Kremer, R. J. (2019). Bioherbicides and nanotechnology: Current status and future trends. *Nano-biopesticides Today and Future Perspectives*, 353–366.

Kulabhusan, P. K., Tripathi, A., & Kant, K. (2022). Gold nanoparticles and plant pathogens: an overview and prospective for biosensing in forestry. *Sensors*, *22*(3), 1259.

Kulkarni, S. (2023). One health implications of agrochemicals and their eco-benign substitutes. In *One health implications of agrochemicals and their sustainable alternatives* (pp. 161–179). Springer Nature Singapore: Singapore.

Kumar, A., Singh, G. A., Gourkhede, P. H., Goutam, P. K., Laxman, T., Pandey, S. K., & Singh, P. (2023). Revolutionizing agriculture: A comprehensive review of nanotechnology applications. *International Journal of Environment and Climate Change*, *13*(11), 3586–3603.

Liang, W., Yu, A., Wang, G., Zheng, F., Hu, P., Jia, J., & Xu, H. (2018). A novel water-based chitosan-La pesticide nanocarrier enhancing defense responses in rice (Oryza sativa L) growth. *Carbohydrate Polymers*, *199*, 437–444.

Liang, Y., Gao, Y., Wang, W., Dong, H., Tang, R., Yang, J., Niu, J., Zhou, Z., Jiang, N., & Cao, Y. (2020). Fabrication of smart stimuli-responsive mesoporous organosilica nano-vehicles for targeted pesticide delivery. *Journal of Hazardous Materials*, *389*, 122075.

Liu, Y., Duan, M., & Yu, Z. (2013). Agricultural landscapes and biodiversity in China. Agriculture, ecosystems & environment, 166, 46–54.

Lord, S. (2024). Data for Background Report 2024-FAO State of Food and Agriculture.

Lowry, G. V., Avellan, A., & Gilbertson, L. M. (2019). Opportunities and challenges for nanotechnology in the agri-tech revolution. *Nature Nanotechnology*, *14*(6), 517–522.

Malik, S. B., Saggu, J. I., Gul, A., Abbasi, B. A., Iqbal, J., Waris, S., & Chalgham, W. (2022). Synthesis and characterization of silver and graphene nanocomposites and their antimicrobial and photocatalytic potentials. *Molecules*, *27*(16), 5184.

McClements, D. J., & Xiao, H. (2017). Is nano safe in foods? Establishing the factors impacting the gastrointestinal fate and toxicity of organic and inorganic food-grade nanoparticles. *npj Science of Food*, *1*(1), 6.

Mishra, J., Singh, R., & Arora, N. K. (2017). Alleviation of heavy metal stress in plants and remediation of soil by rhizosphere microorganisms. *Frontiers in Microbiology*, *8*, 1706.

Mossa, A. T. H., Afia, S. I., Mohafrash, S. M., & Abou-Awad, B. A. (2018). Formulation and characterization of garlic (*Allium sativum* L.) essential oil nanoemulsion and its acaricidal activity on eriophyid olive mites (Acari: Eriophyidae). *Environmental Science and Pollution Research*, *25*, 10526–10537.

Nguyen, T. A. H., Ngo, H. H., Guo, W. S., Zhang, J., Liang, S., Yue, Q. Y., ... & Nguyen, T. V. (2013). Applicability of agricultural waste and by-products for adsorptive removal of heavy metals from wastewater. Bioresource technology, 148, 574–585.

Nuruzzaman, M. D., Rahman, M. M., Liu, Y., & Naidu, R. (2016). Nanoencapsulation, nano-guard for pesticides: A new window for safe application. *Journal of Agricultural and Food Chemistry*, *64*(7), 1447–1483.

Okeke, E. S., Ezeorba, T. P. C., Mao, G., Chen, Y., Feng, W., & Wu, X. (2022). Nano-enabled agrochemicals/materials: Potential human health impact, risk assessment, management strategies and future prospects. *Environmental Pollution*, *295*, 118722.

de Oliveira, J. L., Campos, E. V. R., Goncalves da Silva, C. M., Pasquoto, T., Lima, R., & Fraceto, L. F. (2015). Solid lipid nanoparticles co-loaded with simazine and atrazine: Preparation, characterization, and evaluation of herbicidal activity. *Journal of Agricultural and Food Chemistry, 63*(2), 422–432.

Pandey, D. K., & Mishra, R. (2024). Towards sustainable agriculture: Harnessing AI for global food security. *Artificial Intelligence in Agriculture, 12*, 72–84.

Pang, L., Gao, Z., Feng, H., Wang, S., & Wang, Q. (2019). Cellulose based materials for controlled release formulations of agrochemicals: A review of modifications and applications. *Journal of Controlled Release, 316*, 105–115.

Patel, N., Desai, P., Patel, N., Jha, A., & Gautam, H. K. (2014). Agronanotechnology for plant fungal disease management: A review. *International Journal of Current Microbiology and Applied Sciences, 3*(10), 71–84.

Pathak, V. M., Verma, V. K., Rawat, B. S., Kaur, B., Babu, N., Sharma, A., Dewali, S., Yadav, M., Kumari, R., Singh, S., & Mohapatra, A. (2022). Current status of pesticide effects on environment, human health and it's eco-friendly management as bioremediation: A comprehensive review. *Frontiers in Microbiology, 13*, 962619.

Paul, S.K., Nuruzzaman, M., Correa, T.C. and Naidu, R., 2019. Aluminosilicate nano-skeleton to firm the structure and properties of nano-herbicide formulations.

Pestovsky, Y. S., & Martínez-Antonio, A. (2017). The use of nanoparticles and nanoformulations in agriculture. *Journal of Nanoscience and Nanotechnology, 17*(12), 8699–8730.

Punniyakotti, P., Vinayagam, S., Rajamohan, R., Priya, S. D., Moovendhan, M., & Sundaram, T. (2024). Environmental fate and ecotoxicological behaviour of pesticides and insecticides in non-target environments: Nanotechnology-based mitigation strategies. *Journal of Environmental Chemical Engineering, 12*(5), 113349.

Rafique, M., Sadaf, I., Rafique, M. S., & Tahir, M. B. (2017). A review on green synthesis of silver nanoparticles and their applications. *Artificial Cells, Nanomedicine, and Biotechnology, 45*(7), 1272–1291.

Raj, A., Dubey, A., Malla, M. A., & Kumar, A. (2023). Pesticide pestilence: Global scenario and recent advances in detection and degradation methods. *Journal of Environmental Management, 338*, 117680.

Raliya, R., Saharan, V., Dimkpa, C., & Biswas, P. (2017). Nanofertilizer for precision and sustainable agriculture: Current state and future perspectives. *Journal of Agricultural and Food Chemistry, 66*(26), 6487–6503.

Ramadass, M. and Thiagarajan, P., 2017, November. Effective pesticide nano formulations and their bacterial degradation. In *IOP Conference Series: Materials Science and Engineering* (Vol. 263, No. 2, p. 022050). IOP Publishing.

Rao, J. P., & Geckeler, K. E. (2011). Polymer nanoparticles: Preparation techniques and size-control parameters. *Progress in Polymer Science, 36*(7), 887–913.

Rastogi, A., Tripathi, D. K., Yadav, S., Chauhan, D. K., Živčák, M., Ghorbanpour, M., El-Sheery, N. I., & Brestic, M. (2019). Application of silicon nanoparticles in agriculture. *3 Biotech, 9*, 1–11.

Ren, Z., Cai, T., Wan, Y., Zeng, Q., Li, C., Zhang, J., Ma, K., He, S., Li, J., & Wan, H. (2023). Unintended consequences: Disrupting microbial communities of *Nilaparvata lugens* with non-target pesticides. *Pesticide Biochemistry and Physiology, 194*, 105522.

Roy Choudhury, S., Ghosh, M., & Goswami, A. (2012). Inhibitory effects of sulfur nanoparticles on membrane lipids of *Aspergillus niger*: a novel route of fungistasis. *Current Microbiology, 65*, 91–97.

Saha, S., & Bharadwaj, A. (2023). A step towards smart agriculture using metallic nanostructures. *Plant Stress, 10*, 100216.

Saharan, V., Mehrotra, A., Khatik, R., Rawal, P., Sharma, S. S., & Pal, A. (2013). Synthesis of chitosan based nanoparticles and their in vitro evaluation against phytopathogenic fungi. *International Journal of Biological Macromolecules, 62*, 677–683.

Sahoo, U., Malik, G. C., Banerjee, M., Sahoo, B., & Maitra, S. (2022). Application of nanotechnology in agriculture in India. *International Journal of Natural Sciences, 13*(72), 44422–44429.

Scheringer, M., Praetorius, A., & Goldberg, E. S. (2014). Environmental fate and exposure modeling of nanomaterials. In *Frontiers of nanoscience* (Vol. 7, pp. 89–125). Elsevier.

Schmitz, A., Badgujar, C., Mansur, H., Flippo, D., McCornack, B., & Sharda, A. (2022). Design of a reconfigurable crop scouting vehicle for row crop navigation: a proof-of-concept study. *Sensors, 22*(16), 6203.

Servin, A. D., & White, J. C. (2016). Nanotechnology in agriculture: next steps for understanding engineered nanoparticle exposure and risk. *NanoImpact, 1*, 9–12.

Shakiba, S., Astete, C. E., Paudel, S., Sabliov, C. M., Rodrigues, D. F., & Louie, S. M. (2020). Emerging investigator series: polymeric nanocarriers for agricultural applications: synthesis, characterization, and environmental and biological interactions. Environmental Science: Nano, 7(1), 37–67.

Sharma, A., Kumar, V., Shahzad, B., Tanveer, M., Sidhu, G. P. S., Handa, N., Kohli, S. K., Yadav, P., Bali, A. S., Parihar, R. D., & Dar, O. I. (2019). Worldwide pesticide usage and its impacts on ecosystem. *SN Applied Sciences, 1*, 1–16.

Shehbaz, M. (2023). The future of money: Exploring new trends and technologies. *Qualitative Research Journal for Social Studies, 1*(2), 54–61.

Singh, R. K., Berkvens, R., & Weyn, M. (2021). AgriFusion: An architecture for IoT and emerging technologies based on a precision agriculture survey. IEEE Access, 9, 136253–136283.

Stamatis, N., Hela, D., & Konstantinou, I. (2010). Occurrence and removal of fungicides in municipal sewage treatment plant. *Journal of Hazardous Materials, 175*(1-3), 829–835.

Sun, C., Zeng, Z., Cui, H. and Verheggen, F., 2020. Polymer-based nanoinsecticides: current developments, environmental risks and future challenges-a review. *Biotechnologie, Agronomie, Société et Environnement*, 24.

Tang, Y., Sun, S., Lei, Y., Mo, F., Zhao, J., Li, C., ... Wang, Y. (2024). Assessing the impact of climate and crop diversity on regional greenhouse gas emissions and water demand of cropland. *Agricultural and Forest Meteorology, 355*, 110119.

Tong, Y., Wu, Y., Zhao, C., Xu, Y., Lu, J., Xiang, S., Zong, F., & Wu, X. (2017). Polymeric nanoparticles as a metolachlor carrier: Water-based formulation for hydrophobic pesticides and absorption by plants. *Journal of Agricultural and Food Chemistry, 65*(34), 7371–7378.

Triantafyllidis, V., Mavroeidis, A., Kosma, C., Karabagias, I. K., Zotos, A., Kehayias, G., Beslemes, D., Roussis, I., Bilalis, D., Economou, G., & Kakabouki, I. (2023). Herbicide use in the era of farm to fork: strengths, weaknesses, and future implications. *Water, Air, & Soil Pollution, 234*(2), 94.

Ullah, Z., Gul, F., Iqbal, J., Abbasi, B. A., Kanwal, S., Chalgham, W., & Mahmood, T. (2023). Biogenic synthesis of multifunctional silver oxide nanoparticles (Ag_2ONPs) using Parieteria alsinaefolia delile aqueous extract and assessment of their diverse biological applications. *Microorganisms, 11*(4), 1069.

Usman, M., Farooq, M., Wakeel, A., Nawaz, A., Cheema, S. A., Rehman, H., Ashraf, I., & Sanaullah, M. (2020). Nanotechnology in agriculture: Current status, challenges and future opportunities. *Science of the Total Environment, 721*, 137778.

Vinutha, J. S., Bhagat, D., & Bakthavatsalam, N. (2013). Nanotechnology in the management of polyphagous pest Helicoverpa armigera. *J Acad Indus Res, 1*(10), 606–608.

Wang, Y., Yin, W., & Zeng, J. (2019). Global convergence of ADMM in nonconvex nonsmooth optimization. *Journal of Scientific Computing, 78*(1), 29–63.

Worrall, E. A., Hamid, A., Mody, K. T., Mitter, N., & Pappu, H. R. (2018). Nanotechnology for plant disease management. *Agronomy, 8*(12), 285.

Xiang, Y., Wang, N., Song, J., Cai, D., & Wu, Z. (2013). Micro-nanopores fabricated by high-energy electron beam irradiation: suitable structure for controlling pesticide loss. *Journal of Agricultural and Food Chemistry, 61*(22), 5215–5219.

Xin, X., Judy, J. D., Sumerlin, B. B., & He, Z. (2020). Nano-enabled agriculture: from nanoparticles to smart nanodelivery systems. *Environmental Chemistry, 17*(6), 413–425.

Yadav, A., Yadav, K., Ahmad, R., & Abd-Elsalam, K. A. (2023). Emerging frontiers in nanotechnology for precision agriculture: advancements, hurdles and prospects. *Agrochemicals, 2*(2), 220–256.

Yan, S., Hu, Q., Li, J., Chao, Z., Cai, C., Yin, M., Du, X., & Shen, J. (2019). A star polycation acts as a drug nanocarrier to improve the toxicity and persistence of botanical pesticides. *ACS Sustainable Chemistry & Engineering, 7*(20), 17406–17413.

Zargar, M., Bayat, M., Saquee, F. S., Diakite, S., Ramzanovich, N. M., & Akhmadovich, K. A. (2023). New advances in nano-enabled weed management using poly (Epsilon-Caprolactone)-based nanoherbicides: A review. *Agriculture, 13*(10), 2031.

Zhang, J., Kothalawala, S., & Yu, C. (2023). Engineered silica nanomaterials in pesticide delivery: Challenges and perspectives. *Environmental Pollution, 320*, 121045.

Zhao, X., Cui, H., Wang, Y., Sun, C., Cui, B., & Zeng, Z. (2017). Development strategies and prospects of nano-based smart pesticide formulation. *Journal of Agricultural and Food Chemistry, 66*(26), 6504–6512.

11

Development and Application of Nano-Biosensors for Better Stress Management in Crops

Ilyas Ahmad[1,2]*, Zia Ur Rehman Mashwani*[2]*, Zohaib Younas*[2]*, Tayyaba Yousaf*[2]*, Nazia*[1]*, Ayesha Unzila*[1]*, Zuha Fatima*[1]*, and Waqar Ahmad*[1]

[1] *Department of Botany, Government Degree College no 1, Dera Ismail Khan, KPK, Pakistan*
[2] *Department of Botany, Arid Agriculture university, Rawalpindi, Pakistan*

11.1 Introduction

The world is facing a critical challenge because the demand for food is piling up due to a rapid increase in population, yet crop production has been stagnant in the past few years (Khan et al., 2021). In 2050, the global population is estimated to reach 10 billion, while in 2030, approximately 800 million people will suffer from hunger, and 653 million will be starving (Joint FAO/WHO Expert Committee on Food Additives. Meeting, and World Health Organization, 2011). The world will face significant challenges in ensuring food supply, as the global population is projected to reach 10 billion in 2050 (Ashraf & Javed, 2023). According to the Food and Agriculture Organization (FAO), almost one-third of globally produced food is lost, converting to around 1.3 billion tons annually (2011 estimate) (Kummu et al., 2012). Pre-harvest crop reduction, largely due to diseases, pests, weeds, and abiotic stresses, accounts for 1,051.5 million tons (35% of total losses). Additional losses occur during harvesting and storage, up to 690 million tons (Mesterházy et al., 2020).

Crop yields worldwide are significantly influenced by biotic and abiotic stresses, but the impact of biotic stress on major crops varies by region (Savary et al., 2019). Insect pests result in significant yield losses, ranging from 15% to 20% in major food and cash crops (Rathee & Dalal, 2018). Similarly, yield losses from pathogenic illness are estimated to be between 11 and 30% worldwide (Oerke & Dehne, 2004; Van Esse et al., 2020). Plants that experience abiotic stresses (such as drought, waterlogging, heat, salt, heavy metals, and mineral toxicity) experience morphological, physiological, biochemical, and molecular alterations that negatively impact their growth and development (Figure 11.1) (Atkinson & Urwin, 2012). They have the potential to reduce growth and yield by 50–70% (Acquaah, 2007; Farooq et al., 2009; Savary et al., 2012; Wang et al., 2003; Zahra et al., 2021). Due to abiotic stresses, reactive oxygen species (ROS) are excessively

Nanotechnology Innovations for Food Security and Sustainable Agriculture, First Edition. Edited by Zia Ur Rehman Mashwani, Shah Fahad, Ilyas Ahmad, Shah Saud, Sajid Fiaz, and Taufiq Nawaz.

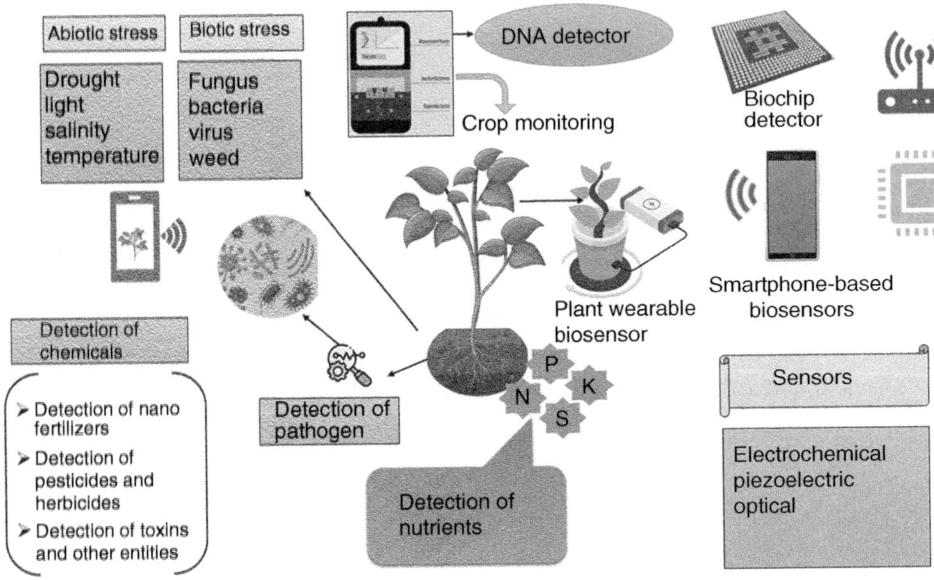

Figure 11.1 Smart sensor-driven systems automate the detection of crop stressors, enable early detection of pathogens and soil nutrient levels, and enable responsive nanofertilizer delivery to enhance crop resilience and yield.

produced, which is the major reason for crop losses (Farooq et al., 2019). According to the global model, CO_2 levels are expected to rise from 400 to 800 ppm (Hofmann et al., 2009; Program, 2009). More than 45% of arid regions face drought conditions (Bot et al., 2000; Thornton et al., 2014). Moreover, contamination from heavy metals (Cr, Cd, As, Pb, Cu, and Hg) decreases respiration, transpiration, photosynthesis, seed germination, and ultimately affects growth, yield, and yield quality (ITPS, FAO, 2015; Manzoor et al., 2018).

11.2 Stress Management for Sustainable Agriculture

Large amounts of food, fiber, and fuel are produced by farms and dairy farms across the United States and are sold at comparatively low prices. Although this abundance is a sign of success, it frequently comes at the expense of the environment, long-term production in agriculture, and public health (Brodt et al., 2011). Every year, the same crop is sown in many fields, which are then usually and effectively mechanically disturbed to remove weeds and absorb agricultural residue. When not in use, the fields are left baren Such activities can contaminate, erode, and otherwise decay the soil. Similar to this, a lot of fertilizer is lost by surface groundwater leaching and surface runoff water, which causes harmful algal blooms and aquatic dead zones, or it escapes into the atmosphere, where it fuels climate change. Large volumes of fertilizer are frequently applied to maximize productivity (Figure 11.1). Another issue is the increasing adoption of herbicide-resistant crops in conjunction with extensive use of herbicides. As a result, herbicide-resistant "superweeds" have emerged, herbicides are carried into nearby farms, and certified organic systems and other farmers now face the additional problem of generating crops that are not herbicide-resistant. The large-scale usage of pesticides is also a danger for human health and the environment when exposed to these chemicals (Bower et al., 2014; Brodt et al., 2011). When combined, these problems focus on how urgently agriculture's sustainability needs to be enhanced (Davis et al., 2012).

Future generations will be less able to produce and flourish if the base of natural resources is diminished by the production of food and fiber. Mesopotamia, the Mediterranean region, the pre-Columbian Southwest United States, and Central America are said to have seen the demise of ancient civilizations due to being heavily stricken by the loss of natural resources as a result of unsustainable forestry and farming methods. The goal of sustainable agriculture is to use natural resources in a way that intensifies their potential for production while reducing negative effects on ecosystems outside of a field's boundaries. Farmers attempt to achieve these objectives, for example, by thinking about how to use already-existing natural processes or how to build their agricultural systems to include essential natural ecosystem activities (Brodt et al., 2011).

To lessen the impact of biotic and abiotic stressors on crop plants, the primary focus has been on using synthetic compounds. According to estimates, up to 3.5 million tons of pesticides were used in 2020 (Zhang, 2018), with a projected worth of USD 103.5 billion. With a consistent 5.5% growth rate, the global pesticide market is expected to reach around USD 107.5 billion and 184 billion in 2023 and 2033, respectively (Pesticides Market Size, Share, Trends & Growth Report, 2033). A 35% reduction in soil respiration, 90% water pollution of agricultural lands, a 70% decrease in insect biomass, a 50% decline in farmland birds, a 30% decline in the honey bee population, and a 42% depletion in species richness were among the health risks and effects of pesticides on nontarget organisms. Additionally, children exposed to pesticides have a 50% chance of developing leukemia, lymphoma, and brain cancer, as well as a 25–30% increased risk of cancer and mental health issues (Ali et al., 2021). It causes difficulties for plants to respond to the combined effects of many stresses. Therefore, a second green revolution is needed to fulfill the world's food needs. Thus, it is necessary to identify some alternative solutions that prioritize human health and environmental sustainability. For achieving the aim of sustainable agriculture, the best alternative solution is nanotechnology as a "key enabling technology" (Parisi et al., 2015; Usman et al., 2020).

Environmental stressors have been recognized in numerous recent studies as significant challenges to 21st-century global food security (Wang & Frei, 2011). Despite the growing demand for agricultural products, several stressors, frequently linked to global warming, pose a threat to agricultural productivity (Challinor et al., 2014; Schlenker & Roberts, 2009; Zhao et al., 2017). To produce biomass, seeds, and yield, plants must adapt to these changing environmental conditions in a physiologically costly manner. Additionally, the combination or modification of various abiotic and biotic stress factors may result in a trade-off between plant responses that are suitable for adaptation to one stress but can increase susceptibility to other stresses (Schön, 2018). It has been discovered that crop plants' growth, development, and productivity are impacted by all of these types of stress (Sanghera et al., 2011).

11.3 Abiotic Stress

According to estimates, ongoing climate change would increase the intensity of abiotic stresses, which could have a more severe impact on crop productivity and serious implications for food security. Abiotic stresses are common in most arable lands worldwide and can cause yield losses of up to 70% in many crops (Tigchelaar et al., 2018; Zörb et al., 2019). Among the main environmental elements that impact crop plant growth, development, and yield are extreme temperatures, water stress (drought and flood), salinity, and heavy metal stress (Vaughan et al., 2018; Zörb et al., 2019). Abiotic stressors that occur frequently and concurrently are more detrimental to food security. These enhance vulnerability to biotic stressors and are not just restricted to abiotic environmental conditions. Since some weeds use water more efficiently than crops, abiotic stressors like drought increase competition between weeds and agricultural plants (Ziska et al., 2010). One of the biggest issues facing agriculture today is abiotic stress. It lowers planted acreage and results in significant losses in crop production globally. This situation gets more complicated as the population grows and the climate changes. By 2050, the world's population is expected to have grown from 7 to 9–10 billion people, necessitating a 60–110% increase in global food production (Rockström et al., 2017). In addition, among other consequences, climate change leads to extreme weather, including drought, severe temperature swings, and excessive precipitation. This affects production and exposes crop species to more abiotic stress. For instance, drought affects 45% of arable land (Bot et al., 2000). Numerous abiotic stressors, including UV, high temperatures, drought, floods, and heavy metals, are frequently experienced by plants. Being sessile, they have developed metabolic and structural adaptations to withstand these conditions, such as increased root area and leaf curling in drought-prone environments; increased synthesis of antioxidants like proline, ascorbic acid, and carotenoids; and increased activity of enzymes involved in ROS scavenging. Oxidative stress, sometimes referred to as a chemical imbalance between pro-oxidants and antioxidants, is brought on by abiotic stress through mechanisms that have been thoroughly reviewed (You & Chan, 2015).

11.3.1 Drought Stress

One of the main things limiting agricultural productivity and significantly impacting crop yield is drought (Escalona et al., 2000; Khan et al., 2013). Plants that are stressed by drought initially react by changing their internal and external structures. The plant develops more slowly and may die as a result of water loss. Plant height reduction, leaf wilting, and changes in leaf number and area are the clear signs of a moisture deficit during the vegetative cycle. Cell enlargement and aging of leaves are intimately linked to plant height, which is significantly impacted by drought. Reduced cell expansion, increased leaf loss, and reduced mitosis during dryness are the primary causes of the decline in plant height. According to Li et al. (2020), plant height in lily and the functional status of leaves—the primary organs for absorption and transpiration—serve as direct indicators of water scarcity. To adapt to drought, plant leaves often take on greater tissue density, greater thickness, and smaller leaf areas (Werner et al., 1999). One of the most obvious characteristics of plant leaves under drought stress is the change in leaf area, which has a direct impact on plant photosynthesis and productivity. Previous research suggests that the accessibility of photoassimilates, canopy temperature, and leaf turgor pressure are the primary causes of changes in plant leaf area (de Freitas et al., 2017). Drought causes a drop in the rate of photosynthesis and leaf turgor pressure, which in turn reduces the area of the leaves (Rucker et al., 1995). In addition to leaves, plant roots are important in drought stress since they are organs that immediately absorb water (Lobet & Draye, 2013). For plants to survive a drought, developed roots can aid in their complete absorption and utilization of the water retained by the soil (Gowda et al., 2011). Thus, in arid regions, the morphological modifications of plant roots are very significant. Root system characteristics, including root density, root hair, and root branches, can have a big impact on a plant's water deficit. Cotton seedling development can be impeded by drought stress, which can also cause fine roots to elongate and thin, limit the lifespan of fine roots of varying diameters, encourage the elongation of root hairs, and hasten their demise (Xiao et al., 2020).

11.3.2 Salinity Stress

A worldwide issue, soil salinity affects about 20% of irrigated land and drastically lowers crop output (Qadir et al., 2014). The global threat of soil salinization impacts 1,100 Mha of soil, or around 7% of the earth's land surface (Sahbeni et al., 2023). Soil salinization is a complex phenomenon that results from both secondary anthropogenic actions and natural geochemical processes (Singh, 2022). Significant losses in agricultural output are caused by the lack of fresh water for irrigation and the continuous deterioration of agricultural lands brought on by salt stress, especially in arid and semiarid regions. Between hundreds and thousands of years ago, soil salinization started. In the past 50 years, it has accelerated the degradation of regions including the San Joaquin Valley in California, the Murray-Darling Basin in Australia, the Yellow River Basin in China, and the Aral Sea Basin in Central Asia (Chang & Brawer Silva, 2014). The primary cause of soil salinization is the buildup of water-soluble salts in the root zone, such as sodium (Na^+), potassium (K^+), chloride (Cl^-), and sulfate (SO_4^{2-}). This results in osmotic fluctuations, which lower the capacity of plant root cells to absorb water from the soil (Kamran et al., 2019; Munns & Tester, 2008; Stavi et al., 2021). Plant cells suffer from hyperionic salt stress, which is caused by the presence of salt ions. Na^+ and Cl^- are thought to be the main ions generating soil salinity among the salinity-inducing water-soluble salts, and excessive Na^+ among cations that exchange causes sodicity (Andrade Foronda, 2022). Salt stress inhibits plant growth, much like a variety of abiotic stresses do, and the rate at which this occurs varies depending on the type of plant, stage of development, and salt content (Yadav et al., 2019). Plants can withstand salt stress by using stunted growth as an adaptive survival mechanism (Munns, 2002). Reduced expression of important regulatory genes involved in cell cycle progression, such as cyclin and cyclin-dependent kinase, may result from salt stress. This could lead to a reduction in the number of cells in the meristem and a growth inhibition that affects the plant's capacity to effectively absorb water and nutrients. When exposed to salt stress, certain plants "panic," react rapidly, and cease to grow. Others, on the other hand, do not react appropriately and run the risk of dying by continuing to develop in extremely salt-stressed environments (Chinnusamy & Zhu, 2003).

11.3.3 Extreme Temperature

The temperature surrounding a plant determines its rate of growth and development, and each species has a minimum, maximum, and optimal temperature range (Hatfield et al., 2011; Schultz, 1964). Temperature increases of more than 5 °C above average can occur during brief periods of extreme weather, lasting a few days. The greatest significant effects on plant productivity would come from extreme occurrences that took place throughout the summer (Kumudini et al., 2014). Crop productivity and the availability of food are greatly impacted by global population increase and major climate change events. Extreme temperatures (cold stress [CS] and heat stress[HS]), like most stresses, can affect a plant's growth and development on its own or in conjunction with other stresses. In the end, they pose serious risks to global food security since they can result in oxidative damage, cellular injury, and even the death of cells after prolonged exposure. Since every plant species has a natural range of ideal temperatures for development and reproduction, drastic changes can quickly have a thermodynamic impact on the intracellular macromolecules (proteins and nucleic acids) and substructures of plant cells (Ruelland & Zachowski, 2010).

11.4 Biotic Stress

Biological stressors on plants include bacteria, viruses, nematodes, weeds, and arachnids. These organisms harm their plant hosts, weaken plants, and have the ability to kill plants. Additionally, they result in agricultural plant losses both before and after harvest (Singla et al., 2016). Biologic stress is an extra difficulty that puts plants under a lot of strain and exacerbates injury by attacking them with pathogens or herbivores (Maron & Crone, 2006; Maron & Kauffman, 2006; Mordecai, 2011; Strauss & Zangerl, 2002). Around the world, fungi, viruses, and bacteria frequently harm agricultural crops, reducing their survival, and biomass production yield, and quality (Culbreath et al., 2008).

11.4.1 Fungal Stress

Plant pathogenic fungi depend on the plants they live in for both carbon and nutrient assimilates, thereby disrupting the plant growth. Fungi frequently infect crops in the field or after harvest, causing significant financial losses for growers and farmers around the world. Numerous fungal genera contain species that harm plants by causing infections, allergies, and toxicity. Both agricultural losses and medical expenses might result in significant financial losses (Bonnick et al., 2007). Globally prevalent, *Colletotrichum* is regarded as one of the most dangerous phytopathogenic fungal genera (Cano et al., 2005). Members of this genus cause field anthracnose, which causes significant financial losses for a variety of crops, including cereals, legumes, coffee, and rubber (Cano et al., 2005).

11.4.2 Pest Attack

Insect herbivores can be broadly classified as either chewing or piercing-and-sucking, and each predicts the type of crop damage they cause: chewing insects can cause potentially massive mechanical damage, while piercing-and-sucking insects cause nutrient loss and frequently vector diseases. Approximately two-thirds of known insect species are phytophagous (use live plants as a food source, excluding pollen and nectar), and insect pests lead to significant global crop losses. Additionally, insects can be classified as either generalists, which use a wide variety of plants as food, or specialists, which have a more specialized range of hosts. Generally speaking, the more specialized an insect is, the better equipped it is to handle specialized host defenses. In reality, things are frequently more complex. For instance, leaf miners might be the larvae of Diptera (Anthomyiidae, Ephydridae, Agromyzidae), Lepidoptera (Gracillariidae, Phyllocnistidae, Nepticulidae, Tischeriidae, Lyonetidae), or Coleoptera (Chrysomelidae, Hispinae, Halticinae), all of which produce nearly identical attack symptoms on plants. Larvae of Lepidoptera (Pyralidae, Noctuidae), Diptera (Muscidae, Anthomyiidae), or Coleoptera (Scarabaeidae) may create "dead-hearts" in graminaceous seedlings.

11.5 Impact of Stress on Plants

11.5.1 Effect on Quality

The effect of abiotic environmental stressors on crop quality has received far less consideration. Crop plants undergo a variety of physiological changes as a result of stress exposure, including modifications to assimilate translocation and photosynthetic gas exchange. (Mohsenzadeh et al., 2006) Abnormalities in evapotranspiration and the uptake of water pattern (Rivelli et al., 2010) impactsthe absorption and transfer of nutrients (Sánchez-Rodríguez et al., 2010) responses of antioxidants (Cai Feng et al., 2011), and apoptosis along with altered gene and enzyme functioning (Kangasjärvi et al., 2005) Such procedures are likely to have an impact on the chemical makeup of crops and, consequently, the quality of agricultural products if they persist for long periods as a result of ongoing stress exposure.

11.5.2 Effect on Growth

Abiotic stressors have a significant negative impact on crop productivity worldwide, resulting in significant yield losses. Agriculture is seriously threatened by the drop in food productivity and quality, which is mostly caused by excessive heat and water shortages (Zandalinas et al., 2021). As a result, reducing agricultural losses brought on by these pressures has become extremely difficult due to climate change, raising concerns about ensuring food security worldwide (Anjum et al., 2011). Optimal and limit temperatures for reproductive activities and grain filling are comparatively lower than those for vegetative growth and development phases, and they are more susceptible to both of these stresses (Prasad & Djanaguiraman, 2014).

11.5.3 Effect on Yield

Abiotic stressors have a significant impact on different crops' reproductive growth and consequently lower their final economic yields. Complex processes like nutrient adaptation and mobility to different reproductive parts, stem reserve accumulation, gametogenesis, fertilization, embryogenesis, endosperm, and seed development are all impacted by drought and high-temperature stress on grain yield. Although crop output can be impacted by these pressures at any stage of growth, the seed-filling stage is essential for determining average seed weight, seed structure, and ultimately the final quantitative and qualitative yield (Sehgal et al., 2018).

11.6 Methods for Sensing Pathogenic Fragment/Stress in Plants

Pathogenic diseases in plants are detected using different methods. Traditional methods for detecting plant stress, such as chlorophyll fluorescence and gas emission analysis, are often limited by their susceptibility to environmental interference, high costs, and inability to detect stress at early stages (Goffart et al., 2008; Li et al., 2010; Muñoz-Huerta et al., 2013; Wu et al., 2007). For instance, fluorescence-based techniques, while useful, are prone to background noise and require complex experimental setups. Similarly, imaging-based approaches, though noninvasive, lack the specificity needed for precise stress identification (Pintó-Marijuan & Munné-Bosch, 2014). In contrast, nano-biosensors have emerged as a groundbreaking solution, offering high sensitivity, real-time monitoring, and the ability to detect stress at the molecular level. Their integration with nanotechnology enables precise, cost-effective, and scalable applications in agriculture, addressing many limitations of conventional methods (Bhagat et al., 2023; Ijaz et al., 2023; Rana et al., 2024; Sheikh et al., 2024).

11.7 Classification of Nano-Biosensors

There are different types of nano-biosensors based on their receptors, transducers, and other components.

11.7.1 Electrochemical Biosensor

Electrochemical biosensors are a type of electrochemical sensor in which a biological component acts as the detector (receiver) (Zhang et al., 2006, 2010; Zhu et al., 2018). The electrochemical biosensor consolidates the analytical power of electrochemical methods with the accurate functionality of the biological component (Sun et al., 2019; Zhang & Ou, 2008). The bioreceptor is bonded to the desirable electrode, and when it interacts with the analyte (the substance being measured), it triggers an electrical signal (such as an amperometric or potentiometric response) that is directly proportional to the analyte concentration (Kordestani & Zhang, 2020; Kordestani et al., 2019; Mousavi et al., 2020). Electrochemical biosensors can be broadly classified into two distinct categories based on the nature of the biomaterial detection process (Jia et al., 2020; Jiang et al., 2020; Wang et al., 2016; Yang et al., 2020a,b; Zhu et al., 2020).

- Catalytic (enzymatic) electrochemical biosensors, in which an enzyme, a cell, or a tissue act as a biological component.
- Propensity biosensors, which utilize the attachment of nucleic acid, antibodies, or membrane receptors.

Enzymes are a crucial group of proteins that act as catalysts for chemical reactions in living systems. They exhibit high efficiency and selectivity in their catalytic functions (Cheng et al., 2016; He et al., 2018a,b; Zhang et al., 2016). Many approaches are used to coat an enzyme on a suitable electrode, including covalent bonding, surface adsorption, polymer trapping, and electrostatic adsorption, to generate an enzyme sensor. This sensor can quantify a particular substrate or inhibitor of a stabilized enzyme with high specificity and effectiveness (Chen et al., 2017a, 2021; Huang et al., 2020; Shi et al., 2017; Wang et al., 2020a). In the production of enzymatic electrochemical biosensors for clinical use, oxide reductase enzymes are the most widely utilized (Bai et al., 2021; Han et al., 2019; Wang et al., 2020b; Zou et al., 2019). Notable examples of enzyme biosensors include glucose sensors based on glucose oxidase and ethanol sensors utilizing alcohol dehydrogenase. However, due to the cramped stability of isolated enzymes, the high cost of some enzymes, and the problems associated with their purification, cells and tissues have been utilized as sources of enzymatic activity in the development of catalytic biosensors (Hu et al., 2014; Singh et al., 2013, 2014; Yang et al., 2014; Zhao et al., 2013). Receptive electrochemical biosensors rely on the selective interaction of biological molecules, including antibodies and polynucleotides, with specific analytes to generate an electrical signal. The remarkable specificity of biochemical bonds, such as DNA pairing or antigen—antibody complex formation, allows for the design of sensors with high sensitivity and selectivity. Immunosensors, a prominent type of affinity sensor, are built on immune responses that involve precise antigen—antibody recognition. These sensors are particularly valuable for identifying and measuring proteins (Gao et al., 2019; Jiang et al., 2013; Song et al., 2009; Xu et al., 2018; Zhang et al., 2019a). The receptor (antigen or antibody) is conjugated with an enzyme, which acts on a substrate to create a product that can be detected through techniques like amperometry or potentiometry. Enzymes such as phosphatase and peroxidase are widely utilized for labeling in electrochemical biosensors. In the case of DNA hybridization-based electrochemical biosensors, the diagnostic ability of nucleic acids, dependent on the pairing of complementary strands, is merged with electrochemical transducers. The probe strand is fixed onto the electrode surface (Chen & Xu, 2020; Deng et al., 2020; Liu et al., 2020a,b; Niu et al., 2020). The presence of target DNA, featuring a sequence complementary to the probe strand, in the sample leads to hybridization, resulting in the generation of a measurable electrical signal (Zhang et al., 2019b, 2020). Double-stranded DNA-modified electrodes in DNA electrochemical biosensors can also be employed to identify small molecules, including drugs or carcinogens, that interact specifically with double-stranded DNA (Walcarius et al., 2013)

11.7.2 Piezoelectric Nanosensors

Piezoelectricity refers to the interplay between mechanical and electrical behaviors in specific solid materials, where mechanical stress induces an electric charge, and conversely, an electric field causes the material to deform. This reversible process represents a fundamental coupling of mechanical and electrical behaviors, enabling innovative applications in various fields (Wang, 2007). The piezoelectric effect has been harnessed for various applications, particularly in the development of advanced sensors. Piezoelectric sensors harness this effect to measure various physical parameters, such as pressure, acceleration, temperature, strain, and force, by converting them into an electrical charge. Recent advancements have resulted in the creation of nanoscale piezoelectric sensors with the capability to measure extremely small forces, down to the nano-newton range (Wang et al., 2006). Piezoelectric nanosensors, although miniature in scale, typically require a charge amplifier that is orders of magnitude larger, at the centimeter scale, to facilitate detection. Despite this, piezoelectric nanosensors have proven valuable in measuring the biomechanics of plants, such as the unique properties of the Venus flytrap (*Dionaea muscipula*) (Volkov et al., 2012, 2013). Understanding the biomechanics, shape-shifting osmotic motors, structures, and mechanosensors of plants provides a rich source of inspiration for developing intelligent materials and versatile structures. These concepts have been utilized in biosensing applications, such as in apple juice detection of staphylococcal enterotoxin B (Maraldo & Mutharasan, 2007) and in investigating plant pathology (Eun et al., 2002).

11.7.3 Chemi-Resistive Sensors

Chemiresistive sensors are used to detect changes caused by the sorption of target molecules onto the sensing material between two electrodes in electrical resistance. Various materials, including semiconductor-based carbon nanotubes and conducting polymers like polypyrrole, can be used as sensing materials. These techniques are particularly functional for gas recognition, with high sensitivity reaching down to the parts-per-trillion range. Chemiresistive sensors have been utilized to detect ethylene, a gaseous phytohormone (Esser et al., 2012; Mirica et al., 2013) and additional organic volatile compounds produced by plants (Liu et al., 2020c; Weerakoon et al., 2011). To investigate plant physiology using electrochemical nanosensors, plant tissues are often homogenized in a known electrolyte solution (Krystofova et al., 2010). Alternatively, plants can be studied indirectly by focusing on molecules involved in plant development and growth present in the soil (Gieling et al., 2005). Recent advancements have enabled the design of nanosensors intended to operate within plants, allowing for in situ analysis. A novel sensor utilizing light-activated nanomaterials, specifically molybdenum-doped bismuth vanadate ($BiVO_4$), was integrated into a photoelectrochemical platform to evaluate antioxidant activity in fruit (Wang et al., 2015). Meanwhile, chemiresistive sensors provide a distinct approach, leveraging changes in electrical resistance to detect target molecules. The integration of these techniques can provide valuable insights into plant physiology and facilitate the detection of key biomolecules.

11.7.4 Surface-Enhanced Raman Scattering Nanosensors

Surface-enhanced Raman scattering (SERS) is a powerful, noninvasive analytical technique that provides insights into the vibrational modes of molecules. This phenomenon occurs when a molecule gains vibrational energy when incident photons drop to a reduced energy state. In 1923, this phenomenon was first described in theory by A. Smekal in 1923 (Smekal (1923) and later experimentally confirmed by C. V. Raman in 1928 (Raman & Krishnan (1928). Raman spectroscopy, including SERS spectroscopy, can be employed to extract information about materials by leveraging the Raman effect. However, the detection of analytes at low concentrations is often hindered by the weak signals produced through Raman scattering. To overcome this limitation, SERS spectroscopy exploits the amplification of Raman signals for molecules adsorbed on nanostructured metallic surfaces (Stiles et al., 2008), achieving enhancements of up to 10^{14}–10^{15}-fold and enabling the detection of single molecules (Kneipp et al., 1997; Nie & Emory, 1997; Stiles et al., 2008). The enhancement of Raman signals in SERS is thought to arise from the interplay of two primary mechanisms: an electromagnetic mechanism and a chemical mechanism. These mechanisms cooperatively amplify the Raman signal of molecules adsorbed on metallic nanostructures, giving rise to the enhanced sensitivity characteristic of SERS (Rodriguez-Lorenzo & Alvarez-Puebla, 2014; Sun et al., 2008a) While the electromagnetic mechanism is considered essential for the SERS effect, the chemical mechanism provides an additional enhancement pathway that enables the technique to achieve ultrasensitive detection limits, making it an extremely powerful analytical tool (Le Ru & Etchegoin, 2009; Moskovits & Suh, 1984). SERS enhancement is attributed to two mechanisms: electromagnetic and chemical. The electromagnetic mechanism of SERS enhancement is rooted in the interaction between incident light and metallic nanostructures, which redistributes the electromagnetic field and amplifies it. The chemical mechanism, on the other hand, involves the interaction between the metal surface and adsorbed molecules, causing changes in the vibrational modes of the molecules, and explains the discrepancies between theoretical predictions and experimental observations of maximum enhancement factors (Kneipp et al., 1997; Nie & Emory, 1997). The chemical mechanism governing SERS involves two primary procedures: charge-transfer procedure and the formation of analyte-–surface complexes between the analyte and the metal (Otto, 2001; Tognaccini et al., 2019) Charge-transfer states entail electronic transitions between the metal's Fermi level and an unoccupied molecular orbital, or vice versa. The formation of surface complexes, on the other hand, modifies the properties of the analyte molecule, potentially facilitating resonance Raman scattering. In plant-based SERS spectra, the majority of peaks are attributed to biomolecules such as adenylated compounds, chlorophyll, flavins, and lipids (Zeiri, 2007). SERS has emerged as a valuable tool for detecting and determining a broad range of biomolecules in plants (Du et al., 2018), including plant hormones (Chen et al., 2017b; Wang et al., 2017). Moreover, its applications extend to food safety diagnostics, where it can be used to detect pesticides (Li et al., 2022; Zhang et al., 2018). With its exceptional sensitivity, SERS can detect analytes at the single-molecule level, making it a powerful tool for various applications. The phenomenon of Raman scattering arises from the inelastic interaction between photons and matter, resulting in a change in energy and direction of the scattered light. The energy difference between the incident and scattered light is attributed to the interaction between photons and the vibrational modes of the material. Electrochemical nanosensors, typically comprising three electrodes, have also been applied in plant sensing studies. Voltammetric and amperometric methods have been used to qualitatively analyze plant responses, demonstrating the utility of these sensors in plant research (Yotova et al., 2013)

11.7.5 Fiber-Optic Biosensors (FOBS) in Plant Nanobionics

Optical fiber-based nanobiosensors (NBSs) offer a novel approach to monitoring plant health by enabling direct measurement of parameters from within the plant tissue. The applications of these sensors extend to various medical fields, where they can be utilized for swift and accurate diagnosis. Their slender profile also makes them suitable for incorporation into hollow instruments, enabling targeted tissue monitoring with reduced invasiveness. NBSs boast several advantages, including non-toxicity, chemical inertness, and suitability for in-plant applications. This enables their safe and effective deployment within plants, where they can be paired with monitoring equipment to track plant health. These sensors are designed to integrate effortlessly with plant monitoring systems and boast a lightweight, user-friendly design. The advancement of fiber-optic NBSs is driven by the dual goals of high performance and biocompatibility, ensuring that the sensors do not disrupt the plant's physiological processes while maintaining their functionality despite potential interference from plant metabolites. Fiber-optic nanobiosensors (NBSs) can be categorized into two main types: extrinsic, where the fiber serves as a conduit for signal transmission, and intrinsic, where the interactions occur directly within the fiber itself. Additionally, fiber-optic biosensors can be differentiated into two groups: minimally invasive NBSs, designed for insertion into plant cavities, and invasive NBSs, intended for integration into plant organs or wood conductive tissue, as reported by Liu et al. (2015).

11.8 Application of Nanosensor in Plants

The agricultural sector requires taking more resilient and sustainable agricultural practices to precisely monitor the inputs and outputs of agricultural commodities. Nanotechnology is playing an important role by proving its high proficiency in monitoring plant health in terms of biotic stresses, abiotic stresses, nutrient deficiencies, and soil conditions (Álvarez et al., 2016; Antonacci et al., 2018; Ghaffar et al., 2020; Salouti & Khadivi Derakhshan, 2020). Nano-biosensors plays important role in evaluating plant health conditions, detecting growing challenges such as plant disease diagnostics and genetic modification and mitigating plant pathology and genetic engineering for efficient agriculture. Early detection of plant diseases and timely prevention are important to ensure enough food supply to the world (Elmer & White, 2018). Nano-biosensors detect important signaling molecules such as ROS, phytohormones such as jasmonic acid, indole butyric acid, ethylene, abscisic acid (ABA), glucose, sucrose, inorganic elements, gaseous ROS, and nitric oxide (Giraldo et al., 2019). These messenger molecules are in the primary response system to arise proper adaptation strategies in plants in response to different types of stresses (Kiegle et al., 2000; Mazars et al., 2010; Mittler, 2017; Suzuki et al., 2013). Various antibody-based nano-biosensor platforms have been designed and employed for the identification of diverse plant pathogens, including a compact, high-capacity biosensor system for the swift detection of plant viruses (Perdikaris et al., 2011). Chemoresistive immuno-sensors utilizing polypyrrole nano-ribbons have been developed for detecting viral plant pathogens (James et al., 2013), while fiber-optic particle plasmon resonance immunosensors based on nanorods have been employed for the identification of orchid viruses (Lin et al., 2014).

Optical detection of analytes in plants is complicated by factors such as tissue thickness and the interference of photosynthetic pigments. Nanosensors, however, are well-suited for this purpose due to their ability to be embedded within plant tissues, facilitating in vivo investigations into cellular signaling and metabolic pathways. Specific examples of nanosensor applications in plant research are explored in the following sections (Table 11.1).

Table 11.1 Methods used for sensing pathogenic stress in plants.

Methods of detection	Description	Efficacy	Shortcoming	Sensitivity	References
PCR (Polymerase chain reaction)	Pathogens are detected by selectively amplifying their DNA using specific primers and repeated heating and cooling cycles.	High selectivity, reliability, sensitivity, and cost-effectiveness, and it is widely recognized and established in the field.	Detailed sample preparation and requires precise temperature control, non-portable.	1–100 fg/μL	Xu et al. (2020a)
DNA amplification Isothermal	The process utilizes specialized primers for DNA amplification without thermal cycling.	Highly selective and reliable, and it eliminates the need for thermal cycling, simplifying the operational process.	Relies on sophisticated primers and a multistep sampling and testing protocol to achieve accurate results.	0.01–1 pg/μL	Aglietti et al. (2019) and Ristaino et al. (2020)
ELISA	This detection method works through the specific binding of an antigen (pathogen protein) to an antibody, utilizing their natural affinity for each other.	It is easy to use, ideal for high-throughput applications, and especially effective for detecting viral antigens.	Time-intensive and may involve elaborate sample preparation and labeling steps, including antigen extraction.	1–100 fg/μL	Charlermroj et al. (2013)
Lateral flow immunoassay	The method relies on colorimetric detection, where the formation of an antigen–antibody complex produces a visible color change.	Easy to carry or transport affordable and easy to handle.	Delivers qualitative data only, needs sample preparation, like extracting antigen proteins.	0.1–1 pg/μL	Anfossi et al. (2013), Panferov et al. (2018), and Song et al. (2014)

(*Continued*)

Table 11.1 (Continued)

Methods of sensing	Mechanism	Efficacy	Disadvantage	References
Raman spectroscopy	Disease detection is based on recognizing chemical changes in plant tissue using molecular signals generated by a laser source.	Easy handling and suitable for particular disease identification.	Susceptible to background fluorescence interference, demands special hardware, and is difficult to implement in field conditions.	Farber and Kurouski (2018), Liu et al. (2020c), and Yeturu et al. (2016)
Chlorophyll fluorescence	Dependent on variations (from stress) in fluorescence resulting from photosynthesis in plants.	It provides insights into photosynthetic efficiency, which can enhance accuracy when used in combination with imaging-based methods.	A time-intensive experimental apparatus is needed, requiring dark adaptation of the sample plants.	Bauriegel and Herppich (2014), Murchie and Lawson (2013), and Pineda et al. (2018)
VOC emission monitoring	Alterations in gaseous emissions from plants are quantified as an indicator of plant health.	This method is ideal for general plant health monitoring, noninvasive, and has the potential for automated continuous monitoring.	Setting up the experiment is difficult, with complicated sampling and testing steps, and the results tend to have low specificity.	Bitas et al. (2013), Cui et al. (2019), Delory et al. (2016), and Li et al. (2019)
Active remote sensing methods	RADAR and LiDAR systems are employed to identify symptomatic morphological changes.	Ideal for large-scale, non-specific plant health monitoring, and LiDAR can also measure parameters like CO_2 concentration and plant water content.	The initial cost is pretty high, the sampling and data processing are complicated, and the results aren't very specific.	Husin et al. (2020) and Lin (2015)
Visible/RGB imaging	Color-based features are identified and extracted, and then classified using computational algorithms.	It uses relatively low-cost hardware (frequently smartphone-based) and is non-invasive.	There is limited potential for pre-symptomatic detection, and the data processing is pretty complex.	Liu et al. (2016, 2019)

11.9 Role of Nano-biosensors in Plant Stress Detection

11.9.1 Nano-biosensors

Nano-biosensors are compact analytical devices that combine nanotechnology and biotechnology to detect and measure biological or chemical molecules. The bimolecular recognition element associated with the transduction element of nanosensors provides rapid, sensitive, and specific detection of target analytes that are displayed as electrical signals or digital outputs, often in real time (Salim et al., 2013). These nanoscale sensors work at nanoscale level to detect environmental fluctuations and changes in chemical and biological reactions. Their ultra-small size causes effective interaction between individual molecules, making them indispensable in a wide range of scientific and environmental applications (Kakarwal & Deshmukh, 2012; Liyakat et al., 2022; Liyakat & Liyakat, 2024; Nerkar et al., 2023a; Swami et al., 2022; Vaijnath et al., 2022). "Miniaturized analytical systems" can be thought of as an easy descriptive term to describe nanosensors that detect analytes or groups of target analytes, including DNA, antibodies, proteins, ions, heavy metals, and volatile organic compounds, and perform specific physicochemical reactions (Salim et al., 2013). The three S's of nano-biosensors—selectivity, stability, and sensitivity—are crucial in evaluating the performance of nano-biosensors (Salim et al., 2013). The primary function of the biosensor is to detect a physiologically specific substance. These substances may be enzymes, proteins, antibodies, and immunological compounds. The biosensor is formed from three components: the detection system known as a bioreceptor, a transducer, and an output system (Kissinger, 2005). The acceptor in biosensors is a biological component that is linked to the converter in several ways (Bakker & Telting-Diaz, 2002; Ge et al., 2014; Ronkainen et al., 2010; Walcarius et al., 2013). This biological element is highly selective for both recognizing analytes and biological interactions (Feng et al., 2020a; Lin et al., 2020; Xie et al., 2018; Zuo et al., 2020). The performance of biosensors is highly dependent upon the way in which the bioreceptor layer attaches to the transducer. The main objective is to set up a close interaction between the biological component and the sensor surface (converter), and to maintain (even enhance) the stability of the biological element (Jiang et al., 2018; Zuo et al., 2015). Target selectivity and identification are essential elements in designing nano-biosensors to carry out accurate molecular recognition of analytes. This accuracy can be obtained by the host–guest relationship between the analyte and the sensors, where the host (chemo-sensors) binds specifically to the guest (analyte) molecule that needs to be recognized. The specificity of the host for the guest molecule is achievable due to specific chemical reactions between the analyte and chemo-sensors in numerous chemical pathways. When the chemo-sensor and analyte interaction is completed, then the chemo-sensor is incorporated into the nano-biosensor, typically a membrane that contacts the transducer. The transducer finds a way to signal that the interaction between host and guest has successfully occurred, and we get electrical, optical, and chemical signals in a readable format (Rodríguez-Sevilla et al., 2014).

11.9.2 Principles of Working

The designation/customization of host molecule that can be used as chemo-sensors in nano-biosensors must meet some criteria and standards, including stability, selectivity, immobilization, signaling, and release of analyte (Figure 11.2).

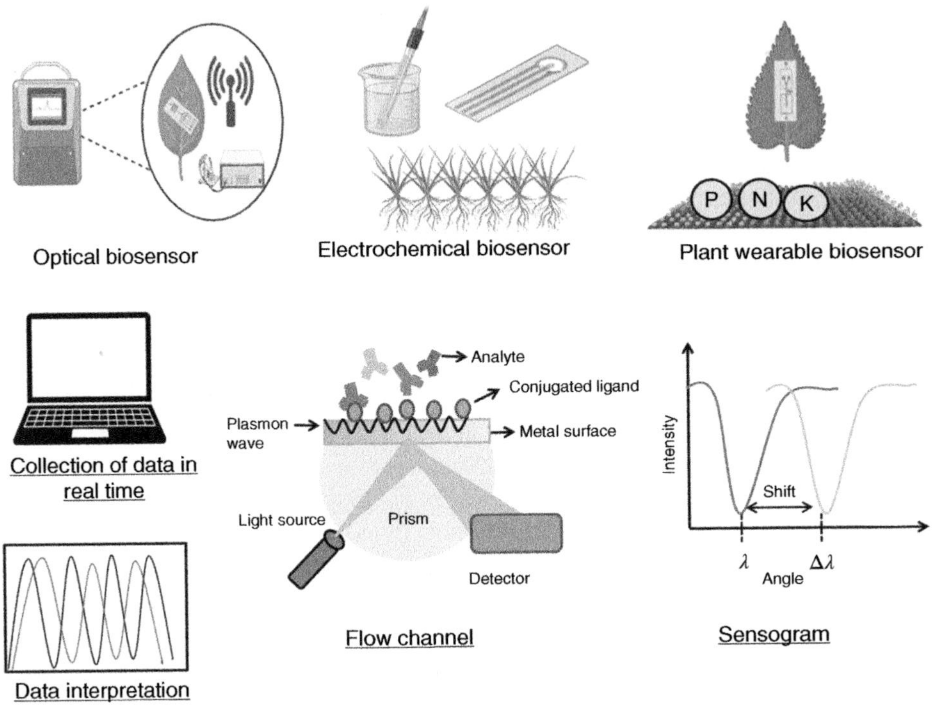

Figure 11.2 Illustrating real-time data collection using wearable biosensors for plants, highlighting optical (plasmon waves, light detection) and electrochemical (analyte-ligand binding) sensing mechanisms. The components include a metal surface, light source, detector, and data interpretation system for continuous monitoring of plant health. Created in BioRender.com.

The host molecule must be resistant to denaturation or degradation for successful binding to the guest molecule. The host molecule must be selective in binding to the guest molecule to prevent unwanted binding to other analytes. The host molecule must be immobilized in a film or membrane in contact with the transducer for enhancing stability, sensitivity, and reusability of nano-biosensors. The host molecule must be able to produce a signal in the transducer that will indicate that a host–guest binding response has occurred. The host molecule must release the analyte after successful detection and signaling so that the host can be used again to detect another analyte. Biological receptors or chemo-sensors used in nano-biosensors are antibodies, aptamers, enzymes, peptides, nucleic acids (DNA/RNA), lectins, membrane receptors, and signaling molecules.

Biological receptors are difficult to handle, and the required conditions can't be optimized efficiently and thus are not practical solutions for many applications. Further advancements in nanotechnology have introduced synthetic receptors; these immobilization environments can be controlled and optimized by aimed molecular designing of nanosensors for a vast range of uses. Genomics, proteomics, and metabolomics provide tools for the synthesis of artificial receptors, and thus the occurrence and recognition of altered metabolic pathways and products can be detected in plant nano-bionics.

Molecular interaction interface biology, bio-interfacial chemistry, and associative mathematical techniques are used for the synthesis of artificial receptors. These synthetic receptor environments are more efficient for accelerated and more accurately targeted selection of analytes. The technique used currently for the synthesis of synthetic receptors is combinatorial array of receptor analysis (CARA) (Fang et al., 2015).

By the development of macrocycles (a class of molecules that consist of a large ring-shaped structure, typically composed of 12 or more atoms), supramolecular chemistry has played its role in the host–guest interaction in nano-sensors. Macrocycles can form complexes with smaller molecules or ions, making them useful for applications such as molecular recognition and sensing. Macrocycles contain internal spaces that have binding sites for encapsulating the guest molecules. The binding site (host cavity) of the macrocycles is designed to fit a specific guest molecule, excluding others that are too large or too small. Here, a question arises: what attracts the guest molecule to the cavity of the host molecule and how is the host–guest assembly stabilized. Macrocycles stabilize the host–guest assembly through non-covalent interactions; these interactions include electrostatic interactions, hydrogen bonding, and dipole–dipole interactions. These interactions are strong enough to hold the complex for the desired period of time, and then the guest (analyte) is released from the host (synthetic receptor) after it has been detected and a signal is produced (da Silva et al., 2013).

For a specific interaction with the analyte the biological layer is responsible, from this interaction the transducer converts the information into a measurable effect (Cui et al., 2021; Wang et al., 2020c; Xu et al., 2020b). For example, the analyte–bioreceptor interaction is converted by mechanical transducers to a change in bending or resonant frequency, by electrochemical transducers into a change in current or potential, and by optical transducers into a change in light frequency or intensity and more. Schematically the biosensor process is shown in Figure 11.1 (Yang et al., 2013). There is a proper reading system for the calculation of physical phenomena such as changes in electrochemical parameters like potential and current, changes in resonant frequency, changes in bending, and changes in optical properties (Feng et al., 2020b; He et al., 2018c; Pan et al., 2020; Wang et al., 2018, 2021; Zhang & Liu, 2019).

11.9.3 Mechanism of Nano-Biosensors

In a nano-biosensor, an electrical device (transducer) is combined with a biological component in a mutually dependent manner, and a bioelectronics analysis system is formed. The first interaction in a nano-biosensor is between a biologically active immobilized receptor (R) and a specific analyte (A) (specific chemical signal). The molecular recognition between the receptor and analyte modifies the arrangements of surface charges, and these alterations produce a signal that can be detected in a readable form in the specific transducer. The molecular recognition element (R) is a crucial component that plays a central role in detecting and analyzing specific substances. This element is composed of diverse biological components, including enzymes, enzymatic systems, antibodies, antigens, receptors, microbial populations, cellular fragments, and signaling molecules. These biological components serve as the foundation for molecular recognition, enabling the detection of various analytes (A) with high specificity and sensitivity (Figure 11.2).

The range of analytes that can be detected and analyzed is vast, encompassing glucose, sugars, amino acids, alcohols, lipids, and nucleotides.

Biosensors can be categorized into three generations. First-generation biosensors involve immobilizing a biocatalyst on the surface of a membrane, which is then attached to a transducer. In second-generation biosensors, the biocatalyst is directly immobilized onto the transducer surface through adsorption or covalent bonding, eliminating the need for a semipermeable membrane. Third-generation biosensors represent a significant advancement, where the biocatalyst is directly linked to a compact electronic device, such as a transistor, enabling miniaturization. The interaction between the biocatalyst, immobilization matrix, and transducer components determines the configuration of the biosensor, including the arrangement of the membrane-electrode-transducer interface (Agrawal et al., 2012).

11.10 Detection of Physiological Responses

11.10.1 Molecular Oxygen Detection

Molecular oxygen functions as the terminal electron acceptor in the electron transport chain during aerobic respiration. Despite the ability of plants, photosynthetic bacteria along with algae, which release oxygen through photolysis in the light-dependent reactions of photosynthesis, atmospheric oxygen remains a key metabolomic precursor for plants. This is because photosynthetic activity is not uniform across tissues and is sometimes inactive (Ast et al., 2012). Many stresses, including flooding, can lead to hypoxia in plant tissues. Differential oxygen abundance is also observed across meristems and organs, with oxygen regulation playing a mechanistic role in plant development (Considine et al., 2017).

Much of the research on oxygen sensing has utilized Clark-type electrodes, where oxygen is chemically reduced to water, producing a measurable current (Clark et al., 1953). Applications of these microelectrodes include monitoring photosynthesis/respiration in potato leaf protoplasts, determining mitochondrial respiration in pea shoots and *Arabidopsis thaliana* leaves, evaluating alternative oxidase function in soybean cotyledons and roots (Bykova et al., 2005; Day et al. 1985; Kearns et al., 1992; Shaw et al., 2017). When put side-by-side with optical sensors, these electrodes show some disadvantages. They damage samples during measurement, need complicated preparation steps, and consume oxygen creating measurement errors in viable cells. Nanosensors can help solve these problems. Currently, two types of nanosensors, electrochemical and optical systems, are being utilized to measure oxygen distribution within tissues. Platinum-coated carbon-quartz electrochemical nanosensors have been utilized to measure dramatic oxygen depletion at *Chara corallina* internode surfaces after cell wall microperforation (Alova et al., 2020). The observed oxygen depletion at the wound site may result from multiple factors, particularly activation of plasma membrane NADPH oxidase and alterations in antioxidant defense mechanisms.

Optical-nanosensors for O_2 possess some properties that make them superior to Clark-type polarographic electrodes, facilitating nanoscale oxygen sensing and large-area imaging. Probes Encapsulated by Biologically Localized Embedding (PEBBLEs) are a leading class of such nanosensors, which incorporate fluorescent dyes within a protective inert matrix to minimize leaching (Buck et al., 2004). Furthermore, the shield maintains structural stability and prevents unwanted interactions with other proteins (Lakowicz, 2006).

A common limitation of PEBBLEs is their red phosphorescence emission, which spectrally overlaps with plant chlorophyll autofluorescence. To overcome this challenge, a microbead-based optical probe utilizing dual-frequency phase modulation was used in plant and algae systems (Elmar Schmälzlin et al., 2005). Nanoparticle O_2 (oxygen) sensors exhibit superb brightness, photostability, and ease of production, and the long-term storage is an additional advantage (Ast et al., 2012). But, their size, ranging between 20 to 600 nm in diameter (Buck et al., 2004), may cause cell injury, thereby limiting their use in living flora (Schmälzlin et al., 2005).

11.10.2 Adenosine Triphosphate Detection

Adenosine triphosphate (ATP) is an organic molecule that drives energy-dependent processes in cells. During metabolism, it is dephosphorylated to adenosine diphosphate or adenosine monophosphate. In plants, ATP synthesis occurs in chloroplasts and mitochondria. A fluorescence resonance energy transfer (FRET)-based optical nanosensorcomprising enhanced *Renilla* luciferase and the fluorescent protein Venus—detected reduced cytoplasmic ATP levels following oligomycin A treatment. Moreover, photosynthetic ATP generation in chloroplast was visualized in transgenic *Arabidopsis* plants by directing the nanosensor to the stroma via transit peptide tagging (Saito et al., 2012).

11.10.3 Calcium(Ca^{2+}) Ions Detection

Calcium ions (Ca^{2+}) play a pivotal role in plant signal transduction pathways, regulating key processes such as stomatal closure, cellular division, and intracellular signaling. In stomatal closure, for example, ABA signaling prompts the release of free Ca^{2+} ions into the cytosol from both extracellular sources and internal stores. This influx of Ca^{2+} reverses the concentration gradient, leading to the efflux of K^+ ions. The subsequent loss of cellular solutes results in reduced turgor pressure, causing the guard cells to become flaccid and the stomatal pores to close. Advanced imaging techniques, such as FRET-based genetically encoded sensors, have enabled high-resolution live-cell visualization of Ca^{2+} dynamics (Krebs et al., 2012). Research on *Lotus japonicus* has identified unique Nod factor-induced Ca^{2+} spiking patterns in both the nucleus and cytosol.

11.10.4 ROS (Reactive Oxygen Species) Detection

Nanosensors are capable of detecting ROS, including superoxide anion ($O2^{\bullet -}$), singlet oxygen ($^1O_2$), hydroxyl radical ($\bullet$OH), and hydrogen peroxide (H_2O_2). These ROS are generated during normal cellular metabolism in organelles such as chloroplasts, mitochondria, peroxisomes, and the apoplast (Apel & Hirt, 2004; Foyer, 2018; Mattila et al., 2015). ROS are integral to various signaling pathways in plants, influencing processes like development, programmed cell death, and stress responses (Neill et al., 2002; Quan et al., 2008). Genetically encoded ratiometric sensors, such as redox-sensitive green fluorescent proteins (roGFPs) (Hanson et al., 2004; Wierer et al., 2012), are widely used to monitor glutathione redox states (García-Quirós et al., 2020; Schnaubelt et al., 2015), while HyPer, a single fluorescent ROS sensor, directly measures H_2O_2 levels. In plants, roGFPs have been extensively employed to determine glutathione redox potential (García-Quirós et al., 2020; Jiang et al., 2006; Rosenwasser et al., 2010; Schwarzländer et al., 2008), and HyPer has been utilized to detect H_2O_2 dynamics in *Arabidopsis* guard cells and roots (Belousov et al., 2006; Costa et al., 2010; Hernández-Barrera et al., 2015). Additionally, modifications of electrochemical sensors with platinum nanoparticles have enabled real-time monitoring of oxidative bursts in single plant protoplasts, providing insights into plant defense mechanisms and signal transduction (Ai et al., 2009).

11.10.5 Plant Hormones Detection

A variety of plant hormones can be identified like strigolactones (Tsuchiya et al., 2015), auxin (Liu et al., 2014a; McLamore et al., 2010), and ethylene (Cristescu et al., 2013; Janssen et al., 2014) by using different types of nanosensors. Plant hormones are the chemical signals that are produced within the plant and are associated with plant's growth and development from embryogenesis also in biotic and abiotic stress tolerance (Bürger & Chory, 2019; Ku et al., 2018; Méndez-Hernández et al., 2019; Pierre-Jerome et al., 2018; Shigenaga & Argueso, 2016; Ullah et al., 2018). Nanosensors enable the potential for in vivo analysis of plant hormone and signaling mechanisms. Strigolactones, root-derived chemical signals, function as phytohormones that govern plant development and play a role in the response of plants to various biological and environmental stressors (Aliche et al., 2020; Umehara et al., 2015). They have been identified as playing roles in:

- Promoting the parasitic organisms to germinate in the roots of host plant (Cook et al., 1966; Cook et al., 1972; Umehara et al., 2015),
- Facilitating plant recognition by symbiotic fungi (Akiyama et al., 2005; Umehara et al., 2015),
- Inhibiting shoot branching in plants (Gomez-Roldan et al., 2008; Umehara et al., 2015).

Striga is one of the parasitic organisms that thrive in the roots of host plants (Cook et al., 1966; Umehara et al., 2015). The optical nanosensor (Yoshimulactone Green) is authorized for the spatiotemporal tracking of strigolactone levels in seed germination of *Striga*. Yoshimulactone Green is recognized by strigolactone receptors and hydrolyzed, producing fluorescent products that can be detected. Beyond facilitating spatiotemporal tracking of strigolactone concentrations, this approach also proved effective in pinpointing dedicated strigolactone receptor (Tsuchiya et al., 2015). Ethylene plays a crucial role in regulating various stages in the life cycle of plants, such as seed sprouting, root formation, flowering, fruit maturity, cellular aging, and reply to both biotic (living organism) and abiotic (like temperature, humidity, and light) stresses (Lin et al., 2009). In agriculture, ethylene is commonly utilized to induce fruit ripening (Wang et al., 2002). Chemiresistive sensors have been effectively employed to detect the gaseous phytohormone ethylene (Esser et al., 2012; Mirica et al., 2013), demonstrating consistent and reliable responses across different fruit types, including bananas, avocados, apples, pears, and oranges.

11.11 Plant Pathogen Detection (Biotic Stress)

Infection caused by Plant pathogens causes significant damage to crops and results in substantial economic losses to the agricultural sector. Currently, the pathogens are detected using PCR-based techniques and microbiological techniques but nano-biosensors propose early and real-time diagnosis of infections, thus facilitating prompt actions to mitigate losses (Bilodeau et al., 2007; Jongman et al., 2020). Real-time, quick, and precise detection of early stages of diseases and infection progression can be achieved using nano-biosensors combined with biomarkers (Khater et al., 2017). Researchers have developed different biosensing techniques, which include auto-sensing, immune-sensing, and electrochemical sensing (Cebula et al., 2019). To investigate infections, researchers routinely use lectins, antibodies, and bacterial receptors. Due to their adaptability to be integrated into biosensors, these receptors are widely used as molecular recognition elements to investigate diseases (El Ichi et al., 2014; Liu et al., 2014b). Globally, 60% agronomic outputs are hindered by plant. To minimize the loss of crops, early detection, evaluation, and identification of pathogens are important (Figure 11.3). Most plant pathology studies work directly or indirectly to prevent catastrophic crop losses (Li et al., 2021).

Quantom dot (QD)-based biosensors have been used to detect cowpea mosaic virus (Medintz et al., 2005), cauliflower mosaic virus (Sun et al., 2008b), *Citrus tristeza* virus (Moreau et al., 2012; Shojaei et al., 2016), grapevine virus A (Tereshchenko et al., 2017), tomato ringspot virus (Zhang et al., 2013), bean pod mottle (Zhang et al., 2013), and Arabis mosaic virus (Zhang et al., 2013). Antibody-functionalized fluorescent silica nanoparticles have been employed as sensors to detect *Xanthomonas axonopodis pv. vesicatoria*, the causative agent of bacterial spot disease in Solanaceae plants (Yao et al., 2009). A gold nanoparticle-enhanced electrochemical sensor enabled the detection of *Pseudomonas syringae* in *A. thaliana* via differential pulse voltammetry (Lau et al., 2017). Mycotoxins are also detected by nanosensors. Mycotoxin residues in corn, wheat, oat, and barley are detected by competitive antibody receptor-based mycosensors (Lattanzio & Nivarlet, 2017; Lattanzio et al., 2012). Flavescence dorée disease of grapevine is detected by gold-functionalized fluorescent DNA oligo-based nanosensor (Firrao et al., 2005). A recent study reported the fabrication of gold nanoparticle-sugar conjugates via functionalized carbohydrates (Yazgan et al., 2021). A sensor for detecting *Penicillium italicum* spores and hyphae in fruit was developed using the gold nanoparticle glycoconjugates (Yazgan et al., 2018). The bacterial strain *Xylella fastidiosa* subsp. pauca CoDiRO is the causative agent of olive quick decline syndrome (OQDS), a disease posing a significant threat to agriculture-dependent economies, such as those in southern Italy. This

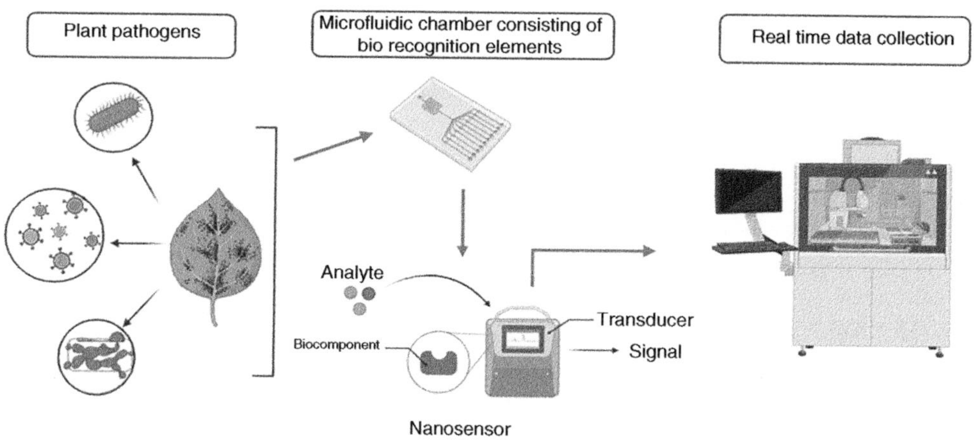

Figure 11.3 Illustratration of biosensing system combining biorecognition elements and transducers to detect plant pathogens through real-time analyte monitoring.

pathogen can also infect other plant species, exhibiting symptoms like leaf scorching and canopy wilting, which may appear months after initial infection. Moreover, some hosts may remain asymptomatic. Therefore, developing sensors for rapid and early detection of infected plants is crucial. Currently, *X. fastidiosa* detection relies on ELISA and PCR methods, as reported by Chiriacò et al. (2018).

11.11.1 Plant Disease Detection Caused by Abiotic Stress

Flooding, water logging, drought, salinity, heat, heavy metals, and other abiotic stresses impose stress on plants in various ways and disrupts plants metabolism and growth patterns. These unfavorable circumstances alter the physio-biochemical responses of plants such as gene expression, membrane permeability, carbohydrate metabolism, stomata closure, photosynthesis, respiration, osmolyte accumulation, membrane composition, hormone regulation, antioxidant enzymes, and generation of free radicals (Figure 11.3), which will affect agricultural efficacy (Husen, 2022). Instant and rapid detection of these physio-biochemical responses of plants stimulated by abiotic stress can aid in achieving sustainable agriculture (Mondal et al., 2022).

11.11.2 Nano-Biosensors in Drought Stress

Drought impacts various physiological activities of plants and imposes multifaceted stress that alters the physical, chemical and molecular attributes of plants (Tiwari et al., 2016). When crops undergo drought stress the first symptoms to be noticed are developmental delay, growth stagnation, and disruption of the metabolic needs of the plant. The plant's tolerance mechanism provides insights into their ability to thrive in unfavorable conditions (Harb et al., 2010).

A novel, low-cost optical platform for measuring soil water content was introduced by Leone et al. (2022). This innovative system utilizes a nano-porous ceramic disc in conjunction with a specially designed optical fiber featuring near-infrared detection. The sensor's design incorporates a Y-shaped bifurcated cable, housing two fibers within a single unit. One fiber connects to a light source, while the other links to a detector. The shared leg enables illumination and collects light from the disc. For soil testing, the sensor is encased in a protective PVC tube, which is then buried in a soil tank. Lan et al. (2020) developed a wearable, graphene-based humidity sensor for plants using capacitive technology. The sensor's interdigital electrode was fabricated on a polyimide film using laser direct writing, and then coated with a graphene oxide (GO) solution. This flexible, GO-based nanosensor can be easily attached to plant leaves without hindering growth. When integrated with wireless devices, the sensor forms a comprehensive monitoring system.

11.11.3 Nano-Biosensors in Extreme Temperature Stress

Temperature stress elevates the concentration of phytohormones such as ABA and indole-3-acetic acid (IAA) in plants. Furthermore, salicylic acid (SA), ethylene (ET), and jasmonates (JAs) have unique roles in response to temperature stress (Siddiqi & Husen, 2019). To monitor plant responses to abiotic stress, researchers have designed an electrochemical biosensor for detecting the phytohormones IAA and SA (Hu et al., 2020). Additionally, Ahmad (2020) employed genetic engineering to develop a pH-stable nanosensor, FLIP-H_2O_2, which exhibits high selectivity for hydrogen peroxide (H_2O_2) while remaining insensitive to other oxidizing agents.

11.11.4 Nano-Biosensors in Salinity Stress

Cultivation failure occurs in water-stressed and sub-humid regions due to an unbalanced concentration of salts. The severity of salinity's harmful effect relies on the concentration of salt and its origin (Shahzad et al., 2019). The toxic concentration of sodium and chloride ions causes devastating water potential in the soil, and water dynamics in plants are disrupted (Munns, 2005). The electrochemical biosensor designed by Hu et al. (2020) can sense very low concentrations of important metabolic signals such as IAA can be detected at 1.99 μM and salicylic acid at 3.30 μM. Notably, the incorporation of multiwall carbon nanotubes (MWNT) and carbon black significantly augmented the sensor's detection capabilities (Mondal et al., 2022). Massachusetts Institute of Technology (MIT) engineers have successfully developed and integrated single-wall carbon nanotube (SWCNT) nanosensors within plant leaves. These innovative sensors detect various stress conditions and transmit data via near-infrared (nIR) optical signals, enabling remote monitoring through smartphones via email and text notifications (Mittal et al., 2017).

11.11.5 Nano-Biosensors in Heavy Metal Stress

Heavy metals are persistent and non-biodegradable and are polluting the environment for centuries. Heavy metals get accumulated in the sediments of coastal wetlands, other smaller and larger water bodies and ultimately become part of food chains and get accumulated in the bodies of living organisms (Numan et al., 2021). The escalating levels of heavy metals in the environment constitute a substantial global risk (Mitra et al., 2022). The incorporation of QDs-based nanomaterials in sensors has revolutionized the detection capabilities of sensors offering improved sensitivity, speed, and cost-effectiveness over traditional methods Wang et al. (2019, 2022) reported biosensors to detect the secretion of biomarkers such as vitronectin-like (VN) to detect plant cell damage through changes in electrochemical reactance. This method also helps in identifying cadmium- and lead-induced stress in different plant species such as *Arabidopsis* and other related species (Guo et al., 2020).

11.11.6 Nano-Biosensors in Detecting Contaminants

Contaminants are detected using nanosensors equipped with recognition elements that selectively interact with target analytes such as copper, cadmium, and other heavy metal ions.(Hairom et al., 2021). Signal is generated when analytes combine with the recognition elements. This triggered response can be measured and detected by nanosensors. The mechanisms underlying nanosensor interactions are diverse, depending on design specifications and intended applications. Certain nanosensors leverage optical phenomena, where shifts in luminescence or absorbance signal the presence of contaminants. Alternatively, others exploit electrochemical transformations, wherein the introduction of target analytes induces measurable changes in electrical conductivity (Hairom et al., 2021; Kumar & Painuli, 2022; Noronha et al., 2019; Potes-Lesoinne et al., 2022)

11.11.7 Early Stress Detection

Crop production is greatly affected worldwide by various abiotic stresses such as salinity, high temperature, flooding, soil erosion, and heavy metal stress (Yadav et al., 2020). To make plants more resilient to stress factors, cutting-edge approaches are necessary, such as developing new technologies and innovative methods to timely detect stress and improve crop yield and growth (Hareesh et al., 2023). For developing these technologies and making more advance approaches, the understanding of plant phenomenon in the face of stress and other external factors is necessary (Galieni et al., 2021). Due to the increasing impact of stressed and unfavorable environment on plants, researchers around the world are interested in making approaches which will lead to early detection and monitoring of plant stress and changes in plant metabolism in response to any biotic or abiotic stress. These approaches include using biomarkers at the cellular and molecular level to detect stress and prevent potential damage to plants (Fernandez et al., 2016; Paes de Melo et al., 2022). ABA, aquaporin, dehydrin, transcription factors, heat shock proteins, antioxidant enzymes, and sRNA are used as biomarkers to detect stress. A wide range of other biomarkers, including phytohormones, enzymes, proteins, and nucleic acids, plays an important role in detecting and monitoring changes in plant metabolism in response to environmental changes even before the occurrence of any physical damage (Alharbi, 2020).

Real-time detection of stress and other plant diseases is important to boost crop productivity and reduce economic losses due to diseases. Early detection of plant diseases and stress is hindered by two hurdles:

1) the need for real-time and cost-effective diagnosis and
2) identifying infections during the early infection stages. To address these hurdles, different innovative approaches and technologies have been devised (John et al., 2022).

The innovation of nanosensors is a groundbreaking approach in the field of plant biotechnology, which enables researchers to detect different biomarkers associated with a plant's physiological and environmental stressors. These nanosensors, integrated with nanomaterials, enable the detection of low concentrations of biomarkers due to their high sensitivity and selectivity, which will otherwise be a challenging task. The recognition of volatile organic compounds produced by plants due to stress can be done by further development in nanosensors, which will provide real-time monitoring of plant health status (Kosgiker, 2018; Liyakat, 2023b; Nikita et al., 2022; Patrick, 1972). Nanosensors are on the verge to revolutionize plant science by enabling the rapid detection of stress and diseases, thereby supporting prompt actions to minimize productivity and economic losses. The combination of nanosensors and wireless technologies can give rise to innovative, data-based agricultural practices that monitor plant health in real-time. Additionally, these sensors can uncover the complexities of plant biochemistry by monitoring metabolites and signaling molecules, offering novel insights into plant–environment interactions (Chinthamu et al., 2023; Kazi, 2023; Kutubuddin et al., 2022; Liyakat, 2023a; Wu et al., 2016). Nanosensors offer a groundbreaking opportunity to unravel the complex responses of plants to environmental stressors, including climate change, nutrient deficiencies, and soil degradation. By identifying specific biochemical changes that occur when plants are stressed, researchers can locate and identify vulnerabilities and opportunities for improvement. This insight can then be harnessed to develop more resilient and adaptable crop varieties through targeted breeding programs or precise genetic modification. As technology continues to evolve and become more affordable, the widespread adoption of nanosensors is poised to revolutionize agricultural practices. This shift could lead to the development of sustainable farming strategies, which are essential for meeting the food demands of a rapidly growing global population. Moreover, the incorporation of nanosensors in plant biology not only enhances our understanding of plant physiology but also paves the way for innovative solutions to pressing agricultural challenges (Karale Aishwarya, 2023; Kazi & Kutubuddin, 2023; Kazi et al., 2023; Liyakat, 2023c,d; Mulani et al., 2023; Nerkar et al., 2023b)

11.12 Conclusion

The development and application of nano-biosensors for plant stress management mark a groundbreaking advancement in agricultural science, offering a transformative solution to one of the most pressing challenges of our time. As global populations grow and climate change intensifies, the need for effective stress management in plants has never been more critical. Stresses, whether biotic, such as pathogens and pests, or abiotic, such as drought, salinity, and extreme temperatures, pose significant threats to plant health, crop yields, and ultimately, global food security. Traditional methods of stress detection, while valuable, often fall short due to their lack of sensitivity, specificity, and real-time monitoring capabilities. These limitations have paved the way for the emergence of nano-biosensors, which leverage the unique properties of nanomaterials to provide precise, rapid, and noninvasive detection of stress responses in plants. Nano-biosensors operate on the principle of converting biological interactions into measurable signals, enabling the detection of stress-related biomarkers at incredibly low concentrations. Their nanoscale dimensions and high surface-to-volume ratio allow for unparalleled sensitivity and specificity, making them ideal for monitoring both physiological and anatomical changes in plants. From optical and electrochemical to mechanical biosensors, each type offers distinct advantages tailored to specific applications, whether it's detecting early signs of drought stress or identifying pathogen infections at the molecular level. The applications of nano-biosensors in agriculture are vast and transformative. They enable real-time monitoring of plant health, allowing farmers and researchers to intervene promptly and mitigate stress before it causes irreversible damage. By providing detailed insights into the molecular and cellular mechanisms of stress responses, nano-biosensors not only enhance our understanding of plant biology but also pave the way for the development of stress-resistant crop varieties. Moreover, their integration with emerging technologies, such as artificial intelligence and the Internet of Things (IoT), holds immense potential for creating smart farming systems that optimize resource use and maximize productivity. In conclusion, nano-biosensors represent a paradigm shift in plant stress management, offering a powerful tool to enhance crop resilience, optimize agricultural practices, and ensure global food security. As we face the growing challenges of a changing climate and increasing food demand, the adoption of nano-biosensors in agriculture will be critical for building a sustainable and resilient future.

11.13 Summary

This chapter explores the transformative potential of nano-biosensors in addressing the critical challenge of plant stress management. With global populations rising and climate change exacerbating environmental stresses, the need for advanced tools to detect and mitigate stress in plants has become increasingly urgent. Stresses ranging from biotic factors like pathogens and pests to abiotic factors such as drought, salinity, and extreme temperatures significantly impact plant health and agricultural productivity. Traditional methods of stress detection, while useful, are often limited by their lack of sensitivity, specificity, and real-time monitoring capabilities. These limitations have driven the development of nano-biosensors, which leverage the unique properties of nanomaterials to provide precise, rapid, and non-invasive detection of stress responses in plants. While the potential of nano-biosensors is immense, challenges such as scalability, cost, and field applicability remain. Future research should focus on addressing these limitations and expanding the use of nano-biosensors for detecting new types of stresses. Collaborative efforts between scientists, engineers, and agricultural experts will be essential to fully realize the benefits of this technology. In summary, nano-biosensors represent a groundbreaking advancement in plant stress management, offering a powerful tool to enhance crop resilience, optimize agricultural practices, and ensure global food security. As we face the growing challenges of a changing climate and increasing food demand, the adoption of nano-biosensors in agriculture will be critical for building a sustainable and resilient future.

References

Acquaah, G. (2007). *Sexual hybridization and wide crosses. Principals of plant genetics and breeding* (pp. 164–180). Oxford: Blackwell Publishing.

Aglietti, C., Luchi, N., Pepori, A. L., Bartolini, P., Pecori, F., Raio, A., ... Santini, A. (2019). Real-time loop-mediated isothermal amplification: an early-warning tool for quarantine plant pathogen detection. *AMB Express*, *9*, 1–14.

Agrawal, R., Satlewal, A., Chaudhary, M., Verma, A., Singh, R., Verma, A., ... Singh, K. (2012). Rapid detection of cadmium-resistant plant growth promotory rhizobacteria: a perspective of ELISA and QCM-based immunosensor. *Journal of Microbiology and Biotechnology*, *22*(6), 849–855.

Ahmad, A. (2020). Construction of a nanosensor for non-invasive imaging of hydrogen peroxide levels in living cells. *Biology*, *9*, 0430. https://doi.org/10.3390/biology9120430

Ai, F., Chen, H., Zhang, S.-H., Liu, S.-Y., Wei, F., Dong, X.-Y., ... Huang, W.-H. (2009). Real-time monitoring of oxidative burst from single plant protoplasts using microelectrochemical sensors modified by platinum nanoparticles. *Analytical Chemistry, 81*(20), 8453–8458.

Akiyama, K., Matsuzaki, K.-i., & Hayashi, H. (2005). Plant sesquiterpenes induce hyphal branching in arbuscular mycorrhizal fungi. *Nature, 435*(7043), 824–827.

Alharbi, R. A. (2020). Proteomics approach and techniques in identification of reliable biomarkers for diseases. *Saudi Journal of Biological Sciences, 27*(3), 968–974.

Ali, S., Ullah, M. I., Sajjad, A., Shakeel, Q., & Hussain, A. (2021). Environmental and health effects of pesticide residues. In *Sustainable agriculture reviews 48: pesticide occurrence, analysis and remediation vol. 2 analysis* (pp. 311–336). Cham: Springer International Publishing.

Aliche, E. B., Screpanti, C., De Mesmaeker, A., Munnik, T., & Bouwmeester, H. J. (2020). Science and application of strigolactones. *New Phytologist, 227*(4), 1001–1011.

Alova, A., Erofeev, A., Gorelkin, P., Bibikova, T., Korchev, Y., Majouga, A., & Bulychev, A. (2020). Prolonged oxygen depletion in microwounded cells of *Chara corallina* detected with novel oxygen nanosensors. *Journal of Experimental Botany, 71*(1), 386–398.

Álvarez, S. P., López, N. E. L., Lozano, J. M., Negrete, E. A. R., & Cervantes, M. E. S. (2016). Plant fungal disease management using nanobiotechnology as a tool. In *Advances and applications through fungal nanobiotechnology* (pp. 169–192). Cham: Springer International Publishing.

Andrade Foronda, D. (2022). Reclamation of a saline-sodic soil with organic amendments and leaching. *Environmental Sciences Proceedings, 16*(1), 56.

Anfossi, L., Di Nardo, F., Giovannoli, C., Passini, C., & Baggiani, C. (2013). Increased sensitivity of lateral flow immunoassay for ochratoxin A through silver enhancement. *Analytical and Bioanalytical Chemistry, 405*, 9859–9867.

Anjum, S. A., Xie, X. Y., Wang, L. C., Saleem, M. F., Man, C., & Lei, W. (2011). Morphological, physiological and biochemical responses of plants to drought stress. *African Journal of Agricultural Research, 6*(9), 2026–2032.

Antonacci, A., Arduini, F., Moscone, D., Palleschi, G., & Scognamiglio, V. (2018). Nanostructured (Bio)sensors for smart agriculture. *TrAC Trends in Analytical Chemistry, 98*, 95–103.

Apel, K., & Hirt, H. (2004). Reactive oxygen species: metabolism, oxidative stress, and signal transduction. *Annual Review of Plant Biology, 55*(1), 373–399.

Ashraf, J., & Javed, A. (2023). Food security and environmental degradation: do institutional quality and human capital make a difference? *Journal of Environmental Management, 331*, 117330.

Ast, C., Schmälzlin, E., Löhmannsröben, H.-G., & Van Dongen, J. T. (2012). Optical oxygen micro-and nanosensors for plant applications. *Sensors, 12*(6), 7015–7032.

Atkinson, N. J., & Urwin, P. E. (2012). The interaction of plant biotic and abiotic stresses: from genes to the field. *Journal of Experimental Botany, 63*(10), 3523–3543.

Bai, B., Guo, Z., Zhou, C., Zhang, W., & Zhang, J. (2021). Application of adaptive reliability importance sampling-based extended domain PSO on single mode failure in reliability engineering. *Information Sciences, 546*, 42–59.

Bakker, E., & Telting-Diaz, M. (2002). Electrochemical sensors. *Analytical Chemistry, 74*(12), 2781–2800.

Bauriegel, E., & Herppich, W. B. (2014). Hyperspectral and chlorophyll fluorescence imaging for early detection of plant diseases, with special reference to *Fusarium* spec. infections on wheat. *Agriculture, 4*(1), 32–57.

Belousov, V. V., Fradkov, A. F., Lukyanov, K. A., Staroverov, D. B., Shakhbazov, K. S., Terskikh, A. V., & Lukyanov, S. (2006). Genetically encoded fluorescent indicator for intracellular hydrogen peroxide. *Nature Methods*, *3*(4), 281–286.

Bhagat, R., Ingle, A. P., & Chen, H. (2023). Nanosensors and nanobiosensors for sustainable agriculture. In *Nanotechnology in agriculture and agroecosystems* (pp. 93–112). Elsevier.

Bilodeau, G., Lévesque, C., De Cock, A., Duchaine, C., Brière, S., Uribe, P., & Hamelin, R. C. (2007). Molecular detection of *Phytophthora ramorum* by real-time polymerase chain reaction using TaqMan, SYBR Green, and molecular beacons. *Phytopathology*, *97*(5), 632–642.

Bitas, V., Kim, H.-S., Bennett, J. W., & Kang, S. (2013). Sniffing on microbes: diverse roles of microbial volatile organic compounds in plant health. *Molecular Plant-Microbe Interactions*, *26*(8), 835–843.

Bonnick, S., Broy, S., Kaiser, F., Teutsch, C., Rosenberg, E., DeLucca, P., & Melton, M. (2007). Treatment with alendronate plus calcium, alendronate alone, or calcium alone for postmenopausal low bone mineral density. *Current Medical Research and Opinion*, *23*(6), 1341–1349.

Bot, A., Nachtergaele, F., & Young, A. (2000). *Land resource potential and constraints at regional and country levels*. Food & Agriculture Org.

Bower, J., Fast, D., Garofolo, F., Gouty, D., Hayes, R., Lowes, S., ... Shoup, R. (2014). *8th GCC: consolidated feedback to US FDA on the 2013 draft FDA guidance on bioanalytical method validation*. Taylor & Francis.

Brodt, S., Six, J., Feenstra, G., Ingels, C., & Campbell, D. (2011). Sustainable agriculture. *Nature Education Knowledge*, *3*(10), 1.

Buck, S. M., Xu, H., Brasuel, M., Philbert, M. A., & Kopelman, R. (2004). Nanoscale probes encapsulated by biologically localized embedding (PEBBLEs) for ion sensing and imaging in live cells. *Talanta*, *63*(1), 41–59.

Bürger, M., & Chory, J. (2019). Stressed out about hormones: how plants orchestrate immunity. *Cell Host & Microbe*, *26*(2), 163–172.

Bykova, N. V., Keerberg, O., Pärnik, T., Bauwe, H., & Gardeström, P. (2005). Interaction between photorespiration and respiration in transgenic potato plants with antisense reduction in glycine decarboxylase. *Planta*, *222*, 130–140.

Cai Feng, C. F., Mei LanJu, M. L., An XiaoLong, A. X., Gao Shun, G. S., Tang Lin, T. L., & Chen Fang, C. F. (2011). Lipid peroxidation and antioxidant responses during seed germination of *Jatropha curcas*. *International Journal of Agriculture and Biology*, *13*(1), 25–30.

Cano, A., Leonard, M. T., & Franz, A. (2005). The significant other version of the Pain Catastrophizing Scale (PCS-S): preliminary validation. *Pain*, *119*(1–3), 26–37.

Cebula, Z., Żołędowska, S., Dziąbowska, K., Skwarecka, M., Malinowska, N., Białobrzeska, W., ... Bogdanowicz, R. (2019). Detection of the plant pathogen *Pseudomonas syringae* pv. *Lachrymans* on antibody-modified gold electrodes by electrochemical impedance spectroscopy. *Sensors*, *19*(24), 5411.

Challinor, A. J., Watson, J., Lobell, D. B., Howden, S. M., Smith, D., & Chhetri, N. (2014). A meta-analysis of crop yield under climate change and adaptation. *Nature Climate Change*, *4*(4), 287–291.

Chang, A. C., & Brawer Silva, D. (2014). *Salinity and drainage in San Joaquin Valley*. California: Springer.

Charlermroj, R., Himananto, O., Seepiban, C., Kumpoosiri, M., Warin, N., Oplatowska, M., ... Elliott, C. T. (2013). Multiplex detection of plant pathogens using a microsphere immunoassay technology. *PLoS One, 8*(4), e62344.

Chen, L., & Xu, X. (2020). Effect evaluation of the long-term care insurance (LTCI) system on the health care of the elderly: a review. *Journal of Multidisciplinary Healthcare, 13*, 863–875.

Chen, Y., He, L., Guan, Y., Lu, H., & Li, J. (2017a). Life cycle assessment of greenhouse gas emissions and water-energy optimization for shale gas supply chain planning based on multi-level approach: case study in Barnett, Marcellus, Fayetteville, and Haynesville shales. *Energy Conversion and Management, 134*, 382–398.

Chen, M., Zhang, Z., Liu, M., Qiu, C., Yang, H., & Chen, X. (2017b). In situ fabrication of label-free optical sensing paper strips for the rapid surface-enhanced Raman scattering (SERS) detection of brassinosteroids in plant tissues. *Talanta, 165*, 313–320.

Chen, Y., Li, J., Lu, H., & Yan, P. (2021). Coupling system dynamics analysis and risk aversion programming for optimizing the mixed noise-driven shale gas-water supply chains. *Journal of Cleaner Production, 278*, 123209.

Cheng, X., He, L., Lu, H., Chen, Y., & Ren, L. (2016). Optimal water resources management and system benefit for the Marcellus shale-gas reservoir in Pennsylvania and West Virginia. *Journal of Hydrology, 540*, 412–422.

Chinnusamy, V., & Zhu, J.-K. (2003). Plant salt tolerance. In *Plant responses to abiotic stress* (pp. 241–270). Springer.

Chinthamu, N., Prasad, M., Chinchawade, A. J., Liyakat, K. K. S., Deepti, K., Karukuri, M., & Kumar, C. M. (2023). Self-secure firmware model for blockchain-enabled IOT environment to embedded system. *European Chemical Bulletin, 12*, S3.

Chiriacò, M. S., Luvisi, A., Primiceri, E., Sabella, E., De Bellis, L., & Maruccio, G. (2018). Development of a lab-on-a-chip method for rapid assay of *Xylella fastidiosa* subsp. pauca strain CoDiRO. *Scientific Reports, 8*(1), 7376.

Clark JR, L. C., Wolf, R., Granger, D., & Taylor, Z. (1953). Continuous recording of blood oxygen tensions by polarography. *Journal of Applied Physiology, 6*(3), 189–193.

Considine, M. J., Diaz-Vivancos, P., Kerchev, P., Signorelli, S., Agudelo-Romero, P., Gibbs, D. J., & Foyer, C. H. (2017). Learning to breathe: developmental phase transitions in oxygen status. *Trends in Plant Science, 22*(2), 140–153.

Cook, C., Whichard, L. P., Turner, B., Wall, M. E., & Egley, G. H. (1966). Germination of witchweed (*Striga lutea* Lour.): isolation and properties of a potent stimulant. *Science, 154*(3753), 1189–1190.

Cook, C., Whichard, L. P., Wall, M., Egley, G. H., Coggon, P., Luhan, P. A., & McPhail, A. (1972). Germination stimulants. II. Structure of strigol, a potent seed germination stimulant for witchweed (*Striga lutea*). *Journal of the American Chemical Society, 94*(17), 6198–6199.

Costa, A., Drago, I., Behera, S., Zottini, M., Pizzo, P., Schroeder, J. I., ... Schiavo, F. L. (2010). H_2O_2 in plant peroxisomes: an in vivo analysis uncovers a Ca^{2+}–dependent scavenging system. *The Plant Journal, 62*(5), 760–772.

Cristescu, S. M., Mandon, J., Arslanov, D., De Pessemier, J., Hermans, C., & Harren, F. J. (2013). Current methods for detecting ethylene in plants. *Annals of Botany, 111*(3), 347–360.

Cui, S., Inocente, E. A. A., Acosta, N., Keener, H. M., Zhu, H., & Ling, P. P. (2019). Development of fast e-nose system for early-stage diagnosis of aphid-stressed tomato plants. *Sensors, 19*(16), 3480.

Cui, D., Li, J., Zhang, X., Zhang, L., Chang, H., & Wang, Q. (2021). Pyrolysis temperature effect on compositions of basic nitrogen species in Huadian shale oil using positive-ion ESI FT-ICR MS and GC-NCD. *Journal of Analytical and Applied Pyrolysis*, *153*, 104980.

Culbreath, A., Kemerait, R., Jr., & Brenneman, T. (2008). Management of leaf spot diseases of peanut with prothioconazole applied alone or in combination with tebuconazole or trifloxystrobin. *Peanut Science*, *35*(2), 149–158.

Davis, K., Nkonya, E., Kato, E., Mekonnen, D. A., Odendo, M., Miiro, R., & Nkuba, J. (2012). Impact of farmer field schools on agricultural productivity and poverty in East Africa. *World Development*, *40*(2), 402–413.

Day, D. A., Neuburger, M., & Douce, R. (1985). Biochemical characterization of chlorophyll-free mitochondria from pea leaves. *Functional Plant Biology*, *12*(3), 219–228.

Delory, B. M., Delaplace, P., Fauconnier, M.-L., & Du Jardin, P. (2016). Root-emitted volatile organic compounds: can they mediate belowground plant-plant interactions? *Plant and Soil*, *402*, 1–26.

Deng, Y., Zhang, T., Clark, J., Aminabhavi, T., Kruse, A., Tsang, D. C., ... Ren, H. (2020). Mechanisms and modelling of phosphorus solid–liquid transformation during the hydrothermal processing of swine manure. *Green Chemistry*, *22*(17), 5628–5638.

Du, S., Yu, C., Tang, L., & Lu, L. (2018). Applications of SERS in the detection of stress-related substances. *Nanomaterials*, *8*(10), 757.

El Ichi, S., Leon, F., Vossier, L., Marchandin, H., Errachid, A., Coste, J., ... Fournier-Wirth, C. (2014). Microconductometric immunosensor for label-free and sensitive detection of Gram-negative bacteria. *Biosensors and Bioelectronics*, *54*, 378–384.

Elmer, W., & White, J. C. (2018). The future of nanotechnology in plant pathology. *Annual Review of Phytopathology*, *56*(1), 111–133.

Escalona, J. M., Flexas, J., & Medrano, H. (2000). Stomatal and non-stomatal limitations of photosynthesis under water stress in field-grown grapevines. *Functional Plant Biology*, *27*(1), 87–87.

Esser, B., Schnorr, J. M., & Swager, T. M. (2012). Selective detection of ethylene gas using carbon nanotube-based devices: utility in determination of fruit ripeness. *Angewandte Chemie International Edition*, *51*(23), 5752–5756.

Eun, A. J.-C., Huang, L., Chew, F.-T., Fong-Yau Li, S., & Wong, S.-M. (2002). Detection of two orchid viruses using quartz crystal microbalance-based DNA biosensors. *Phytopathology*, *92*(6), 654–658.

Fang, X., Chen, J., Dai, L., Ma, H., Zhang, H., Yang, J., ... Yan, C. (2015). Proteomic dissection of plant responses to various pathogens. *Proteomics*, *15*(9), 1525–1543.

Farber, C., & Kurouski, D. (2018). Detection and identification of plant pathogens on maize kernels with a hand-held Raman spectrometer. *Analytical Chemistry*, *90*(5), 3009–3012.

Farooq, M., Wahid, A., Kobayashi, N., Fujita, D., & Basra, S. M. (2009). Plant drought stress: effects, mechanisms and management. In *Sustainable agriculture* (pp. 153–188). Dordrecht: Springer Netherlands.

Farooq, M. A., Niazi, A. K., Akhtar, J., Farooq, M., Souri, Z., Karimi, N., & Rengel, Z. (2019). Acquiring control: the evolution of ROS-induced oxidative stress and redox signaling pathways in plant stress responses. *Plant Physiology and Biochemistry*, *141*, 353–369.

Feng, Q., Li, Y., Wang, N., Hao, Y., Chang, J., Wang, Z., ... Wang, L. (2020a). A biomimetic nanogenerator of reactive nitrogen species based on battlefield transfer strategy for enhanced immunotherapy. *Small*, *16*(25), 2002138.

Feng, S., Lu, H., Tian, P., Xue, Y., Lu, J., Tang, M., & Feng, W. (2020b). Analysis of microplastics in a remote region of the Tibetan Plateau: implications for natural environmental response to human activities. *Science of the Total Environment, 739*, 140087.

Fernandez, O., Urrutia, M., Bernillon, S., Giauffret, C., Tardieu, F., Le Gouis, J., ... Gibon, Y. (2016). Fortune telling: metabolic markers of plant performance. *Metabolomics, 12*, 1–14.

Firrao, G., Moretti, M., Rosquete, M. R., Gobbi, E., & Locci, R. (2005). Nanobiotransducer for detecting flavescence dorée phytoplasma. *Journal of Plant Pathology, 87*(2), 101–107.

Foyer, C. H. (2018). Reactive oxygen species, oxidative signaling and the regulation of photosynthesis. *Environmental and Experimental Botany, 154*, 134–142.

de Freitas, A. R., Monteiro, C. B., Pedrosa, M. V. B., de Lima, P. A. M., & Alexandre, R. S. (2017). Effects of sewage sludge and length on initial growth of guava seedlings. *African Journal of Agricultural Research, 12*(49), 3415–3420.

Galieni, A., D'Ascenzo, N., Stagnari, F., Pagnani, G., Xie, Q., & Pisante, M. (2021). Past and future of plant stress detection: an overview from remote sensing to positron emission tomography. *Frontiers in Plant Science, 11*, 609155.

Gao, T., Li, C., Zhang, Y., Yang, M., Jia, D., Jin, T., ... Li, R. (2019). Dispersing mechanism and tribological performance of vegetable oil-based CNT nanofluids with different surfactants. *Tribology International, 131*, 51–63.

García-Quirós, E., Alché, J. D., Karpinska, B., & Foyer, C. H. (2020). Glutathione redox state plays a key role in flower development and pollen vigour. *Journal of Experimental Botany, 71*(2), 730–741.

Ge, X., Asiri, A. M., Du, D., Wen, W., Wang, S., & Lin, Y. (2014). Nanomaterial-enhanced paper-based biosensors. *TrAC Trends in Analytical Chemistry, 58*, 31–39.

Ghaffar, N., Farrukh, M. A., & Naz, S. (2020). Applications of nanobiosensors in agriculture. In *Nanoagronomy* (pp. 179–196). Cham: Springer International Publishing.

Gieling, T. H., Van Straten, G., Janssen, H., & Wouters, H. (2005). ISE and Chemfet sensors in greenhouse cultivation. *Sensors and Actuators B: Chemical, 105*(1), 74–80.

Global Growth Insights. (2025, October 6). *Pesticides market size, share, trends & growth report, 2033*. https://www.globalgrowthinsights.com/market-reports/pesticides-market-116195

Giraldo, J. P., Wu, H., Newkirk, G. M., & Kruss, S. (2019). Nanobiotechnology approaches for engineering smart plant sensors. *Nature Nanotechnology, 14*(6), 541–553.

Goffart, J., Olivier, M., & Frankinet, M. (2008). Potato crop nitrogen status assessment to improve N fertilization management and efficiency: past–present–future. *Potato Research, 51*, 355–383.

Gomez-Roldan, V., Fermas, S., Brewer, P. B., Puech-Pagès, V., Dun, E. A., Pillot, J.-P., ... Portais, J.-C. (2008). Strigolactone inhibition of shoot branching. *Nature, 455*(7210), 189–194.

Gowda, V. R., Henry, A., Yamauchi, A., Shashidhar, H., & Serraj, R. (2011). Root biology and genetic improvement for drought avoidance in rice. *Field Crops Research, 122*(1), 1–13.

Guo, Z., Kang, Y., Liang, S., & Zhang, J. (2020). Detection of Hg(II) in adsorption experiment by a lateral flow biosensor based on streptavidin-biotinylated DNA probes modified gold nanoparticles and smartphone reader. *Environmental Pollution, 266*, 115389.

Hairom, N. H. H., Soon, C. F., Mohamed, R. M. S. R., Morsin, M., Zainal, N., Nayan, N., ... Harun, N. H. (2021). A review of nanotechnological applications to detect and control surface water pollution. *Environmental Technology & Innovation, 24*, 102032.

Han, X., Chen, N., Yan, J., Liu, J., Liu, M., & Karellas, S. (2019). Thermodynamic analysis and life cycle assessment of supercritical pulverized coal-fired power plant integrated with No. 0 feedwater pre-heater under partial loads. *Journal of Cleaner Production*, *233*, 1106–1122.

Hanson, G. T., Aggeler, R., Oglesbee, D., Cannon, M., Capaldi, R. A., Tsien, R. Y., & Remington, S. J. (2004). Investigating mitochondrial redox potential with redox-sensitive green fluorescent protein indicators. *Journal of Biological Chemistry*, *279*(13), 13044–13053.

Harb, A., Krishnan, A., Ambavaram, M. M., & Pereira, A. (2010). Molecular and physiological analysis of drought stress in *Arabidopsis* reveals early responses leading to acclimation in plant growth. *Plant Physiology*, *154*(3), 1254–1271.

Hareesh, P. S., Resmi, T. R., Sheela, M. N., & Makeshkumar, T. (2023). Cassava mosaic disease in South and Southeast Asia: current status and prospects. *Frontiers in Sustainable Food Systems*, *7*, 1086660.

Hatfield, J. L., Boote, K. J., Kimball, B. A., Ziska, L., Izaurralde, R. C., Ort, D., & Wolfe, D. (2011). Climate impacts on agriculture: implications for crop production. *Agronomy Journal*, *103*(2), 351–370.

He, L., Chen, Y., & Li, J. (2018a). A three-level framework for balancing the tradeoffs among the energy, water, and air-emission implications within the life-cycle shale gas supply chains. *Resources, Conservation and Recycling*, *133*, 206–228.

He, L., Chen, Y., Zhao, H., Tian, P., Xue, Y., & Chen, L. (2018b). Game-based analysis of energy-water nexus for identifying environmental impacts during Shale gas operations under stochastic input. *Science of the Total Environment*, *627*, 1585–1601.

He, L., Shen, J., & Zhang, Y. (2018c). Ecological vulnerability assessment for ecological conservation and environmental management. *Journal of Environmental Management*, *206*, 1115–1125.

Hernández-Barrera, A., Velarde-Buendía, A., Zepeda, I., Sanchez, F., Quinto, C., Sánchez-Lopez, R., ... Cardenas, L. (2015). Hyper, a hydrogen peroxide sensor, indicates the sensitivity of the *Arabidopsis* root elongation zone to aluminum treatment. *Sensors*, *15*(1), 855–867.

Hofmann, D. J., Butler, J. H., & Tans, P. P. (2009). A new look at atmospheric carbon dioxide. *Atmospheric Environment*, *43*(12), 2084–2086.

Hu, C., Yang, D.-P., Zhu, F., Jiang, F., Shen, S., & Zhang, J. (2014). Enzyme-labeled Pt@BSA nanocomposite as a facile electrochemical biosensing interface for sensitive glucose determination. *ACS Applied Materials & Interfaces*, *6*(6), 4170–4178.

Hu, Y., Wang, X., Wang, C., Hou, P., Dong, H., Luo, B., & Li, A. (2020). A multifunctional ratiometric electrochemical sensor for combined determination of indole-3-acetic acid and salicylic acid. *RSC Advances*, *10*(6), 3115–3121.

Huang, Z., Zheng, H., Guo, L., & Mo, D. (2020). Influence of the position of artificial boundary on computation accuracy of conjugated infinite element for a finite length cylindrical shell. *Acoustics Australia*, *48*(2), 287–294.

Husen, A. (2022). *Environmental pollution and medicinal plants*. CRC Press.

Husin, N. A., Khairunniza-Bejo, S., Abdullah, A. F., Kassim, M. S., Ahmad, D., & Azmi, A. N. (2020). Application of ground-based LiDAR for analysing oil palm canopy properties on the occurrence of basal stem rot (BSR) disease. *Scientific Reports*, *10*(1), 6464.

Ijaz, M., Khan, F., Ahmed, T., Noman, M., Zulfiqar, F., Rizwan, M., & Li, B. (2023). Nanobiotechnology to advance stress resilience in plants: current opportunities and challenges. *Materials Today Bio*, *22*, 100759.

ITPS, FAO. (2015). Status of the world's soil resources (SWSR): Main report. *Food and Agriculture Organization of the United Nations and intergovernmental technical panel on soils*, 650.

James, D., Varga, A., & Sanderson, D. (2013). Genetic diversity of *Plum pox virus*: strains, disease and related challenges for control. *Canadian Journal of Plant Pathology*, *35*(4), 431–441.

Janssen, S., Schmitt, K., Blanke, M., Bauersfeld, M.-L., Wöllenstein, J., & Lang, W. (2014). Ethylene detection in fruit supply chains. *Philosophical Transactions of the Royal Society A: Mathematical, Physical and Engineering Sciences*, *372*(2017), 20130311.

Jia, L., Liu, B., Zhao, Y., Chen, W., Mou, D., Fu, J., ... Zhao, L. (2020). Structure design of MoS_2@Mo_2C on nitrogen-doped carbon for enhanced alkaline hydrogen evolution reaction. *Journal of Materials Science*, *55*, 16197–16210.

Jiang, K., Schwarzer, C., Lally, E., Zhang, S., Ruzin, S., Machen, T., ... Feldman, L. (2006). Expression and characterization of a redox-sensing green fluorescent protein (reduction-oxidation-sensitive green fluorescent protein) in *Arabidopsis*. *Plant Physiology*, *141*(2), 397–403.

Jiang, Q., Wang, G., Jin, S., Li, Y., & Wang, Y. (2013). Predicting human microRNA-disease associations based on support vector machine. *International Journal of Data Mining and Bioinformatics*, *8*(3), 282–293.

Jiang, Q., Shao, F., Gao, W., Chen, Z., Jiang, G., & Ho, Y.-S. (2018). Unified no-reference quality assessment of singly and multiply distorted stereoscopic images. *IEEE Transactions on Image Processing*, *28*(4), 1866–1881.

Jiang, D., Chen, F.-X., Zhou, H., Lu, Y.-Y., Tan, H., Yu, S.-J., ... Jin, Z.-B. (2020). Bioenergetic crosstalk between mesenchymal stem cells and various ocular cells through the intercellular trafficking of mitochondria. *Theranostics*, *10*(16), 7260.

John, S. A., Chattree, A., Ramteke, P. W., Shanthy, P., Nguyen, T. A., & Rajendran, S. (2022). Nanosensors for plant health monitoring. In *Nanosensors for smart agriculture* (pp. 449-–461). Elsevier.

Joint FAO/WHO Expert Committee on Food Additives. Meeting, and World Health Organization. (2011). *Evaluation of certain contaminants in food: seventy-second [72nd] report of the Joint FAO/WHO Expert Committee on Food Additives*. World Health Organization.

Jongman, M., Carmichael, P. C., & Bill, M. (2020). Technological advances in phytopathogen detection and metagenome profiling techniques. *Current Microbiology*, *77*(4), 675–681.

Kakarwal, S., & Deshmukh, D. R. (2012). Performance analysis of face recognition by principal component analysis and feed forward neural network. *International Journal of Engineering Innovations and Research*, *1*(1), 40–45.

Kamran, M., Parveen, A., Ahmar, S., Malik, Z., Hussain, S., Chattha, M. S., ... Chen, J.-T. (2019). An overview of hazardous impacts of soil salinity in crops, tolerance mechanisms, and amelioration through selenium supplementation. *International Journal of Molecular Sciences*, *21*(1), 148.

Kangasjärvi, J., Jaspers, P., & Kollist, H. (2005). Signalling and cell death in ozone-exposed plants. *Plant, Cell & Environment*, *28*(8), 1021–1036.

Karale Aishwarya, A. (2023). Smart billing cart using RFID, YOLO and deep learning for mall administration. *International Journal of Instrumentation and Innovation Sciences*, *8*(2).

Kazi, V. (2023). Deep learning, YOLO and RFID based smart billing handcart. *Journal of Communication Engineering & Systems, 13*(1), 1–8.

Kazi, S., & Kutubuddin, K. (2023). Fruit grading, disease detection, and an image processing strategy. *Journal of Image Processing and Artificial Intelligence, 9*(2), 17–34.

Kazi, S., Shaikh, M., & Kutubuddin, K. (2023). Machine learning in the production process control of metal melting. *Journal of Advancement in Machines, 8*(2), 29–35.

Kearns, A., Whelan, J., Young, S., Elthon, T. E., & Day, D. A. (1992). Tissue-specific expression of the alternative oxidase in soybean and siratro. *Plant Physiology, 99*(2), 712–717.

Khan, M. A., Iqbal, M., Akram, M., Ahmad, M., Hassan, M. W., & Jamil, M. (2013). Recent advances in molecular tool development for drought tolerance breeding in cereal crops: a review. *Zemdirbyste-Agriculture, 100*(3), 325–334.

Khan, M., Khan, A. U., Hasan, M. A., Yadav, K. K., Pinto, M. M., Malik, N., ... Sharma, G. K. (2021). Agro-nanotechnology as an emerging field: a novel sustainable approach for improving plant growth by reducing biotic stress. *Applied Sciences, 11*(5), 2282.

Khater, M., De La Escosura-Muñiz, A., & Merkoçi, A. (2017). Biosensors for plant pathogen detection. *Biosensors and Bioelectronics, 93*, 72–86.

Kiegle, E., Moore, C. A., Haseloff, J., Tester, M. A., & Knight, M. R. (2000). Cell-type-specific calcium responses to drought, salt and cold in the *Arabidopsis* root. *The Plant Journal, 23*(2), 267–278.

Kissinger, P. T. (2005). Biosensors-a perspective. *Biosensors and Bioelectronics, 20*(12), 2512–2516.

Kneipp, K., Wang, Y., Kneipp, H., Perelman, L. T., Itzkan, I., Dasari, R. R., & Feld, M. S. (1997). Single molecule detection using surface-enhanced Raman scattering (SERS). *Physical Review Letters, 78*(9), 1667.

Kordestani, H., & Zhang, C. (2020). Direct use of the Savitzky–Golay filter to develop an output-only trend line-based damage detection method. *Sensors, 20*(7), 1983.

Kordestani, H., Zhang, C., & Shadabfar, M. (2019). Beam damage detection under a moving load using random decrement technique and Savitzky–Golay filter. *Sensors, 20*(1), 243.

Kosgiker, G. M. (2018). Machine learning-based system, food quality inspection and grading in food industry. *International Journal of Food and Nutritional Sciences, 11*(10), 723–730.

Krebs, M., Held, K., Binder, A., Hashimoto, K., Den Herder, G., Parniske, M., ... Schumacher, K. (2012). FRET-based genetically encoded sensors allow high-resolution live cell imaging of Ca^{2+} dynamics. *The Plant Journal, 69*(1), 181–192.

Krystofova, O., Trnkova, L., Adam, V., Zehnalek, J., Hubalek, J., Babula, P., & Kizek, R. (2010). Electrochemical microsensors for the detection of cadmium(II) and lead(II) ions in plants. *Sensors, 10*(6), 5308–5328.

Ku, Y.-S., Sintaha, M., Cheung, M.-Y., & Lam, H.-M. (2018). Plant hormone signaling crosstalks between biotic and abiotic stress responses. *International Journal of Molecular Sciences, 19*(10), 3206.

Kumar, C., & Painuli, R. (2022). Nanomaterials for electrochemical treatment of pollutants in water. In *Applications of advanced nanomaterials in water treatment* (pp. 165–184). CRS Press.

Kummu, M., De Moel, H., Porkka, M., Siebert, S., Varis, O., & Ward, P. J. (2012). Lost food, wasted resources: global food supply chain losses and their impacts on freshwater, cropland, and fertiliser use. *Science of the Total Environment, 438*, 477–489.

Kumudini, S., Andrade, F. H., Boote, K., Brown, G., Dzotsi, K., Edmeades, G., ... Hammer, G. (2014). Predicting maize phenology: intercomparison of functions for developmental response to temperature. *Agronomy Journal*, *106*(6), 2087–2097.

Kutubuddin, K., Halli, U., & Kosgiker, G. (2022). Voltage sag mitigation using DVR and ultra capacitor. *Journal of Semiconductor Devices and Circuits*, *9*(3), 21–31p.

Lakowicz, J. R. (2006). *Principles of fluorescence spectroscopy*. Springer.

Lan, L., Le, X., Dong, H., Xie, J., Ying, Y., & Ping, J. (2020). One-step and large-scale fabrication of flexible and wearable humidity sensor based on laser-induced graphene for real-time tracking of plant transpiration at bio-interface. *Biosensors and Bioelectronics*, *165*, 112360.

Lattanzio, V. M., & Nivarlet, N. (2017). Multiplex dipstick immunoassay for semiquantitative determination of *Fusarium mycotoxins* in oat. In *Oat: methods and protocols* (pp. 137–142). New York, NY: Springer New York.

Lattanzio, V. M., Nivarlet, N., Lippolis, V., Della Gatta, S., Huet, A.-C., Delahaut, P., ... Visconti, A. (2012). Multiplex dipstick immunoassay for semi-quantitative determination of *Fusarium mycotoxins* in cereals. *Analytica Chimica Acta*, *718*, 99–108.

Lau, H. Y., Wu, H., Wee, E. J., Trau, M., Wang, Y., & Botella, J. R. (2017). Specific and sensitive isothermal electrochemical biosensor for plant pathogen DNA detection with colloidal gold nanoparticles as probes. *Scientific Reports*, *7*(1), 38896.

Le Ru, E., & Etchegoin, P. (2009). EM enhancements and plasmon resonances: examples and discussion. In *Principles of surface-enhanced Raman spectroscopy* (pp. 299–365). Elsevier Amsterdam.

Leone, M., Consales, M., Passeggio, G., Buontempo, S., Zaraket, H., Youssef, A., ... Cusano, A. (2022). Fiber optic soil water content sensor for precision farming. *Optics & Laser Technology*, *149*, 107816.

Li, Y., Chen, D., Walker, C., & Angus, J. (2010). Estimating the nitrogen status of crops using a digital camera. *Field Crops Research*, *118*(3), 221–227.

Li, Z., Paul, R., Ba Tis, T., Saville, A. C., Hansel, J. C., Yu, T., ... Wei, Q. (2019). Non-invasive plant disease diagnostics enabled by smartphone-based fingerprinting of leaf volatiles. *Nature Plants*, *5*(8), 856–866.

Li, W., Wang, Y., Zhang, Y., Wang, R., Guo, Z., & Xie, Z. (2020). Impacts of drought stress on the morphology, physiology, and sugar content of Lanzhou lily (*Lilium davidii* var. *unicolor*). *Acta Physiologiae Plantarum*, *42*, 1–11.

Li, Z., Liu, Y., Hossain, O., Paul, R., Yao, S., Wu, S., ... Wei, Q. (2021). Real-time monitoring of plant stresses via chemiresistive profiling of leaf volatiles by a wearable sensor. *Matter*, *4*(7), 2553–2570.

Li, H., Merkl, P., Sommertune, J., Thersleff, T., & Sotiriou, G. A. (2022). SERS Hotspot engineering by aerosol self-assembly of plasmonic Ag nanoaggregates with tunable interparticle distance. *Advanced Science*, *9*(22), 2201133.

Lin, Y. (2015). LiDAR: an important tool for next-generation phenotyping technology of high potential for plant phenomics? *Computers and Electronics in Agriculture*, *119*, 61–73.

Lin, Z., Zhong, S., & Grierson, D. (2009). Recent advances in ethylene research. *Journal of Experimental Botany*, *60*(12), 3311–3336.

Lin, H.-Y., Huang, C.-H., Lu, S.-H., Kuo, I.-T., & Chau, L.-K. (2014). Direct detection of orchid viruses using nanorod-based fiber optic particle plasmon resonance immunosensor. *Biosensors and Bioelectronics*, *51*, 371–378.

Lin, J., Cai, X., Liu, Z., Liu, N., Xie, M., Zhou, B., ... Guo, Z. (2020). Anti-liquid-interfering and bacterially antiadhesive strategy for highly stretchable and ultrasensitive strain sensors based on Cassie-Baxter wetting state. *Advanced Functional Materials*, *30*(23), 2000398.

Liu, J. T., Hu, L. S., Liu, Y. L., Chen, R. S., Cheng, Z., Chen, S. J., ... Huo, K. F. (2014a). Real-time monitoring of auxin vesicular exocytotic efflux from single plant protoplasts by amperometry at microelectrodes decorated with nanowires. *Angewandte Chemie International Edition*, *53*(10), 2643–2647.

Liu, X., Zhang, N., Bing, T., & Shangguan, D. (2014b). Carbon dots based dual-emission silica nanoparticles as a ratiometric nanosensor for Cu^{2+}. *Analytical Chemistry*, *86*(5), 2289–2296.

Liu, L.-h., Zhou, X.-h., & Shi, H.-c. (2015). Portable optical aptasensor for rapid detection of mycotoxin with a reversible ligand-grafted biosensing surface. *Biosensors and Bioelectronics*, *72*, 300–305.

Liu, T., Chen, W., Wu, W., Sun, C., Guo, W., & Zhu, X. (2016). Detection of aphids in wheat fields using a computer vision technique. *Biosystems Engineering*, *141*, 82–93.

Liu, L., Wang, R., Xie, C., Yang, P., Wang, F., Sudirman, S., & Liu, W. (2019). PestNet: an end-to-end deep learning approach for large-scale multi-class pest detection and classification. *IEEE Access*, *7*, 45301–45312.

Liu, Y., Xiao, H., Hao, Y., Ye, L., Jiang, X., Wang, H., & Sun, X. (2020a). Diagnosis of citrus greening using Raman spectroscopy-based pattern recognition. *Journal of Applied Spectroscopy*, *87*, 150–158.

Liu, Y., Zhang, B., Feng, Y., Lv, X., Ji, D., Niu, Z., ... Fan, Y. (2020b). Development of 340-GHz transceiver front end based on GaAs monolithic integration technology for THz active imaging array. *Applied Sciences*, *10*(21), 7924.

Liu, Y., Xu, T., Liu, Y., Gao, Y., & Di, C. (2020c). Wear and heat shock resistance of Ni-WC coating on mould copper plate fabricated by laser. *Journal of Materials Research and Technology*, *9*(4), 8283–8288.

Liyakat, K. K. S. (2023a). Analysis for field distribution in optical waveguide using linear Fem method. *Journal of Optical communication Electronics*, *9*(1), 23–28.

Liyakat, K. K. S. (2023b). IoT-based weather information prototype using WeMos. *Journal of Control and Instrumentation Engineering*, *9*(1), 10–22.

Liyakat, K. K. S. (2023c). IoT based smart healthcare monitoring. *Liberation of Creativity: Navigating New Frontiers in Multidisciplinary Research*, *2*, 456–477.

Liyakat, K. K. S. (2023d). IoT based substation health monitoring. *Magnification of Research: Advanced Research in Social Sciences and Humanities*, *2*, 160–171.

Liyakat, K. S. S., & Liyakat, K. K. S. (2024). Nanosensors in agriculture field: a study. *International Journal of Applied Nanotechnology*, *10*(02), 12–22p.

Liyakat, K. K. S., Warhade, N. S., Pol, R. S., Jadhav, H. M., & Mulani, A. O. (2022). Yarn quality detection for textile industries using image processing. *Journal Of Algebraic Statistics*, *13*(3), 3465–3472.

Lobet, G., & Draye, X. (2013). Novel scanning procedure enabling the vectorization of entire rhizotron-grown root systems. *Plant Methods*, *9*, 1–11.

Manzoor, J., Sharma, M., & Wani, K. A. (2018). Heavy metals in vegetables and their impact on the nutrient quality of vegetables: a review. *Journal of Plant Nutrition*, *41*(13), 1744–1763.

Maraldo, D., & Mutharasan, R. (2007). Detection and confirmation of staphylococcal enterotoxin B in apple juice and milk using piezoelectric-excited millimeter-sized cantilever sensors at 2.5 fg/mL. *Analytical Chemistry*, *79*(20), 7636–7643.

Maron, J. L., & Crone, E. (2006). Herbivory: effects on plant abundance, distribution and population growth. *Proceedings of the Royal Society B: Biological Sciences, 273*(1601), 2575–2584.

Maron, J. L., & Kauffman, M. J. (2006). Habitat-specific impacts of multiple consumers on plant population dynamics. *Ecology, 87*(1), 113–124.

Mattila, H., Khorobrykh, S., Havurinne, V., & Tyystjärvi, E. (2015). Reactive oxygen species: reactions and detection from photosynthetic tissues. *Journal of Photochemistry and Photobiology B: Biology, 152*, 176–214.

Mazars, C., Thuleau, P., Lamotte, O., & Bourque, S. (2010). Cross-talk between ROS and calcium in regulation of nuclear activities. *Molecular Plant, 3*(4), 706–718.

McLamore, E. S., Diggs, A., Calvo Marzal, P., Shi, J., Blakeslee, J. J., Peer, W. A., & Porterfield, D. M. (2010). Non-invasive quantification of endogenous root auxin transport using an integrated flux microsensor technique. *The Plant Journal, 63*(6), 1004–1016.

Medintz, I. L., Sapsford, K. E., Konnert, J. H., Chatterji, A., Lin, T., Johnson, J. E., & Mattoussi, H. (2005). Decoration of discretely immobilized cowpea mosaic virus with luminescent quantum dots. *Langmuir, 21*(12), 5501–5510.

Méndez-Hernández, H. A., Ledezma-Rodríguez, M., Avilez-Montalvo, R. N., Juárez-Gómez, Y. L., Skeete, A., Avilez-Montalvo, J., ... Loyola-Vargas, V. M. (2019). Signaling overview of plant somatic embryogenesis. *Frontiers in Plant Science, 10*, 77.

Mesterházy, Á., Oláh, J., & Popp, J. (2020). Losses in the grain supply chain: causes and solutions. *Sustainability, 12*(6), 2342.

Mirica, K. A., Azzarelli, J. M., Weis, J. G., Schnorr, J. M., & Swager, T. M. (2013). Rapid prototyping of carbon-based chemiresistive gas sensors on paper. *Proceedings of the National Academy of Sciences of the United States of America, 110*(35), E3265–E3270.

Mitra, D., Saritha, B., Janeeshma, E., Gusain, P., Khoshru, B., Nouh, F. A. A., ... Rabari, A. (2022). Arbuscular mycorrhizal fungal association boosted the arsenic resistance in crops with special responsiveness to rice plant. *Environmental and Experimental Botany, 193*, 104681.

Mittal, R., Debs, L. H., Patel, A. P., Nguyen, D., Patel, K., O'Connor, G., ... Eshraghi, A. A. (2017). Neurotransmitters: the critical modulators regulating gut–brain axis. *Journal of Cellular Physiology, 232*(9), 2359–2372.

Mittler, R. (2017). ROS are good. *Trends in Plant Science, 22*, 11–19.

Mohsenzadeh, S., Malboobi, M., Razavi, K., & Farrahi-Aschtiani, S. (2006). Physiological and molecular responses of *Aeluropus lagopoides* (Poaceae) to water deficit. *Environmental and Experimental Botany, 56*(3), 314–322.

Mondal, R., Dam, P., Chakraborty, J., Paret, M. L., Katı, A., Altuntas, S., ... Mandal, A. K. (2022). Potential of nanobiosensor in sustainable agriculture: the state-of-art. *Heliyon, 8*(12), e12207.

Mordecai, E. A. (2011). Pathogen impacts on plant communities: unifying theory, concepts, and empirical work. *Ecological Monographs, 81*(3), 429–441.

Moreau, A. L., Janissen, R., Santos, C. A., Peroni, L. A., Stach-Machado, D. R., de Souza, A. A., ... Cotta, M. A. (2012). Highly-sensitive and label-free indium phosphide biosensor for early phytopathogen diagnosis. *Biosensors and Bioelectronics, 36*(1), 62–68.

Moskovits, M., & Suh, J. (1984). Surface selection rules for surface-enhanced Raman spectroscopy: calculations and application to the surface-enhanced Raman spectrum of phthalazine on silver. *The Journal of Physical Chemistry, 88*(23), 5526–5530.

Mousavi, A. A., Zhang, C., Masri, S. F., & Gholipour, G. (2020). Structural damage localization and quantification based on a CEEMDAN Hilbert transform neural network approach: a model steel truss bridge case study. *Sensors, 20*(5), 1271.

Mulani, A. O., Patil, R. M., & Liyakat, K. K. S. (2023). Modelo de apariencia discriminatorio para un sólido seguimiento en línea de múltiples objetivos. *Télématique: Revista Electrónica de Estudios Telemáticos*, *22*(1), 24–43.

Munns, R. (2002). Comparative physiology of salt and water stress. *Plant, Cell & Environment*, *25*(2), 239–250.

Munns, R. (2005). Genes and salt tolerance: bringing them together. *New Phytologist*, *167*(3), 645–663.

Munns, R., & Tester, M. (2008). Mechanisms of salinity tolerance. *Annual Review of Plant Biology*, *59*(1), 651–681.

Muñoz-Huerta, R. F., Guevara-Gonzalez, R. G., Contreras-Medina, L. M., Torres-Pacheco, I., Prado-Olivarez, J., & Ocampo-Velazquez, R. V. (2013). A review of methods for sensing the nitrogen status in plants: advantages, disadvantages and recent advances. *Sensors*, *13*(8), 10823–10843.

Murchie, E. H., & Lawson, T. (2013). Chlorophyll fluorescence analysis: a guide to good practice and understanding some new applications. *Journal of Experimental Botany*, *64*(13), 3983–3998.

Neill, S. J., Desikan, R., Clarke, A., Hurst, R. D., & Hancock, J. T. (2002). Hydrogen peroxide and nitric oxide as signalling molecules in plants. *Journal of Experimental Botany*, *53*(372), 1237–1247.

Nerkar, P. M., Liyakat, K. K. S., Dhaware, B. U., & Liyakat, K. S. S. (2023a). Predictive data analytics framework based on heart healthcare system (HHS) using machine learning. *Journal of Advanced Zoology*, *44*, 3673–3686.

Nerkar, P. M., Shinde, S. S., Liyakat, K. K. S., Desai, S., & Kazi, S. S. L. (2023b). Monitoring fresh fruit and food using IoT and machine learning to improve food safety and quality. *Tuijin Jishu/Journal of Propulsion Technology*, *44*(3), 2927–2931.

Nie, S., & Emory, S. R. (1997). Probing single molecules and single nanoparticles by surface-enhanced Raman scattering. *Science*, *275*(5303), 1102–1106.

Nikita, S., Sanika, C., Nagveni, K., Sakshi, K., & Kazi, K. (2022). Announcement system in Bus. *Journal of Image Processing and Intelligent remote sensing*, *2*(6).

Niu, Z., Zhang, B., Wang, J., Liu, K., Chen, Z., Yang, K., ... Ji, D. (2020). The research on 220GHz multicarrier high-speed communication system. *China Communications*, *17*(3), 131–139.

Noronha, V. T., Aquino, Y. M., Maia, M. T., & Freire, R. M. (2019). Sensing of water contaminants: from traditional to modern strategies based on nanotechnology. In *Nanomaterials applications for environmental matrices* (pp. 109–150). Elsevier.

Numan, A., Gill, A. A., Rafique, S., Guduri, M., Zhan, Y., Maddiboyina, B., ... Dang, N. N. (2021). Rationally engineered nanosensors: a novel strategy for the detection of heavy metal ions in the environment. *Journal of Hazardous Materials*, *409*, 124493.

Oerke, E.-C., & Dehne, H.-W. (2004). Safeguarding production-losses in major crops and the role of crop protection. *Crop Protection*, *23*(4), 275–285.

Otto, A. (2001). Theory of first layer and single molecule surface enhanced Raman scattering (SERS). *physica status solidi (a)*, *188*(4), 1455–1470.

Paes de Melo, B., Carpinetti, P. D. A., Fraga, O. T., Rodrigues-Silva, P. L., Fioresi, V. S., de Camargos, L. F., & Ferreira, M. F. D. S. (2022). Abiotic stresses in plants and their markers: a practice view of plant stress responses and programmed cell death mechanisms. *Plants*, *11*(9), 1100.

Pan, D., Xia, X.-X., Zhou, H., Jin, S.-Q., Lu, Y.-Y., Liu, H., ... Jin, Z.-B. (2020). COCO enhances the efficiency of photoreceptor precursor differentiation in early human embryonic stem cell-derived retinal organoids. *Stem Cell Research & Therapy*, *11*, 1–12.

Panferov, V. G., Safenkova, I. V., Byzova, N. A., Varitsev, Y. A., Zherdev, A. V., & Dzantiev, B. B. (2018). Silver-enhanced lateral flow immunoassay for highly-sensitive detection of potato leafroll virus. *Food and Agricultural Immunology, 29*(1), 445–457.

Parisi, C., Vigani, M., & Rodríguez-Cerezo, E. (2015). Agricultural nanotechnologies: what are the current possibilities? *Nano Today, 10*(2), 124–127.

Patrick, E. A. (1972). *Fundamentals of pattern recognition.* Englewood Cliffs, NJ: Prentice-Hall.

Perdikaris, A., Vassilakos, N., Yiakoumettis, I., Kektsidou, O., & Kintzios, S. (2011). Development of a portable, high throughput biosensor system for rapid plant virus detection. *Journal of Virological Methods, 177*(1), 94–99.

Pierre-Jerome, E., Drapek, C., & Benfey, P. N. (2018). Regulation of division and differentiation of plant stem cells. *Annual Review of Cell and Developmental Biology, 34*(1), 289–310.

Pineda, M., Pérez-Bueno, M. L., & Barón, M. (2018). Detection of bacterial infection in melon plants by classification methods based on imaging data. *Frontiers in Plant Science, 9*, 164.

Pintó-Marijuan, M., & Munné-Bosch, S. (2014). Photo-oxidative stress markers as a measure of abiotic stress-induced leaf senescence: advantages and limitations. *Journal of Experimental Botany, 65*(14), 3845–3857.

Potes-Lesoinne, H. A., Ramirez-Alvarez, F., Perez-Gonzalez, V. H., Martinez-Chapa, S. O., & Gallo-Villanueva, R. C. (2022). Nanomaterials for electrochemical detection of pollutants in water: a review. *Electrophoresis, 43*(1–2), 249–262.

Prasad, P. V., & Djanaguiraman, M. (2014). Response of floret fertility and individual grain weight of wheat to high temperature stress: sensitive stages and thresholds for temperature and duration. *Functional Plant Biology, 41*(12), 1261–1269.

Program, U. G. C. R. (2009). *Global climate change impacts in the United States.* Cambridge University Press.

Qadir, M., Quillérou, E., Nangia, V., Murtaza, G., Singh, M., Thomas, R. J., ... Noble, A. D. (2014). Economics of salt-induced land degradation and restoration. *Paper presented at the Natural Resources Forum, 38*(4), 282–295.

Quan, L. J., Zhang, B., Shi, W. W., & Li, H. Y. (2008). Hydrogen peroxide in plants: a versatile molecule of the reactive oxygen species network. *Journal of Integrative Plant Biology, 50*(1), 2–18.

Raman, C. V., & Krishnan, K. S. (1928). A new type of secondary radiation. *Nature, 121*(3048), 501–502.

Rana, K., Tripathi, R. M., & Pudake, R. N. (2024). Advances in the fabrication of nanosensors for management of abiotic stress in crop plants. In *Innovative methods in horticultural crop improvement: biosensors and nanosensors* (pp. 75–97). Springer.

Rathee, M., & Dalal, P. (2018). Emerging insect pests in Indian agriculture. *Indian Journal of Entomology, 80*(2), 267–281.

Ristaino, J. B., Saville, A. C., Paul, R., Cooper, D. C., & Wei, Q. (2020). Detection of *Phytophthora infestans* by loop-mediated isothermal amplification, real-time LAMP, and droplet digital PCR. *Plant Disease, 104*(3), 708–716.

Rivelli, A. R., De Maria, S., Pizza, S., & Gherbin, P. (2010). Growth and physiological response of hydroponically-grown sunflower as affected by salinity and magnesium levels. *Journal of Plant Nutrition, 33*(9), 1307–1323.

Rockström, J., Williams, J., Daily, G., Noble, A., Matthews, N., & Gordon, L. (2017). Sustainable intensification of agriculture for human prosperity and global sustainability. *Ambio, 46*, 4–17.

Rodriguez-Lorenzo, L., & Alvarez-Puebla, R. (2014). Surface-enhanced Raman scattering (SERS) nanoparticle sensors for biochemical and environmental sensing. In *Nanosensors for chemical and biological applications* (pp. 197–230). Elsevier.

Rodríguez-Sevilla, E., Ramírez-Silva, M.-T., Romero-Romo, M., Ibarra-Escutia, P., & Palomar-Pardavé, M. (2014). Electrochemical quantification of the antioxidant capacity of medicinal plants using biosensors. *Sensors, 14*(8), 14423–14439.

Ronkainen, N. J., Halsall, H. B., & Heineman, W. R. (2010). Electrochemical biosensors. *Chemical Society Reviews, 39*(5), 1747–1763.

Rosenwasser, S., Rot, I., Meyer, A. J., Feldman, L., Jiang, K., & Friedman, H. (2010). A fluorometer-based method for monitoring oxidation of redox-sensitive GFP (roGFP) during development and extended dark stress. *Physiologia Plantarum, 138*(4), 493–502.

Rucker, K., Kvien, C., Holbrook, C., & Hook, J. (1995). Identification of peanut genotypes with improved drought avoidance traits. *Peanut Science, 22*(1), 14–18.

Ruelland, E., & Zachowski, A. (2010). How plants sense temperature. *Environmental and Experimental Botany, 69*(3), 225–232.

Sahbeni, G., Ngabire, M., Musyimi, P. K., & Székely, B. (2023). Challenges and opportunities in remote sensing for soil salinization mapping and monitoring: a review. *Remote Sensing, 15*(10), 2540.

Saito, K., Chang, Y., Horikawa, K., Hatsugai, N., Higuchi, Y., Hashida, M., ... Nagai, T. (2012). Luminescent proteins for high-speed single-cell and whole-body imaging. *Nature Communications, 3*(1), 1262.

Salim, W. W. A. W., Zeitchek, M. A., Hermann, A. C., Ricco, A. J., Tan, M., Selch, F., ... ul Haque, A. (2013). Multi-analyte biochip (MAB) based on all-solid-state ion-selective electrodes (ASSISE) for physiological research. *Journal of Visualized Experiments: JoVE, 74*, 50020.

Salouti, M., & Khadivi Derakhshan, F. (2020). Biosensors and nanobiosensors in environmental applications. In *Biogenic nano-particles and their use in agro-ecosystems* (pp. 515–591). Singapore: Springer Singapore.

Sánchez-Rodríguez, E., del Mar Rubio-Wilhelmi, M., Cervilla, L. M., Blasco, B., Rios, J. J., Leyva, R., ... Ruiz, J. M. (2010). Study of the ionome and uptake fluxes in cherry tomato plants under moderate water stress conditions. *Plant and Soil, 335*, 339–347.

Sanghera, G. S., Wani, S. H., Hussain, W., & Singh, N. (2011). Engineering cold stress tolerance in crop plants. *Current Genomics, 12*(1), 30.

Savary, S., Ficke, A., Aubertot, J.-N., & Hollier, C. (2012). Crop losses due to diseases and their implications for global food production losses and food security. *Food Security, 4*(4), 519–537.

Savary, S., Willocquet, L., Pethybridge, S. J., Esker, P., McRoberts, N., & Nelson, A. (2019). The global burden of pathogens and pests on major food crops. *Nature Ecology & Evolution, 3*(3), 430–439.

Schlenker, W., & Roberts, M. J. (2009). Nonlinear temperature effects indicate severe damages to US crop yields under climate change. *Proceedings of the National Academy of Sciences of the United States of America, 106*(37), 15594–15598.

Schmälzlin, E., Van Dongen, J. T., Klimant, I., Marmodée, B., Steup, M., Fisahn, J., & Löhmannsröben, H.-G. (2005). An optical multifrequency phase-modulation method using microbeads for measuring intracellular oxygen concentrations in plants. *Biophysical Journal, 89*(2), 1339–1345. https://doi.org/10.1529/biophysj.105.063453

Schnaubelt, D., Queval, G., Dong, Y., Diaz-Vivancos, P., Makgopa, M. E., Howell, G., ... Foyer, C. H. (2015). Low glutathione regulates gene expression and the redox potentials of the nucleus and cytosol in *Arabidopsis thaliana*. *Plant, Cell & Environment, 38*(2), 266–279.

Schön, C. (2018). BayKlimaFit-strategies for the adaptation of crop plants to climate change.

Schultz, T. (1964). *De la agricultura*. Editorial Alianza.

Schwarzländer, M., Fricker, M., Müller, C., Marty, L., Brach, T., Novak, J., ... Meyer, A. (2008). Confocal imaging of glutathione redox potential in living plant cells. *Journal of Microscopy, 231*(2), 299–316.

Sehgal, A., Sita, K., Siddique, K. H., Kumar, R., Bhogireddy, S., Varshney, R. K., ... Nayyar, H. (2018). Drought or/and heat-stress effects on seed filling in food crops: impacts on functional biochemistry, seed yields, and nutritional quality. *Frontiers in Plant Science, 9*, 1705.

Shahzad, H., Ullah, S., Iqbal, M., Bilal, H. M., Shah, G. M., Ahmad, S., ... Ahmad, I. (2019). Salinity types and level-based effects on the growth, physiology and nutrient contents of maize (*Zea mays*). *Italian Journal of Agronomy, 14*(4), 1326.

Shaw, D. S., Meitha, K., Considine, M. J., & Foyer, C. H. (2017). Mitochondrial respiration and oxygen tension. In J. G. Kapuganti (Ed.), *Plant respiration and internal oxygen: methods and protocols* (pp. 97–113). New York, NY: Springer New York.

Sheikh, Z. N., Raina, S., Bhat, U. H., Sharma, N., Saffeullah, P., Mohammad, F., ... Islam, S. (2024). Deciphering the potential role of nano-based biosensors in growth, development and stress management of crop plants. *South African Journal of Botany, 167*, 542–553.

Shi, K., Tang, Y., Liu, X., & Zhong, S. (2017). Non-fragile sampled-data robust synchronization of uncertain delayed chaotic Lurie systems with randomly occurring controller gain fluctuation. *ISA Transactions, 66*, 185–199.

Shigenaga, A. M., & Argueso, C. T. (2016). No hormone to rule them all: Interactions of plant hormones during the responses of plants to pathogens. *Paper presented at the Seminars in Cell & Developmental Biology, 56*, 174–189.

Shojaei, T. R., Salleh, M. A. M., Sijam, K., Rahim, R. A., Mohsenifar, A., Safarnejad, R., & Tabatabaei, M. (2016). Fluorometric immunoassay for detecting the plant virus Citrus tristeza using carbon nanoparticles acting as quenchers and antibodies labeled with CdTe quantum dots. *Microchimica Acta, 183*, 2277–2287.

Siddiqi, K. S., & Husen, A. (2019). Plant response to jasmonates: current developments and their role in changing environment. *Bulletin of the National Research Centre, 43*(1), 1–11.

da Silva, C. P., Franzoi, A. C., Fernandes, S. C., Dupont, J., & Vieira, I. C. (2013). Development of biosensor for phenolic compounds containing PPO in β-cyclodextrin modified support and iridium nanoparticles. *Enzyme and Microbial Technology, 52*(4–5), 296–301.

Singh, A. (2022). Soil salinity: a global threat to sustainable development. *Soil Use and Management, 38*(1), 39–67.

Singh, J., Roychoudhury, A., Srivastava, M., Chaudhary, V., Prasanna, R., Lee, D. W., ... Malhotra, B. (2013). Highly efficient bienzyme functionalized biocompatible nanostructured nickel ferrite–chitosan nanocomposite platform for biomedical application. *The Journal of Physical Chemistry C, 117*(16), 8491–8502.

Singh, J., Roychoudhury, A., Srivastava, M., Solanki, P. R., Lee, D. W., Lee, S. H., & Malhotra, B. (2014). A dual enzyme functionalized nanostructured thulium oxide based interface for biomedical application. *Nanoscale, 6*(2), 1195–1208.

Singla, J., Krattinger, S., Wrigley, C., Faubion, J., Corke, H., & Seetharaman, K. (2016). *Biotic stress resistance genes in wheat* (pp. 388–392). Academic Press. https://doi.org/10.1016/B978-0-12-394437-5.00229-1

Smekal, A. (1923). Zur quantentheorie der dispersion. *Naturwissenschaften, 11*(43), 873–875.

Song, Y., Cui, K., Wang, L., & Chen, S. (2009). The electrodeposition of Ag nanoparticles on a type I collagen-modified glassy carbonelectrode and their applications as a hydrogen peroxide sensor. *Nanotechnology, 20*(10), 105501.

Song, S., Liu, N., Zhao, Z., Njumbe Ediage, E., Wu, S., Sun, C., ... Wu, A. (2014). Multiplex lateral flow immunoassay for mycotoxin determination. *Analytical Chemistry, 86*(10), 4995–5001.

Stavi, I., Thevs, N., & Priori, S. (2021). Soil salinity and sodicity in drylands: a review of causes, effects, monitoring, and restoration measures. *Frontiers in Environmental Science, 9*, 712831.

Stiles, P. L., Dieringer, J. A., Shah, N. C., & Van Duyne, R. P. (2008). Surface-enhanced Raman spectroscopy. *Annual Review of Analytical Chemistry, 1*, 601–626.

Strauss, S. Y., & Zangerl, A. R. (2002). Plant-insect interactions in terrestrial ecosystems. *Plant-Animal Interactions: An Evolutionary Approach, 2002*, 77–106.

Sun, L., Song, Y., Wang, L., Guo, C., Sun, Y., Liu, Z., & Li, Z. (2008a). Ethanol-induced formation of silver nanoparticle aggregates for highly active SERS substrates and application in DNA detection. *The Journal of Physical Chemistry C, 112*(5), 1415–1422.

Sun, W., Zhong, J., Qin, P., & Jiao, K. (2008b). Electrochemical biosensor for the detection of cauliflower mosaic virus 35 S gene sequences using lead sulfide nanoparticles as oligonucleotide labels. *Analytical Biochemistry, 377*(2), 115–119.

Sun, L., Su, Z., Xia, Y., Zhang, C., & Li, C. (2019). Superwide-range fiber bragg grating displacement sensor based on an eccentric gear: principles and experiments. *Journal of Aerospace Engineering, 32*(1), 04018129.

Suzuki, N., Miller, G., Salazar, C., Mondal, H. A., Shulaev, E., Cortes, D. F., ... Schlauch, K. (2013). Temporal-spatial interaction between reactive oxygen species and abscisic acid regulates rapid systemic acclimation in plants. *The Plant Cell, 25*(9), 3553–3569.

Swami, D., Thamake, S. S., Ubale, N. S., Lokhande, P. V., & Liyakat, D. K. K. S. (2022). Sending notification to someone missing you through smart watch. *International Journal of Information Technology and Computer Engineering (IJITC), 2*(8), 19–24.

Tereshchenko, A., Fedorenko, V., Smyntyna, V., Konup, I., Konup, A., Eriksson, M., ... Bechelany, M. (2017). ZnO films formed by atomic layer deposition as an optical biosensor platform for the detection of *Grapevine virus* A-type proteins. *Biosensors and Bioelectronics, 92*, 763–769.

Thornton, P. K., Ericksen, P. J., Herrero, M., & Challinor, A. J. (2014). Climate variability and vulnerability to climate change: a review. *Global Change Biology, 20*(11), 3313–3328.

Tigchelaar, M., Battisti, D. S., Naylor, R. L., & Ray, D. K. (2018). Future warming increases probability of globally synchronized maize production shocks. *Proceedings of the National Academy of Sciences of the United States of America, 115*(26), 6644–6649.

Tiwari, S., Lata, C., Chauhan, P. S., & Nautiyal, C. S. (2016). *Pseudomonas putida* attunes morphophysiological, biochemical and molecular responses in *Cicer arietinum* L. during drought stress and recovery. *Plant Physiology and Biochemistry, 99*, 108–117.

Tognaccini, L., Ricci, M., Gellini, C., Feis, A., Smulevich, G., & Becucci, M. (2019). Surface enhanced Raman spectroscopy for in-field detection of pesticides: a test on dimethoate residues in water and on olive leaves. *Molecules, 24*(2), 292.

Tsuchiya, Y., Yoshimura, M., Sato, Y., Kuwata, K., Toh, S., Holbrook-Smith, D., ... Kinoshita, T. (2015). Probing strigolactone receptors in *Striga hermonthica* with fluorescence. *Science, 349*(6250), 864–868.

Ullah, A., Manghwar, H., Shaban, M., Khan, A. H., Akbar, A., Ali, U., ... Fahad, S. (2018). Phytohormones enhanced drought tolerance in plants: a coping strategy. *Environmental Science and Pollution Research, 25*, 33103–33118.

Umehara, M., Cao, M., Akiyama, K., Akatsu, T., Seto, Y., Hanada, A., ... Yamaguchi, S. (2015). Structural requirements of strigolactones for shoot branching inhibition in rice and *Arabidopsis*. *Plant and Cell Physiology, 56*(6), 1059–1072.

Usman, M., Farooq, M., Wakeel, A., Nawaz, A., Cheema, S. A., ur Rehman, H., ... Sanaullah, M. (2020). Nanotechnology in agriculture: current status, challenges and future opportunities. *Science of the Total Environment, 721*, 137778.

Vaijnath, S. P., Siddheshwar, M. P., Pradip, M. S., & Liayakat, D. K. K. S. (2022). Smart safty device for women. *International Journal of Aquatic Science, 13*(1), 556–560.

Van Esse, H. P., Reuber, T. L., & van der Does, D. (2020). Genetic modification to improve disease resistance in crops. *New Phytologist, 225*(1), 70–86.

Vaughan, M. M., Block, A., Christensen, S. A., Allen, L. H., & Schmelz, E. A. (2018). The effects of climate change associated abiotic stresses on maize phytochemical defenses. *Phytochemistry Reviews, 17*, 37–49.

Volkov, A. G., Murphy, V. A., Clemmons, J. I., Curley, M. J., & Markin, V. S. (2012). Energetics and forces of the *Dionaea muscipula* trap closing. *Journal of Plant Physiology, 169*(1), 55–64.

Volkov, A. G., Harris, S. L., II, Vilfranc, C. L., Murphy, V. A., Wooten, J. D., Paulicin, H., ... Markin, V. S. (2013). Venus flytrap biomechanics: forces in the *Dionaea muscipula* trap. *Journal of Plant Physiology, 170*(1), 25–32.

Walcarius, A., Minteer, S. D., Wang, J., Lin, Y., & Merkoçi, A. (2013). Nanomaterials for bio-functionalized electrodes: recent trends. *Journal of Materials Chemistry B, 1*(38), 4878–4908.

Wang, Z. L. (2007). Nanopiezotronics. *Advanced Materials, 19*(6), 889–892.

Wang, Y., & Frei, M. (2011). Stressed food–the impact of abiotic environmental stresses on crop quality. *Agriculture, Ecosystems & Environment, 141*(3–4), 271–286.

Wang, K. L.-C., Li, H., & Ecker, J. R. (2002). Ethylene biosynthesis and signaling networks. *The Plant Cell, 14*(suppl_1), S131–S151.

Wang, W., Vinocur, B., & Altman, A. (2003). Plant responses to drought, salinity and extreme temperatures: towards genetic engineering for stress tolerance. *Planta, 218*, 1–14.

Wang, X., Zhou, J., Song, J., Liu, J., Xu, N., & Wang, Z. L. (2006). Piezoelectric field effect transistor and nanoforce sensor based on a single ZnO nanowire. *Nano Letters, 6*(12), 2768–2772.

Wang, L., Han, D., Ni, S., Ma, W., Wang, W., & Niu, L. (2015). Photoelectrochemical device based on Mo-doped $BiVO_4$ enables smart analysis of the global antioxidant capacity in food. *Chemical Science, 6*(11), 6632–6638.

Wang, Y., Li, C., Zhang, Y., Yang, M., Li, B., Jia, D., & Mao, C. (2016). Experimental evaluation of the lubrication properties of the wheel/workpiece interface in minimum quantity lubrication (MQL) grinding using different types of vegetable oils. *Journal of Cleaner Production, 127*, 487–499.

Wang, F., Gu, X., Zheng, C., Dong, F., Zhang, L., Cai, Y., ... Zhang, Z. (2017). Ehrlich reaction evoked multiple spectral resonances and gold nanoparticle hotspots for Raman detection of plant hormone. *Analytical Chemistry, 89*(17), 8836–8843.

Wang, M., Yang, L., Hu, B., Liu, J., He, L., Jia, Q., ... Zhang, Z. (2018). Bimetallic NiFe oxide structures derived from hollow NiFe Prussian blue nanobox for label-free electrochemical biosensing adenosine triphosphate. *Biosensors and Bioelectronics, 113*, 16–24.

Wang, X., Cheng, M., Yang, Q., Wei, H., Xia, A., Wang, L., ... Huang, X. (2019). A living plant cell-based biosensor for real-time monitoring invisible damage of plant cells under heavy metal stress. *Science of the Total Environment, 697*, 134097.

Wang, Q., Liu, B., & Wang, Z. (2020a). Investigation of heat transfer mechanisms among particles in horizontal rotary retorts. *Powder Technology, 367*, 82–96.

Wang, X.-F., Gao, P., Liu, Y.-F., Li, H.-F., & Lu, F. (2020b). Predicting thermophilic proteins by machine learning. *Current Bioinformatics, 15*(5), 493–502.

Wang, P., Yao, T., Li, Z., Wei, W., Xie, Q., Duan, W., & Han, H. (2020c). A superhydrophobic/electrothermal synergistically anti-icing strategy based on graphene composite. *Composites Science and Technology, 198*, 108307.

Wang, B., Song, Z., & Sun, L. (2021). A review: comparison of multi-air-pollutant removal by advanced oxidation processes–industrial implementation for catalytic oxidation processes. *Chemical Engineering Journal, 409*, 128136.

Wang, X., Kong, L., Zhou, S., Ma, C., Lin, W., Sun, X., ... Wang, P. (2022). Development of QDs-based nanosensors for heavy metal detection: a review on transducer principles and in-situ detection. *Talanta, 239*, 122903.

Weerakoon, K. A., Shu, J. H., & Chin, B. A. (2011). A chemiresistor sensor with a poly3-hexylthiophene active layer for the detection of insect infestation at early stages. *IEEE Sensors Journal, 11*(7), 1617–1622.

Werner, C., Correia, O., & Beyschlag, W. (1999). Two different strategies of mediterranean macchia plants to avoid photoinhibitory damage by excessive radiation levels during summer drought. *Acta Oecologica, 20*(1), 15–23.

Wierer, S., Peter, S., Elgass, K., Mack, H.-G., Bieker, S., Meixner, A. J., ... Schleifenbaum, F. (2012). Determination of the in vivo redox potential by one-wavelength spectro-microscopy of roGFP. *Analytical and Bioanalytical Chemistry, 403*, 737–744.

Wu, J., Wang, D., Rosen, C. J., & Bauer, M. E. (2007). Comparison of petiole nitrate concentrations, SPAD chlorophyll readings, and QuickBird satellite imagery in detecting nitrogen status of potato canopies. *Field Crops Research, 101*(1), 96–103.

Wu, Y., Jin, J., & Xing, X. (2016). *Multi-target tracking method with flexibility of illumination change*. Paper presented at the 2016 IEEE International Conference on Consumer Electronics-China (ICCE-China).

Xiao, S., Liu, L., Zhang, Y., Sun, H., Zhang, K., Bai, Z., ... Li, C. (2020). Fine root and root hair morphology of cotton under drought stress revealed with RhizoPot. *Journal of Agronomy and Crop Science, 206*(6), 679–693.

Xie, J., Wen, D., Liang, L., Jia, Y., Gao, L., & Lei, J. (2018). Evaluating the validity of current mainstream wearable devices in fitness tracking under various physical activities: comparative study. *JMIR mHealth and uHealth, 6*(4), e9754.

Xu, M., Li, T., Wang, Z., Deng, X., Yang, R., & Guan, Z. (2018). Reducing complexity of HEVC: a deep learning approach. *IEEE Transactions on Image Processing, 27*(10), 5044–5059.

Xu, T., Yao, Z., Liu, J., Zhang, H., Din, G. M. U., Zhao, S., ... Gao, L. (2020a). Development of droplet digital PCR for the detection of *Tilletia laevis*, which causes common bunt of wheat, based on the SCAR marker derived from ISSR and real-time PCR. *Scientific Reports, 10*(1), 16106.

Xu, P., Lu, W., Zhang, J., & Zhang, L. (2020b). Efficient hydrolysis of ammonia borane for hydrogen evolution catalyzed by plasmonic Ag@Pd core–shell nanocubes. *ACS Sustainable Chemistry & Engineering, 8*(33), 12366–12377.

Yadav, S. P., Bharadwaj, R., Nayak, H., Mahto, R., Singh, R. K., & Prasad, S. K. (2019). Impact of salt stress on growth, productivity and physicochemical properties of plants: a review. *International Journal of Chemical Studies, 7*(2), 1793–1798.

Yadav, S., Modi, P., Dave, A., Vijapura, A., Patel, D., & Patel, M. (2020). Effect of abiotic stress on crops. *Sustainable Crop Production, 3*(17), 5–16.

Yang, G.-H., Zhou, Y.-H., Wu, J.-J., Cao, J.-T., Li, L.-L., Liu, H.-Y., & Zhu, J.-J. (2013). Microwave-assisted synthesis of nitrogen and boron co-doped graphene and its application for enhanced electrochemical detection of hydrogen peroxide. *RSC Advances, 3*(44), 22597–22604.

Yang, Y., Asiri, A. M., Du, D., & Lin, Y. (2014). Acetylcholinesterase biosensor based on a gold nanoparticle–polypyrrole–reduced graphene oxide nanocomposite modified electrode for the amperometric detection of organophosphorus pesticides. *Analyst, 139*(12), 3055–3060.

Yang, J., Li, S., Wang, Z., Dong, H., Wang, J., & Tang, S. (2020a). Using deep learning to detect defects in manufacturing: a comprehensive survey and current challenges. *Materials, 13*(24), 5755.

Yang, Y., Liu, J., Yao, J., Kou, J., Li, Z., Wu, T., ... Sun, H. (2020b). Adsorption behaviors of shale oil in kerogen slit by molecular simulation. *Chemical Engineering Journal, 387*, 124054.

Yao, K. S., Li, S., Tzeng, K., Cheng, T. C., Chang, C. Y., Chiu, C., & Lin, Z. (2009). Fluorescence silica nanoprobe as a biomarker for rapid detection of plant pathogens. *Advanced Materials Research, 79*, 513–516.

Yazgan, I., Zhang, J., Kariuki, V., Akgul, A., Cronmiller, L. E., Akgul, A., ... Eshun, G. (2018). Selective sensing and imaging of penicillium italicum spores and hyphae using carbohydrate–lectin interactions. *ACS Sensors, 3*(3), 648–654.

Yazgan, I., Osonga, F. J., Miller, R. M., Kariuki, V. M., Zhang, J., Feng, J., ... Sadik, O. A. (2021). Greener one-pot synthesis of gold nanoparticle glycoconjugates using functionalized sugars. *ACS Agricultural Science & Technology, 1*(4), 379–389.

Yeturu, S., Jentzsch, P. V., Ciobotă, V., Guerrero, R., Garrido, P., & Ramos, L. A. (2016). Handheld Raman spectroscopy for the early detection of plant diseases: *Abutilon mosaic* virus infecting *Abutilon* sp. *Analytical Methods, 8*(17), 3450–3457.

Yotova, L., Yaneva, S., & Marinkova, D. (2013). Biomimetic nanosensors for determination of toxic compounds in food and agricultural products. *Journal of Chemical Technology & Metallurgy, 48*(3).

You, J., & Chan, Z. (2015). ROS regulation during abiotic stress responses in crop plants. *Frontiers in Plant Science, 6*, 1092.

Zahra, N., Hafeez, M. B., Shaukat, K., Wahid, A., Hussain, S., Naseer, R., & Farooq, M. (2021). Hypoxia and anoxia stress: plant responses and tolerance mechanisms. *Journal of Agronomy and Crop Science, 207*(2), 249–284.

Zandalinas, S. I., Fritschi, F. B., & Mittler, R. (2021). Global warming, climate change, and environmental pollution: recipe for a multifactorial stress combination disaster. *Trends in Plant Science, 26*(6), 588–599.

Zeiri, L. (2007). SERS of plant material. *Journal of Raman Spectroscopy: An International Journal for Original Work in all Aspects of Raman Spectroscopy, Including Higher Order Processes, and also Brillouin and Rayleigh Scattering, 38*(7), 950–955.

Zhang, W. (2018). Global pesticide use: profile, trend, cost/benefit and more. *Proceedings of the International Academy of Ecology and Environmental Sciences, 8*(1), 1.

Zhang, J., & Liu, B. (2019). A review on the recent developments of sequence-based protein feature extraction methods. *Current Bioinformatics, 14*(3), 190–199.

Zhang, C., & Ou, J. (2008). Control structure interaction of electromagnetic mass damper system for structural vibration control. *Journal of Engineering Mechanics, 134*(5), 428–437.

Zhang, C. W., Ou, J. P., & Zhang, J. Q. (2006). Parameter optimization and analysis of a vehicle suspension system controlled by magnetorheological fluid dampers. *Structural Control and Health Monitoring: The Official Journal of the International Association for Structural Control and Monitoring and of the European Association for the Control of Structures, 13*(5), 885–896.

Zhang, C., Li, L., & Ou, J. (2010). Swinging motion control of suspended structures: principles and applications. *Structural Control and Health Monitoring, 17*(5), 549–562.

Zhang, M., Chen, W., Chen, X., Zhang, Y., Lin, X., Wu, Z., & Li, M. (2013). Multiplex immunoassays of plant viruses based on functionalized upconversion nanoparticles coupled with immunomagnetic separation. *Journal of Nanomaterials, 2013*(1), 317437.

Zhang, B., Xu, D., Liu, Y., Li, F., Cai, J., & Du, L. (2016). Multi-scale evapotranspiration of summer maize and the controlling meteorological factors in north China. *Agricultural and Forest Meteorology, 216*, 1–12.

Zhang, P., Wang, L. M., Zheng, D. W., Lin, T. F., Wei, X. D., Liu, X. Y., & Wang, H. Q. (2018). Surface-enhanced Raman spectroscopic analysis of N6-benzylaminopurine residue quantity in sprouts with gold nanoparticles. *Journal of Environmental Science and Health, Part B, 53*(9), 561–566.

Zhang, H., Qiu, Z., Cao, J., Abdel-Aty, M., & Xiong, L. (2019a). Event-triggered synchronization for neutral-type semi-Markovian neural networks with partial mode-dependent time-varying delays. *IEEE Transactions on Neural Networks and Learning Systems, 31*(11), 4437–4450.

Zhang, B., Ji, D., Fang, D., Liang, S., Fan, Y., & Chen, X. (2019b). A novel 220-GHz GaN diode on-chip tripler with high driven power. *IEEE Electron Device Letters, 40*(5), 780–783.

Zhang, B., Niu, Z., Wang, J., Ji, D., Zhou, T., Liu, Y., & Fan, Y. (2020). Four-hundred gigahertz broadband multi-branch waveguide coupler. *IET Microwaves, Antennas & Propagation, 14*(11), 1175–1179.

Zhao, Y., Zhang, W., Lin, Y., & Du, D. (2013). The vital function of Fe_3O_4@Au nanocomposites for hydrolase biosensor design and its application in detection of methyl parathion. *Nanoscale, 5*(3), 1121–1126.

Zhao, C., Liu, B., Piao, S., Wang, X., Lobell, D. B., Huang, Y., ... Ciais, P. (2017). Temperature increase reduces global yields of major crops in four independent estimates. *Proceedings of the National Academy of Sciences of the United States of America, 114*(35), 9326–9331.

Zhu, L., Zhang, C., Guan, X., Uy, B., Sun, L., & Wang, B. (2018). The multi-axial strength performance of composited structural BCW members subjected to shear forces. *Steel & Composite Structures, 27*(1), 75–87.

Zhu, S., Wang, X., Zheng, Z., Zhao, X.-E., Bai, Y., & Liu, H. (2020). Synchronous measuring of triptolide changes in rat brain and blood and its application to a comparative pharmacokinetic study in normal and Alzheimer's disease rats. *Journal of Pharmaceutical and Biomedical Analysis, 185*, 113263.

Ziska, L. H., Tomecek, M. B., & Gealy, D. R. (2010). Competitive interactions between cultivated and red rice as a function of recent and projected increases in atmospheric carbon dioxide. *Agronomy Journal, 102*(1), 118–123.

Zörb, C., Geilfus, C. M., & Dietz, K. J. (2019). Salinity and crop yield. *Plant Biology, 21*, 31–38.

Zou, Q., Xing, P., Wei, L., & Liu, B. (2019). Gene2vec: gene subsequence embedding for prediction of mammalian N6-methyladenosine sites from mRNA. *RNA, 25*(2), 205–218.

Zuo, C., Chen, Q., Tian, L., Waller, L., & Asundi, A. (2015). Transport of intensity phase retrieval and computational imaging for partially coherent fields: the phase space perspective. *Optics and Lasers in Engineering, 71*, 20–32.

Zuo, C., Li, J., Sun, J., Fan, Y., Zhang, J., Lu, L., ... Chen, Q. (2020). Transport of intensity equation: a tutorial. *Optics and Lasers in Engineering, 135*, 106187.

12

Nano-Based Applications for Veterinary and Dairy Production

Saba Fatima[1], Asghar Khan[2], Arfan Yousaf[2], Sadaf Anees[2], Asma Ayoob[2], and Ramzan Khan[2]

[1] *Animal Quarantine Department, Ministry of National Food Security and Research, Islamabad, Pakistan*
[2] *Department of Clinical Studies, Faculty of Veterinary and Animal Sciences, Arid Agriculture University, Rawalpindi, Pakistan*

12.1 Introduction

Nanotechnology has recently emerged as a cutting-edge tool across various fields, including dairy production and veterinary medicine. Nanoparticles have paved the way for revolutionary development in therapeutics, diagnostics, and disease prevention in animals. This is due to their unique features like rapid solubility, enhanced activity, and greater surface area to volume ratio. A strong potential has been shown by nanotechnology in increased vaccine efficacy, high nutrient absorption, rapid drug delivery, and fight against antimicrobial resistance (AMR). Nanotechnology has emerged as an attractive field of research in livestock and veterinary sciences (Algharib et al., 2020; Han et al., 2020).

Nanotechnology, the term "nano," which comes from the Greek word for "dwarf," is frequently used in connection with terminology such as nanometer, nanobot, and nanotechnology. It also referred to as "nanotech," the study and application of materials and technologies smaller than 100 nm in size. Because of their small size, minor modifications in their chemical and physical properties can result in more reactivity and solubility than bigger particles (Ali et al., 2021). Nanotechnology has been widely studied for their potential to replace conventional therapeutic methods and enhance efficacy of many drugs. For example, efficacy of gold nanorods has been investigated in photothermal therapy (PTT) to treat tumors in cats. This study has demonstrated an alternate noninvasive tumor treatment in animals (Abdoon et al., 2015). Furthermore, silver nanoparticles have been successfully used to treat vancomycin-resistant *Staphylococcus aureus* (VRSA) in bovine mastitis (Algharib et al., 2020). This study has paved the way to combat AMR in mastitis-causing pathogens, a major challenge in dairy sector.

Similarly, nanovaccines have earned significant attention due to their ability to increase immune responses and provision of sustained antigen release. Nanoparticles have been used as delivery systems and vaccine adjuvants for certain diseases. A study has been conducted

Nanotechnology Innovations for Food Security and Sustainable Agriculture, First Edition. Edited by Zia Ur Rehman Mashwani, Shah Fahad, Ilyas Ahmad, Shah Saud, Sajid Fiaz, and Taufiq Nawaz.

on use of nanoparticles encapsulating antigen as nanovaccine against cystic echinococcosis. This research has demonstrated stability and rapid delivery to mucosal tissues, leading to boosted immunogenicity (Silvarrey et al., 2024). Additionally, another immunological research has been conducted to modulate immune response in horses that were suffering from recurrent airway obstruction (RAO). This study highlights the potential of nanoparticle-bound CpG oligonucleotides as immunomodulatory agents (Klier et al., 2015).

In livestock sector, nanoparticles have shown promising results as nutritional supplements by enhancing production and improving animal health. For instance, nanoselenium (Nano-Se) have shown top notch results as an antioxidant and has increased selenium concentration in milk in dairy cows (Han et al., 2020). Similarly, Nano-Se has been studied to improve rumen fermentation, microbial protein synthesis, and nutrient digestibility in sheep (Shi et al., 2011). Furthermore, curcumin nanoparticles have also shown promising results in promoting gut health, immunity, and growth performance in poultry (Reda et al., 2020). Nanotechnology-based approaches offer the best replacement for conventional therapies that are facing challenges such as AMR, low drug solubility, and reduced bioavailability. The emergence of nanovaccine is especially a field of research that should be prioritized in developing countries where disease control in livestock is important for public health and food security.

Despite potential advantages of nanotechnology in dairy production and veterinary medicine, several challenges remain to be resolved. Safety concerns and side effects of nanoparticles are a major obstacle in way to clinical application of nanotechnology. The high reactivity and tiny size may interact with biological systems in harmful way, leading to immune dysregulation, oxidative stress, and cytotoxicity (Bogdanchikova et al., 2016). Furthermore, residual accumulation of nanoparticles in animal's tissues and their effects on animal and public health are the areas of concern that should be thoroughly evaluated. Quality control and standardization are also major challenges in the use of nanotechnology in clinical application. Ethical considerations pertaining to the use of nanoparticles in animals also require attention. Ensuring the welfare of animals and minimizing potential risks associated with nanotechnology are crucial factors. Furthermore, careful regulation and assessment should be adopted to check the impact of nanoparticles on environment, if released unintentionally into ecosystem.

The incorporation of nanotechnology into dairy production and veterinary medicine presents a number of opportunities for improving animal health, welfare, and productivity. However, dealing with the challenges linked with the safety, cost, standardization, regulation, and ethical concerns is crucial for the efficient use of these technologies. In future, researchers should focus on improving the safety profiles of nanoparticles usage and establishing regulatory guidelines to ensure their effective and safe use in the veterinary clinical applications. Additionally, facilitating collaboration between industry stakeholders, researchers, and regulatory agencies will be important in developing the field of veterinary nanotechnology.

12.2 Nanotechnology in Veterinary Medicine

Nanotechnology has revolutionized the diagnosis, prevention, and treatment of animal diseases. The following are key areas where nanotechnology has revolutionized veterinary medicine.

12.2.1 Diagnosis

Nanotechnology has surfaced as an emerging tool in diagnostics, giving innovative solutions for the accurate and early diagnosis of various diseases. Nanoparticles have unique chemical, physical, and optical properties that make them extraordinarily effective in diagnosis. Nanostructures, including cantilevers, nanowires, liposomes, quantum dots, and dendrimers, have shown remarkable potential in increasing diagnostic capabilities. Nanowires are highly sensitive tools that can detect chemical and biological species at even extremely low concentrations. The integration of nanowires into sensor devices helps to interact with target molecules. These integrated sensors offer innovative applications in diagnostics due to their ability to provide multiplexed detection of disease markers and real-time monitoring, thus enhancing the early diagnosis of diseases (Zheng et al., 2005; Zhang & Ning, 2012). A study by Chang et al. (2020) developed a highly sensitive silicon nanowire field-effect transistor (SiNW-FET) designed for detecting cardiac troponin I, which is an important marker of heart injury in hamsters. This novel approach demonstrated nanowire potential for real-time, highly sensitive diagnosis of cardiac injury (Chang et al., 2020. Cantilever-based nanosensors, another emerging technology, use microscopic flexible beams for the detection of specific disease biomarkers, making it a cost-effective and efficient diagnostic technique. Additionally, it reinforces rapid screening of multiple targets simultaneously, offering remarkable advancements in diagnostic specificity and sensitivity (Majumdar, 2002; Yue et al., 2004). A novel nanoscale genosensor was developed using an atomic force microscope cantilever functionalized with complementary strands of miRNA-203 and miRNA-205, both of which are potential biomarkers for detecting metastasis in head and neck squamous cell carcinomas, particularly in cervical lymph nodes (Andrade et al., 2019).

Quantum dots (QDs) are nanoparticles that are semiconductors and are known for their extraordinary optical properties. Their stability and brightness make them suitable for detecting and imaging biological targets. QDs have facilitated in-depth visualization of disease processes due to their role in labeling tissues, cells, and single molecules. Additionally, they play a role in the detection of multiple biomarkers simultaneously (Alivisatos, 1996; Gao et al., 2004). Quantum dots (QDs) have been used to increase the sensitivity of the immunochromatographic assays for diagnosing brucellosis. When combined with the gold nanoparticles, QDs enhanced the signal by two to three times compared to QDs alone and more than nine times compared to gold nanoparticles alone. This method enhanced diagnosis sensitivity to 98% of ELISA sensitivity, providing a rapid, effective, and portable diagnostic tool (Sotnikov et al., 2020). Liposomes, which are lipid-based nanoscale vesicles, are used as imaging agents due to their ability to encapsulate imaging molecules and biocompatibility. They serve as carriers in imaging techniques like computed tomography (CT) and magnetic resonance imaging (MRI) (Lasic, 1998; Martina et al., 2005). Quantum dots are used in theragnostic liposomes to improve diagnosis and treatment of cancer. QDs, due to their intense signals and high photostability, are incorporated into bilayers of liposome for targeted drug delivery and imaging. Studies have shown enhanced cellular uptake and in vivo bioimaging when QDs are integrated into liposomes, making them auspicious tools for future cancer therapy and diagnostics (Muthu & Feng, 2013).

Dendrimers are highly branched, nanoscale tree-like polymers with precise structural control, making them appropriate for diagnostic applications. Due to the capacity of these nanoparticles to amplify signals and bind multiple ligands effectively, they have played a significant role in diagnostic assays (Svenson & Tomalia, 2005; Tomalia et al., 2007). Dendrimers are being investigated as potential carriers for the contrast agents in imaging techniques. Their molecular structures allow for better imaging and targeting of tissues such as tumors and blood vessels. However, issues like production consistency, drug uniformity, and purity need to be resolved before clinical application (Krause et al., 2000).

The emerging field of nanotechnology may detect cancer biomarkers at even low concentrations and target deep-seated cancer cells (Hajjafari et al., 2024). Nanoparticles are also valuable in the development of highly sensitive biosensors for early diagnosis of Alzheimer's disease (AD) by detecting localized protein deposition (Panghal and Flora, 2024).

12.2.2 Prevention

Nanovaccines are increasingly recognized as a groundbreaking strategy for combating animal pathogens, offering innovative solutions to the challenges that plague conventional vaccines. Traditional vaccines, such as inactivated or live-attenuated types, often fall short due to limited protection duration, suboptimal immune responses, and safety concerns (Brito et al., 2013). Additionally, these vaccines are not always effective against rapidly evolving pathogens, newly emerging diseases, or chronic infections. Nanotechnology, with its capacity to engineer nanoparticles possessing unique properties like high surface area, controlled release, and enhanced antigen presentation, offers a promising alternative for vaccine development (Pati et al., 2018).

Various nanoplatforms, including virus-like particles (VLPs), liposomes, virosomes, and polymeric nanoparticles, are being explored for their potential to improve vaccine efficacy (Brito et al., 2013; Al-Halifa et al., 2019). VLPs, which structurally resemble natural viruses but lack genetic material, are highly immunogenic and safe. Their repetitive antigenic structures effectively stimulate both humoral and cellular immune responses, making them ideal candidates for vaccination (Pati et al., 2018). Liposomes and virosomes, which are lipid-based nanoparticles, have demonstrated considerable potential as antigen carriers. They enhance immunogenicity by encapsulating antigens, nucleic acids, and adjuvants, thereby providing versatile platforms for vaccine development (Al-Halifa et al., 2019).

Polymeric nanoparticles, particularly those made from polylactic-co-glycolic acid (PLGA), chitosan, and polyethyleneimine, offer additional benefits such as biodegradability, biocompatibility, and targeted delivery. These properties make them attractive candidates for improving vaccine delivery systems (Pati et al., 2018). The mechanism underlying the effectiveness of nanovaccines involves enhancing antigen delivery to antigen-presenting cells (APCs), such as dendritic cells. By engineering nanoparticles to interact with APCs via receptor-mediated endocytosis, researchers can promote efficient antigen processing and presentation to T cells. This stimulation triggers both innate and adaptive immune responses, facilitating the generation of memory cells that provide long-term protection (Brito et al., 2013; Pati et al., 2018).

Furthermore, nanovaccines show promise in enhancing mucosal immunity by improving antigen stability and bioavailability when administered through noninvasive routes like oral or nasal delivery (Al-Halifa et al., 2019). Their nanoscale size allows them to be effectively taken up by dendritic cells and other immune cells, resulting in improved cross-presentation of antigens via major histocompatibility complex molecules. This process leads to robust activation of both CD^{8+} cytotoxic T cells and CD^{4+} helper T cells, promoting a balanced Th1 and Th2 immune response (Pati et al., 2018). Despite their immense potential, the development of nanovaccines is not without challenges. Critical issues include optimizing formulation stability, ensuring reproducible manufacturing processes, and addressing safety concerns related to nanoparticle toxicity. Additionally, navigating the regulatory landscape for nanovaccine-based products presents a complex hurdle, requiring rigorous assessment of their efficacy and safety before approval.

Nevertheless, nanovaccines continue to hold tremendous promise in veterinary vaccinology by offering enhanced antigen delivery, superior immunogenicity, and sustained immune responses. Continued research and innovation in this field could effectively overcome the limitations associated with traditional vaccines, providing powerful tools for managing infectious diseases in animals and contributing to a One Health approach to global health (Celis-Giraldo et al., 2021). Recent advancements have demonstrated the potential of various nanovaccine platforms against diverse animal pathogens. For example, Pei et al. (2023) developed a self-assembling nanovaccine incorporating the Foot-and-Mouth Disease Virus (FMDV) VP1 protein and truncated flagellin (nFLiC) covalently linked to a self-assembling protein (i301). This formulation exhibited remarkable thermal stability, efficient cellular uptake, and robust immune response induction, comparable to traditional inactivated vaccines, making it a strong candidate for FMDV vaccination.

Similarly, Qamar et al. (2021) investigated copper oxide nanoparticles (CuONPs) synthesized from *Momordica charantia* fruit extract as an adjuvant for a live Newcastle Disease (ND) vaccine. The CuONP-enhanced vaccine provided 80% protection against the R2B NDV strain in chicks, significantly higher than the 60% protection achieved by the standard live vaccine. It also exhibited increased hemagglutinin inhibition (HI) titers and improved hematological and biochemical parameters, underscoring its potential as a potent vaccine delivery system. Furthermore, Acevedo-Villanueva et al. (2021) examined the efficacy of a chitosan nanoparticle (CNP) vaccine containing *Salmonella enteritidis* outer-membrane-proteins (OMPs) and flagellin proteins administered in ovo to broilers. The CNP-vaccinated birds demonstrated elevated IgY and IgA antibody levels, alongside reduced *Salmonella cecal* loads by 0.62 and 0.85 $\log_{10}$ CFU/g at first and second weeks postchallenge, respectively, compared to controls. Importantly, no adverse effects on hatchability, growth performance, or immune response were observed, highlighting CNP as a promising candidate for efficient mass vaccination against *Salmonella* in poultry.

12.2.3 Treatment

Photothermal therapy (PTT), which employs near-infrared (NIR) light to induce heat for therapeutic purposes, has emerged as a promising approach for antibacterial treatment. However, its effectiveness is often hindered by excessive temperature levels that can cause damage to normal cells. To address this limitation, Zhu et al. (2023) developed a novel hydrogel composed of a mixture of sodium alginate and AgNPs-modified polydopamine nanoparticles (NPs). This hydrogel demonstrated effective antibacterial properties at low temperatures, offering a safer alternative for treating bacterial infections. In addition, Chang et al. (2023) successfully formulated a nanocomposite hydrogel functional wound dressing by combining AgNPs with PTT. This innovative wound dressing was found to significantly alleviate inflammation and promote the healing of infected skin wounds, highlighting the potential of integrating nanotechnology with PTT for enhanced therapeutic outcomes.

Moreover, Ahn et al. (2023) synthesized Ag–Te NPs using Te nanorods, demonstrating their potential in breast cancer treatment through both in vitro and in vivo studies. By leveraging photothermal conversion, these nanoparticles effectively ablated hyperthermic cancer cells while also exhibiting antibacterial activity against breast tumor bacterial cells, illustrating their dual therapeutic benefits. Another promising technique, photodynamic therapy (PDT), involves irradiating lesions with a specific wavelength of laser light, activating photosensitive drugs selectively accumulated within the affected tissue. The resulting photochemical reaction destroys the targeted lesion. Shivashankarappa et al. demonstrated the efficacy of PDT for melanoma treatment by coupling AgNPs with photosensitizer drug molecules. This approach enabled selective induction of apoptosis or necrosis in cancer cells and allowed for targeted drug delivery to specific sites (Shivashankarappa and Sanjay 2019).

Polymeric nanoparticles are also gaining attention for their potential in cancer treatment through enhanced drug delivery systems. Zhong et al. reported the codelivery of sorafenib (SORA) and crizotinib (CRIZ) using poly(ethylene glycol)-poly(e-caprolactone)-poly (ethylene glycol) (PEG–PCL–PEG) nanoparticles. This formulation exhibited improved apoptosis rates and reduced tumor growth in lung cancer models. Furthermore, the nanoparticles provided sustained drug release, enhanced cellular uptake, and minimized systemic toxicity, making them a promising therapeutic option for lung cancer treatment (Zhong et al., 2021). The application of dopamine-loaded nanoparticles has shown considerable potential in treating Parkinson's disease (PD). Monge-Fuentes et al. utilized albumin/poly (D,L-lactic-*co*-glycolic acid) (PLGA) nanosystems loaded with dopamine (DA) in a 6-hydroxydopamine (6-OHDA) PD mouse model. These nanoparticles successfully crossed the blood-brain barrier (BBB) and replenished dopamine in the nigrostriatal pathway, resulting in significant improvements in motor coordination, balance, and sensorimotor function compared to traditional treatments. The findings suggest that albumin/PLGA nanoparticles could provide an innovative and efficient approach for PD therapy (Monge-Fuentes et al., 2021).

Furthermore, tellurium nanoparticles (TeNPs) synthesized by *Streptomyces graminisoli* have demonstrated remarkable antibacterial activity against methicillin-resistant *Staphylococcus aureus* (MRSA). With a mean size of 21.4 nm and exhibiting rod and rosette forms, these TeNPs showed an inhibition zone of 24 ± 0.7 mm and a minimum inhibitory concentration (MIC) of 50 mg/mL during in vitro testing. Additionally, in vivo studies on infected rats revealed improved survival rates, reduced bacterial counts in organs, and effective modulation of inflammatory cytokines when treated with TeNPs alone or combined with vancomycin, suggesting their potential as effective therapeutic agents against MRSA (Abed et al., 2023). In the realm of antifungal therapy, gold nanoparticles (AuNPs) synthesized using olive leaf extract have demonstrated significant efficacy against *Candida albicans*, a causative agent of cutaneous candidiasis. Kareem et al. reported that AuNPs exhibited a MIC of 40.77 ng/mL, effectively inhibiting fungal growth. In vivo experiments on mice showed that AuNPs provided a faster cure, achieving complete recovery within three days compared to four days for nystatin-treated mice. These findings underscore the potential of green-synthesized AuNPs as a safe and effective alternative treatment for cutaneous candidiasis (Kareem et al., 2021).

12.3 Therapeutics and Drug Delivery

12.3.1 Targeted Drug Delivery

Nanotechnology has led to the creation of innovative drug delivery methods using nanoparticles for treating posterior segment diseases. Nanoparticles (NPs), first and foremost, offer the potential for targeted drug delivery. They can be engineered with precision to selectively target tissues or cells, including photoreceptor cells or the retinal pigment epithelium (RPE). Moreover, nanocarriers help to minimize side effects by safeguarding the drug against degradation in vivo, enhancing absorption and distribution, and increasing drug bioavailability, thus reducing the required dosage for effective treatment. Nanomedicines can prolong drug release, increasing the therapeutic impact lifetime and perhaps eliminating the requirement for frequent dosage, which can significantly improve patient adherence. In addition, these nanomedicines can carry various therapeutic ingredients such as small-molecule drugs, biologics, and genes. They provide noninvasive delivery options such as eye drops or ointments, while also serving imaging functionalities (Wu et al., 2024). Poly(lactic-*co*-glycolic acid) nanoparticles (PLGA-NPs) are promising systems for controlled drug release and targeted delivery. These nanoparticles enhance therapeutic efficacy by modulating drug release kinetics, improving pharmacokinetics, and reducing off-target toxicity. They have shown potential in treating chronic diseases such as cancer, arthritis, and neurological disorders. PLGA-NPs' surface functionalization facilitates targeted delivery, enhancing drug accumulation at specific sites while minimizing systemic exposure. Despite their benefits, challenges like formulation stability, scalability, and regulatory approval remain (Zeb et al., 2022). Janus nanoparticles composed of dextran and bismuth selenide (Dex-BSe) have demonstrated significant efficacy against drug-resistant biofilms. These nanoparticles exploit the photothermal properties of BSe and the biofilm-targeting ability of dextran to penetrate and disperse biofilms effectively. Additionally, under NIR light, Dex-BSe nanoparticles exhibit enhanced antibacterial activity due to their self-propelled active motion, resulting in efficient biofilm removal. In vivo studies confirmed their ability to eradicate *S. aureus* and MRSA biofilms in wound and abscess models, highlighting their potential for clinical applications (Liu et al., 2023). κ-carrageenan-coated zinc oxide nanoparticles (ZnO-CR NPs) have demonstrated potential in targeted drug delivery and apoptosis induction in oral cancer. Synthesized using *Kappaphycus alvarezii*, these nanoparticles exhibited effective free radical scavenging, substantial antimicrobial activity against pathogens like *S. aureus* and *C. albicans*, and a dose-dependent anticancer effect on KB cells. The ZnO-CR NPs significantly reduced cell viability by inducing apoptosis via modulation of BCL-2, BAX, and P53 gene expression, making them a promising candidate for oral cancer treatment (Marunganathan et al., 2024).

12.3.2 Bioavailability

Nanoparticulate drug delivery systems (NDDS) have gained attention for improving the oral bioavailability of drugs, which is often limited by harsh gastrointestinal (GI) conditions, poor permeability, and extensive pre-systemic metabolism. NDDS can enhance drug absorption by protecting drugs from premature degradation, increasing intracellular penetration, and promoting drug uptake. Factors like nanoparticle size, surface properties, and shape significantly influence their interaction with biological pathways. For instance, nanoparticles smaller than 200 nm can be more effectively absorbed by the GI tract, while PEGylation and other surface modifications improve mucus penetration and reduce recognition by efflux transporters. Nanoparticles with hydrophilic coatings or specific ligands can enhance cellular uptake and drug accumulation, demonstrating their potential in oral drug delivery (Wang et al., 2020). Solid lipid nanoparticles (SLNs) and nanostructured lipid carriers (NLCs) are lipid-based carrier systems designed to enhance the oral bioavailability of drugs with poor intestinal permeability. These nanoparticles effectively encapsulate lipophilic and hydrophilic drugs, providing controlled release and protection from degradation. SLNs and NLCs can enhance absorption by increasing drug solubility, promoting lymphatic uptake, and inhibiting efflux transporters. Their ease of manufacturing, biocompatibility, and scalability make them attractive candidates for oral drug delivery systems targeting Class IV drugs and protein/peptide drugs (Talegaonkar & Bhattacharyya, 2019).

Selenium nanoparticles (SeNPs) synthesized using *Bacillus cereus* demonstrated enhanced bioavailability compared to bulk selenium. The biosynthesis process involved reducing sodium selenate with *B. cereus* cultures, producing spherical SeNPs with a mean diameter of 170 nm. In vivo studies on mice revealed higher absorption rates of SeNPs than bulk selenium, with improved excretion rates, suggesting reduced toxicity. These findings highlight the potential of SeNPs in veterinary and medical applications due to their enhanced bioavailability and reduced toxicity (Pouri et al., 2017). Silver nanoparticles (AgNPs) administered orally to rats showed significant accumulation in the liver and kidneys, with the liver being the primary target organ. Fecal excretion was the main elimination route, while urinary excretion was negligible. Imaging studies revealed that AgNPs penetrate the liver but remain mostly in the cortex of the kidneys. Speciation analysis indicated the formation of Ag(I) complexes due to AgNP oxidation, particularly metallothionein complexes in the kidneys. The study highlights the potential toxicity of AgNPs from prolonged exposure (Jiménez-Lamana et al., 2014). Enhancing the bioavailability of medicinal substances is critical in veterinary medicine, as many medicines have low solubility and absorption. Nanotechnology improves drug bioavailability in veterinary medicine by increasing solubility, stability, and targeted delivery. Nanoparticles encapsulate pharmaceuticals, preventing breakdown and improving absorption (Osama et al., 2020).

Nanotechnology addresses these issues by increasing medicine solubility, stability, and absorption. For example, polymeric nanoparticles and lipid-based nanocarriers enhance hydrophobic drug absorption, resulting in improved therapeutic benefits, and preserving them from enzymatic breakdown and increasing bioavailability. These carriers may interact with cell membranes, which facilitates the direct drug delivery into cells. Liposomes' physicochemical features, such as lipid content and surface charge, can be modified to improve medication delivery and release profiles (Chariou et al., 2020). In addition, polymeric nanoparticles have been used to increase the bioavailability of veterinary medicines. These nanoparticles can improve the solubility of weakly water-soluble medications and protect them against degradation, which leads to enhanced therapeutic outcomes (Carvalho et al., 2020).

12.3.3 Controlled and Sustained Release Systems

Smart sustained-release nanoparticles are designed to provide controlled and prolonged drug delivery, particularly for cancer treatment. These nanoparticles respond to specific physiological triggers or external stimuli to release drugs at the desired time and site. Common nanocarrier systems include liposomes, dendrimers, micelles, and polymeric nanoparticles, each offering advantages such as enhanced drug retention, targeted delivery, and reduced side effects. Despite their potential, challenges like manufacturing reproducibility, scalability, and regulatory approval remain significant hurdles for clinical translation (Bai et al., 2022). Stimuli-responsive nanoparticles (srNPs) have been developed for controlled drug delivery and synergistic cancer immunotherapy. These nanoparticles respond to physiological triggers like pH, glutathione (GSH), enzymes, and reactive oxygen species (ROS), or external stimuli such as light, ultrasound, and magnetic fields. By exploiting the unique tumor microenvironment (TME), srNPs enable targeted drug release, enhance immune responses, and improve therapeutic efficacy. They have shown promise in converting "cold" tumors into "hot" ones, facilitating immune cell infiltration and activation for robust cancer treatment. Despite their potential, challenges in manufacturing, stability, and clinical translation remain (Zhang et al., 2021). Mesoporous silica nanoparticles (MSNs) have emerged as a promising platform for controlled-release drug delivery due to their unique structural features, including high surface area, large pore volume, and excellent biocompatibility. These nanoparticles are designed to enhance the efficacy and safety of therapeutic agents by providing targeted and controlled release mechanisms. The release of drugs from MSNs is commonly triggered by internal or external stimuli. Internal stimuli include redox conditions, enzymatic activity, and pH variations, while external stimuli consist of light, ultrasound, magnetic fields, and temperature, which can be categorized as either internal or external (Karimi et al., 2016).

The ability of MSNs to be functionalized with various molecular gatekeepers or nanovalves enhances their performance in controlled-release systems. These gatekeepers, which can be polymers, peptides, or inorganic nanostructures, effectively seal the pores of the nanoparticles, preventing premature drug release until the appropriate stimulus is applied (Karimi et al., 2016). For instance, redox-responsive MSNs utilize disulfide bond-linked gatekeepers that are cleaved under high concentrations of intracellular glutathione, commonly found in cancer cells, leading to targeted drug release (Karimi et al., 2016). Enzyme-responsive MSNs are designed to respond to specific enzymes overexpressed in diseased tissues, while pH-responsive MSNs exploit the acidic environment of tumor cells to trigger drug release (Karimi et al., 2016).

Additionally, external stimuli such as light and magnetic fields provide noninvasive methods for controlled drug release. NIR light is particularly advantageous due to its deep tissue penetration and minimal damage to surrounding tissues. Magnetic-responsive MSNs are incorporated with magnetic cores, allowing targeted drug release upon exposure to external magnetic fields (Karimi et al., 2016). Temperature-responsive MSNs, which release their payload when subjected to hyperthermia, are another effective strategy for controlled delivery, particularly in tumor environments where localized heating is achievable (Karimi et al., 2016). The integration of these stimulus-responsive mechanisms makes MSNs highly versatile and efficient for drug delivery applications, especially in cancer therapy where selective targeting and minimal side effects are critical. Further studies focusing on optimizing the functionalization and targeting capabilities of MSNs are essential to enhance their clinical translation (Karimi et al., 2016).

12.3.4 Toxicity

Nanotechnology offers significant advantages in various biomedical applications, but concerns about nanoparticle toxicity continue to be a critical issue. Due to their small size, nanoparticles can easily penetrate cells and tissues, which, while beneficial for targeted drug delivery, also presents potential health risks. Studies have highlighted several adverse effects associated with nanoparticle exposure, including oxidative damage, genotoxicity, and immune system disruption. When nanoparticles accumulate in organs such as the liver, lungs, and kidneys, they can cause long-term harm and compromise organ function (Elalfy et al., 2018). The ability of nanoparticles to trigger inflammatory responses is particularly concerning. Once inside the body, nanoparticles can interact with cellular structures, leading to immune dysregulation and chronic inflammation. Such disruptions in the immune system may weaken the body's natural defenses, making it more vulnerable to various diseases. Additionally, evidence suggests that prolonged exposure to nanoparticles may result in DNA damage, which raises concerns about their potential carcinogenicity. The genotoxic effects of nanoparticles are attributed to their capacity to generate ROS, which can interfere with cellular processes and induce mutations.

Another critical issue associated with nanoparticles is bioaccumulation. When nanoparticles are introduced into the body, their high reactivity and ability to cross biological barriers can result in their accumulation within vital organs over time. This buildup can lead to toxicological effects, impairing the normal functioning of organs and potentially causing irreversible damage (Xuan et al., 2023). These findings underscore the importance of thoroughly assessing the safety and biocompatibility of nanoparticles before their widespread use in veterinary and medical applications. Developing standardized guidelines for evaluating nanoparticle toxicity and implementing strategies to minimize potential risks are essential steps to ensure their safe and effective use.

12.3.5 Theragnostic

Theragnostic, an emerging field of nanotechnology, combines diagnostic and therapeutic capabilities into a single nanoparticle-based system, offering immense potential for advancing veterinary medicine. By integrating real-time imaging, disease monitoring, and targeted drug delivery, theragnostic nanoparticles enable precise and efficient treatment strategies that significantly enhance therapeutic accuracy (Bai et al., 2018). This dual-purpose approach is particularly valuable in addressing complex veterinary challenges, including viral infections, inflammatory disorders, and cancers in animals.

The fundamental mechanism of theragnostic nanomedicine involves utilizing nanoparticles designed to carry both imaging agents and therapeutic compounds. This integrated system allows for the simultaneous diagnosis and treatment of diseases, making it a highly efficient tool for veterinary applications (Meena et al., 2018). In canine oncology, for example, magnetic and gold nanoparticles have been employed to accurately visualize tumors while concurrently administering chemotherapy in a controlled manner. This approach improves treatment efficacy while minimizing damage to surrounding healthy tissues. Moreover, the application of theragnostic nanoparticles in veterinary practice offers several key advantages. One of the most significant benefits is the ability to monitor treatment efficacy in real time, which helps reduce the risk of developing drug resistance and minimizes adverse effects associated with conventional therapies (Bilgili & Uysal, 2019). Additionally, theragnostic systems are particularly effective in managing conditions like mastitis in dairy cows, where early detection and targeted therapy can substantially decrease economic losses and promote animal welfare.

Overall, theragnostic represents a groundbreaking advancement in veterinary nanomedicine, providing innovative solutions for diagnosing and treating various animal diseases with enhanced precision and efficiency.

12.3.6 Cross-Biological Barriers

Nanoparticles have significantly transformed drug delivery systems by enabling medications to overcome biological barriers that traditionally limit therapeutic efficacy. One of the most remarkable advantages of nanotechnology in veterinary medicine is its ability to enhance drug delivery efficiency by traversing obstacles such as the BBB and the intestinal lining (Zhang et al., 2021). These advancements are particularly valuable in improving treatment options for neurological and systemic diseases in animals, where conventional therapies often fall short. The BBB, characterized by its selective permeability, poses a substantial challenge for treating neurological disorders in animals. Conventional medications often struggle to penetrate this barrier, limiting their therapeutic potential and efficacy. However, nanoparticles like liposomes and dendrimers have demonstrated promising results in facilitating targeted drug delivery to the brain. By enhancing drug permeability and stability, these nanoparticles have improved the treatment of various neurological conditions, offering new hope for more effective therapies.

Similarly, the gastrointestinal barrier presents a significant challenge, particularly for orally administered drugs. Nanocarriers have proven essential in protecting medications from enzymatic degradation within the stomach, thereby enhancing their oral bioavailability. This enhanced absorption is especially beneficial for antibiotics, probiotics, and vaccines, where efficient delivery is critical for achieving optimal therapeutic outcomes (Grabrucker et al., 2016).

Overall, the application of nanoparticles in overcoming biological barriers represents a critical advancement in veterinary medicine. By enabling precise and efficient drug delivery to previously inaccessible sites, nanotechnology offers new possibilities for treating complex animal diseases more effectively and safely.

12.3.7 Shelf Life of Therapeutic Agents

Nanotechnology has significantly improved the shelf life and stability of therapeutic agents in veterinary medicine by preventing degradation and preserving bioactivity over extended periods (Hassan et al., 2020). This advancement is particularly essential for enhancing the efficacy and reliability of various veterinary drugs, vaccines, and therapeutic compounds, ensuring their effectiveness even under challenging storage conditions. One of the most effective strategies for enhancing drug stability involves nanoparticle-based encapsulation. In this approach, medications are enclosed within liposomes, polymeric nanoparticles, or SLNs, which act as protective carriers that shield the drugs from environmental degradation (Youssef et al., 2019). This encapsulation process helps maintain the therapeutic properties of drugs, even when exposed to harsh conditions that would otherwise compromise their efficacy.

Moreover, nanotechnology offers promising solutions for extending vaccine viability, especially in resource-limited areas where maintaining proper cold storage is difficult. Nanoparticles serving as vaccine carriers significantly reduce the dependence on refrigeration, allowing stable storage under adverse conditions. This enhancement improves vaccine accessibility and effectiveness in remote locations, making it an invaluable tool for improving animal health management (Bilgili & Uysal, 2019). Additionally, the development of extended-release formulations using nanomaterials has proven effective in providing controlled and sustained therapeutic effects. By preventing premature drug degradation, these formulations reduce the frequency of administration, resulting in more efficient treatment protocols and improved animal health (Bai et al., 2018). Such advancements highlight the transformative potential of nanotechnology in veterinary medicine by enhancing the stability, efficacy, and accessibility of therapeutic agents.

12.3.8 Antimicrobial Activity

The increasing prevalence of bacteria, fungi, and viruses, coupled with the growing resistance of pathogens to conventional antimicrobial agents, necessitates the exploration of alternative solutions (Hwangbo et al., 2019). Nanomaterials have emerged as promising candidates for combating microbial resistance due to their multiple mechanisms of action against pathogens, which significantly reduce the likelihood of resistance development. Various studies have demonstrated the high bioactivity of nanoparticles made from copper (Zare Khafri et al., 2018), silver (Safarpoor et al., 2018), nickel (Hashem et al., 2022), and selenium (Dinc et al., 2022). Their remarkable efficacy is primarily attributed to their small size and high surface-to-volume ratio, which enhances their interaction with microbial cell membranes.

Despite the demonstrated efficacy of nanoparticles, the precise mechanisms by which they exert their antimicrobial effects remain incompletely understood. It is believed that nanoparticles may produce ROS, resulting in decreased concentrations of antioxidant compounds and progressive oxidative damage to essential cellular components, including lipids, DNA, and proteins. Excessive ROS production disrupts redox homeostasis, thereby inducing oxidative stress, which compromises membrane integrity, protein functionality, and DNA structure. The generation of various ROS, such as hydrogen peroxide, superoxide radicals, and hydroxyl radicals, is highly dependent on the chemical nature of the nanoparticles used (Verma and Dutta, 2017).

Nanoparticles, particularly silver, copper, and CNPs, have shown extensive antibacterial, antifungal, and antiviral activities, making them valuable alternatives to traditional antibiotics. Silver nanoparticles (AgNPs) are especially effective at disrupting bacterial cell membranes, resulting in cell death. Practical applications of AgNPs, such as silver nanoparticle-based wound dressings for dairy cows, have demonstrated significant reductions in infection rates, underscoring their potential use in veterinary medicine (Algharib et al., 2020). CNPs have also shown efficacy against multidrug-resistant (MDR) bacteria, particularly in poultry farming. Research indicates that these nanoparticles can effectively reduce the reliance on conventional antibiotics in chicken farms, thereby promoting more sustainable and efficient disease management practices (Şenel & McClure, 2004).

In addition to their direct antimicrobial effects, nanoparticles enhance the effectiveness of standard antibiotics by facilitating more efficient drug delivery. Metal nanoparticles, such as zinc oxide, gold, copper, silver, and titanium dioxide, have been shown to improve the efficacy of conventional antibiotics, thereby reducing dosage requirements and minimizing the risk of resistance development (Smekalova et al., 2016). Silver nanoparticles, in particular, have demonstrated exceptional antibacterial, antifungal, and antiviral properties, making them valuable in treating various animal diseases, including mastitis, respiratory infections, and wound infections (Frippiat et al., 2025). Furthermore, nanoparticles are increasingly being integrated into nanocoatings for biosecurity applications. These coatings, incorporated into disinfectants, wound dressings, and animal enclosures, significantly reduce bacterial contamination and enhance hygiene standards (El-Sayed & Kamel, 2020).

12.3.9 Anti-inflammatory Agent

Zinc oxide (ZnO) nanoparticles synthesized under different conditions demonstrated remarkable antibacterial activity, particularly when produced using ultrasound-assisted methods. This sample exhibited the highest antibacterial efficacy, with a 99% inhibition rate against *S. aureus* and an 80% inhibition rate against *Pseudomonas aeruginosa*. The superior antibacterial performance is largely attributed to the differences in bacterial cell wall composition, making ZnO nanoparticles especially effective against Gram-positive microorganisms. Moreover, the ultrasound-assisted ZnO nanoparticles displayed impressive anti-inflammatory activity, achieving a 93% inhibition rate, which surpassed the effectiveness of the reference drug, diclofenac. This enhanced anti-inflammatory property highlights the potential of ZnO nanoparticles as a powerful alternative for treating infections and inflammatory conditions in veterinary medicine (Lopez et al., 2023).

12.4 Nanotechnology in Parasitology

Parasites pose a constant and unacceptable threat to the well-being of millions of individuals in tropical and subtropical regions, as well as domesticated animals worldwide, leading to significant human suffering and economic losses. Parasitology, the study of parasites and their interactions with hosts, was once believed to be resolved through medications, vaccinations, and preventative measures. However, this underestimated the complexity of host–parasite interactions (Cox, 2009). Nanotechnology has emerged as a transformative tool in various scientific disciplines, including veterinary parasitology, opening new avenues for diagnosing, preventing, and treating parasitic diseases in animals (Muraleedharan & Chhabra, 2018).

Life-threatening verminous aneurysms caused by the destructive parasite *Strongylus vulgaris* in horses are difficult to identify in time. A highly sensitive Nano-Gold ELISA has been developed for diagnosis, with no cross-reactivity to sera from animals infected with other helminths (Baghdadi et al., 2024). This nanotechnology-enhanced ELISA provides highly sensitive and targeted methods for parasite diagnosis using signal-amplifying gold nanoparticles, revolutionizing parasitological diagnostics and improving animal health management. Gold nanoparticles and quantum dots are utilized for imaging modalities to visualize parasites in different tissues and organs, including Toxocara (Jahanmahin & Borji, 2023) and Schistosoma (Qadeer et al., 2022). Nanoparticles have been employed for bioimaging to locate and track Leishmania parasites in host tissues, enabling early detection and monitoring of disease progression (Tiwari et al., 2023). In clinical sciences, food safety, and drug analysis, screen-printing electrode biosensors have gained significant interest. These biosensors detect minimal concentrations of analytes by identifying variations in voltage, current, and conductance from immunological reactions. Nanotechnology has improved existing diagnostic methods and introduced novel approaches, leading to the development of highly sensitive and adaptable diagnostic devices using nanoparticles (Sadr et al., 2023).

Nanobiosensors are now used to diagnose Schistosoma (Mutro Nigo et al., 2020) and Toxocara (Jahanmahin & Borji, 2023). Research on nanobiosensors for early hydatid cyst detection has demonstrated their potential to offer noninvasive, rapid, and precise identification methods, improving the sensitivity and specificity of diagnostic techniques (Sadr et al., 2023). A sensor utilizing a chitosan-coated carbon paste and gold nanoparticles showed a detection limit of 0.06 μg/mL for *Schistosoma japonicum* antigen detection (Mutro Nigo et al., 2020). Gold nanoparticles and aminated magnetic particles have also been utilized for the highly sensitive detection of *Schistosoma mansoni* DNA using electrochemical techniques (Mutro Nigo et al., 2020).

Fiber optic spectrometry-based biosensors have been investigated for detecting *Trypanosoma* infections, demonstrating the potential of spectrometric biosensors for noninvasive diagnostics (Zhao et al., 2025).

A nanovaccine containing *Toxoplasma gondii*'s ribosomal P2 protein has been developed, eliciting strong humoral responses in animal models, providing a promising strategy for toxoplasmosis prophylaxis (Yu et al., 2022). Nanoparticle-based vaccines incorporating *Leishmania infantum* antigens have demonstrated significant efficacy in reducing parasite burdens in hamster livers and spleens, offering exciting prospects for visceral leishmaniasis control (González et al., 2023). An innovative subunit nanovaccine against *Trypanosoma cruzi* has shown protection in mouse models, significantly reducing parasite presence in tissues post-rechallenge (Chowdhury et al., 2020). Polymeric nanoparticles coated with *Giardia lamblia* membranes were developed as a multivalent nanovaccine, inducing robust adaptive immunity in mice and effectively preventing *G. lamblia* infection upon intranasal administration (Zhou et al., 2022).

Silver, gold, and polymeric nanoparticles are being investigated as alternatives or adjuncts to conventional antiparasitic treatments. Silver nanoparticles (AgNPs) exhibit broad-spectrum antimicrobial activity, disrupting parasite cellular structures via oxidative stress and interaction with sulfur-containing proteins or DNA. They have been tested for efficacy against Leishmania and Plasmodium species (Rai, 2012). Gold nanoparticles (AuNPs) are biocompatible and possess modifiable surface properties that facilitate targeted drug delivery. Functionalized AuNPs have shown promising results against *T. cruzi* and *Plasmodium falciparum* by improving drug transport to parasite cells (Dykman & Khlebtsov, 2011).

Biodegradable polymeric nanoparticles serve as efficient drug carriers, enhancing the solubility and stability of antiparasitic medications. These nanoparticles enable controlled release and targeted drug delivery, reducing toxicity and improving treatment efficacy against parasites such as Leishmania (Bajwa et al., 2022). Recent advances in nanotechnology have also introduced photothermal and PDTs for parasitic infections. Gold and silver nanoparticles are being explored for their potential in targeted heat-based or light-activated therapies to eliminate parasites without harming host tissues (Tiwari et al., 2023).

12.5 Nanotechnology in Dairy Production

Nowadays, the dairy and food industries are undergoing a tremendous technological and socioeconomic revolution. A classic illustration is the rapid shift in product availability and the expanding range of consumer choices. Customers are increasingly looking for fresh, authentic, appropriate, and enjoyable food options (Chavada, 2016). Conversely, global consumers are favoring products with superior quality and safety as well as novel packaging materials that amplify their desirable qualities (Amrita Poonia, 2019). Similarly, the ban of ionophore group antibiotics in animal nutrition has led to greater interest in biotechnology techniques to improve animal health and production (Gilbert, 2012). Therefore, adoption of innovative and advanced technologies is critical for remaining competitive and maintaining leadership in the food and food processing sectors. Likewise, to meet the rising customer demand for natural, minimally processed, and high-quality products, the use of nanotechnology in dairy industry is set to bring a major breakthrough. This technology help in producing high-quality, safe dairy products and developing advanced packaging material with improved qualities (Girigoswami et al., 2021).

Encapsulating feed additives that involves coating or trapping at nanoscale enhances their utilization and easy recognition by animal's body. These feed additives having enhanced resistance to heat, spoilage, and light while preventing degradation by proteases and digestive enzymes (Wang et al., 2022). Encapsulation ensures the availability of nutrients to specific target sites and not broken down too early in the rumen that allows them to reach small intestines for better absorption. Likewise, the ability of animals to turn feed into energy is enhanced by the nanosized feed additives, which maximize nutrient absorption and utilization, improving growth and feed conversion rates. Therefore, this enhances the effectiveness of feed additives in animal nutrition, promoting better growth and animal health (Dumlu, 2024).

Studies have shown that incorporating nanofeed additives into rations has been shown to improve meat and egg quality in chickens, as well as milk yield and other parameters in cattle and sheep (Gelaye, 2024; Dong et al., 2022). Similarly, a study was conducted on zinc nanoparticles in milk yield enhancement in goats, which shows a remarkable improvement in milk production. This study also clarifies that the improvement in performance is due to the high intake of dry matter and protein degradation. Similarly, the use of zinc nanoparticles supplementation increases the absorption of nutrients and enhances the activity of digestive enzymes, gut health and immune response. Which in turn contributed in enhancement of milk production and improvement in feed efficiency. Additionally, it also aids in lowering gut inflammation, supporting overall growth performance. Conversely, nanoparticles also show a tremendous decline in somatic cell counts in cows with subclinical mastitis, which ultimately enhances milk production (Mansour et al., 2024).

Similarly, studies shown that calves supplemented with ZnO nanoparticles exhibited higher post-weaning intake of dry matter, enhanced nutrients digestibility, and elevated blood zinc levels as compare to the ones receiving conventional zinc oxide. Additionally, both forms of zinc improve rumen volatile fatty acid (VFA) concentrations, reduced rumen ammonia levels, boosted antioxidant capacity in blood, and lowered the occurrence of diarrhea and pneumonia in calves (Abdollahi et al., 2020). Milk, an essential raw material in the dairy industry, includes significant levels of carbohydrates, fat, and proteins, making it susceptible to spoiling. Oxidative and hydrolytic processes, triggered by enzymes from milk and psychotropic microorganisms, can cause unwanted changes in milk and dairy products. Additionally, milk is a perfect environment for hazardous bacteria such as *Listeria monocytogenes*, *S. aureus*, *Salmonella*, and toxic strains of *Escherichia coli*, creating a considerable risk of contamination and loss of quality in dairy products as shown in Figure 12.1 (Brandelli et al., 2023; Claeys et al., 2013; Fusco, 2020).

Nanotechnology can be utilized to develop nanosized delivery methods for vitamins, minerals, and other nutrients, increasing their absorption and effectiveness in dairy

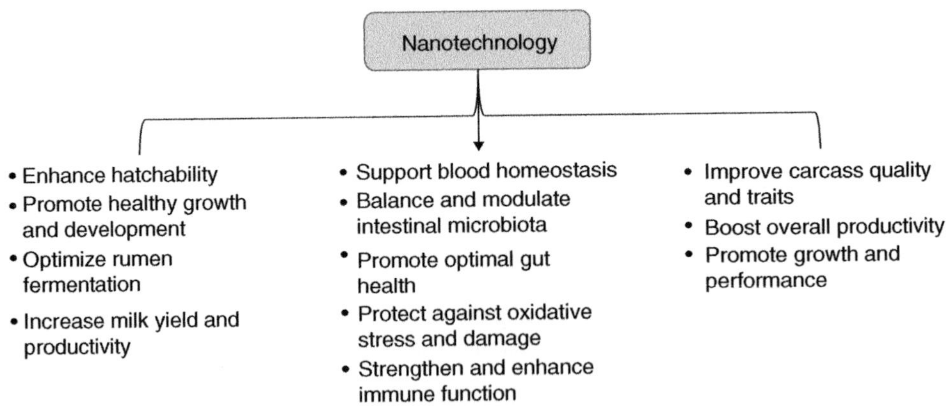

Figure 12.1 Beneficial effects of nanoparticles in animal growth and health.

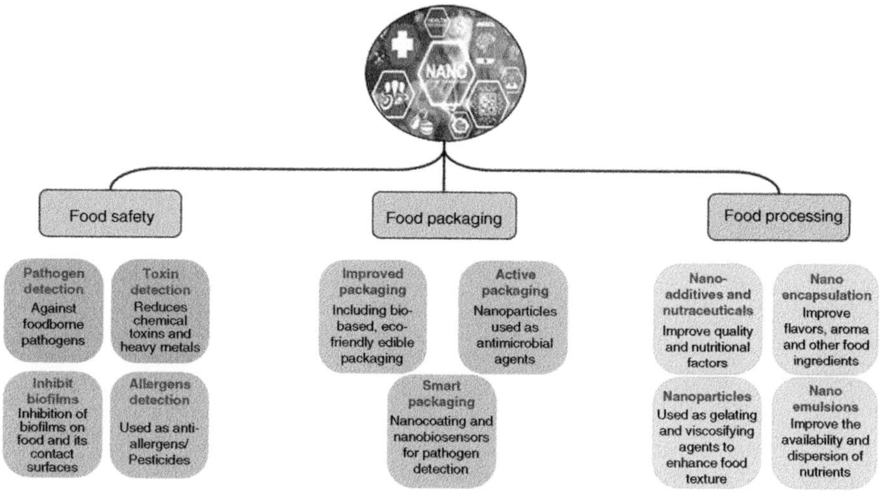

Figure 12.2 Applications of nanotechnology in dairy industry. *Source:* wladimir1804/Adobe Stock Photos.

products such as milk, cheese, and yogurt. It can also be used in packaging to monitor and maintain the quality of dairy products. They are additionally permitted to find contaminants or spoilage microorganisms in food, hence improving food safety. Similarly, it enhances the consistency, creaminess, and flavor of dairy products by altering the composition of fat molecules at the nanoscale, resulting in a more pleasant taste and mouthfeel. Likewise, nanotechnology is gaining much attraction to create antimicrobial coatings for dairy packaging, which aid to prevent spoiling and improve shelf life by inhibiting bacteria development (Sonea, et al., 2021). Food safety, quality control, and traceability could all be greatly enhanced by nanobiosensors, an emerging technology. These sensors provide increased precision in food analysis, processing, packaging, and safety by detecting and measuring chemicals at the nanoscale as shown in Figure 12.2 (Guruprasath et al., 2024).

12.6 Benefits of Nano-Based Applications

Nanotechnology has revolutionized various industries by offering enhanced efficiency, precision, and sustainability across numerous applications. This cutting-edge technology involves manipulating materials at the atomic or molecular level, enabling the design of nanoparticles with unique structures that provide remarkable advantages in several fields as shown in Figure 12.3.

12.6.1 Food Industry

In the food industry, nanotechnology plays a crucial role in enhancing product quality, safety, and shelf life. By reducing materials to a nanoscale, nanoparticles exhibit superior penetrability, chemical activity, enzymatic activation, and catalytic properties compared to

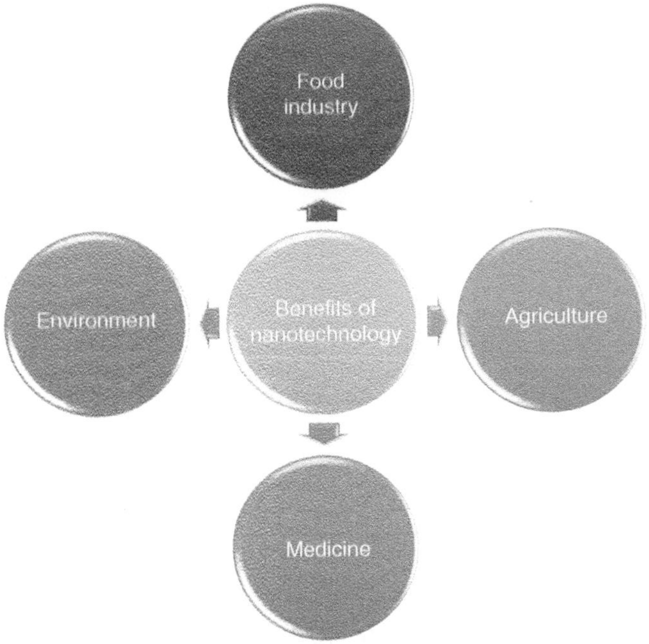

Figure 12.3 Benefits of nanotechnology in various fields.

their bulk counterparts (Biswas et al., 2022). These attributes make them invaluable for developing innovative coating materials aimed at preserving agricultural products. For instance, nanoparticles are incorporated into edible coatings to extend the freshness of fruits and vegetables by preventing microbial growth, reducing the effects of harmful radiation, and improving mechanical and functional properties (Yadollahi et al., 2010; Pudake et al., 2019). Studies have shown that nanocoatings effectively enhance food safety during storage by acting as antimicrobial agents and maintaining organoleptic and nutritional qualities (Arroyo et al., 2020; Zambrano-Zaragoza et al., 2020; Suhag et al., 2022).

12.6.2 Agricultural Advancements

Nanotechnology has significantly transformed modern agriculture through the development of nanofertilizers, designed to enhance nutrient use efficiency and overcome the limitations of traditional fertilization methods. Nanofertilizers such as nano-zinc have demonstrated increased nutrient bioavailability, resulting in improved crop growth, yield, and nutritional quality. For example, the application of nano-zinc fertilizers has been shown to enhance zinc efficiency and biofortification in wheat, contributing to better plant health and nutritional value (Munir et al., 2024). Moreover, combining nano-zinc with plant growth-promoting bacteria has proven effective in enhancing the biochemical and metabolic attributes of crops like maize, especially in challenging environments like the tropical Cerrado. These advancements underscore the potential of nanofertilizers to revolutionize agriculture by providing targeted, efficient nutrient delivery systems and promoting sustainable farming practices.

12.6.3 Environmental Protection

Nanoparticles play a pivotal role in environmental protection through innovative technologies aimed at pollution remediation, prevention, and natural process enhancement. One of the most promising applications involves pollution remediation, particularly using iron nanoparticles for degrading harmful contaminants like chlorinated hydrocarbons, heavy metals, and pesticides. The small size and high surface area of nanoparticles allow them to be transported with groundwater flow, facilitating in situ remediation of pollutants without excavation. This process mainly involves oxidation-reduction reactions that transform hazardous substances into less toxic forms (Masciangioli & Zhang, 2003; Zhang et al. 1998; Ponder et al., 2000). Nanotechnology also enhances bioremediation efforts, with magnetite nanoparticles coated on bacteria improving desulfurization activity and enabling easier separation and reuse of bacterial cells, thus enhancing efficiency and cost-effectiveness (Li et al., 2009). Furthermore, nanoparticles like titanium dioxide (TiO_2) and zinc oxide (ZnO) are effective for photoremediation, where photocatalytic degradation of organic contaminants occurs under ultraviolet or visible light (Kamat & Meisel, 2003).

12.6.4 Medicine

In veterinary medicine, nanotechnology offers groundbreaking advancements in disease diagnosis, prevention, and treatment. The unique properties of nanoparticles, including high surface area, reactivity, and enhanced solubility, contribute to their efficacy in various therapeutic applications. Nanoparticles have shown particular promise in improving drug delivery systems by increasing bioavailability and providing controlled release mechanisms. For example, gold nanorods have been successfully applied in PTT for treating tumors in cats, providing a noninvasive alternative to conventional treatments (Abdoon et al., 2015). Similarly, silver nanoparticles have demonstrated potent antimicrobial activity against vancomycin-resistant *S. aureus* in bovine mastitis, highlighting their potential in combating AMR (Algharib et al., 2020).

Furthermore, nanotechnology plays a vital role in vaccine development through the use of nanovaccines incorporating VLPs, liposomes, and polymeric nanoparticles. These nanovaccines enhance immunogenicity and provide sustained antigen release, proving effective in inducing protective immune responses against various animal pathogens, including viral infections like foot-and-mouth disease and bacterial pathogens such as Salmonella (Celis-Giraldo et al., 2021). Additionally, nanoparticles contribute to nutritional supplementation and growth improvement in livestock. Nano-Se has demonstrated excellent antioxidant properties, increasing selenium concentrations in milk from dairy cows and enhancing rumen fermentation in sheep (Han et al., 2020; Shi et al., 2011). Similarly, curcumin nanoparticles have been shown to promote gut health, immunity, and growth performance in poultry (Reda et al., 2020).

Nanoparticles are also employed to enhance the efficiency of conventional antibiotics, reducing dosage requirements and minimizing resistance development. Metal nanoparticles such as zinc oxide, silver, copper, and titanium dioxide have demonstrated efficacy against various pathogens, providing alternative antimicrobial strategies (Smekalova et al., 2016). Nanocoatings incorporated into disinfectants, wound dressings, and animal enclosures further improve biosecurity by reducing bacterial contamination and enhancing hygiene (El-Sayed & Kamel, 2020). However, despite their numerous benefits, nanoparticles pose safety concerns due to their high reactivity and small size, which can disrupt biological systems, cause oxidative stress, and potentially result in cytotoxicity (Bogdanchikova et al., 2016). Moreover, their accumulation in tissues and potential environmental impact require thorough evaluation to ensure safe application in veterinary medicine.

12.7 Case Studies and Real-World Applications

12.7.1 Antihyperglycemic Activity of Green Silver Nanoparticles in Diabetic Rats

Virgen-Ortiz et al. (2015) conducted a comprehensive study to evaluate the antihyperglycemic properties of biocompatible silver nanoparticles synthesized through green chemistry using *Rumex hymenosepalus* extracts. The experiment was carried out on male Wistar rats, in which diabetes was experimentally induced via streptozotocin (STZ) injection to simulate hyperglycemia. The green silver nanoparticles (GSNPs) produced through this eco-friendly approach had an average diameter of 9 nm and were effectively stabilized by polyphenols present in the extract. The concentration of the synthesized nanoparticles was determined to be 1.07 mg/mL, ensuring consistency in the administered dosage. The experimental design consisted of two groups of rats: the first group included diabetic rats induced with STZ to establish hyperglycemia, while the second group comprised diabetic rats treated with GSNPs. The treated rats received a low dose of GSNPs (150 μg/kg body weight) through intraperitoneal injection over a period of nine days. The results were promising, revealing a significant reduction in fasting blood glucose levels by approximately 50%, from an initial hyperglycemic state of 315.2 ± 36.1 to 159 ± 15 mg/dL. Moreover, the glucose tolerance test demonstrated an enhanced capacity for glucose regulation in GSNP-treated diabetic rats, suggesting improved metabolic control. The antihyperglycemic activity observed in the treated rats was primarily attributed to increased glucokinase activity and elevated GLUT-2 expression in hepatic tissues, indicating a potential mechanism through which GSNPs exert their therapeutic effects. This study provides valuable insights into the potential of green silver nanoparticles synthesized from *R. hymenosepalus* extracts as effective therapeutic agents for managing diabetes. The significant reduction in blood glucose levels achieved with low doses of GSNPs underscores their potential applicability in veterinary medicine for treating diabetic conditions. Furthermore, the use of green synthesis methods enhances the biocompatibility and safety profile of the nanoparticles, making them a promising candidate for future clinical applications.

12.7.2 Nanoparticles for Treatment of Bovine *Staphylococcus aureus* Mastitis

Algharib et al. (2020) conducted an in-depth review exploring the application of various nanoparticles to enhance the treatment of *S. aureus* mastitis in dairy cattle. *S. aureus* is recognized as a major pathogen responsible for causing severe bovine mastitis, a condition that leads to significant economic losses in the dairy industry. Unfortunately, traditional antibiotic therapies often prove ineffective due to challenges such as AMR, intracellular persistence, and biofilm formation, all of which contribute to the pathogen's resilience and persistence. The review highlighted several types of nanoparticles, including liposomes, polymeric nanoparticles (such as chitosan and PLGA), SLNs, nanogels, and metal nanoparticles, particularly silver nanoparticles. These nanoparticles employ a variety of mechanisms to combat *S. aureus* infections. Among their beneficial effects are enhanced intracellular drug delivery, disruption of bacterial cell membranes, inhibition of biofilm formation, and increased antibacterial activity against drug-resistant strains of *S. aureus*. Notably, silver nanoparticles have demonstrated significant efficacy against VRSA by enhancing bacterial cell death when used in combination with vancomycin. This synergy between silver nanoparticles and vancomycin offers a promising strategy for overcoming resistance issues. Additionally, liposome-based systems loaded with ciprofloxacin exhibited a high loading efficiency of approximately 45% and effectively inhibited biofilm formation, a critical factor in the persistence of *S. aureus* infections. CNPs were found to exhibit enhanced antibacterial efficacy against biofilms and intracellular infections, highlighting their potential use in tackling hard-to-treat infections. Similarly, SLNs demonstrated improved antibacterial effects when combined with vancomycin, with particle sizes ranging from 95 to 100 nm. Enrofloxacin-loaded SLNs further enhanced treatment outcomes by increasing intracellular accumulation by 70–150 times compared to free drugs, providing a more efficient delivery system for combating intracellular pathogens.

Moreover, copper-based nanogels were also successfully employed for the treatment of clinical mastitis, showcasing the versatility of nanoparticle applications in veterinary medicine. Collectively, these findings underscore the potential of nanoparticles as a powerful tool for treating *S. aureus* mastitis in dairy cattle. By addressing key challenges such as drug resistance and enhancing the efficacy of existing antibiotics, nanoparticles offer a promising and innovative approach to improving animal health and productivity.

12.7.3 Silver Nanoparticles for Treatment of Canine Distemper

Bogdanchikova et al. (2016) conducted a groundbreaking study to evaluate the therapeutic potential of silver nanoparticles (AgNPs) in treating canine distemper, a highly lethal viral disease caused by the canine distemper virus (CDV). The study aimed to assess the efficacy of AgNPs in managing both non-neurological and neurological cases of CDV, acknowledging that disease stage significantly influences treatment outcomes. The AgNPs used in the study were composed of 6% metallic silver and 94% polyvinylpyrrolidone (PVP) solution, with particle sizes ranging from 0.5 to 100 nm and a concentration of 1.8 mg/mL. Treatment protocols were tailored according to disease severity: non-neurological cases were administered a dosage of 1 mL/kg daily for 7–15 days, while neurological cases received a higher dosage of 1 mL/kg twice daily for 25–28 days. The results demonstrated exceptional success in treating non-neurological cases, achieving a remarkable 90% recovery rate, with complete recovery observed within 10–15 days. These findings suggest that AgNPs possess considerable therapeutic potential for early stage CDV infections. However, treating neurological cases proved to be significantly more challenging. Only 6% of the dogs in the neurological group recovered, with just 1 of 15 dogs showing complete recovery without long-term sequelae. In contrast, the untreated control group exhibited a 100% mortality rate, emphasizing the severity of the disease and the urgent need for effective therapeutic interventions. While the study highlights the promising efficacy of AgNPs in treating non-neurological canine distemper, it also underscores the need for further research to improve therapeutic outcomes in advanced neurological cases. These findings provide a critical foundation for future studies aimed at optimizing nanoparticle-based therapies for combating CDV and enhancing clinical recovery rates in affected animals.

12.7.4 Photothermal Therapy Using Gold Nanorods for Treatment of Mammary Gland Tumor in a Cat

Abdoon et al. (2015) investigated the use of PTT employing gold nanorods (GNRs) for the treatment of mammary gland adenocarcinoma Grade II in a seven-year-old Shirazi cat. The treatment protocol involved intratumoral injection of GNRs at a dosage of 75 μg/kg body weight, followed by exposure to an 808 nm laser for 10 minutes. This procedure was administered twice, with a 15-day interval between treatments. The results were promising, with 60% tumor ablation observed 15 days after the initial injection. Complete tumor remission was achieved 31 days after the first injection, as confirmed by ultrasound scanning, which demonstrated the total ablation of the tumor. Importantly, blood tests and liver and kidney function analyses indicated no signs of toxicity, suggesting that the treatment was well-tolerated. An exceptional outcome of this study was the restoration of reproductive function in the treated cat. Remarkably, the cat became pregnant within two months following tumor remission and successfully delivered three healthy kittens. This case provides compelling evidence of the potential of gold nanorods as a noninvasive, effective cancer treatment that preserves reproductive functions. Summary of case studies is given in Figure 12.4.

12.7.5 Nanovaccine for Cystic Echinococcosis in Dogs

Silvarrey et al. (2024) developed an innovative oral nanovaccine targeting *Echinococcus granulosus*, utilizing polymeric nanoparticles formulated with PLGA and Eudragit-RSPO to encapsulate EgFABP1, a fatty acid-binding protein derived from the parasite. The primary objective of this approach is to interrupt the transmission cycle of *E. granulosus* by immunizing dogs, which serve as the definitive hosts. The nanoparticles demonstrated a high antigen encapsulation efficiency exceeding 50%, with particle sizes averaging 208.22 ± 2.38 nm, making them suitable for efficient M-cell internalization. Furthermore, the nanoparticles exhibited a positive zeta potential of 32.75 mV, enhancing their stability and cellular uptake. Crucially, the integrity of the antigen remained stable when exposed to artificial gastric and intestinal fluids, indicating that the formulation could effectively protect the antigen during oral administration. While these findings are promising, further in vivo studies are necessary to evaluate the immunogenicity and protective efficacy of this nanovaccine. Nevertheless, this innovative approach presents a valuable strategy for reducing the transmission of *E. granulosus* to humans by targeting dogs through oral immunization.

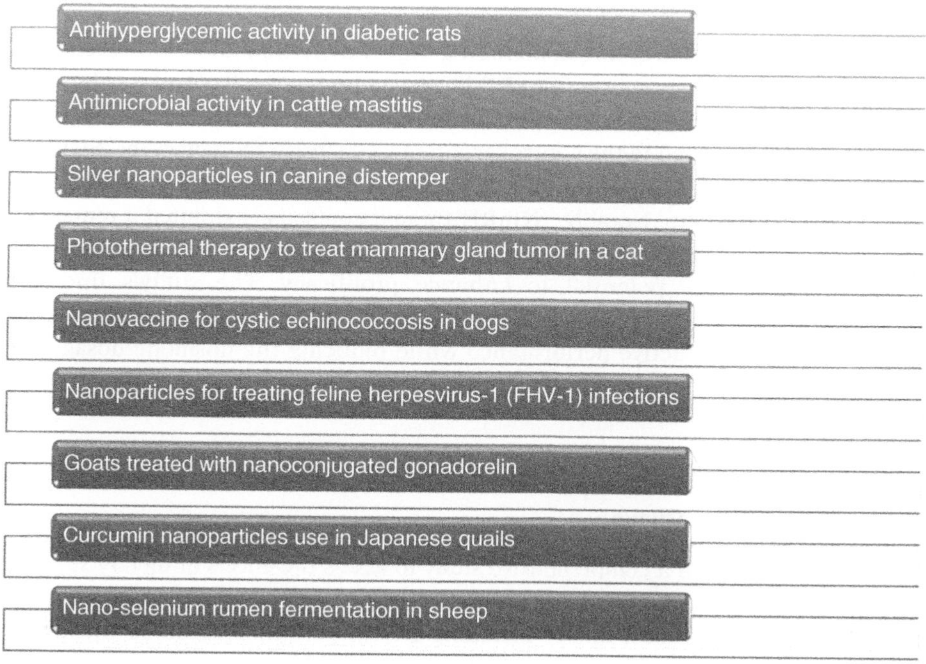

Figure 12.4 Case studies and real-world applications.

12.7.6 LTAC Nanoparticles for Treating Feline Herpesvirus-1 Infections

Lappin et al. (2023) conducted an insightful study examining the therapeutic potential of immune-stimulatory LTAC (liposome-dual TLR agonist complexes) nanoparticles for managing Feline Herpesvirus-1 (FHV-1) surface ocular infections in cats. The research included both naturally infected cats and experimental infection models to comprehensively evaluate the efficacy of this novel treatment. The LTAC nanoparticles demonstrated a robust antiviral response by inducing type I interferon production, effectively suppressing FHV-1 replication in vitro. This immune-stimulatory mechanism significantly reduced viral replication, suggesting a promising strategy for controlling FHV-1 infections.

12.7.7 Reproductive Performance of Goats Treated with Nanoconjugated Gonadorelin

Hashem and Sallam (2020) conducted a detailed investigation into the reproductive performance of goats treated with gonadorelin conjugated with chitosan-TPP nanoparticles (NGnRH), aiming to enhance bioavailability and efficacy while minimizing hormone dosage. The study involved 33 multiparous Zaraiebi goats divided into three groups: control (placebo), GnRH (50 μg/mL gonadorelin), and NGnRH (12.5 μg/mL chitosan-TPP conjugated gonadorelin). The NGnRH formulation demonstrated impressive properties, with an entrapment efficiency of 91.2%, particle size of 93.91 nm, and a zeta potential of 11.6 mV, suggesting excellent stability and potential for effective delivery.

Results revealed that NGnRH-treated goats exhibited significantly enhanced reproductive performance compared to controls, including higher ovulation rates, improved prolificacy, and elevated serum progesterone (P4) levels. Although the P4 level was highest in the GnRH group (12.38 ng/mL), it was also considerably elevated in the NGnRH group (8.10 ng/mL) compared to the control group (4.78 ng/mL). Moreover, the NGnRH treatment achieved an 81.81% conception rate by day 45, slightly lower than the 90.90% observed in the GnRH group. Notably, NGnRH-treated goats also exhibited a remarkable increase in the twinning rate (88.88%) compared to the control group (42.85%), highlighting the formulation's potential to enhance prolificacy. The study's findings underscore the advantages of NGnRH, demonstrating that the nanoformulation can effectively enhance reproductive performance while reducing the hormone dosage by 75% compared to conventional methods.

12.7.8 Biological Curcumin Nanoparticles for Growth and Health Improvement in Japanese Quails

Reda et al. (2020) conducted a comprehensive study to assess the effects of dietary supplementation with biological curcumin nanoparticles (CNPs) on various physiological and performance parameters in Japanese quails, including growth performance, blood chemistry, immunity, antioxidant status, and cecal microbiota. The experiment involved 270 quails divided into six groups, each receiving different concentrations of nanocurcumin (0.1, 0.2, 0.3, 0.4, and 0.5 g/kg diet) over a five-week period. The results demonstrated a significant enhancement in body weight gain, feed conversion ratio (FCR), and immune function, with the most favorable FCR (2.86) observed in the group supplemented with 0.4 g/kg of nanocurcumin. Additionally, dietary inclusion of nanocurcumin positively influenced serum biochemistry, as evidenced by increased levels of total protein, albumin, globulin, and immunoglobulins (IgG and IgM), particularly in the group receiving 0.2 g/kg of nanocurcumin.

12.7.9 Elemental Nano-Selenium for Feed Digestibility and Rumen Fermentation in Sheep

Shi et al. (2011) conducted a meticulous investigation to evaluate the effects of elemental Nano-Se supplementation on feed digestibility, rumen fermentation, and urinary purine derivatives in sheep. The study utilized eight male ruminally cannulated sheep arranged in a replicated 4×4 Latin square design, receiving Nano-Se at varying doses of 0.3, 3.0, and 6.0 g/kg dry matter (DM), while the control group received no supplementation. The findings revealed that Nano-Se supplementation significantly enhanced nutrient digestibility, rumen fermentation, and microbial protein synthesis, with the most favorable outcomes observed at a dosage of 3.0 g/kg DM. Specifically, Nano-Se administration increased total VFAs from 73.63 mM in the control group to 77.72 mM in the Nano-Se (NS)-med group, indicating improved fermentation efficiency. Additionally, propionate production was elevated, while ammonia nitrogen concentration was reduced, suggesting enhanced nitrogen utilization. The effective degradability of neutral detergent fiber and crude protein (CP) derived from soybean meal was also improved, reflecting better-feed breakdown and absorption. Furthermore, total purine derivatives (PD) excretion was significantly increased, indicating heightened microbial protein synthesis, which is essential for optimizing rumen microbial activity.

12.7.10 Nanoselenium Supplementation for Lactating Dairy Cows

Han et al. (2020) conducted a well-designed study to compare the effects of Nano-Se supplementation with traditional sodium selenite on lactation performance, milk selenium concentration, and selenoprotein (Sel) gene expression in lactating dairy cows. The experiment involved twelve multiparous Holstein cows randomly assigned to two groups: one receiving 0.30 mg Se/kg of dry matter (DM) from sodium selenite and the other supplemented with Nano-Se. The trial lasted 30 days, during which various physiological and molecular parameters were assessed to determine the efficacy of Nano-Se. The findings revealed that Nano-Se supplementation significantly enhanced milk selenium concentration, increasing it from 23.19 to 30.60 μg/L, while also boosting milk glutathione peroxidase (GSH-Px) activity. Furthermore, plasma selenium levels and GSH-Px activity were markedly higher in cows receiving Nano-Se compared to those supplemented with sodium selenite. Molecular analysis of the mammary gland demonstrated that Nano-Se supplementation substantially upregulated the expression of various selenoprotein genes, including GPX1, GPX2, GPX4, TXNRD2, TXNRD3, SelW, SelT, SelK, and SelF. These genes are critical for maintaining antioxidant defenses and overall cellular health.

12.8 Challenges and Considerations

While nanotechnology presents tremendous potential for advancing veterinary medicine, several critical challenges and considerations must be addressed to ensure its successful application. These challenges encompass safety concerns, standardization issues, cost, scalability, regulatory barriers, ethical considerations, and the necessity for comprehensive clinical studies. One of the most pressing challenges is the potential toxicity of nanoparticles. Although numerous studies have demonstrated their therapeutic benefits, concerns remain regarding their long-term safety and biocompatibility. Nanoparticles may induce oxidative stress, cytotoxicity, and inflammatory responses depending on their composition, size, and dosage. For instance, silver nanoparticles (AgNPs) have proven effective in treating canine distemper and *S. aureus* infections in cattle (Bogdanchikova et al., 2016; Algharib et al., 2020). However, the risk of nanoparticle accumulation in tissues and potential cytotoxic effects on nontarget cells necessitates further investigation. Likewise, Nano-Se supplementation has shown significant benefits in enhancing feed digestibility and antioxidant status in sheep and dairy cows (Shi et al., 2011; Han et al., 2020), but the potential risks associated with prolonged exposure remain unclear. Standardization and quality control also pose substantial challenges. Nanoparticle synthesis requires meticulous control over factors such as particle size, shape, surface charge, and coating materials, all of which critically influence biological activity, biodistribution, and toxicity. Inconsistencies in nanoparticle synthesis can lead to unreliable results, limiting reproducibility and hindering progress. For example, the biological activity of curcumin nanoparticles used in poultry studies may vary depending on formulation techniques and particle size (Reda et al., 2020). Establishing standardized protocols for nanoparticle production is essential to ensure consistency, reliability, and comparability across studies. Moreover, the high cost and scalability of nanoparticle production present additional obstacles. While laboratory-scale methods are effective for research purposes, transitioning to commercial-scale manufacturing demands substantial investment in infrastructure and resources. The use of gold nanorods for cancer therapy in a cat (Abdoon et al., 2015) exemplifies this challenge; despite their remarkable therapeutic efficacy, the high cost of gold nanomaterials limits their broader application. Therefore, developing cost-effective synthesis methods without compromising quality is critical for making nanotechnology-based treatments more accessible in veterinary medicine. Regulatory approval remains another significant barrier. The absence of well-defined regulatory frameworks for nanoparticle-based veterinary therapies hampers their commercialization. Although promising results have been reported, such as the use of gelatin nanoparticle-bound CpG-ODN for treating RAO in horses (Klier et al., 2015), there are currently insufficient guidelines for ensuring their safety and efficacy. Regulatory bodies must establish rigorous protocols for evaluating the quality, safety, and therapeutic efficacy of nanoparticle-based products before their approval for veterinary use. Ethical considerations also warrant careful attention. The application of nanoparticles in animals raises concerns related to animal welfare, environmental impact, and public perception. Ensuring that the benefits of nanoparticle-based therapies outweigh potential harm is essential for ethical justification. Additionally, the potential environmental impact of nanoparticles, particularly if excreted or unintentionally released, requires thorough assessment. Another crucial challenge is the need for comprehensive clinical studies. Although many nanoparticles have demonstrated efficacy in vitro or through small-scale animal models, large-scale clinical trials are essential to validate these findings and establish robust guidelines for their application. For instance, LTAC nanoparticles have shown promise in treating FHV-1 infections by reducing viral shedding and enhancing immune response (Lappin et al., 2023). Similarly, nanovaccines aimed at reducing the transmission of *E. granulosus* to humans by targeting dogs have shown potential (Silvarrey et al., 2024). However, further studies are needed to confirm their safety and efficacy under field conditions. Addressing these challenges requires a collaborative, multidisciplinary approach involving veterinary researchers, toxicologists, material scientists, regulatory agencies, and industry stakeholders. Developing standardized protocols, conducting long-term safety studies, and establishing clear regulatory guidelines will be essential to advance veterinary nanotechnology. Furthermore, improving public awareness and understanding of the benefits and risks associated with nanoparticle-based treatments is crucial for promoting their acceptance and successful integration into veterinary practices.

12.9 Conclusion

Nanotechnology has emerged as a revolutionary tool in veterinary medicine and dairy production, offering groundbreaking advancements in diagnostics, therapeutics, disease prevention, and animal welfare. The application of nanoparticles has proven effective in enhancing drug delivery, improving vaccine efficacy, combating AMR, and facilitating precise diagnostic tools.

Moreover, nano-based systems have shown remarkable potential in promoting feed utilization, enhancing animal growth, and improving overall health. However, despite these promising benefits, challenges such as safety concerns, standardization issues, cost-effectiveness, and regulatory barriers remain significant obstacles to their widespread application. The potential toxicity of nanoparticles, their accumulation in tissues, and their impact on nontarget cells highlight the need for thorough safety evaluations before clinical applications. Additionally, the lack of standardized protocols and quality control measures complicates the reproducibility and scalability of nanoparticle production. Addressing these challenges will require collaborative efforts among researchers, regulatory agencies, and industry stakeholders to establish clear guidelines for the safe and effective use of nanotechnology in veterinary medicine. Future research should focus on optimizing nanoparticle formulations, enhancing their biocompatibility, and ensuring their efficacy under real-world conditions. The development of cost-effective manufacturing methods and robust clinical trials will be crucial for advancing nanotechnology from experimental research to practical veterinary applications. Ultimately, the integration of nanotechnology in veterinary medicine holds great promise for improving animal health, welfare, and productivity while ensuring sustainable practices in the dairy and livestock industries.

References

Abdollahi, M., Rezaei, J., & Fazaeli, H. (2020). Performance, rumen fermentation, blood minerals, leukocyte and antioxidant capacity of young Holstein calves receiving high-surface ZnO instead of common ZnO. *Archives of Animal Nutrition*, *74*(3), 189–205.

Abdoon, A. S., Salama, A. A., & Taha, A. E. (2015). Photothermal therapy using gold nanorods for treatment of mammary gland tumor in a cat. *Photodiagnosis and Photodynamic Therapy*, *12*(2), 325–328.

Abed, N. N., Abou El-Enain, I. M. M., Helal, E. E. H., & Yosri, M. (2023). Novel biosynthesis of tellurium nanoparticles and investigation of their activity against common pathogenic bacteria. *Journal of Taibah University Medical Sciences*, *18*(2), 400–412. https://doi.org/10.1016/j.jtumed.2022.10.006

Acevedo-Villanueva, K., Renu, S., Gourapura, R., & Selvaraj, R. (2021). Efficacy of a nanoparticle vaccine administered in-ovo against Salmonella in broilers. *PLoS One*, *16*(4), e0247938. https://doi.org/10.1371/journal.pone.0247938

Ahn, H., Kang, S., Kang, K., Lee, D. N., Min, D. H., & Jang, H. (2023). Rod-to-sphere elemental reconstruction of biocompatible Ag(2)Te-Ag(4.53)-Te(3) nanoparticles for triple negative breast cancer photo-nanotherapy. *Journal of Materials Chemistry B*, *11*(23), 5142–5150.

Al-Halifa, S., Gauthier, L., Arpin, D., Bourgault, S., & Archambault, D. (2019). Nanoparticle-based vaccines against respiratory viruses. Frontiers in immunology, 10, 22–spiepr DHYP

Algharib, S. A., Dawood, A., & Xie, S. (2020). Nanoparticles for treatment of bovine *Staphylococcus aureus* mastitis. *Frontiers in Veterinary Science*, *7*, 373.

Ali, A., Ijaz, M., Khan, Y. R., Sajid, H. A., Hussain, K., Rabbani, A. H., & Ahmed, I. (2021). Role of nanotechnology in animal production and veterinary medicine. *Tropical Animal Health and Production*, *53*, 1–14.

Alivisatos, A. P. (1996). Semiconductor clusters, nanocrystals, and quantum dots. *Science, 271*(5251), 933–937.

Andrade, M. A., Rodrigues, L. F., Ierich, J. C., Melendez, M. E., Carvalho, A. L., de Carvalho, A. C., Steffens, C., Abdalla, F. C., Oliveira, O. N., & Leite, F. L. (2019). A nanomechanical genosensor using functionalized cantilevers to detect the cancer biomarkers miRNA-203 and miRNA-205. *IEEE Sensors Journal, 20*(6), 2860–2867.

Arroyo, B. J., Bezerra, A. C., Oliveira, L. L., Arroyo, S. J., de Melo, E. A., & Santos, A. M. P. (2020). Antimicrobial active edible coating of alginate and chitosan add ZnO nanoparticles applied in guavas (*Psidium guajava* L). *Food Chemistry, 309*, 125566.

Baghdadi, H. B., Abdelsalam, M., & Attia, M. M. (2024). Diagnostic innovations in equine parasitology: A Nanogold-ELISA for sensitive serodiagnosis of migratory *Strongylus vulgaris* larvae infections. *BMC Veterinary Research, 20*(1), 579.

Bai, D. P., Lin, X. Y., Huang, Y. F., & Zhang, X. F. (2018). Theranostics aspects of various nanoparticles in veterinary medicine. *International Journal of Molecular Sciences, 19*(11), 3299.

Bai, X., Smith, Z. L., Wang, Y., Butterworth, S., & Tirella, A. (2022). Sustained drug release from smart nanoparticles in cancer therapy: A comprehensive review. *Micromachines, 13*(1623). https://doi.org/10.3390/mi13101623

Bajwa, H. U. R., Khan, M. K., Abbas, Z., Riaz, R., Rehman, T. U., Abbas, R. Z., & Alouffi, A. (2022). Nanoparticles: Synthesis and their role as potential drug candidates for the treatment of parasitic diseases. *Life, 12*(5), 750.

Bilgili, A., & Uysal, M. H. (2019). Use areas of nanotechnology in veterinary medicine. *Journal of Food Science and Engineering, 9*, 201–216.

Biswas, R., Alam, M., Sarkar, A., Haque, M. I., Hasan, M. M., & Hoque, M. (2022). Application of nanotechnology in food: Processing, preservation, packaging and safety assessment. *Heliyon, 8*(11), e11795.

Bogdanchikova, N., Khanadeev, V., Kladko, D., & Balakin, K. (2016). Silver nanoparticles for treatment of canine distemper. *International Journal of Nanotechnology, 12*(9-11), 805–816.

Brandelli, A., Lopes, N. A., & Pinilla, C. M. B. (2023). Nanostructured antimicrobials for quality and safety improvement in dairy products. *Foods, 12*(13), 2549.

Brito, L. G., Rocha, R. B., Barbieri, F. D. S., Ribeiro, E. S., Vendrami, F. B., Souza, G. C., ... Oliveira, M. C. (2013). Babesia bovis infection in cattle in the southwestern Brazilian Amazon. *Ticks and Tick-borne Diseases, 4*(1-2), 78–82.

Carvalho, M. R., Reis, R. L., & Oliveira, J. M. (2020). Dendrimer nanoparticles for colorectal cancer applications. *Journal of Materials Chemistry B, 8*(6), 1128–1138.

Celis-Giraldo, C. T., López-Abán, J., Muro, A., Patarroyo, M. A., & Manzano-Román, R. (2021). Nanovaccines against animal pathogens: The latest findings. *Vaccine, 9*(9), 988. https://doi.org/10.3390/vaccines9090988

Chang, S.-M., Palanisamy, S., Wu, T.-H., Chen, C.-Y., Cheng, K.-H., Lee, C.-Y., Yuan, S.-S. F., & Wang, Y.-M. (2020). Utilization of silicon nanowire field-effect transistors for the detection of a cardiac biomarker, cardiac troponin I, and their applications involving animal models. *Scientific Reports, 10*(1), 22027. https://doi.org/10.1038/s41598-020-78829-7

Chang, R., Zhao, D., Zhang, C., Liu, K., He, Y., Guan, F., et al. (2023). Nanocomposite multifunctional hyaluronic acid hydrogel with photothermal antibacterial and antioxidant properties for infected wound healing. *International Journal of Biological Macromolecules, 226*, 870–884.

Chariou, P. L., Ortega-Rivera, O. A., & Steinmetz, N. F. (2020). Nanocarriers for the delivery of medical, veterinary, and agricultural active ingredients. *ACS Nano, 14*(3), 2678–2701.

Chavada, P. (2016). Novel application of nanotechnology in dairy and food industry: Nano inside. *International Journal of Agriculture Sciences, 2016*, 0975–3710.

Chowdhury, I. H., Lokugamage, N., & Garg, N. J. (2020). Experimental nanovaccine offers protection against repeat exposures to *Trypanosoma cruzi* through activation of polyfunctional T cell response. *Frontiers in Immunology, 11*, 595039. https://doi.org/10.3389/fimmu.2020.595039

Claeys, W. L., Cardoen, S., Daube, G., De Block, J., Dewettinck, K., Dierick, K., & Herman, L. (2013). Raw or heated cow milk consumption: Review of risks and benefits. *Food Control, 31*(1), 251–262.

Cox, F. E. (2009). *Modern parasitology: A textbook of parasitology*. John Wiley & Sons.

Dinc, S. K., Vural, O. A., Kayhan, F. E., & San Keskin, N. O. (2022). Facile biogenic selenium nanoparticle synthesis, characterization and effects on oxidative stress generated by UV in microalgae. *Particuology, 70*, 30–42. https://doi.org/10.1016/J.PARTIC.2021.12.005

Dong, Y., Zhang, K., Han, M., Miao, Z., Liu, C., & Li, J. (2022). Low level of dietary organic trace minerals improved egg quality and modulated the status of eggshell gland and intestinal microflora of laying hens during the late production stage. *Frontiers in Veterinary Science, 9*, 920418.

Dumlu, B. (2024). Importance of nano-sized feed additives in animal nutrition. *Journal of Agricultural Production, 5*(1), 55–72.

Dykman, L. A., & Khlebtsov, N. G. (2011). Gold nanoparticles in biology and medicine: Recent advances and prospects. *Acta Naturae, 3*(2), 34–55.

Elalfy, M. M., El-hadidy, M. G., & Abouelmagd, M. M. (2018). Update of usefulness and adverse effects of nanoparticles on animals and human health. *Journal of Veterinary Medicine and Health, 2*, 106.

El-Sayed, A., & Kamel, M. (2020). Advanced applications of nanotechnology in veterinary medicine. *Environmental Science and Pollution Research, 27*(16), 19073–19086.

Frippiat, T., Art, T., & Delguste, C. (2025). Silver nanoparticles as antimicrobial agents in veterinary medicine: Current applications and future perspectives. *Nanomaterials, 15*(3), 202.

Fusco, V., Chieffi, D., Fanelli, F., Logrieco, A. F., Cho, G. S., & Kabisch, J. (2020). & Franz, C. M. Microbial quality and safety of milk and milk products in the 21st century. Comprehensive Reviews in Food Science and Food Safety, 19(4), 2013–spiepr DHYP2049.

Gao, X., Cui, Y., Levenson, R. M., Chung, L. W. K., & Nie, S. (2004). In vivo cancer targeting and imaging with semiconductor quantum dots. *Nature Biotechnology, 22*(8), 969–976.

Gelaye, Y. (2024). Application of nanotechnology in animal nutrition: Bibliographic review. *Cogent Food & Agriculture, 10*(1), 2290308.

Gilbert, N. (2012). Rules tighten on use of antibiotics on farms: clampdown aims to stop spread of drug-resistant microbes. *Nature, 481*(7380), 125–126.

Girigoswami, A., Ghosh, M. M., Pallavi, P., Ramesh, S., & Girigoswami, K. (2021). Nanotechnology in detection of food toxins–focus on the dairy products. *Biointerface Research in Applied Chemistry, 11*(6), 14155–14172.

González, M. A. C., Gonçalves, A. A. M., Ottino, J., Leite, J. C., Resende, L. A., Melo-Júnior, O. A., Silveira, P., Cardoso, M. S., Fujiwara, R. T., Bueno, L. L., Santos, R. L., Carvalho, T. F., Garcia, G. M., Paes, P. R. O., Galdino, A. S., Chávez-Fumagalli, M. A., Melo, M. M., Silveira-Lemos, D., Martins-Filho, O. A., Dutra, W. O., ... Giunchetti, R. C. (2023). Vaccination with formulation of nanoparticles loaded with *Leishmania amazonensis* antigens confers protection against experimental visceral Leishmaniasis in hamster. *Vaccine*, *11*(1), 111. https://doi.org/10.3390/vaccines11010111

Grabrucker, A. M., Ruozi, B., Belletti, D., Pederzoli, F., Forni, F., Vandelli, M. A., & Tosi, G. (2016). Nanoparticle transport across the blood brain barrier. *Tissue Barriers*, *4*(1), e1153568.

Guruprasath, N., Sankarganesh, P., Adeyeye, S. A. O., Babu, A. S., & Parthasarathy, V. (2024). Review on emerging applications of nanobiosensor in food safety. *Journal of Food Science*.

Hajjafari, A., Sadr, S., Rahdar, A., Bayat, M., Lotfalizadeh, N., Dianaty, S., Rezaei, A., Moghaddam, S. P., Hajjafari, K., Simab, P. A., Kharaba, Z., Borji, H., & Pandey, S. (2024). Exploring the integration of nanotechnology in the development and application of biosensors for enhanced detection and monitoring of colorectal cancer. *Inorganic Chemistry Communications, 112409.*

Han, D., Zhang, C., & Zhang, Z. (2020). Nano-selenium supplementation for lactating dairy cows. *Biological Trace Element Research*, *196*(1), 80–86.

Hashem, A. H., Al Abboud, M. A., Alawlaqi, M. M., Abdelghany, T. M., & Hasanin, M. (2022). Synthesis of nanocapsules based on biosynthesized nickel nanoparticles and potato starch: Antimicrobial, antioxidant, and anticancer activity. *Starch-Starke*, *74*, 2100165. https://doi.org/10.1002/STAR.20210.165

Hashem, N. M., & Sallam, S. M. (2020). Reproductive performance of goats treated with nanoconjugated gonadorelin. *Animal Reproduction Science*, *218*, 106482. https://doi.org/10.1016/j.anireprosci.2020.106482

Hassan, A. A., Mansour, M. K., El Hamaky, A. M., El Ahl, R. M. S., & Oraby, N. H. (2020). Nanomaterials and nanocomposite applications in veterinary medicine. In *Multifunctional hybrid nanomaterials for sustainable agri-food and ecosystems* (pp. 583–638). Elsevier.

Hwangbo, M., Claycomb, E. C., Liu, Y., Alivio, T. E. G., Banerjee, S., & Chu, K. H. (2019). Effectiveness of zinc oxide-assisted photocatalysis for concerned constituents in reclaimed wastewater: 1,4-Dioxane, trihalomethanes, antibiotics, antibiotic resistant bacteria (ARB), and antibiotic resistance genes (ARGs). *Science of the Total Environment*, *649*, 1189–1197. https://doi.org/10.1016/j.scitotenv.2018.08.360

Jahanmahin, A., & Borji, H. (2023). Nanotechnology-based approaches for the treatment of toxocariasis: A prospective review. *Journal of Veterinary Physiology and Pathology*, *2*(2), 12–19.

Jiménez-Lamana, J., Laborda, F., Bolea, E., Abad-Álvaro, I., Castillo, J. R., Bianga, J., He, M., Bierla, K., Mounicou, S., Ouerdane, L., Gaillet, S., Rouanet, J. M., & Szpunar, J. (2014). An insight into silver nanoparticles bioavailability in rats. *Metallomics*, *6*(12), 2242–2249. https://doi.org/10.1039/c4mt00200h

Kamat, P. V., & Meisel, D. (2003). Nanoscience opportunities in environmental remediation. *Comptes Rendus. Chimie*, *6*(8–10), 999–1007.

Kareem, H. A., Samaka, H. M., & Abdulridha, W. M. (2021). Evaluation of the effect of the gold nanoparticles prepared by green chemistry on the treatment of cutaneous candidiasis. *Current Medical Mycology*, *7*(1), 1–5. https://doi.org/10.18502/cmm.7.1.61.6

Karimi, M., Mirshekari, H., Aliakbari, M., Sahandi-Zangabad, P., & Hamblin, M. R. (2016). Smart mesoporous silica nanoparticles for controlled-release drug delivery. *Nanotechnology Reviews, 5*(2), 195–207. https://doi.org/10.1515/ntrev-2015-0057

Klier, J., Lehmann, B., Failing, K., & Reinhold, P. (2015). Nanoparticulate CpG immunotherapy in RAO-affected horses: Phase I and IIa clinical trials. *Journal of Veterinary Internal Medicine, 29*(1), 320–328.

Krause, W., Hackmann-Schlichter, N., Maier, F. K., & Müller, R. (2000). Dendrimers in diagnostics. *Topics in Current Chemistry, 210*, 262–304.

Lappin, M. R., Blondeau, J., & Jensen, W. A. (2023). LTAC nanoparticles for treating feline herpesvirus-1 infections. *Frontiers in Veterinary Science, 10*, 741028. https://doi.org/10.3389/fvets.2023.741028

Lasic, D. D. (1998). Novel applications of liposomes. *Trends in Biotechnology, 16*(7), 307–321.

Li, N., Ma, L., Wang, J., Zheng, L., Liu, J., Duan, Y., ... Xie, Y. (2009). Interaction between nano-anatase TiO2 and liver DNA from mice in vivo. *Nanoscale Research Letters, 5*(1), 108.

Liu, Z., Guo, K., Yan, L., Zhang, K., Wang, Y., Ding, X., Zhao, N., & Xu, F.-J. (2023). Janus nanoparticles targeting extracellular polymeric substance achieve flexible elimination of drug-resistant biofilms. *Nature Communications, 14*(1), 5132. https://doi.org/10.1038/s41467-023-40830-9

Lopez-Miranda, J. L., Molina, G. A., González-Reyna, M. A., España-Sánchez, B. L., Esparza, R., Silva, R., & Estévez, M. (2023). Antibacterial and anti-inflammatory properties of ZnO nanoparticles synthesized by a green method using sargassum extracts. *International Journal of Molecular Sciences, 24*(2), 1474.

Majumdar, A. (2002). Bioassays based on molecular nanomechanics. *Disease Markers, 18*(2), 167–174.

Mansour, H., Aboamer, A. A., Agamy, R., Ali, S. M., Mohamed, M. Y., Abd El-Aziz, M. E., & Farag, I. M. (2024). Effect of zinc oxide and selenium nanoparticles on milk production efficiency and related gene expression in Egyptian Baladi Goats. *Egyptian Journal of Chemistry.*

Martina, M. S., Fortin, J. P., Menager, C., Clays, K., & Laurent, S. (2005). Generation of superparamagnetic liposomes as efficient MRI contrast agents. *Journal of the American Chemical Society, 127*(30), 10676–10680.

Marunganathan, V., Kumar, M. S. K., Kari, Z. A., Giri, J., Shaik, M. R., Shaik, B., & Guru, A. (2024). Marine-derived κ-carrageenan-coated zinc oxide nanoparticles for targeted drug delivery and apoptosis induction in oral cancer. *Molecular Biology Reports, 51*(89). https://doi.org/10.1007/s11033-023-09146-1

Masciangioli, T., & Zhang, W. X. (2003). Peer reviewed: Environmental technologies at the nanoscale.

Meena, N., Sahni, Y., Thakur, D., & Singh, R. (2018). Applications of nanotechnology in veterinary therapeutics. *Journal of Entomology and Zoology, 6*(2), 167–175.

Monge-Fuentes, V., Mayer, A. B., Lima, M. R., Geraldes, L. R., Zanotto, L. N., Moreira, K. G., Martins, O. P., Piva, H. L., Felipe, M. S., Amaral, A. C., Bocca, A. L., Tedesco, A. C., & Mortari, M. R. (2021). Dopamine-loaded nanoparticle systems circumvent the blood–brain barrier restoring motor function in a mouse model for Parkinson's disease. *Scientific Reports, 11*(1), 15185. https://doi.org/10.1038/s41598-021-94175-8

Munir, A., Salah, M. A., Ali, M., Ali, B., Saleem, M. H., Samarasinghe, K. G. B. A., & Anas, M. (2024). Advancing agriculture: Harnessing smart nanoparticles for precision fertilization. *BioNanoScience, 14*(4), 3846–3863.

Muraleedharan, K., & Chhabra, M. (2018). Nanotechnology applications and potential in parasitology: An overview. *Vet. Immunol. Biotechnol, 1*, 25–40.

Muthu, M. S., & Feng, S. S. (2013). Theranostic liposomes for cancer diagnosis and treatment: Current development and pre-clinical success. *Expert Opinion on Drug Delivery, 10*(2), 151–155.

Mutro Nigo, M., Salieb-Beugelaar, G. B., Battegay, M., Odermatt, P., & Hunziker, P. (2020). Schistosomiasis: From established diagnostic assays to emerging micro/nanotechnology-based rapid field testing for clinical management and epidemiology. *Precision Nanomedicine, 3*(1), 439–458.

Osama, E., El-Sheikh, S. M., Khairy, M. H., & Galal, A. A. (2020). Nanoparticles and their potential applications in veterinary medicine. *Journal of Advanced Veterinary Research, 10*(4), 268–273.

Panghal, A., & Flora, S. J. S. (2024). Nanotechnology in the diagnostic and therapy for Alzheimer's disease. *Biochimica et Biophysica Acta (BBA) - General Subjects, 1868*(3), 130559.

Pati, R., Shevtsov, M., & Sonawane, A. (2018). Nanoparticle vaccines against infectious diseases. *Frontiers in Immunology, 9*, 2224.

Pei, C., Dong, H., Teng, Z., Wei, S., Zhang, Y., Yin, S., Tang, J., Sun, S., & Guo, H. (2023). Self-assembling nanovaccine fused with flagellin enhances protective effect against foot-and-mouth disease virus. *Vaccine, 11*(1675). https://doi.org/10.3390/vaccines11111675

Poonia, A. (2019). Recent trends in nanomaterials used in dairy industry. *Nanoscience for Sustainable Agriculture*, 375–396.

Ponder, S. M., Darab, J. G., & Mallouk, T. E. (2000). Remediation of Cr (VI) and Pb (II) aqueous solutions using supported, nanoscale zero-valent iron. Environmental science & technology, 34(12), 2564–spiepr DHYP2569.

Pouri, S., Motamedi, H., Honary, S., & Kazeminezhad, I. (2017). Biological synthesis of selenium nanoparticles and evaluation of their bioavailability. *Brazilian Archives of Biology and Technology, 60*, e170452. https://doi.org/10.1590/1678-4324-2017160452

Pudake, R. N., Chauhan, N., & Kole, C. (2019). *Nanoscience for sustainable agriculture*. Springer eBooks. https://doi.org/10.1007/978-3-319-97852-9

Qadeer, A., Ullah, H., Sohail, M., Safi, S. Z., Rahim, A., Saleh, T. A., & Horky, P. (2022). Potential application of nanotechnology in the treatment, diagnosis, and prevention of schistosomiasis. *Frontiers in Bioengineering and Biotechnology, 10*, 1013354.

Qamar, H., Chauhan, D. K., Owais, M., & Hussain, T. (2021). Protective efficacy of novel nanoparticle-based vaccine against Newcastle disease. *Biochemical and Cellular Archives, 21*(2), 5173–5181. https://connectjournals.com/03896.2021.21.5173

Rai, B. M. (2012). Silver nanoparticles as nano-antimicrobials: Bioactivity, benefits and bottlenecks. In M. Rai, A. Yadav, & N. Cioffi (Eds.), *Nano-antimicrobials: Progress and prospects* (pp. 211). Springer.

Reda, F. M., El-Saadony, M. T., Elnesr, S. S., Alagawany, M., & Tufarelli, V. (2020). Biological curcumin nanoparticles for growth and health improvement in Japanese quails. *Animals, 10*(6), 754.

Sadr, S., Lotfalizadeh, N., Abbasi, A. M., Soleymani, N., Hajjafari, A., Roohbaksh Amooli Moghadam, E., & & Borji, H. (2023). Challenges and prospective of enhancing hydatid cyst chemotherapy by nanotechnology and the future of nanobiosensors for diagnosis. *Tropical Medicine and Infectious Disease*, *8*(11), 494.

Safarpoor, M., Ghaedi, M., Asfaram, A., Yousefi-Nejad, M., Javadian, H., Zare Khafri, H., & Bagherinasab, M. (2018). Ultrasound-assisted extraction of antimicrobial compounds from *Thymus daenensis* and *Silybum marianum*: Antimicrobial activity with and without the presence of natural silver nanoparticles. *Ultrasonics Sonochemistry*, *42*, 76–83.

Şenel, S., & McClure, S. J. (2004). Potential applications of chitosan in veterinary medicine. *Advanced Drug Delivery Reviews*, *56*(10), 1467–1480.

Shi, L., Xun, W., Yue, W., Zhang, C., Ren, Y., & Shi, L. (2011). Elemental nano-selenium for feed digestibility and rumen fermentation in sheep. *Animal Feed Science and Technology*, *163*(1-2), 136–142.

Shivashankarappa, A., & Sanjay, K. R. (2019). Photodynamic therapy on skin melanoma and epidermoid carcinoma cells using conjugated 5- aminolevulinic acid with microbial synthesised silver nanoparticles. *Journal of Drug Targeting*, *27*(4), 434–441.

Silvarrey, C., Fernández, S., Díaz, A., & Gutiérrez, J. (2024). Nanovaccine for cystic echinococcosis in dogs. *Veterinary Parasitology*, *309*, 109820.

Smekalova, M., Aragon, V., Panacek, A., Prucek, R., Zboril, R., & Kvitek, L. (2016). Enhanced antibacterial effect of antibiotics in combination with silver nanoparticles against animal pathogens. *The Veterinary Journal*, *209*, 174–179.

Sonea, C., Lupusoru, M., & Opris, I. (2021). Augmentation of Nutrition by Nanotechnology. *Modern Approaches to Augmentation of Brain Function*, 415–431.

Sotnikov, D. V., Barshevskaya, L. V., Zherdev, A. V., Eskendirova, S. Z., Mukanov, K. K., Mukantayev, K. K., Ramankulov, Y. M., & Dzantiev, B. B. (2020). Immunochromatographic system for serodiagnostics of cattle brucellosis using gold nanoparticles and signal amplification with quantum dots. *Applied Sciences*, *10*(3), 738. https://doi.org/10.3390/app10.30738

Suhag, R., Kumar, R., Dhiman, A., Sharma, A., Prabhakar, P. K., Gopalakrishnan, K., Kumar, R., & Singh, A. (2022). Fruit peel bioactives, valorisation into nanoparticles and potential applications: A review. *Critical Reviews in Food Science and Nutrition*, *63*(24), 6757–6776. https://doi.org/10.1080/10408398.2022.2043237

Svenson, S., & Tomalia, D. A. (2005). Dendrimers in biomedical applications—Reflections on the field. *Advanced Drug Delivery Reviews*, *57*(15), 2106–2129.

Talegaonkar, S., & Bhattacharyya, A. (2019). Potential of lipid nanoparticles (SLNs and NLCs) in enhancing oral bioavailability of drugs with poor intestinal permeability. *AAPS PharmSciTech*, *20*(121). https://doi.org/10.1208/s12249-019-1337-8

Tiwari, R., Gupta, R. P., Singh, V. K., Kumar, A., Rajneesh, M., & P., Sundar, S., Gautam, V., & Kumar, R. (2023). Nanotechnology-based strategies in parasitic disease management: from prevention to diagnosis and treatment. *ACS Omega*, *8*(45), 42014–42027. https://doi.org/10.1021/acsomega.3c04587

Tomalia, D. A., Reyna, L. A., & Svenson, S. (2007). Dendrimers as multi-purpose nanodevices for oncology drug delivery and diagnostic imaging. *Biochemical Society Transactions*, *35*(1), 61–67.

Verma, S., & Dutta, R. K. (2017). Enhanced ROS generation by ZnO-ammonia modified graphene oxide nanocomposites for photocatalytic degradation of trypan blue dye and 4-nitrophenol. *Journal of Environmental Chemical Engineering, 5*, 4776–4787. https://doi.org/10.1016/J.JECE.2017.08.026

Virgen-Ortiz, J. J., Sánchez-Mendoza, M. E., & Rincón-Sánchez, A. R. (2015). Antihyperglycemic activity of green silver nanoparticles in diabetic rats. *Journal of Nanomedicine & Nanotechnology, 6*(2), 1–8. https://doi.org/10.4172/2157-7439.1000272

Wang, K., Lu, X., Lu, Y., Wang, J., Lu, Q., Cao, X., & Yang, Z. (2022). Nanomaterials in animal husbandry: research and prospects. *Frontiers in Genetics, 13*, 915911.

Wang, Y., Pi, C., Feng, X., Hou, Y., Zhao, L., & Wei, Y. (2020). The influence of nanoparticle properties on oral bioavailability of drugs. *International Journal of Nanomedicine, 15*, 6295–6310. https://doi.org/10.2147/IJN.S257269

Wu, Y., Li, X., Fu, X., Huang, X., Zhang, S., Zhao, N., Ma, X., Saiding, Q., Yang, M., Tao, W., Zhou, X., & Huang, J. (2024). Innovative nanotechnology in drug delivery systems for advanced treatment of posterior segment ocular diseases. *Advanced Science, 11*(32), 2403399.

Xuan, Y., Zhang, W., Zhu, X., & Zhang, S. (2023). An updated overview of some factors that influence the biological effects of nanoparticles. *Frontiers in Bioengineering and Biotechnology, 11*, 1254861.

Yadollahi, A., Arzani, K., & Khoshghalb, H. (2010). The role of nanotechnology in horticultural crops postharvest management. *Acta Horticulturae, 875*, 49–56. https://doi.org/10.17660/actahortic.2010.875.4

Youssef, F. S., El-Banna, H. A., Elzorba, H. Y., & Galal, A. M. (2019). Application of some nanoparticles in the field of veterinary medicine. *International Journal of Veterinary Science and Medicine, 7*(1), 78–93.

Yu, Z., He, K., Cao, W., Aleem, M. T., Yan, R., Xu, L., Song, X., & Li, X. (2022). Nano vaccines for *T. gondii* ribosomal P2 protein with nanomaterials as a promising DNA vaccine against toxoplasmosis. *Frontiers in Immunology, 13*, 839489. https://doi.org/10.3389/fimmu.2022.839489\

Yue, M., Lin, H., Dedrick, D. E., & Stenken, J. A. (2004). A 2-D microcantilever array for multiplexed biomolecular analysis. *Journal of Microelectromechanical Systems, 13*(2), 290–299.

Zambrano-Zaragoza, M. L., Quintanar-Guerrero, D., Del Real, A., González-Reza, R. M., Cornejo-Villegas, M. A., & Gutiérrez-Cortez, E. (2020). Effect of nano-edible coating based on beeswax solid lipid nanoparticles on strawberry's preservation. *Coatings, 10*, 253. https://doi.org/10.3390/coatings10.30253

Zare Khafri, H., Ghaedi, M., Asfaram, A., Javadian, H., & Safarpoor, M. (2018). Synthesis of CuS and ZnO/Zn$(OH)_2$ nanoparticles and their evaluation for in vitro antibacterial and antifungal activities. *Applied Organometallic Chemistry, 32*. https://doi.org/10.1002/AOC.4398

Zeb, A., Gul, M., Nguyen, T.-T.-L., & Maeng, H.-J. (2022). Controlled release and targeted drug delivery with poly(lactic-co-glycolic acid) nanoparticles: Reviewing two decades of research. *Journal of Pharmaceutical Investigation, 52*(683–724). https://doi.org/10.1007/s40005-022-00584-w

Zhang, W. X., Wang, C. B., & Lien, H. L. (1998). Treatment of chlorinated organic contaminants with nanoscale bimetallic particles. Catalysis today, 40(4), 387–spiepr DHYP395.

Zhang, J., Lin, Y., Lin, Z., Wei, Q., Qian, J., Ruan, R., Jiang, X., Hou, L., Song, J., Ding, J., & Yang, H. (2021). Stimuli-responsive nanoparticles for controlled drug delivery in synergistic cancer immunotherapy. *Advanced Science*, *9*(5), 2103444. https://doi.org/10.1002/advs.202103444

Zhang, G. J., & Ning, Y. (2012). Silicon nanowire biosensor and its applications in disease diagnostics: A review. *Analytica Chimica Acta*, *749*, 1–15.

Zhao, Y., Zhang, L., Wang, A., & Zhou, D. (2025). Biosensor technology: Advances and applications in livestock infectious disease diagnosis. *Veterinary Sciences*, *12*(1), 23. https://doi.org/10.3390/vetsci12010023

Zheng, G., Patolsky, F., Cui, Y., Wang, W. U., & Lieber, C. M. (2005). Multiplexed electrical detection of cancer markers with nanowire sensor arrays. *Nature Biotechnology*, *23*(10), 1294–1301.

Zhong, T., Liu, X., Li, H., & Zhang, J. (2021). Co-delivery of sorafenib and crizotinib encapsulated with polymeric nanoparticles for the treatment of in vivo lung cancer animal model. *Drug Delivery*, *28*(1), 2108–2118. https://doi.org/10.1080/10.17544.2021.1979129

Zhou, J., Miyamoto, Y., Ihara, S., Kroll, A. V., Nieskens, N., Tran, V. N., Hanson, E. M., Fang, R. H., Zhang, L., & Eckmann, L. (2022). Codelivery of antigens and adjuvant in polymeric nanoparticles coated with native parasite membranes induces protective mucosal immunity against *Giardia lamblia*. *The Journal of Infectious Diseases*, *226*(2), 319–323. https://doi.org/10.1093/infdis/jiac074

Zhu, H., Cheng, X., Zhang, J., Wu, Q., Liu, C., & Shi, J. (2023). Constructing a selfhealing injectable SABA/Borax/PDA@AgNPs hydrogel for synergistic low-temperature photothermal antibacterial therapy. *Journal of Materials Chemistry B*, *11*(3), 618–630.

13

Role of Nanotechnology for Better Food Preservation

Muhammad Nauman Khan[1,2], Barkat Ullah[1], Nasir Assad[3], Marzia Batool Laila[3], Tooba[3], Sana Wahab[4], Alevcan Kaplan[5], Amjad Ali[6], Shah Fahad[7], and Syed Mukaram Shah[8]

[1] *Department of Botany, Islamia College Peshawar, Peshawar, Pakistan*
[2] *Biology Laboratory, University Public School, University of Peshawar, Peshawar, Pakistan*
[3] *Institute of Chemistry, University of Sargodha, Sargodha, Pakistan*
[4] *Department of Plant Sciences, Quaid-i-Azam University, Islamabad, Pakistan*
[5] *Department of Crop and Animal Production, Sason Vocational School, Batman University, Batman, Turkey*
[6] *Department of Sustainable Crop Production, Università Cattolica Del Sacro Cuore, Piacenza, Italy*
[7] *Department of Agronomy, Abdul Wali Khan University, Pakistan*
[8] *Centre of Plant Biodiversity, University of Peshawar, Peshawar, Pakistan*

13.1 Introduction to Nanotechnology in Food Preservation

Nutrition is essential for maintaining human life and promoting growth. Food can be consumed in its natural form or after undergoing processing to supply energy and facilitate growth (Michel et al., 2024). The problem of food wastage has become a major concern worldwide in recent times. The process of food manufacturing and consumption results in significant food waste. Food deprivation is prevalent as a result of inadequate supply chains, increasing populations, and the impacts of climate change (Sahoo et al., 2024). The overall food waste generated consisted of 20% from production, 1% from processing, 19% from distribution, and 60% from consumption. Key factors contributing to food waste encompass cooking shrinkage, manufacturing challenges, supply chain disruptions, increasing consumer demands, fluctuating weather conditions, soil runoff, and regulatory constraints (Hanis and Fernando, 2024). As the population continues to increase, there is an escalating demand among consumers for food that is fresh, healthy, and nutritious (Lami et al., 2024). All varieties of food, upon being harvested, begin to show indications of deterioration. As a result, food wastage has become a notable issue in every sector of food processing. An effective strategy to tackle the issues surrounding food wastage involves the implementation of food preservation techniques (Heydari, 2024).

The goal of food preservation is to increase the goods' shelf life without compromising its nutritional value, safety, or sensory characteristics; this is essential for ensuring the availability and safety of food (Wang et al., 2024). Refrigeration, freezing, canning, and the use

Nanotechnology Innovations for Food Security and Sustainable Agriculture, First Edition. Edited by Zia Ur Rehman Mashwani, Shah Fahad, Ilyas Ahmad, Shah Saud, Sajid Fiaz, and Taufiq Nawaz.

of chemical preservatives are examples of traditional preservation technologies that have been crucial in guaranteeing the availability of food (Rahman, 2020). But these methods often cause people to worry about the effects on the environment and public health, such as how much energy they consume, how much greenhouse gas they release, and whether or not they leave behind dangerous chemical residues (Lisboa et al., 2024). The food sector must embrace new, environmentally friendly preservation methods that are in line with worldwide initiatives to lessen environmental impacts and safeguard public health if sustainability is to maintain its growing importance. Methods and developments that successfully increase food's shelf life while decreasing its environmental impact and improving its quality and safety are at the heart of the sustainable food preservation idea (Lisboa et al., 2024).

Nanotechnology presents an innovative approach to overcoming the challenges associated with traditional food preservation techniques. Through the application of nanoparticles (NPs), new strategies can be formulated to improve food safety and quality, all while reducing environmental consequences (Bajpai et al., 2018). The remarkable mesoscopic properties make NPs highly attractive for the other applications, such as huge surface area, highly reactive surface area, small particle size, great strength, quantum effects, and ductility (Gorde et al., 2024). Due to their unique properties, which distinguish them from the bulk material in terms of physiochemical and biological properties, research began on their synthesis, characterization, applications, and evaluations. In another study, scientists used a novel small modular reactor to produce radioactive isotopes used in imaging, diagnostic techniques, and radiation therapy (Eker et al., 2024). The results of these studies have significantly inspired scientific innovation, which has contributed to significant development and revolution in the agri-food industry in the past decades. Nanotechnology applications in food systems have led to the development of many novel products with better characteristics, including texture, flavor, sensory properties, and stability. Packaging materials that incorporate antimicrobial NPs show promise in preventing bacterial growth and extending the shelf life of food without the need for excessive refrigeration or chemical preservatives (Gupta et al., 2024). In addition, nanocoatings can be used on food commodities to form a protective barrier against spoilage and oxidation, thus prolonging shelf life and maintaining freshness. Solutions based on nanotechnology will help reduce energy consumption, lower greenhouse gas emissions, and eliminate harmful chemicals, which align with the principles of sustainable food preservation (Thakur et al., 2024a).

13.2 Principles of Nanotechnology in Food Preservation

Man has been striving to perfect and create better methods of food preservation since prehistoric times. There has been a long way for mankind to go from the days of attempting to keep food fresh by burying it in damp caves to the refrigeration methods of the modern period. Cold streams and cellars would also be useful for preserving food. Modern methods of drying and fermenting have their roots in techniques developed about 10,000 years ago (Jurić et al., 2024). Humans have long relied on a variety of preservation techniques, including fermentation, salting, sundrying, roasting, oven baking, smoking, steaming, curing, pickling, canning, bottling, jellying, irradiation, carbonation, and the use of chemical or artificial preservatives. There was a lack of thorough understanding of the scientific process behind any of these approaches, despite their shared goal of reducing or eliminating the disease-causing organism's rate of reproduction. Other popular methods for food preservation include osmotic inhibition and vacuum preservation. The presence and usage of the aforementioned preservation techniques in the Greek, Roman, and Egyptian civilizations are supported by archaeological evidence. To prevent food from going bad, the ancient Egyptians would sundry it (Zhou et al., 2024a). To preserve food from spoilage caused by microbes, the Romans came up with the concept of pickling. However, the addition of honey or sugar to preservation methods was a Greek invention that caused food to become jelly-like. By creating a rudimentary type of artificial refrigeration in 1784, William Cullen achieved the first technologically inventive advance in food preservation techniques (Zhou et al., 2024a). In the early 1800s, people began salting and preserving food to extend its shelf life (Zhou et al., 2024a). Some others have also made important contributions to preservation techniques; for example, in 1809, Nicolas Appert (Olawore et al., 2024) developed a vacuum bottling method to supply food to French troops; this method helped pave the way for tinning and, later, canning, which were developed by Peter Durand in 1810. Louis Pasteur developed the process of pasteurization in 1862, which allowed for the preservation of milk, wine, and beer (Jambholkar et al., 2024). But these were rudimentary ways, and they couldn't keep food for much longer. A better and more long-lasting method of food preservation was required.

The word "nano" is often associated with concepts like "extremely small" or "related to atoms" after the majority of people (García et al., 2010). The "nanoscience" discipline is born out of merging this idea with scientific methods. It finds use in every area of science and technology. Improved food storage, longer shelf life, pollution monitoring, and the incorporation of antibacterial or health supplement components are just a few of the many uses for nanotechnology in the food science area. Preharvest and postharvest difficulties in agricultural production have been significantly mitigated by the use of nanotechnology in food preservation (Sahoo et al., 2021).

In the field of functional foods, where bioactive components are prone to degradation and eventual deactivation in hostile settings, the use of nanoencapsulation proved important in increasing food product shelf life. With the help of this method, deterioration processes are successfully slowed down or prevented altogether till the product reaches its ultimate destination (Chowdhury et al., 2024). As an additional benefit, the inclusion of edible nanocoatings onto a variety of food items serves as a barrier to the exchange of gases and moisture. These coatings not only impart colors, flavors, antioxidants, enzymes, and anti-browning agents, but they also contribute to the enhanced shelf life of produced foods, even after the packaging has been opened. This is because these coatings supply all of these things (Jurić et al., 2024). In addition, nanosensors that are included into packaging make it possible to monitor essential aspects of food quality in real time. These aspects include temperature, humidity, and chemical concentrations. With this preventative method, it is possible to identify the deterioration of food at an earlier stage, which enables prompt interventions and helps to avoid food waste (Kotekar et al., 2024). In conclusion, the performance of packaging materials is greatly improved when NPs are incorporated into the manufacture of the materials. These substances have the potential to enhance their barrier qualities, so rendering them more resistant to the entrance of oxygen, light, and moisture. Furthermore, several NPs have natural antimicrobial capabilities, which further restrict the growth of microorganisms and lengthen the shelf life of the food that has been packed. These new uses of nanotechnology not only contribute to the creation of food items that are safer and fresher, but they also help to reduce the amount of food that is wasted and the environmental effect that is involved with the manufacturing and distribution of food. Nanomaterial and their applications in food technology are shown in Table 13.1.

Table 13.1 Applications of nanomaterial in food technology.

Nanostructured materials/ particles	Applications in food technology	Activity	References
Zinc oxide	The antibacterial properties of food preservation packaging are highlighted by its characteristics, which are used to keep fresh foods from spoiling and to keep them from adhering together during storage.	An effective antimicrobial; the mechanical characteristics of the synthesized nanomaterial were unaffected by exposure to UV-A; activated oxygen scavenging materials are used to prevent oxygen transport inside the packing containers.	(Mizielińska et al., 2018)
Titanium dioxide	To limit oxygen transfer in packing containers, materials that scavenge activated oxygen are used; the mechanical properties of the produced nanomaterial were unaffected by exposure to UV-A.	An effective antimicrobial; the mechanical characteristics of the produced nanomaterial were unaffected by exposure to UV-A; materials that absorb activated oxygen are used to prevent oxygen transfer in containers for storage.	(Kumar et al., 2020)
Copper-based	With the use of active packaging, food goods may be kept fresh for longer and protected from harmful germs and microbes that cause deterioration.	Used to prevent the spread of infectious diseases; their large surface area allowed them to bind to cell membranes, amplifying their antimicrobial effects; they also pass water vapor, act as a barrier, are resistant to heat and UV rays, and are effective against microbes.	(Kumar et al., 2021)
Silver-based	By controlling the growth of harmful bacteria and other microbes, as well as by increasing the food's shelf life, active packaging is an essential component in food preservation.	Problems with yellowing, lack of transparency, and heat stability; improved mechanical and barrier qualities; greater antioxidant efficacy; and characteristic antimicrobial activity against bacterial strains, including Gram-positive and Gram-negative.	(Biswas et al., 2022)

13.3 Nanocoatings for Food Preservation

The market value of fresh produce is contingent upon its capacity to endure various physical, physiological, microbiological, and biochemical stresses throughout the storage process. Ensuring a clean environment throughout the harvesting and processing stages, minimizing enzyme activity and moisture levels, and meticulously controlling storage temperature and humidity are essential measures for prolonging the shelf life of the product (Chaudhary et al., 2020). Drying, fermenting, pickling, and freezing represent some of the traditional methods employed for the preservation of produce. In comparison to conventional methods for extending the shelf life of food, nano-edible coatings present a more sustainable and versatile alternative.

After harvest, fruits and vegetables release carbon dioxide into the atmosphere as a result of their respiration process. The application of the coating leads to a decrease in ethylene synthesis and a shift toward predominantly anaerobic respiration, necessitating reduced oxygen levels. This approach effectively prolongs shelf life by nearly twofold, while also enhancing freshness, firmness, and nutritional value through the reduction of physiological water loss and the deceleration of ripening (Gupta et al., 2022). The characteristics and quantity of the coating significantly influence the interior environment, subsequently impacting the levels of oxygen and carbon dioxide, as well as weight loss. These coatings function by establishing a semipermeable protective layer on the fruit's surface, thereby minimizing quality losses. The metabolic functions of fruits are notably hindered by physical barriers, which control gas exchange and water vapor movement – two critical processes necessary for preserving the freshness of fruits. A reduction in respiration inhibits the plant's ability to absorb oxygen and release carbon dioxide, thus delaying the ripening process. Additionally, it helps retain moisture levels and keep the fruit from getting too mushy. A thin, biodegradable layer can be applied to the surface of produce that will help slow the rate of ripening and spoiling. This is accomplished by regulating gas exchange, minimizing moisture loss, and offering antimicrobial defense. Therefore, this edible coating using materials such as nano-ZnO, SiO_2, TiO_2, and Ag NPs can effectively increase the shelf life of the fruit better than common coatings without these materials.

Nanocoatings are a safe alternative for food and beverages, are environmentally friendly, and do not need any special equipment to be applied. This approach is a holistic, inexpensive option for controlling postharvest losses and extending shelf life. Apart from protective functions, nanocoatings can also be doped with active substances. In addition, fruits and vegetables contain bioactive compounds that play a key role in improving health and strengthening the immune system. There are many bioactive compounds, including micronutrients, antioxidants, polyphenols, flavoring agents, pigments, probiotics and prebiotics, omega-3 fatty acids, and many other nutraceuticals in films and coatings. Common issues include the bioavailability, solubility, and stability of these substances. Through the application of nanotechnology, particularly nanoformulations, enhancements to these compounds can be achieved (Nile et al., 2020). Nanomaterials enhance the bioavailability, solubility, and biological activity of encapsulated components due to their reduced particle sizes and increased surface areas per unit mass (Haris et al., 2023). Recent findings indicate that encapsulating essential oils in a nanocoating enhances their biological effectiveness by expanding their surface area, although it diminishes their influence on sensory attributes. This occurs because essential oils can effectively mask the odor or flavor of the coating. According to the findings of de Oliveira Filho et al. (2021), NPs have the potential to enhance bioavailability and reduce interactions with other components of the food matrix, even when employing smaller amounts of bioactive material. This occurs due to their enhancement of the transport of active elements through biological membranes. The bioaccessibility of nonencapsulated resveratrol, due to its poor solubility, is approximately 48.1%. Leena et al., 2020 demonstrated that zein nanofibers infused with resveratrol enhanced bioaccessibility by 1.4 times relative to the original form, resulting in an encapsulation percentage of 67.6%. The application of the nanocoating, leveraging its antioxidant properties, led to a notable reduction in weight loss, enhanced color retention, and effectively inhibited browning in apple slices.

The nano-edible film and coating technology to incorporate probiotics can be considered a considerable progress. Probiotics are mainly composed of a mixture of different types of bacteria, and they are often found in fermented dairy products including milk, cheese, puddings, fruit drinks, and yoghurts. One of the most important methods to improve food stability and safety is to use probiotic edible films in food matrices. Such films prevent the growth of spoilage microorganisms either by outcompeting spoilage microorganisms or by producing antimicrobial substances in the case of probiotics. They also serve as a great way to introduce probiotics into a person's diet. One of the fundamental properties of nano-edible films is the right of live probiotics to target the intestines, which might enable health benefits for a long time. NPs of probiotics and prebiotics have the potential to increase their bioavailability, enhance their release mechanism, and improve their survival rate. According to the previously mentioned study by Zabihollahi et al. (2020), a series of probiotic nanofilms were prepared based on carboxymethyl cellulose with cellulose nanofibers (CNFs) and inulin as a prebiotic to enhance the mechanical properties of the nanofilms and survival of probiotics. The addition of CNF to the film resulted in significant improvements in its mechanical properties, water resistibility and stability. Antimicrobial activity against *Escherichia coli* was also markedly enhanced following treatment with the inulin prebiotic in vivo in the probiotic *Lactobacillus plantarum*, with a 36% increase in probiotic storage viability observed. These effects of this probiotic were demonstrated in the film. Probiotic-supplemented broth prevented pathogen growth and produced low-molecular-weight material (e.g., ethanol, hydrogen peroxide, and organic acids [e.g., lactic and acetic acid] as well as diacetyl and acetaldehyde). The chicken fillets showed a shelf life of 6 days, while the control sample had a shelf life of only 3 days. *Lactobacillus acidophilus* was used as a synthesis agent to create the carbon dots (CDs), which were then incorporated into a bacterial nanocellulose film. Introducing these probiotic CDs has significantly enhanced the film's antibacterial and UV shielding efficacy and inhibits development of *Escherichia coli* and *Listeria monocytogenes*. When incorporated into the cellulose matrix, the nanopaper possessed excellent antibacterial properties, increased flexibility, and potential for UV-blocking effects. These coatings thus protect delicate fruits from microbial contamination and UV exposure, extending their freshness (Kousheh et al., 2020).

13.4 Nanoencapsulation Techniques

In order to do nanoencapsulation, nanocapsules are utilized. The term "nanoencapsulation" refers to a method of nanopackaging substances using nanoemulsification, nanostructuration, and nanocomposite processes. The end result is a product with properties like controlled release of the core. Improved bioavailability and efficacy, improved stability, protection from oxidation, taste creation, controlled release in response to moisture or pH, sequential delivery of numerous active ingredients, altered flavor character, prolonged organoleptic perception, and ease of handling are just a few of the many advantages they offer (del Rosario Herrera-Rivera et al., 2024). These nanovesicular systems are characterized by a core–shell architecture, where the medication is contained in a reservoir or a cavity encased in a polymer membrane or coating (Altemimi et al., 2024). Molecular dispersions, as well as liquids or solids containing the active ingredient, can all fit within the cavity. A key role for nanocapsules in food preservation is the delivery of the desired component while simultaneously entrapping the smell and undesirable components. The biological system experiences an improvement in the bioavailability of the drug when the dietary supplements are carried across the gastrointestinal tract in nanocapsules. Nanoprecipitation, emulsion–diffusion, double emulsification, emulsion–coacervation, polymer coating, and layer-by-layer assembly are the six fundamental methods for building nanocapsules (Zhou et al., 2024b). One key distinction between nanoemulsions and regular emulsions is that the latter does not alter the visual look of the food item when included in it. Pesticides, fertilizers, and vaccinations may all be delivered to plants in the form of these nanocapsules. They also increase the food's nutritional value by delivering lipophilic health supplements such as vitamins, minerals, fatty acids, and growth hormones (Vasanthakumaran et al., 2024). The primary advantage of encapsulation is that it shields the concealed component, allowing for its accurate delivery to the target regardless of environmental factors. One type of nanoencapsulant is the liposome, which is a nanobased carrier. Nanoliposomes aid in the system's multicomponent-regulated and targeted distribution. Nutraceuticals, nutrients, vitamins, enzymes, antimicrobials, and additives are known to be delivered by these (Duncan and Gaspar, 2011). Electrospinning zein fibers coated with gallic acid is a novel encapsulation technology that is the subject of ongoing study (Abouelnour et al., 2024). Before the lipids reach their destination, zein fiber prevents their degradation in the system. Quite surprisingly, the food packaging business can make extensive use of this groundbreaking technique. Due to the enhanced solubility and specificity of the encapsulated components, lipid-based encapsulation techniques outperform alternative encapsulation methods. By preventing the component from coming into significant contact with food material, this approach ensures that both the food's natural characteristics and the component's ability to function inside the biological system remain unaffected. Colloidosomes are similar in appearance; they are tiny, capsule-like structures with hollow interiors, and their size is less than a quarter of a human cell (Francis et al., 2024). Many people think that the shell contains multiple components and can act as a transporter for medications and dietary supplements in the body. Nanocochleates also aid in enhancing the processed food's quality. The micronutrients are stabilized when these nanocoils round them. Phosphatidyl serine, phosphatidic acid, dioleoyl phosphatidyl serine, phosphatidyl choline, phosphatidylethanolamine, diphosphatidylglycerol, dioleoyl phosphatidic acid, dimyristoyl phosphatidyl serine, dipalmitoyl phosphatidyl glycerol, and distearoyl phosphatidyl serine are some of the soy phospholipids that make it up (Biswas et al., 2022). With the goal of developing vaccinations that can control the body's immunological response, nanoencapsulation of probiotics is another new area where nanotechnology is proving its worth. The probiotics are effectively transported to the digestive system and kept viable through the use of nanoencapsulation technology. According to many research, starch-like NPs aid in lipid body preservation and are also efficiently distributed to their intended sites inside biological systems (De Azeredo, 2009). Antioxidant nanoencapsulated delivery systems include archaeosomes. They are made from lipids found in archaebacterial membranes. It is well-known that these lipids are both heat- and stress-resistant. In addition, there have been results indicating that α-tocopherol NPs in fat droplets can prevent milk from deteriorating (Chaudhry et al., 2008). Canola active oil, produced by a firm in Haifa, Israel, known as Shemen, is one of the few commercially available food items that have discovered a purpose as a material for nanoencapsulation (Singh et al., 2024). The remaining food items are made in the United States, and the brands and names of these products include the following: Fortified Fruit Juice, made by High Vive, uses nanoencapsulation to deliver fortified vitamins, lycopene, theanine, and sun-active iron; NanoResveratrol, made by Life Enhancement, is a plant-based lipid encased in a natural phospholipid shell, like phosphatidylcholine; Health Plus International's Spray for Life Vitamin Supplements, made by Health Plus, uses nanoencapsulation to deliver fortified vitamin beverages; Jamba Juice Hawaii's Daily Boost is a nanoencapsulated product containing bioactive components or vitamin-fortified beverages; and The Color Emulsion is made by Wild Flavors and is used to create β-carotenal, apocarotenal, or paprika nanoemulsions. RBC Life Sciences Inc.'s Nanoceuticals Slim Shake Chocolate employs nanoclusters to enhance the flavor profile of the shake while eliminating the need for added sugar (Zhan et al., 2024). Nanotea represents yet another nanoencapsulated offering developed by Qinhuangdao Taiji Ring Nano-Products Co. Ltd., China (Rhim et al., 2013).

13.5 Nanomaterials in Food Packaging

Packaging plays a crucial role in the food industry by preserving the sensory and nutritional qualities of food throughout storage. Food packaging fulfills two essential roles: preventing spoilage and safeguarding the delicate bioactive components of food from environmental influences and physical harm. There is considerable enthusiasm for utilizing nanotechnology to enhance food packaging solutions. Functional NPs enhance the barrier properties, thermal stability, strength, and durability of packaging materials, thereby contributing to the prolonged shelf life of food items (Gupta et al., 2024). Incorporating NPs into conventional food packaging materials, including films and containers, exemplifies one approach. Alternatively, the development of nanocomposite multilayer packaging materials can be accomplished by combining organic, inorganic, and hybrid nanocoatings using techniques such as immersion, spraying, or rubbing. The application of nanotechnology facilitates the creation of dynamic, smart, and improved packaging solutions. This development greatly enhances physical and mechanical characteristics, provides antibacterial and antioxidant features, boosts UV absorption, and supports the monitoring and management of food conditions (Ashfaq et al., 2022). Numerous forecasts suggest that intelligent packaging is poised for the most rapid expansion among all smart packaging elements, attributed to its economical, engaging, and accessible features (Cunha and Mazieri, 2024).

NPs improve the mechanical and physical characteristics of food packaging, aiding in factors like biodegradability, strength, flexibility, thermal resistance, UV absorption, and resistance to water vapor and oxygen. The addition of metallic oxides to polymers greatly enhances their mechanical characteristics, barrier effectiveness, and light transmission capabilities (Ashfaq et al., 2022). Arfat and colleagues created films made from biocomposites that incorporate fish skin gelatin and Ag–Cu NPs. The results showed that the films displayed improved resistance to heat and UV radiation, exhibited greater mechanical strength, and successfully prevented the growth of both Gram-positive (*L. monocytogenes*) and Gram-negative (*Salmonella typhimurium*) bacteria (Arfat et al., 2017). Cellulose NPs exhibit significant effectiveness in blocking oxygen flow while permitting water to permeate. This property indicates a promising application for enhancing the oxygen barrier performance of polymers when integrated within moisture-resistant matrices (Wu et al., 2021). The investigation conducted by Tabatabaei et al. revealed improvements in tensile strength, water vapor barrier, and oxygen barrier in gelatin-k-carrageenan biocomposites that incorporate nano-SiO_2 (Tabatabaei et al., 2018).

Findings indicate that nanoclay enhances barrier properties against gas and UV light (Tan et al., 2024). Rostamzad et al. developed a protein film that incorporates transglutaminase and nanoclay. A study demonstrated that even a minimal addition of nanoclay enhanced the mechanical and physical properties of the fish protein film (Rostamzad et al., 2016). Khalaj et al. synthesized polypropylene nanocomposites by employing nanoclay and nanoclay modified with iron NPs. The incorporation of 2% by weight montmorillonite into plain polypropylene led to a decrease in oxygen and water vapor permeabilities by 22% and 33%, respectively, when compared to the control sample. The analysis of plain polypropylene versus polypropylene infused with 0.2% iron NPs revealed a notable improvement, showcasing a 55% rise in oxygen permeability and a 77% rise in water vapor permeability (Khalaj et al., 2016).

To enhance food packaging, the application of nanocoating is being implemented. Food can be covered with thin films or layers composed of either edible or nonedible materials. An intriguing method that demonstrates potential for prolonging the shelf life of food products is the application of edible coatings. Various biopolymers, such as lipid proteins and polysaccharides, are utilized in the formulation of edible coatings. Composites and nanocomposites were created to address the growing demand for food packaging that integrates multiple functionalities into one material. This is due to the fact that coatings composed of a single polymer offer only one characteristic, which might be insufficient to adequately maintain the quality of the food. Although nanofillers improve the color, mechanical properties, and barrier characteristics of edible films compared to standard options, further investigation is required for them to rival synthetic films (Ashfaq et al., 2022). Dash and colleagues developed films incorporating starch and pectin along with titanium dioxide (TiO_2). The mechanical and oxygen barrier characteristics were enhanced by NPs of TiO_2. The UV impermeability of the starch–pectin-blended films was observed to increase with a higher concentration of TiO_2 NPs (Xie et al., 2023).

The practice of incorporating specific elements into the packaging material or the package headspace to improve the properties of the packing polymer is referred to as active packaging (Nian et al., 2024). These elements could encompass oxygen scavengers, ethylene absorbers, antioxidants, carbon dioxide emitters, and antibacterial agents. Antimicrobial drugs hinder bacterial growth by disrupting their cellular structure or obstructing their metabolic pathways (Nian et al., 2024). Reactive oxygen species (ROS), generated in part by NPs, impede DNA replication and ATP production, resulting in cell damage or cell death.

The development of active packaging featuring robust antibacterial properties could significantly improve through the use of gelatin films that incorporate NPs, including nanoclay or metal NPs. Fasihnia et al. developed a nanocomposite film using polyethylene and organoclay NPs (Fasihnia et al., 2018). The combination of the two organoclays was found to significantly inhibit the growth of both *Staphylococcus aureus* and *Escherichia coli*. The development of film for active packaging was carried out by Yang et al. utilizing polyvinyl alcohol, chitosan, and lignin NPs. The film exhibited antimicrobial and antioxidant properties against the Gram-negative bacteria *Xanthomonas arboricola* pv. pruni and *Erwinia carotovora* subsp. carotovora. The packaging can be utilized in the advancement of active packaging for meals that are both fresh and rich in fat (Yang et al., 2016). To develop active packaging, Ramesh et al. synthesized films using polyvinyl alcohol, incorporating fennel seed oil and cellulose NPs into the mixture. The film exhibited improved antibacterial and antioxidant properties (Ramesh and Radhakrishnan, 2019).

The highly reactive unsaturated hydrocarbon molecule referred to as gaseous ethylene is responsible for the ripening and quality degradation of perishable produce. A promising approach to addressing the issue of the limited shelf life of perishable food items is the use of active packaging that includes an ethylene scavenger. The phrase "ethylene scavenger" denotes a category of compounds that can either "capture and contain" ethylene or "take up water via the chemical interaction between two substances" (Gaikwad et al., 2020). Zhang et al. developed a film that demonstrates outstanding ethylene scavenging, antioxidant, and antibacterial properties, utilizing chitosan, TiO_2 NPs, and black plum peel extract (Zhang et al., 2019).

The food sector might benefit greatly from nanostructures and nanomaterials due to their many desirable characteristics, including their tiny size, high surface-to-volume ratio, and quantum effects. NPs of metals and oxides (e.g., silver, zinc, copper, titanium dioxide, and magnesium oxide) with inherent antibacterial capabilities have recently attracted a lot of interest from the food industry. Among other things, research has shown that protecting food from dangerous and spoilage-inducing microorganisms using antimicrobial NPs in active packaging extends its storage life and improves consumer convenience (Hoseinnejad et al., 2018; Sharma et al., 2020). Strength, weight reduction, and oxygen permeability are all improved when active antimicrobial NPs are mixed with the packaging material. To put it simply, intelligent packaging materials are those "materials and articles that monitor the condition of packaged food or the environment surrounding the food" according to a 2004 definition by the European Commission. Throughout the supply chain, intelligent packaging – also called smart packaging – uses high-tech devices such as barcodes, radio frequency identification tags, sensors, and indicators to communicate, sense, record, and send information about the food's safety, quality, and history (Yao et al., 2024). Advantages and disadvantages of nanomaterials are shown in Table 13.2.

Table 13.2 Advantages and disadvantages of nanomaterial in food packing.

Materials	Advantages	Disadvantages	Properties	References
Silver	• Silver significantly improves mechanical properties, barrier performance, and antimicrobial functions. • It can substantially reduce flammability while maintaining the transparency of the polymer matrix.	• The presence of Ag nanoparticles in containers effectively suppressed microbial growth for a period extending to 10 days. • Extending the shelf life of packaged food products reduces waste production.	• Packaging that inhibits the growth of microbes. • Systems that remove ethylene and carbon dioxide. • Systems that scavenge oxygen from the air. • Silver's mechanisms include adhering to bacterial cell surfaces, damaging their DNA, penetrating their cells, releasing more antimicrobial $Ag+$ ions, and creating cell membrane pits.	(Khaydarov et al., 2009)
Gold	• Polyethylene terephthalate (PET) bottles with a covering will be more efficient in preserving fragile drinks like beer or juice.	• When the accumulation of gold nanoparticles becomes permanent, leading to a slow fading of the red color in time–temperature integrators, the freezing phenomena occurs.	• The influence of different parameters on the lower critical solution temperature (LCST) at a solid–water interface is investigated via the use of gold nanoparticles.	(Sperling et al., 2008)
Clay nanoparticle	• Using nanoclays has demonstrated to lower carbon dioxide permeability by 50% and moisture by 30% correspondingly. • Meat products' reduced lipid oxidation results in longer shelf life.	The interplay between low specific gravity and high aspect ratio leads to diminished compatibility. As a result, packing cannot be completed.	• Originally designed primarily to improve UV and gas barrier properties, clay nanoparticle materials also provide heat resistance, stability, strength, and stiffness; hence, they find use in food packaging. By means of optimal morphological arrangement under influence of both enthalpic and entropic elements, clay nanoparticles improve their characteristics.	(Seghar et al., 2011)
Titanium dioxide	• The addition of TiO_2 to polylactic acid can prolong the shelf life of mangoes by up to 15 days. • TiO_2 is characterized by its inert and nontoxic properties.	• The phenomenon of rancidity arises from the movement of TiO_2 into the lipid matrix.	• TiO_2 surface hydroxyl radicals and reactive oxygen species were produced when fatty acids and polyunsaturated phospholipids in microbial cell membranes were oxidized. The results showed that the rate of color recovery was directly related to the concentration of oxygen.	(Weir et al., 2012)
Zinc oxide	• As disinfectants, photocatalytic agents and UV blockers mostly rely on zinc oxide. • Zinc oxide contains several mineral components really extensively. • Zinc oxide has rather strong antibacterial action.	• Zinc oxide manifests on the skin as an opaque white layer. • The application of zinc oxide leads to the generation of byproducts that are detrimental to the environment.	• Zinc enhances the effectiveness of the antibiotics, antibacterial agents, and antifungal properties. • Zinc oxide nanoparticles have shown promise as a food packaging material due to their antibacterial properties and their capacity to inhibit bacterial contamination in processed poultry meat, particularly in relation to *Salmonella typhimurium* and *Staphylococcus aureus*.	(Mirzaei and Darroudi, 2017)

13.6 Antimicrobial Nanomaterials

The food industry is experiencing a significant transformation due to the rise of antimicrobial nanomaterials. These advanced materials, designed at the nanoscale, demonstrate strong antimicrobial properties against a diverse range of microorganisms, including bacteria, fungi, and viruses. The distinct features, including elevated surface area-to-volume ratios and increased reactivity, allow for effective inhibition of microbial growth, thus prolonging the shelf life of food items and reducing the likelihood of foodborne illnesses. This method presents significant potential for transforming food safety and security, while guaranteeing a dependable and nutritious food supply for consumers around the globe (Rai et al., 2022).

Silver NPs (Ag NPs) have become a significant tool in combating foodborne pathogens. The antimicrobial capabilities arise from a complex mechanism: Ag NPs compromise bacterial cell walls, obstruct essential cellular functions such as DNA replication and protein synthesis, trigger oxidative stress, and deactivate important enzymes. This complex attack, combined with their substantial surface area-to-volume ratio, allows Ag NPs to efficiently target a broad range of microorganisms, greatly improving food safety and prolonging shelf life. Research indicates that the integration of Ag NPs into biodegradable films such as chitosan or cellulose markedly prolongs the shelf life of fresh produce by suppressing fungal growth. Furthermore, active packaging systems embedded with Ag NPs have demonstrated efficacy in managing microbial growth in meat products through the controlled release of silver ions. The combination of Ag NPs with other antibacterial nanomaterials, such as graphene oxide (GO), enhances their antibiotic activities through synergistic effects. The encapsulation of the bacterial cell membrane within its 2D layered GO structure has the potential to trigger oxidative stress in the basal plane, resulting in membrane damage (Díez-Pascual, 2018). The sample with 43% Ag demonstrated the highest antibacterial performance, achieving an efficacy of nearly 100% against *S. aureus* and *Pseudomonas aeruginosa*. The proposed theory for the antibacterial action suggests that the GO encapsulates the bacteria, while the toxicity of Ag effectively eliminates them.

Ag NPs are one of the most prominent strategies used against foodborne pathogens. The antimicrobial properties are due to multistep mechanism as shown in Figure 13.1. Ag NPs damage the bacterial cell walls, inhibit the vital cellular processes including DNA replication and protein synthesis, induce oxidative stress, and inhibit several key enzymes (Hashmi et al., 2024; Jabbar et al., 2023; Assad et al., 2023). Their high surface area-to-volume ratio facilitates an effective attack against a multitude of microorganisms, thus greatly enhancing food safety and extending food shelf life. Studies suggest that the incorporation of Ag NPs into biodegradable films including chitosan or cellulose significantly extends the shelf life of fresh fruits and vegetables by inhibiting fungal growth, and Ag NP-embedded active packaging systems have been reported to be effective in controlling the growth of microorganisms in meat products via controlled release of silver ions (More et al., 2023). The synergy by using Ag NPs along with other antibacterial nanomaterials (e.g., graphene oxide [GO]) enables the increase in the antibiotic activities of Ag NPs. The 2D layered GO structure can cover a bacterial cell membrane to the strain, possibly inducing oxidative stress at the basal plane and hurting the membrane. Díez-Pascual (2018) investigated the antibacterial properties of Ag–GO nanocomposites covalently grafted by

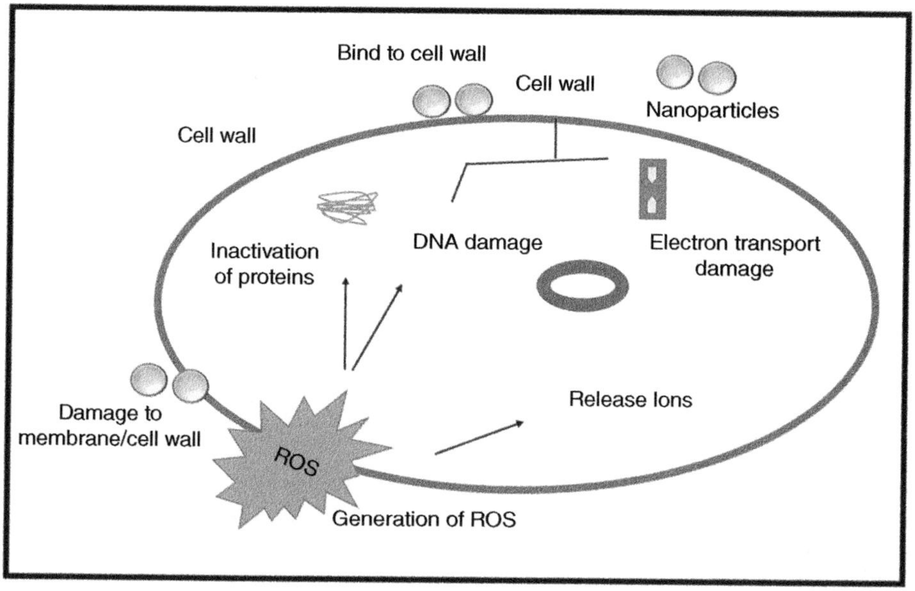

Figure 13.1 Mechanism of nanomaterial against antimicrobial agents.

means of thiol groups in the absence of reducing agents and stabilizers. Among all the samples, 43% Ag showed the best antibacterial property with an efficacy of nearly 100% for both *S. aureus* and *P. aeruginosa*.

Titanium dioxide NPs (TiO_2 NPs) have received much attention in the field of food preservation due to their strong photocatalytic activity. TiO_2 NPs generate ROS under light irradiation, particularly ultraviolet (UV) light, which can effectively inactivate a range of microorganisms, including bacteria, fungi, and viruses. It has promising applications in producing self-disinfecting food packaging materials and antimicrobial coatings in food contact surfaces. Also, using TiO_2 NPs in food packaging films could enhance the barrier properties of food packaging films toward oxygen, moisture, and light to improve the shelf life of packaged foods greatly. However, the release route of TiO_2 NPs into the food chain and their environmentally toxicological outcomes must be critically assessed (Thakur et al., 2024b).

ZnO NPs have surfaced as noteworthy antimicrobial agents in the food industry, presenting a range of benefits (Yu et al., 2024). Additional antimicrobial efficacy is aided because their small sizes allow for more interaction of the surface area with microorganisms. ZnO NPs can easily be incorporated into various types of food packaging materials such as films and coatings to even edible coatings, providing a convenient and efficient means of antimicrobial delivery. Researches show shelf life of perishable food can be significantly prolonged through packaging materials that inhibit spoilage microorganisms in the food because of active packaging with ZnO NPs, which can potentially reduce food waste as well as promote food safety. Conducting thorough safety assessments and ensuring that the use of ZnO NPs in food complies with relevant regulations and guidelines is important (Sirelkhatim et al., 2015; Noor et al., 2025).

Alongside the extensively studied nanomaterials like silver and zinc oxide, other nanomaterials display promising antimicrobial properties for food preservation. Due to its ability to generate ROS and disrupt the membrane of microbial cells, copper oxide NPs (CuO NPs) have well-characterized antimicrobial effects. Chitosan NPs as a natural polymer have an antimicrobial effect on different microorganisms like bacterial and fungal. Its diversity has potential applications in food preservation, including as novel antimicrobial polymeric packaging materials and advanced food coatings. However, in-depth safety assessments and regulatory aspects are needed to ensure the safe and efficient use of these novel technologies in the foods industry (Ma et al., 2022).

13.7 Nanosensors for Food Quality Monitoring

A sensor is a device that can detect changes in the physical stimulus and convert them into measurable and recordable waves. Nanosensors are tiny sensors that rely on measurements taken at the nanoscale for their accuracy. Pesticides, spoiled food, poisons, foodborne diseases, and off-putting aromas and flavors may all be detected with their help. They can also track the passage of time, the temperature, and the oxygen saturation levels. Furthermore, nanosensors are essential for the detection of harmful pathogens in food. There has been a worldwide uptick in cases of food poisoning due to tainted foods. About 250 distinct diseases have been associated with eating food that is contaminated with bacteria, viruses, and parasites (Ghorbian et al., 2025). Healthcare and social and economic stability are both hit hard by the management of these interrelated diseases. To ensure the quality and safety of food, it is vital to regularly test it for microbial contamination. In the realm of disease detection and analysis, particular nanosensors represent an innovative technology that has received considerable acclaim worldwide.

Due to the quick loss of vitamins and antioxidants in food during the process of manufacturing and storage, the food industry has added nanosensors for their detection. Experiment with ionic liquid nanocomposites containing single-walled and multi-walled carbon nanotubes shows food enzymes produced using nanotechnology in milk, wheat flour, and fruit juices. It gives information on the ripening of plantains, bananas, and other fruits by determining the level of sucrose, glucose, and ascorbic acid through the use of nanosensors (Barbosa, 2025). Toxins are produced by microbes as a defensive strategy. Gold NPs modified with anti-aflatoxin antibodies were employed to identify these toxins in food samples (Kotekar et al., 2024). Carbon nanotubes–based luminous biosensors have shown the capacity to detect marine toxin peroxide in the flesh of marine mussels (Pteriomorphia). Gilent magnetic beads with anti-aflatoxin M1 antibody were used to capture aflatoxin M1, which was reported in milk samples and gold nanoprobes (Ji et al., 2024). The issue of food adulteration is increasingly recognized as a critical global health concern, presenting substantial threats to human well-being. Food adulterants can result in symptoms including vomiting, diarrhea, paralysis, or even fatal consequences (Anagaw et al., 2024). Contaminants may encompass various items, including foods, beverages, sweets, and those with high protein levels. Melamine, recognized as an adulterant, was identified through the use of gold NPs (in picomoles) (Kotekar et al., 2024).

Pathogenic microorganisms present in food play a crucial role in the deterioration of food quality and the emergence of food-related diseases. The identification of bacterial deoxyribonucleic acid (DNA) or bacterial cells functions as a technique to detect these microorganisms. Matthews et al., 2025 reported that, although it is more laborious and requires more accurate handling than other methods, the application of NPs produced acceptable results for detection of bacteria and DNA separation. Likewise, multiparametric magneto-fluorescent nanosensors have been developed by Kotekar et al. (2024) to allow for sensitive and rapid detection for *E. coli* O157:H7 infection in <1 hour (Kotekar et al., 2024). The magnetic NPs demonstrated enhanced effectiveness in isolating *E. coli* O157:H7 from ground beef and flesh samples by utilizing magnetic NP–antibody conjugates (Kotekar et al., 2024).

The pathogenic bacterium *Listeria monocytogenes* was isolated from milk samples using magnetic iron oxide NPs. The polymerase chain reaction (PCR) was then utilized to measure the DNA levels (Balakrishna et al., 2024). In the case of PCR-based microbial identification, 16S ribosomal ribonucleic acid (rRNA) is an essential selective marker. In contrast, 16S rRNA microarray has limitations such as high cost and poor sensitivity, while NP-based identification has potential to be precise and reliable. There are ratiometric sensors that combine organic fluorescent dyes, quantum dot arrays, and rare-earth metal ions for the detection of copper (II) ions in vegetable and fruit samples (Chelliah et al., 2021). Recent advancements have enabled the detection of arsenic in water using optical dual-mode nanosensors that integrate gold NPs and carbon quantum dots.

13.8 Environmental and Safety Considerations

A significant variety of products arising from the evolving field of nanotechnology poses important risks to public health. The introduction of synthesized NPs in the food chain can result in the bioaccumulation of hazardous contaminants in foodstuffs that are dangerous to humans and the environment. Many components of foods have naturally occurring nanostructures. NPs have been incorporated in food packaging to enhance barrier and mechanical properties of traditional and bio-based materials. Further, they have been adapted to take on sophisticated, functional tasks, like embedding active or intelligent elements in a package. Such elements should also be capable of transferring or gaining substances from or into the packaged food or surrounding atmosphere and must be placed on the food package so as to convey important information on storage and use condition (Dudefoi et al., 2018). Consuming foods that have been in contact with such nanopackaging might create an exposure route, which could be a serious health threat. This is particularly important with regard to toxicity and ecotoxicity as particle nanomaterials can migrate from packaging into food. This is due to poor packaging performance. Nanostructured materials have beneficial properties as antimicrobials and antioxidants, but it is also proposed that their increased generation of ROS could lead to oxidative stress or cellular damage1213 (Fu et al., 2014). Nanotoxicity may cause DNA damage, excessive cell stimulation, alterations in cell motility, toxicity, cytotoxicity, apoptosis, and possibly the development of cancer. A range of factors can influence this impact, including the toxicity of the NPs used, the specific packing matrix applied, the degree of migration that occurred, and the absorption rate of the particular food item (Smaoui et al., 2023). The growing body of evidence concerning health risks associated with NP absorption is becoming increasingly evident. In addition to the usual risks and hazards, nano-based food items may cause health problems owing to bioaccumulation, increased activity, and overconsumption (Erisen and Uludag, 2024). One way to evaluate the hazards of nanostructured materials is to look at how they enter the body, how they are deposited, and how they move around inside (Zubair et al., 2024). There has to be further research and risk assessment into the potential security risks posed by prolonged exposure to these compounds by inhalation or skin absorption, particularly in regards to the possibility of cytotoxicity and toxicity in the long run (Kim et al., 2024). Research on the potential toxicity of nanomaterials is limited, although some studies have shown that NPs from packaging materials can leach into food, which people then eat (del Rosario Herrera-Rivera et al., 2024). However, it is possible for these NPs to build up in several human organs, including the digestive tract, small intestine, liver, kidneys, lungs, and spleen. In addition, problems including renal illness, liver damage, and lung damage might result from only one oral intake of ZnO NPs. NPs may easily enter the bloodstream after being ingested via the gastrointestinal tract because they can cross biological barriers.

The impact of NPs in food on human health seems to have been the subject of little investigation. A comprehensive analysis of a nanomaterial's chemical makeup, physical and chemical characteristics, interactions with living organisms, and exposure levels is necessary for risk assessment. It is crucial to address several important issues (problem formulation) before beginning an assessment of a nanomaterial intended for application in the food and feed chain. To confirm that a material is a nanomaterial, its physicochemical properties must be determined (Erisen and Uludag, 2024). The strength of nanoscale characteristics in a nanomaterial is inherently linked to the unique features or phenomena that the nanomaterial displays. It is expected that a nanomaterial will exhibit behavior akin to that of its non-nanomaterial counterpart, if such a counterpart exists, in situations where it loses these characteristics due to degradation or disintegration. The main issues related to the safety of ingested nanomaterials encompass food contamination and various potential hazards, components of food, and the likelihood of NPs migrating to organs and other tissues. Furthermore, there is a concern regarding the possible local adverse effects in the gastrointestinal tract, which has been thoroughly investigated using in vitro digestion models (Baig et al., 2021). In light of this, the majority of safety concerns related to orally ingested nanomaterials center on those that may bypass the digestive system. The models can accurately simulate the conditions found in the gastrointestinal system, including the mouth, stomach, and gut. The degree of physiological representation is the primary feature that differentiates these models from one another. The models exhibit considerable variation, ranging from simplistic to highly detailed, with options for either static or dynamic configurations. They may include enzymes, bile salts, and other elements as required, or exclude them altogether (Zhou and McClements, 2022). Additionally, the physiological responses replicated can differ among various models, such as those simulating feeding versus starvation, or contrasting infants with adults. The Food and Agriculture Organisation of the United Nations (FAO), the World Health Organisation (WHO), and the European Union (EU) have all highlighted the importance of improving safety protocols related to NPs in food

(WHO, 2010). A recent revision by the European Food Safety Authority (EFSA) intends to provide a solid basis for the assessment of risks resulting from the application of nanoscience and nanotechnology along the food and feed chain, with human and animal health in focus. EFSA is responsible for multiple fields of application: novel foods, food and feed additives, food contact materials, and pesticides. These amended guidelines, titled Scientific Committee Guidance on Nano Risk Assessment (SC Guidance on Nano-RA, SC, 2021–2023), include several scientific studies that provide key evidence to support the physicochemical properties of nanomaterials, exposure assessment, hazard characterization, and applications in various fields of nanomedicine (SC, 2021) (Committee et al., 2021). Increasing concerns have emerged regarding NPs, NP fractions, and materials with nanoscale properties, as these could potentially pose risks to consumers. The Scientific Committee on Consumer Safety (SCCS) has published a Scientific Consultation (SCCS, 2019) that emphasizes the features of NPs which may pose safety issues. These NPs may be evaluated within the framework of risk assessment. The Consultation pointed out the lack of stringent regulations for assessing the overall safety issues related to certain NPs, emphasizing several characteristics that should raise alarms about the safety of the nanomaterial. This decision was made in recognition of the scarce recommendations available. A grading strategy was formulated during the Consultation, in accordance with expert insights, to assign a grade that clearly reflects the potential risks associated with nanomaterials (Isibor, 2024). The findings suggest that degradation, genotoxicity, accumulation, and immunotoxicity are essential elements of toxicokinetics and human hazard evaluation, likely shaped by the distinct properties of nanomaterials. According to Dekkers et al. (2016), prioritizing these issues in the hazard evaluation of nanomaterials is crucial due to their significant impact (Dekkers et al., 2016). Before incorporating nanotechnology into the food system, it is vital to concentrate on the safety assessment of food or food products to reduce the toxicity and health risks linked to NPs. The NPs are required to meet the following criteria:

- Establish a reliable approach for detecting and quantifying NPs in food products or culinary items.
- Develop a set of guidelines for the application of NPs in food products or food items using a standardized framework.
- Evaluate the maximum levels of NP exposure for consumers arising from the intake of food products or food items. Develop a pharmacokinetic model for dietary NPs present in food products.

NPs can migrate into packaged food, but comprehensive migration tests and risk assessments are insufficient. This migration is the transfer of low-molecular-weight components from the packaging to the food material, and it can be described by Fick's second law, which determines diffusion. Therefore, migration studies are an integral part of the development of new advanced food packaging materials, to evaluate whether undesirable or harmful substances could migrate into food in general or specific use conditions (Biswas et al., 2022). The investigation found that most experimental studies did not sufficiently address the possible migration of NPs from food packaging materials to food products. This gap is probably due to the absence of precise analytical tools suitable for identifying

NPs at low concentrations and minuscule dimensions. Studies that report no evidence of migration during a given time should provide information on detection limits (i.e., fugitive concentrations, and detection limits should be provided in terms of particle mass or number concentration as a function of particle size). The behavior of migration may also be influenced by the antibacterial, antimicrobial, water-retaining, and protective characteristics of NPs, aspects that should also be integrated into these studies (Alehosseini et al., 2019). Unfortunately, the extensive use and distribution of exhaled nitric oxide (ENO) have not been rigorously investigated regarding their toxicity and potential hazards. This is a significant fact that has not garnered the attention it warrants. ENO can be tailored to address particular requirements and selected for its low toxicity to humans. They can be utilized right away for the detection of pollutants. The combination of these distinct attributes establishes ENO as a credible solution for tackling environmental pollution (Anvar et al., 2021). It is crucial to specify particular safety standards, such as the permissible limits or constraints of NPs employed in food-related uses. Moreover, environmental regulations may present obstacles concerning the utilization of nanomaterials in food packaging, given that these materials will eventually necessitate disposal. A comprehensive collection of data has been gathered concerning the regulations that oversee the use of nanotechnology in the food sector. As a result, additional examination is necessary to assess the possible toxicity of nanostructured materials utilized in the food sector. Moreover, experts in the domain, together with users, require an extensive worldwide structure for knowledge exchange that empowers them to tackle and resolve different facets of implementation, consumption, disposal, and long-term impacts. This will advance the development of applications and research within the food industry.

13.9 Challenges and Limitations

The advancements made in the application of nanotechnology in the food industry have been significant; nonetheless, the progress concerning nanostructures within this field has been somewhat constrained. While nanotechnology holds considerable promise for creating groundbreaking products and processes in the food industry, several challenges need to be tackled. A growing number of individuals are voicing concerns about the safety of nanotechnology products designed for human consumption and use (Jagtiani, 2022). The progress in food nanotechnology research is advancing rapidly. Therefore, it is crucial to conduct a comprehensive analysis of the possible health risks to humans before initiating the commercial production of nanofood products. At present, there is a lack of established guidelines for evaluating the acceptability of nanomaterials found in food. The primary obstacle involves creating food delivery systems that guarantee safety for human consumption while also maintaining economic feasibility in production. The transfer and integration of NPs from packaging materials into food pose a significant challenge that must be tackled to ensure that foods provide adequate nutrition. The behavior of materials at the nanoscale shows significant differences, but our understanding of how to explore these variations is still quite limited. A more profound understanding of the nanoscale behaviors and toxicological impacts of nanomaterials will result in improved safety standards for their practical application (Biswas et al., 2022).

Moreover, it is also challenging to study the migration of NPs when relying on purely predictive models that only considered diffusion-based migrations. A molecular mechanism involving three subprocesses is generally found for nanoparticle (NP) migration: (i) due to a concentration gradient, the molecule penetrates the polymer to reach the foodstuffs; (ii) when the food touches the packaging, the molecule detaches from the polymer and is absorbed by the foodstuff; and (iii) due to a concentration difference, the molecule diffuses within the foodstuff (Bott et al., 2014). However, there is also undeniable lack of understanding when it came to the possible release mechanisms for the identified NPs. The varying effects of the physiochemical characteristics for gastrointestinal absorption of NPs – such as morphology, composition, surface properties, charge, aggregation state, and functional ingredients – shows that the question of the risk for consumers for translocating NPs originating from food packaging is likely to be more complex than other studies. These are certainly areas where more research is needed going forward. It is also important to study the implications of NPs and potential risks, as well as toxicological and environmental aspects. New discovery showed that NPs are able to penetrate biological barriers, which enables such particles enter into cells and even organs (Khan et al., 2019). The synthesis of NPs through various chemical processes can lead to adverse effects, resulting in the generation of toxic by-products that are detrimental to the environment and contribute to contamination. Therefore, in addition to public demand and popularity, it is crucial to implement a comprehensive risk assessment strategy, biosafety measures, regulatory frameworks, and address public concerns during the food processing, production, packaging, and consumption of nano-based food products. This phenomenon can be attributed to the increasing acceptance of nano-enhanced food items. Furthermore, it is crucial to perform both in vitro and in vivo studies on the interactions of NPs with biological organisms prior to the development and production of antimicrobial NPs that provide sustainable alternatives. This analysis must be finalized before the materials can be produced. The Food and Drug Administration has concluded that the distinct attributes and features presented by nanotechnology should be included in evaluations of the safety, effectiveness, public health implications, or regulatory status of products utilizing nanotechnology (Osman et al., 2024).

Meanwhile, the European Commission has developed extensive legislation aimed at recognizing technological advancements that utilize nanostructured materials, which display physical and chemical properties that are markedly different from those observed at larger scales. This framework discusses NPs alongside smart and active materials, emphasizing the significance of these developments. The measures implemented by the European Commission are outlined later. In this context, the limited understanding of their potential toxicity via oral exposure has resulted in nanomaterials receiving less attention than the cutaneous or inhalation routes. This relates to the finding that oral exposure is the primary mode of administration. The introduction of laws specifically designed for nanotechnology, focused on tackling the various safety issues linked to this technology, has the potential to greatly impact the entire food processing sector. The presence of NPs in dietary substances can lead to biological effects, including oxidative stress responses, protein denaturation, and DNA damage. At present, a universally applicable standardized safety evaluation technique for NPs, especially those associated with food and nanotechnology, does not exist. The regulation of NP usage in food is of utmost significance. This entails formulating risk assessments and safety management strategies that are in harmony with the aims and objectives of modern toxicity testing (Rauscher et al., 2017).

13.10 Conclusion and Recommendations

Nanotechnology has revolutionized the food industry by introducing novel techniques for sustaining nourishment, wrapping consumables, and enhancing protection. The properties of NPs, such as their heightened surface-to-volume proportion, amplified reactivity, and powerful antimicrobial influence – have enabled noteworthy improvements in prolonging shelf life, minimizing decay, and confirming nutritional quality with certainty. Groundbreaking applications like nanocoatings, nanoencapsulation, and nanosensors have illustrated remarkable efficiency in maintaining freshness, monitoring nourishment quality in real time, and identifying contaminants with exceptional sensitivity. While these innovations are impressive, fully incorporating nanotechnology into the nutritional supply chain presents notable risks to health and the environment. Concerns regarding NP migration from coverings into nourishment, prospective bioaccumulation in human tissues over the long term, and far-reaching ecological impacts remain critically important. Current investigations highlight substantial gaps in danger evaluation protocols, detection methods, and administrative directives, emphasizing the urgent necessity for exhaustive studies and stringent evaluations to guarantee protection.

To maximize the benefits of nanotechnology in the food industry while ensuring safety and sustainability, it is crucial to undertake comprehensive risk assessments through thorough toxicological studies and standardized methodologies for rigorously evaluating NP migration, bioaccumulation, and toxicity within ecosystems. Regulatory frameworks must be strengthened by cautiously establishing clear, evidence-based guidelines harmonized to international standards while mandating transparent front-of-package labeling of all nano-enhanced food products. Advancements in detection technologies should prioritize developing next-generation analytical tools capable of detecting NPs at ever-lower concentrations and leveraging nanosensors for real-time, in-field food quality monitoring. Promoting sustainable nanotechnology practices necessarily involves encouraging the judicious use of biodegradable, eco-friendly materials with minimal embodied carbon footprints, supporting imaginative recyclable packaging solutions designed for disassembly, and conducting comprehensive life cycle assessments to proactively minimize adverse environmental impacts. Public engagement and education efforts must aim to raise critical awareness of both nanotechnology's promising benefits and its inherent risks to informed consumers so that they can make choices aligned with their values. Lastly, fostering collaborative, interdisciplinary networks among food scientists, engineers, toxicologists, environmental researchers, regulators, and other stakeholders will help address multifaceted challenges associated with safety, innovation, ethics, and compliance while supporting solution-oriented research that integrates food science, nanotechnology, toxicology, and environmental studies. Responsible and sustainable use of nanotechnology will prove pivotal in sculpting the future landscape of global food systems, preservation techniques, and safety standards.

References

Abouelnour, M. A., Abd EL-Aziz, M. A., Osman, K. M., Fathy, I. N., Tayeh, B. A., & Elfakharany, M. E. (2024). Recycling of marble and granite waste in concrete by incorporating nano alumina. *Construction and Building Materials, 411*, 134456.

Alehosseini, A., Gómez-Mascaraque, L. G., Martínez-Sanz, M., & López-Rubio, A. (2019). Electrospun curcumin-loaded protein nanofiber mats as active/bioactive coatings for food packaging applications. *Food Hydrocolloids, 87*, 758–771.

Altemimi, A. B., Farag, H. A. M., Salih, T. H., Awlqadr, F. H., Al-Manhel, A. J. A., Vieira, I. R. S., & Conte-Junior, C. A. (2024). Application of nanoparticles in human nutrition: A review. *Nutrients, 16*(5), 636.

Anagaw, Y. K., Ayenew, W., Limenh, L. W., Geremew, D. T., Worku, M. C., Tessema, T. A., Simegn, W., & Mitku, M. L. (2024). Food adulteration: Causes, risks, and detection techniques. *SAGE Open Medicine, 12*, 20503121241250184.

Anvar, A. A., Ahari, H., & Ataee, M. (2021). Antimicrobial properties of food nanopackaging: A new focus on foodborne pathogens. *Frontiers in Microbiology, 12*, 690706.

Arfat, Y. A., Ahmed, J., Hiremath, N., Auras, R., & Joseph, A. (2017). Thermo-mechanical, rheological, structural and antimicrobial properties of bionanocomposite films based on fish skin gelatin and silver-copper nanoparticles. *Food Hydrocolloids, 62*, 191–202.

Ashfaq, A., Khursheed, N., Fatima, S., Anjum, Z., & Younis, K. (2022). Application of nanotechnology in food packaging: Pros and Cons. *Journal of Agriculture and Food Research, 7*, 100270.

Assad, N., Naeem-ul-Hassan, M., Sher, M., Alatawi, F. S., Alatawi, M. S., Omran, A. M., Jame, R. M., Adnan, M., Khan, M. N., & Ali, B. (2023). Aconitum lycoctonum L.(Ranunculaceae) mediated biogenic synthesis of silver nanoparticles as potential antioxidant, anti-inflammatory, antimicrobial and antidiabetic agents. *BMC Chemistry, 17*(1).

Baig, N., Kammakakam, I., & Falath, W. (2021). Nanomaterials: A review of synthesis methods, properties, recent progress, and challenges. *Materials Advances, 2*(6), 1821–1871.

Bajpai, V. K., Kamle, M., Shukla, S., Mahato, D. K., Chandra, P., Hwang, S. K., Kumar, P., Huh, Y. S., & Han, Y.-K. (2018). Prospects of using nanotechnology for food preservation, safety, and security. *Journal of Food and Drug Analysis, 26*(4), 1201–1214.

Balakrishna, K., Lakshminarayan, V., Ganeshkumar, N., & Kingston, J. (2024). Listeria monocytogenes in the food chain–recent advancements in technologies for rapid detection and identification: Identification of Listeria monocytogenes from foods. *Journal of Microbiology, Biotechnology and Food Sciences*, e11628–e11628.

Barbosa, J. B. (2025). *Microbial Contamination and Food Safety (Volume I)*. MDPI.

Biswas, R., Alam, M., Sarkar, A., Haque, M. I., Hasan, M. M., & Hoque, M. (2022). Application of nanotechnology in food: Processing, preservation, packaging and safety assessment. *Heliyon, 8*(11).

Bott, J., Störmer, A., & Franz, R. (2014). A model study into the migration potential of nanoparticles from plastics nanocomposites for food contact. *Food Packaging and Shelf Life, 2*(2), 73–80.

Chaudhary, S., Kumar, S., Kumar, V., & Sharma, R. (2020). Chitosan nanoemulsions as advanced edible coatings for fruits and vegetables: Composition, fabrication and developments in last decade. *International Journal of Biological Macromolecules, 152*, 154–170.

Chaudhry, Q., Scotter, M., Blackburn, J., Ross, B., Boxall, A., Castle, L., Aitken, R., & Watkins, R. (2008). Applications and implications of nanotechnologies for the food sector. *Food Additives and Contaminants, 25*(3), 241–258.

Chelliah, R., Wei, S., Daliri, E. B.-M., Rubab, M., Elahi, F., Yeon, S.-J., Jo, K. H., Yan, P., Liu, S., & Oh, D. H. (2021). Development of nanosensors based intelligent packaging systems: Food quality and medicine. *Nanomaterials, 11*(6), 1515.

Chowdhury, S., Kar, K., & Mazumder, R. (2024). Exploration of different strategies of nanoencapsulation of bioactive compounds and their ensuing approaches. *Future Journal of Pharmaceutical Sciences, 10*(1), 72.

Committee, E. S., More, S., Bampidis, V., Benford, D., Bragard, C., Halldorsson, T., Hernández-Jerez, A., Hougaard Bennekou, S., Koutsoumanis, K., & Lambré, C. (2021). Guidance on risk assessment of nanomaterials to be applied in the food and feed chain: Human and animal health. *EFSA Journal, 19*(8), e06768.

Cunha, K. C. T., & Mazieri, M. R. (2024). Intelligent packaging and value generating: Technological development opportunities based on Patent Analysis. *World Patent Information, 76*, 102258.

De Azeredo, H. M. (2009). Nanocomposites for food packaging applications. *Food Research International, 42*(9), 1240–1253.

de Oliveira Filho, J. G., Miranda, M., Ferreira, M. D., & Plotto, A. (2021). Nanoemulsions as edible coatings: A potential strategy for fresh fruits and vegetables preservation. *Food, 10*(10), 2438.

Dekkers, S., Oomen, A. G., Bleeker, E. A., Vandebriel, R. J., Micheletti, C., Cabellos, J., Janer, G., Fuentes, N., Vázquez-Campos, S., & Borges, T. (2016). Towards a nanospecific approach for risk assessment. *Regulatory Toxicology and Pharmacology, 80*, 46–59.

del Rosario Herrera-Rivera, M., Torres-Arellanes, S. P., Cortés-Martínez, C. I., Navarro-Ibarra, D. C., Hernández-Sánchez, L., Solis-Pomar, F., Pérez-Tijerina, E., & Román-Doval, R. (2024). Nanotechnology in food packaging materials: Role and application of nanoparticles. *RSC Advances, 14*(30), 21832–21858.

Díez-Pascual, A. M. (2018). *Antibacterial activity of nanomaterials*. MDPI.

Dudefoi, W., Villares, A., Peyron, S., Moreau, C., Ropers, M.-H., Gontard, N., & Cathala, B. (2018). Nanoscience and nanotechnologies for biobased materials, packaging and food applications: New opportunities and concerns. *Innovative Food Science & Emerging Technologies, 46*, 107–121.

Duncan, R., & Gaspar, R. (2011). Nanomedicine (s) under the microscope. *Molecular Pharmaceutics, 8*(6), 2101–2141.

Eker, F., Duman, H., Akdaşçi, E., Bolat, E., Sarıtaş, S., Karav, S., & Witkowska, A. M. (2024). A comprehensive review of nanoparticles: From classification to application and toxicity. *Molecules, 29*(15), 3482.

Erisen, D. E., & Uludag, K. (2024). Estimating biosafety of biodegradable biomedical materials from in vitro ion tolerance parameters and toxicity of nanomaterials in brain. In *Transformative Approaches to Patient Literacy and Healthcare Innovation*. IGI Global.

Fasihnia, S. H., Peighambardoust, S. H., & Peighambardoust, S. J. (2018). Nanocomposite films containing organoclay nanoparticles as an antimicrobial (active) packaging for potential food application. *Journal of Food Processing and Preservation, 42*(2), e13488.

Francis, D. V., Abdalla, A. K., Mahakham, W., Sarmah, A. K., & Ahmed, Z. F. (2024). Interaction of plants and metal nanoparticles: Exploring its molecular mechanisms for sustainable agriculture and crop improvement. *Environment International, 108859*.

Fu, P. P., Xia, Q., Hwang, H.-M., Ray, P. C., & Yu, H. (2014). Mechanisms of nanotoxicity: Generation of reactive oxygen species. *Journal of Food and Drug Analysis*, *22*(1), 64–75.

Gaikwad, K. K., Singh, S., & Negi, Y. S. (2020). Ethylene scavengers for active packaging of fresh food produce. *Environmental Chemistry Letters*, *18*, 269–284.

García, M., Forbe, T., & Gonzalez, E. (2010). Potential applications of nanotechnology in the agro-food sector. *Food Science and Technology*, *30*, 573–581.

Ghorbian, M., Ghobaei-Arani, M., Babaei, M. R., & Ghorbian, S. (2025). Nanotechnology and nanosensors in personalized healthcare: A comprehensive review. *Sensing and Bio-Sensing Research*, *100740*.

Gorde, P. M., Dash, D. R., Singh, S. K., & Singha, P. (2024). Advancements in sustainable food packaging: A comprehensive review on utilization of nanomaterials, machine learning and deep learning. *Sustainable Chemistry and Pharmacy*, *39*, 101619.

Gupta, R. K., Abd El Gawad, F., Ali, E. A., Karunanithi, S., Yugiani, P., & Srivastav, P. P. (2024). Nanotechnology: Current applications and future scope in food packaging systems. *Measurement: Food*, *13*, 100131.

Gupta, V., Biswas, D., & Roy, S. (2022). A comprehensive review of biodegradable polymer-based films and coatings and their food packaging applications. *Materials*, *15*(17), 5899.

Hanis, M. H., & Fernando, Y. (2024). Smart logistics solutions for reducing food waste: A case of D nipah catering. *International Journal of Industrial Management*, *18*(1), 11–21.

Haris, M., Hussain, T., Mohamed, H. I., Khan, A., Ansari, M. S., Tauseef, A., Khan, A. A., & Akhtar, N. (2023). Nanotechnology–A new frontier of nano-farming in agricultural and food production and its development. *Science of the Total Environment*, *857*, 159639.

Hashmi, S. S., Ibrahim, M., Adnan, M., Ullah, A., Khan, M. N., Kamal, A., Iqbal, M., Kaplan, A., Assad, N., & Elshikh, M. S. (2024). Green synthesis of silver nanoparticles from *Olea europaea* L. extracted polysaccharides, characterization, and its assessment as an antimicrobial agent against multiple pathogenic microbes. *Open Chemistry*, *22*(1), 20240016.

Heydari, M. (2024). Cultivating sustainable global food supply chains: A multifaceted approach to mitigating food loss and waste for climate resilience. *Journal of Cleaner Production*, *442*, 141037.

Hoseinnejad, M., Jafari, S. M., & Katouzian, I. (2018). Inorganic and metal nanoparticles and their antimicrobial activity in food packaging applications. *Critical Reviews in Microbiology*, *44*(2), 161–181.

Isibor, P. O. (2024). Regulations and policy considerations for nanoparticle safety. In *Environmental nanotoxicology: Combatting the minute contaminants*. Springer.

Jabbar, A., Abbas, A., Assad, N., Naeem-ul-Hassan, M., Alhazmi, H. A., Najmi, A., Zoghebi, K., Al Bratty, M., Hanbashi, A., & Amin, H. M. (2023). A highly selective Hg 2+ colorimetric sensor and antimicrobial agent based on green synthesized silver nanoparticles using Equisetum diffusum extract. *RSC Advances*, *13*(41), 28666–28675.

Jagtiani, E. (2022). Advancements in nanotechnology for food science and industry. *Food Frontiers*, *3*(1), 56–82.

Jambholkar, P. C., Choudhari, S. G., & Sharma, M. (2024). Louis Pasteur: A legacy unmasked. *Cureus*, *16*(8), e68080.

Ji, Y., Wang, R., & Zhao, H. (2024). Toward sensitive and reliable immunoassays of marine biotoxins: From rational design to food analysis. *Journal of Agricultural and Food Chemistry*, *72*(29), 16076–16094.

Jurić, M., Bandić, L. M., Carullo, D., & Jurić, S. (2024). Technological advancements in edible coatings: Emerging trends and applications in sustainable food preservation. *Food Bioscience*, 103835.

Khalaj, M.-J., Ahmadi, H., Lesankhosh, R., & Khalaj, G. (2016). Study of physical and mechanical properties of polypropylene nanocomposites for food packaging application: Nano-clay modified with iron nanoparticles. *Trends in Food Science & Technology*, *51*, 41–48.

Khan, I., Saeed, K., & Khan, I. (2019). Nanoparticles: Properties, applications and toxicities. *Arabian Journal of Chemistry*, *12*(7), 908–931.

Khaydarov, R., Khaydarov, R., Estrin, Y., Evgrafova, S., Scheper, T., Endres, C., & Cho, S. (2009). *Silver nanoparticles: Environmental and human health impacts. Nanomaterials: Risks and benefits* (pp. 287–297). Springer.

Kim, S. H., Bae, S., Sung, Y. W., & Hwang, Y. S. (2024). Effects of particle size on toxicity, bioaccumulation, and translocation of zinc oxide nanoparticles to bok choy (*Brassica chinensis* L.) in garden soil. *Ecotoxicology and Environmental Safety*, *280*, 116519.

Kotekar, G., Gaikwad, S., & Nigam, A. (2024). Macro world in nano hands: Nano revolutions in medicine and food processing with the aid of nanosensors. *Sensors International, 100291.*

Kousheh, S. A., Moradi, M., Tajik, H., & Molaei, R. (2020). Preparation of antimicrobial/ultraviolet protective bacterial nanocellulose film with carbon dots synthesized from lactic acid bacteria. *International Journal of Biological Macromolecules*, *155*, 216–225.

Kumar, A., Choudhary, A., Kaur, H., Mehta, S., & Husen, A. (2021). Metal-based nanoparticles, sensors, and their multifaceted application in food packaging. *Journal of Nanobiotechnology*, *19*(1), 256.

Kumar, S., Mukherjee, A., & Dutta, J. (2020). Chitosan based nanocomposite films and coatings: Emerging antimicrobial food packaging alternatives. *Trends in Food Science & Technology*, *97*, 196–209.

Lami, O., Mesías Díaz, F. J., Martín González, A., Hernández Matamoros, A., Escribano Sánchez, M. & Martínez Carrasco, F. 2024. Can fruit be more sustainable? A study on consumer preferences towards the use of natural preservatives in cherries.

Leena, M. M., Yoha, K., Moses, J., & Anandharamakrishnan, C. (2020). Edible coating with resveratrol loaded electrospun zein nanofibers with enhanced bioaccessibility. *Food Bioscience*, *36*, 100669.

Lisboa, H. M., Pasquali, M. B., dos Anjos, A. I., Sarinho, A. M., de Melo, E. D., Andrade, R., Batista, L., Lima, J., Diniz, Y., & Barros, A. (2024). Innovative and sustainable food preservation techniques: Enhancing food quality, safety, and environmental sustainability. *Sustainability*, *16*(18), 8223.

Ma, X., Zhou, S., Xu, X., & Du, Q. (2022). Copper-containing nanoparticles: Mechanism of antimicrobial effect and application in dentistry-a narrative review. *Frontiers in Surgery, 9(905892.*

Matthews, K. R., Kniel, K. E., & Critzer, F. J. (2025). *Food microbiology: An introduction*. John Wiley & Sons.

Michel, M., Eldridge, A. L., Hartmann, C., Klassen, P., Ingram, J., & Meijer, G. W. (2024). Benefits and challenges of food processing in the context of food systems, value chains and sustainable development goals. *Trends in Food Science & Technology, 104703.*

Mirzaei, H., & Darroudi, M. (2017). Zinc oxide nanoparticles: Biological synthesis and biomedical applications. *Ceramics International*, *43*(1), 907–914.

Mizielińska, M., Kowalska, U., Jarosz, M., & Sumińska, P. (2018). A comparison of the effects of packaging containing nano ZnO or polylysine on the microbial purity and texture of cod (Gadus morhua) fillets. *Nanomaterials*, *8*(3), 158.

More, P. R., Pandit, S., Filippis, A. D., Franci, G., Mijakovic, I., & Galdiero, M. (2023). Silver nanoparticles: Bactericidal and mechanistic approach against drug resistant pathogens. *Microorganisms*, *11*(2), 369.

Nian, L., Wang, M., Sun, X., Zeng, Y., Xie, Y., Cheng, S., & Cao, C. (2024). Biodegradable active packaging: Components, preparation, and applications in the preservation of postharvest perishable fruits and vegetables. *Critical Reviews in Food Science and Nutrition*, *64*(8), 2304–2339.

Nile, S. H., Baskar, V., Selvaraj, D., Nile, A., Xiao, J., & Kai, G. (2020). Nanotechnologies in food science: Applications, recent trends, and future perspectives. *Nano-Micro Letters*, *12*, 1–34.

Noor, A., Ahmad, N., Ali, A., Ali, M., Iqbal, M., Khan, M. N., Laila, M. B., Shah, S. N., Kaplan, A., & Ercişli, S. (2025). Green synthesis of silver-doped ZnO nanoparticles from Adiantum venustum D. Don (Pteridaceae): Antimicrobial and antioxidant evaluation. *Journal of Basic Microbiology*, *e2400543*.

Olawore, O., Ogunmola, M., & Desai, S. (2024). Engineered nanomaterial coatings for food packaging: Design, manufacturing, regulatory, and sustainability implications. *Micromachines*, *15*(2), 245.

Osman, A. I., Zhang, Y., Farghali, M., Rashwan, A. K., Eltaweil, A. S., Abd El-Monaem, E. M., Mohamed, I. M., Badr, M. M., Ihara, I., & Rooney, D. W. (2024). Synthesis of green nanoparticles for energy, biomedical, environmental, agricultural, and food applications: A review. *Environmental Chemistry Letters*, *22*(2), 841–887.

Rahman, M. S. (2020). Food preservation: An overview. *Handbook of Food Preservation*, 7–18.

Rai, A., Sharma, V. K., Jain, A., Sharma, M., Pandey, A., Singh, H. B., Gupta, V. K., & Singh, B. N. (2022). Microbe-fabricated nanoparticles as potent biomaterials for efficient food preservation. *International Journal of Food Microbiology*, *379*, 109833.

Ramesh, S., & Radhakrishnan, P. (2019). Cellulose nanoparticles from agro-industrial waste for the development of active packaging. *Applied Surface Science*, *484*, 1274–1281.

Rauscher, H., Rasmussen, K., & Sokull-Klüttgen, B. (2017). Regulatory aspects of nanomaterials in the EU. *Chemie Ingenieur Technik*, *89*(3), 224–231.

Rhim, J.-W., Park, H.-M., & Ha, C.-S. (2013). Bio-nanocomposites for food packaging applications. *Progress in Polymer Science*, *38*(10–11), 1629–1652.

Rostamzad, H., Paighambari, S. Y., Shabanpour, B., Ojagh, S. M., & Mousavi, S. M. (2016). Improvement of fish protein film with nanoclay and transglutaminase for food packaging. *Food Packaging and Shelf Life*, *7*, 1–7.

Sahoo, A., Dwivedi, A., Madheshiya, P., Kumar, U., Sharma, R. K., & Tiwari, S. (2024). Insights into the management of food waste in developing countries: With special reference to India. *Environmental Science and Pollution Research*, *31*(12), 17887–17913.

Sahoo, M., Vishwakarma, S., Panigrahi, C., & Kumar, J. (2021). Nanotechnology: Current applications and future scope in food. *Food Frontiers*, *2*(1), 3–22.

Seghar, S., Azem, S., & Aït Hocine, N. (2011). Effects of clay nanoparticles on the mechanical and physical properties of unsaturated polyester. *Advanced Science Letters*, *4*(11–12), 3424–3430.

Sharma, R., Jafari, S. M., & Sharma, S. (2020). Antimicrobial bio-nanocomposites and their potential applications in food packaging. *Food Control*, *112*, 107086.

Singh, N., Kumar, B., Usman, U. L., & Susan, M. A. B. H. (2024). Nano revolution: Exploring the frontiers of nanomaterials in science, technology, and society. *Nano-Structures & Nano-Objects, 39*, 101299.

Sirelkhatim, A., Mahmud, S., Seeni, A., Kaus, N. H. M., Ann, L. C., Bakhori, S. K. M., Hasan, H., & Mohamad, D. (2015). Review on zinc oxide nanoparticles: Antibacterial activity and toxicity mechanism. *Nano-Micro Letters, 7*, 219–242.

Smaoui, S., Chérif, I., Hlima, H. B., Khan, M. U., Rebezov, M., Thiruvengadam, M., Sarkar, T., Shariati, M. A., & Lorenzo, J. M. (2023). Zinc oxide nanoparticles in meat packaging: A systematic review of recent literature. *Food Packaging and Shelf Life, 36*, 101045.

Sperling, R. A., Gil, P. R., Zhang, F., Zanella, M., & Parak, W. J. (2008). Biological applications of gold nanoparticles. *Chemical Society Reviews, 37*(9), 1896–1908.

Tabatabaei, R. H., Jafari, S. M., Mirzaei, H., Nafchi, A. M., & Dehnad, D. (2018). Preparation and characterization of nano-SiO_2 reinforced gelatin-k-carrageenan biocomposites. *International Journal of Biological Macromolecules, 111*, 1091–1099.

Tan, Y., Yang, Q., Zheng, M., Sarwar, M. T., & Yang, H. (2024). Multifunctional nanoclay-based hemostatic materials for wound healing: A review. *Advanced Healthcare Materials, 13*(6), 2302700.

Thakur, D., Rana, P., Singh, S. K., Bakshi, M., Kumar, S., & Singh, S. (2024a). Nanoemulsion edible coating for shelf-life improvement and quality control in perishable products. *Plant Nano Biology*, 100114.

Thakur, N., Thakur, N., Kumar, A., Thakur, V. K., Kalia, S., Arya, V., Kumar, A., Kumar, S., & Kyzas, G. Z. (2024b). A critical review on the recent trends of photocatalytic, antibacterial, antioxidant and nanohybrid applications of anatase and rutile TiO_2 nanoparticles. *Science of the Total Environment*, 169815.

Vasanthakumaran, M., Ramesh, M., Murugan, K., Hema, T., Rajaganesh, R., & Hwang, J.-S. (2024). Developmental toxicity, biochemical and biomarker in the zebrafish (*Danio rerio*) embryo exposed to biosynthesized cadmium oxide nanoparticles. *Chemosphere, 369*, 143851.

Wang, J., Zhao, F., Huang, J., Li, Q., Yang, Q., & Ju, J. (2024). Application of essential oils as slow-release antimicrobial agents in food preservation: Preparation strategies, release mechanisms and application cases. *Critical Reviews in Food Science and Nutrition, 64*(18), 6272–6297.

Weir, A., Westerhoff, P., Fabricius, L., Hristovski, K., & Von Goetz, N. (2012). Titanium dioxide nanoparticles in food and personal care products. *Environmental Science & Technology, 46*(4), 2242–2250.

WHO, G. S. (2010). FAO/WHO Expert meeting on the application of nanotechnologies in the food and agriculture sectors: Potential food safety implications. Meeting report.

Wu, F., Misra, M., & Mohanty, A. K. (2021). Challenges and new opportunities on barrier performance of biodegradable polymers for sustainable packaging. *Progress in Polymer Science, 117*, 101395.

Xie, Q., Liu, X., Zhang, Y., & Liu, G. (2023). Development and characterization of a new potato starch/watermelon peel pectin composite film loaded with TiO_2 nanoparticles and microencapsulated *Lycium barbarum* leaf flavonoids and its use in the Tan mutton packaging. *International Journal of Biological Macromolecules, 252*, 126532.

Yang, W., Owczarek, J., Fortunati, E., Kozanecki, M., Mazzaglia, A., Balestra, G. M., Kenny, J. M., Torre, L., & Puglia, D. (2016). Antioxidant and antibacterial lignin nanoparticles in polyvinyl alcohol/chitosan films for active packaging. *Industrial Crops and Products*, *94*, 800–811.

Yao, Q.-B., Huang, F., Lu, Y.-H., Huang, J.-M., Ali, M., Jia, X.-Z., Zeng, X.-A., & Huang, Y.-Y. (2024). Polysaccharide-based food packaging and intelligent packaging applications: A comprehensive review. *Trends in Food Science & Technology*, 104390.

Yu, Y., Alkasir, R., Farrukh, A., Riaz, N., Rasool, A., Kaleem, I., Khattak, S. H., Bangash, S. A. K., Khan, M. N., & Wahab11, S. (2024). Synergistic antibacterial potential of Zno-Nps with different antibiotics against multidrug-resistant *Escherichia coli* and *Pseudomonas aeruginosa*. *Polish Journal of Environmental Studies*, *33*(4), 4893–4902.

Zabihollahi, N., Alizadeh, A., Almasi, H., Hanifian, S., & Hamishekar, H. (2020). Development and characterization of carboxymethyl cellulose based probiotic nanocomposite film containing cellulose nanofiber and inulin for chicken fillet shelf life extension. *International Journal of Biological Macromolecules*, *160*, 409–417.

Zhan, Z., Feng, Y., Zhao, J., Qiao, M., & Jin, Q. (2024). Valorization of seafood waste for food packaging development. *Food*, *13*(13), 2122.

Zhang, X., Liu, Y., Yong, H., Qin, Y., Liu, J., & Liu, J. (2019). Development of multifunctional food packaging films based on chitosan, TiO_2 nanoparticles and anthocyanin-rich black plum peel extract. *Food Hydrocolloids*, *94*, 80–92.

Zhou, H., & McClements, D. J. (2022). Recent advances in the gastrointestinal fate of organic and inorganic nanoparticles in foods. *Nanomaterials*, *12*(7), 1099.

Zhou, Q., Lan, W., & Xie, J. (2024a). Phenolic acid-chitosan derivatives: An effective strategy to cope with food preservation problems. *International Journal of Biological Macromolecules*, *254*, 127917.

Zhou, Y., Wang, Y., Peijnenburg, W., Vijver, M. G., Balraadjsing, S., Dong, Z., Zhao, X., Leung, K. M., Mortensen, H. M., & Wang, Z. (2024b). Application of machine learning in nanotoxicology: A critical review and perspective. *Environmental Science & Technology*, *58*(34), 14973–14993.

Zubair, M., Riaz, M., Kiani, M. N., Aslam, H. M., Fatima, A., Sultan, H., & Rehman, A. U. (2024). Application of nanotechnology for targeted drug delivery and nontoxicity. *International Journal of General Practice Nursing*, *2*(2), 57–67.

14

Nanoencapsulation Approaches in Food Processing and Packaging

Sharjeel Haider[1], Nosheen Mirza[2], Muhammad Anwar-ul-Haq[3], Khadija Bibi[4], Venuste Munyaneza[1], Ayaz Ali[1], Iftikhar Ali Ahmed[1], Muhammad Mehran[1], Sidra Sohail[5], Asma Zafar[6], Dua e Zainab[7], and Sara Rauf[1]

[1] *College of Resources and Environment/Microelement Research Center/Key Laboratory of Arable Land Conservation (Middle and Lower Reaches of Yangtze River), Ministry of Agriculture and Rural Affairs, Huazhong Agricultural University, Wuhan, China*
[2] *Department of Soil and Environmental Sciences, Ghazi University, Dera Ghazi Khan, Pakistan*
[3] *Institute of Soil and Environmental Sciences, University of Agriculture Faisalabad, Pakistan*
[4] *Department of Botany, Faculty of Sciences, Ghazi University, Dera Ghazi Khan, Pakistan*
[5] *State Key lab of soil erosion and dryland farming on the loess plateau, Institute of Soil and Water Conservation, Northwest A&F University, China*
[6] *College of Natural Resources and Environment, Northwest A&F University, Yangling, Shaanxi, PR China*
[7] *Key Laboratory of Green and Low Carbon Agriculture on Dryland in Northwest China, Ministry of Agriculture and Rural Affairs, College of Natural Resource and Environment, Northwest A & F University, China*

14.1 Introduction to Nanoencapsulation

The basic life components, DNA along with amino acids sugars and hormones, exist in the nanometer ranges (Weiss et al., 2006). People have engineered nanomaterials through inspiration from the natural structures that they observed, in order to advance the progress of human civilization. Richard Feynman established nanostructures during 1959 while Nario Taniguchi first used the word nanotechnology for submicron particle management in 1974. The term "nano" refers to a magnitude of 10–9 m (Quintanilla-Carvajal et al., 2010). The British Standards Institution defined nanotechnology as its description of designing and building materials along with their characterization and production along with their applications, which use nanoscale size and shape manipulation (Bawa et al., 2005). Scientific research in the last few decades has found its most promising field in nanotechnology. Nanotechnology focuses on making materials using processes for production and processing, which results in products smaller than 1,000 nm in size (Sanguansri and Augustin, 2006). Smaller-sized nanoparticles within the nanoscale dimension boost their surface exposure level, followed by a drastic rise in their chemical response as well as significant transformations in electrical and optical properties. The characteristics of nanoscale materials provide numerous distinctive applications in multiple fields (Neethirajan and Jayas, 2011). Silicon chips have been developed through nanotechnology techniques for more than 20 years. Nano sized material

Nanotechnology Innovations for Food Security and Sustainable Agriculture, First Edition. Edited by Zia Ur Rehman Mashwani, Shah Fahad, Ilyas Ahmad, Shah Saud, Sajid Fiaz, and Taufiq Nawaz.

development enables revolutionary progress in computers and electronics that transformed worldwide operations and reshaped human living possibilities.

A process called encapsulation led to the development of particles between a few nanometers to a few millimeters in size through substances enclosed inside other substances. Basically, the entrapped substance exists as a liquid form, though solids or gases are also possible. Nanoencapsulation is a process that forms a thin layer of protection against environmental degradation and retains the nutritional and functional qualities of bioactive compounds in nanoscale capsules by employing fats, starches, dextrins, alginates, protein and lipid materials as encapsulation materials. The main reason for using encapsulation is the fact that some nutrients do not remain in the food for a significant amount of time or may react with the other food components, causing undesirable effects. Two encapsulation approaches exist: under micro- and nanoscales. Researchers describe microencapsulation as a method that enhances food nutrient persistence and manages release timing for specific consumption stages and during gut transit (microencapsulation of vitamin). The protective shield of nanoencapsulation defends bioactive food components against adverse environmental factors, while at the same time improving solubility and odor masking, as well as strengthening the absorption of functional ingredients that show limited bioavailability. Drug delivery operates primarily through nanoencapsulation and conjugation as its main methodologies. Nanoparticles enhance both drug-encapsulating capabilities and delivery precision, which leads to improved treatment effects. Medical researchers apply nanoparticle technology, which enables them to place new genes and replace faulty genes into patients' genetic makeup (Dinh et al., 2024). A suitable nanocarrier system helps protect bioactive compounds by using chitosan, zein, and alginate as resistance components against gastrointestinal tract enzymes through various delivery methods, which include association colloids, nanoparticles, nanoemulsions, nanofibers/nanotubes, nanolaminates. The selection of an encapsulation method depends on two primary criteria related to core and wall material properties: (i) core nature and (ii) wall material characteristics such as size, thickness, solubility, permeability, and delivery speed. These production processes can be categorized as chemical (emulsion and interfacial polymerization), physical–chemical (emulsification and coacervation), and physical–mechanical methods (spray-drying/spray-cooling/spray-congealing/prilling, freeze-drying, electrodynamic methods, and extrusion) (Perinelli et al., 2020). Some applications consolidate different methods, where homogenization produces emulsions before converting them to dry powder through spray-drying together with freeze-drying methods (da Silva Barbosa et al., 2021). Research shows that spray-drying dominates flavor encapsulation, with approximately 80–90% applications, whereas spray-chilling and melt extrusion and melt injection, respectively, follow with 5–10% and 2–3% usage and 2% applications (Gupta et al., 2016).

14.2 Principles of Nanoencapsulation

14.2.1 Size and Surface Properties of Nanocarriers

Nanocarriers' success in nano-encapsulation relies heavily on their size characteristics together with their surface attributes. Nanocarrier dimensions between 1 and 1000 nm regulate their particle surface extent, their dissolution capacity, and their biological systemic relationships. Nanocarriers that are small display higher ratios between their surface area and volume so they penetrate biological environments better and improve the availability of their encapsulated substances. Nanocarrier stability together with their behavior depends significantly on surface attributes, which include zeta potential, surface hydrophobicity, and functional group composition (Duan and Li, 2013; Powers et al., 2007; Singh and Lillard Jr, 2009). Nanoparticle dispersion benefits from stable surface charges, which stop their agglomeration (Cooley et al., 2018). Drug delivery systems can be engineered with surface modifications of ligands or antibodies to enhance cell and tissue specificity in targeted drug delivery.

Nanocarrier surface attributes govern their availability in the body as well as their stability together with cellular uptake potential and distribution within the system (Duan and Li, 2013; Honary and Zahir, 2013). The zeta (ζ) potential of nanocarriers serves two purposes since it reveals surface charges that produce electrostatic interactions and helps identify proper coating materials (Jahanshahi and Babaei, 2008). An electrical current applied during recording with laser Doppler velocimetry demonstrates the movement of nanocarriers to determine their zeta potential (Clogston and Patri, 2010). Because of its accuracy and sensitivity along with versatility, electrophoretic light scattering has become the usually adopted technique. The measurement of zeta potential remains an approach that shows high sensitivity to the ionic strength level and pH values. The measurement requires solution dilution prior to the start. Interpreted data from nanocarrier solutions containing mixed oppositely charged carriers can become unreliable (Joye and McClements, 2014).

Hydrophobic properties of nanocarriers can be measured through three techniques: adsorption probe method, hydrophobic interaction chromatography, and contact angle measurements and biphasic partitioning. X-ray photon correlation spectroscopy also helps identify specific chemical groups on the nanocarrier surface for hydrophobicity prediction (Scholes et al., 1999).

14.2.2 Biocompatibility and Biodegradability

The key aspect of nanoencapsulation involves deploying readily degradable yet compatible biological materials to build nanocarriers. Nanocarriers exhibit biocompatibility when they can stay in biological environments without producing toxic reactions. A nanocarrier that breaks down safely within the body without creating any dangerous residues belongs to the category of biodegradable materials. Lipids together with polymers and proteins represent preferred materials that undergo body metabolism while remaining nontoxic. The safe delivery of nanoencapsulated substances depends heavily on this principle because drug delivery applications require extended time-based safety. Due to their degradation ability, biodegradable nanocarriers operate best for minimizing environmental harm throughout several years of agricultural use together with other potential applications (Witika et al., 2020). Biodegradable polymers usually demonstrate biocompatibility properties that make it possible to use electronics in implantable biomedical applications. The majority of biodegradable polymers do not suffice for certain application demands that require flexibility, electrical conductivity, dielectric properties, and gas and water vapor blocking properties. Current research focuses on nanocomposite preparation through biopolymer-nanofiller combinations meant to add functional characteristics without compromising biodegradability and biocompatibility (Liu et al., 2019).

14.2.3 Release Mechanism

The release mechanism stands as one of the fundamental principles in nano-encapsulation since it specifies the timing and process of releasing encapsulated substances from nano-carriers. Controlled release represents a method that delivers the contained substance through a slow, gradual process over an extended period to minimize high levels of concentration in the bloodstream while also decreasing the severity of side effects. Drug delivery systems highly benefit from sustained and homogeneous drug release because it enhances treatment results together with lowering side effect risks. The design of nanocarriers allows engineers to specify triggerable release functionality. External factors, including pH and temperature as well as enzymes and other stimuli, enable controlled release of encapsulated substances from the delivery system. The acidic conditions of tumors can be used to trigger chemotherapy drug release from nanocarriers designed for cancer treatment. The ability to trigger drug release proves exceptionally beneficial when scientists need to target particular tissue or cellular locations for effective therapy. The random core movement known as (i) diffusion is one of five mechanisms of release, where a concentration gradient drives the process. In this process, the release of the active agent depends on various factors, including the physical–chemical characteristics of both the core and the matrix, as well as the ratio between them (Wang et al., 2020). (ii) Swelling, where differences in solvent concentration cause the whole shell structure to swell with increased pore size, making it difficult to maintain capsule integrity and causing core molecule release (Lengyel et al., 2019). (iii) Fragmentation occurs when the matrix is disrupted by physical, chemical, or biological stresses, and in this process, the amount of core released depends on the magnitude of the stress as well as the shape and size of the resulting capsule fragments (McClements, 2014). (iv) Erosion is a process that can be caused by various factors, including temperature, pH, enzymes, and mechanical stimulation. Surface erosion beside bulk erosion represent two distinct ways by which capsules degrade during the process (Wischke and Schwendeman, 2011). (v) Osmotic ressure and leaching occurs where difference in osmotic potential or solvent penetration into the capsule lead to dissolution and migration of the active core into the surrounding medium. This process is strongly influenced by the solubility of the active agent, the permeability of the capsule wall, and the nature of the surrounding environment (Gouin, 2004). Moreover, surface erosion leads to capsule surface degradation and bulk erosion leads to complete capsule breakdown. The drug release process starts from either the point of application surface or from the breached surface (Fredenberg et al., 2011).

14.2.4 Types of Nanocarriers

Nanoencapsulation makes use of nanocarriers that function as small-sized materials for encapsulating active components while enhancing their protection characteristics along with stability and delivery precision. Nanocarriers exist in three main categories, including lipids, polymers, and their combination as hybrid systems. The selection of nanocarriers during nanoencapsulation technology determines the overall performance and steady implementation of the included substance, alongside affecting its delivery characteristics. Specified applications require unique properties found within three major nanocarrier types: lipid-based, polymer-based, and inorganic nanocarriers. The drug delivery system consisting of liposomes and solid lipid nanoparticles (SLNs) functions as an effective carrier for hydrophobic and hydrophilic substances because they maintain high biocompatibility and offer straightforward usage for biomedical and pharmaceutical applications (Kyriakoudi et al., 2021). Various nanocarriers built from polymers, including nanoparticles, micelles, and dendrimers, present adjustable properties for their size, surface composition, and biodegradability based on treatment needs (Bahrami et al., 2020; Masiero et al., 2021). There is expanding use of inorganic nanocarriers made from silica and gold nanoparticles, because they provide both chemical versatility and chemical resistance properties, as well as high drug-loading capability suitable for diagnostic imaging and drug delivery applications (Alshawwa et al., 2022; Yu et al., 2018).

14.2.5 Targeted Delivery

Targeted delivery works as a principle to provide accurate delivery of the encapsulated substance to certain locations and cell types. Nanocarriers possess design capabilities that enable them to detect and adhere to distinctive target-cell receptors or antigens, thus delivering the encapsulated substance exclusively to its intended destination (Jahangirian et al., 2017; Lam et al., 2017). The technique finds critical application in cancer drug administration because it directs medication delivery directly to cancer cells while reducing harm to other body systems and improving treatment effectiveness. Nanocarriers receive ligands or antibodies for surface attachment to achieve active targeting, since these bound elements specifically recognize target cell receptors (Kamaly et al., 2012; Shi et al., 2011). Passive targeting occurs when nanocarriers use their dimensions and surface properties for tissue accumulation through enhanced permeability and retention (EPR). Through targeted delivery, medical interventions become more effective while minimizing adverse effects, as the treatment specifically reaches its intended target area (Carmeliet and Jain, 2000).

14.3 Methods of Nanoencapsulation

14.3.1 Solvent Evaporation Method

The bioactive substance and polymer get dissolved together with a volatile organic solvent for this process. The solvent evaporation process results in the encapsulated substance becoming trapped within the polymer matrix when the solvent creates vacuum pressure or under spray drying conditions (Figure 14.1). The microencapsulation process relies heavily on solvent evaporation techniques since their implementation in recent times. The technological group of solvent evaporation techniques splits into two main sections: emulsification solvent evaporation and extraction. The original method has resulted in multiple technological variations, which were developed in recent years. Laboratory technicians can use solvent evaporation techniques to create microsphere morphologies and other features that precisely match delivery needs when transporting peptides and vaccines using carrier biomaterials. The research area received comprehensive attention regarding method

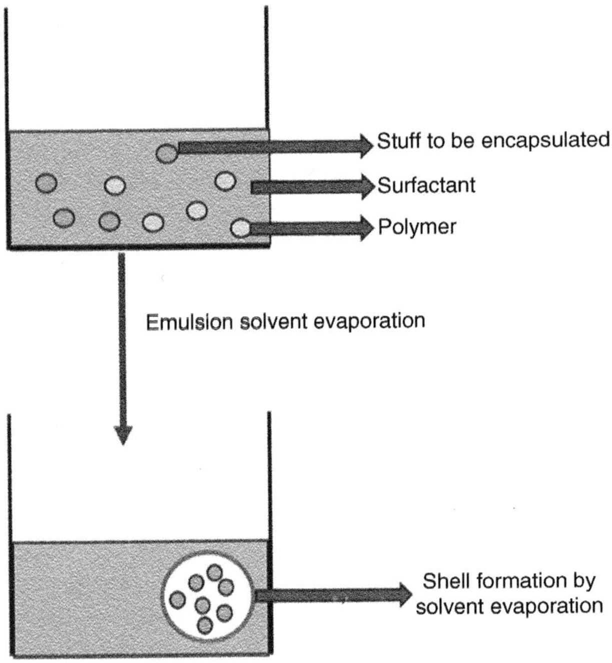

Figure 14.1 Schematic representation of the solvent evaporation method for nanoencapsulation.

variations for solvent evaporation alongside core materials, coat materials, and emulsion stabilizers with multiple process variables discussed extensively. The drug targeting technology presents promising prospects to pharmaceutical scientists while also creating specific obstacles regarding scale-up challenges, together with non-organic solvent use and biodegradable alternative polymer use, in addition to implementing hybrid delivery technologies from multiple encapsulation methods to preserve peptide stability during production and post-process stability of microspheres (Hwisa et al., 2013; Li et al., 2008; Maqbool and Noreen, 2019; Staff et al., 2013).

14.3.2 Nanoprecipitation (Solvent-Displacement Method)

The combination of a solution that contains a polymer and a bioactive substance with water as a non-solvent through rapid mixing triggers polymer precipitation, which encloses the bioactive substance (Figure 14.2). Scientists consider nanoprecipitation to be a reproducible, easy method for producing submicron-sized polymer particles. Multidisciplinary research works have focused on improving technique reproducibility because industry applications require more suitable processes. The preparation of different polymeric particulate carriers through the nanoprecipitation process continues to be a common method (Miladi et al., 2016). Industrial production of nanoparticles benefits from flash nanoprecipitation as a competent manufacturing process. Hydrophobic materials and hydrophilic chemical compounds usually are encapsulated through nanoprecipitation. Expert studies

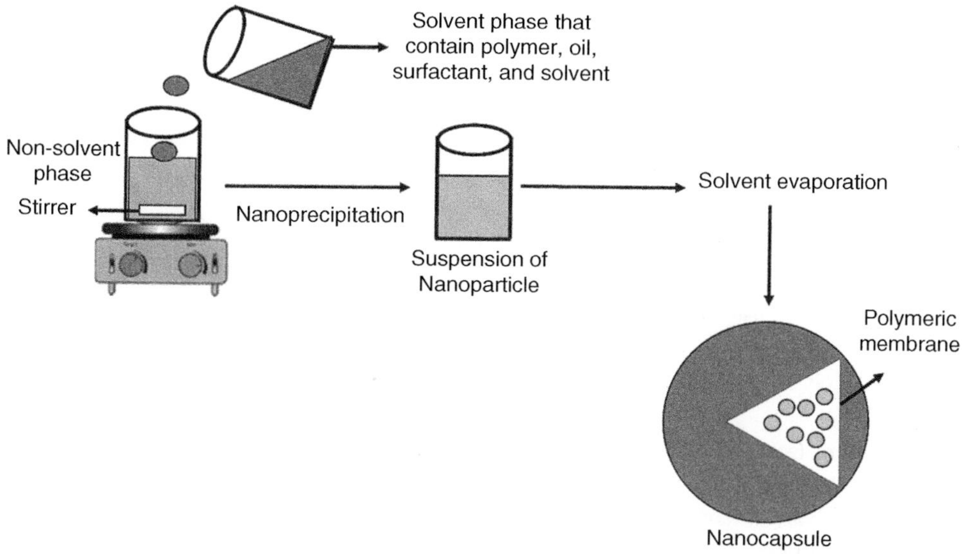

Figure 14.2 Illustration of the nanoprecipitation technique for nanoencapsulation.

have demonstrated that nanoprecipitation serves effectively as a suitable encapsulation method for natural compounds.

The particle characteristics of nanoparticles depend significantly on all elements involved in both processing methods and formulation development within the nanoprecipitation technique (Figure 14.2). Nanoparticles intended for in vivo studies have used both biodegradable and nonbiodegradable polymers for their preparation. The biodistribution patterns of active-loaded nanoparticles have been studied through literature research after using various administration routes to different body regions (Rivas et al., 2017).

14.3.3 Electrospinning Method

A polymer solution transforms into fine fibers through the application of electric field in the electrospinning process (Figure 14.3). The fiber production method enables researchers to embed bioactive substances inside the fibers when they are created. The electrospinning technique manages (bio)polymeric solutions or melts through high electrical voltages to produce suitable fibrous micro- and nanostructures by optimizing process variables post-production (Figure 14.3). Electrospinning alongside its related process called electrospraying demonstrates potential for being a new method to deliver food compounds because this method enables the use of aqueous solutions while working at room temperature without needing coagulation reactions to produce nanoscale and microscale matrices. Nanofibers (which have diameters under 1 μm) are regularly produced as a standard manufacturing practice.

The food industry shows strong research interest in nanofiber materials from electrospinning due to their potential application in controlled delivery systems. By using the food-grade, compatible room-temperature process, industry can efficiently encapsulate bioactives by

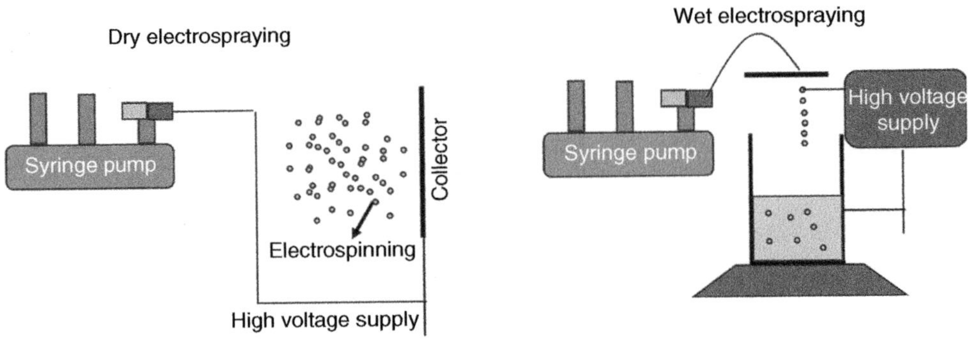

Figure 14.3 Overview of the dry electrospraying method for nanoencapsulation.

protecting them from denaturation and achieving improved stability. The development of electrospun fibers presents promising prospects to improve the design quality alongside performance advancements of products and delivery frameworks for supplementary food compounds (Ghorani and Tucker, 2015; Rostamabadi et al., 2020; Shetty et al., 2022).

14.3.4 High-Pressure Homogenization

A solution containing a bioactive substance and a stabilizing agent moves through a small orifice when submerged in high-pressure conditions during high-pressure homogenization (HPH) processing. Nanoparticles emerge through the breakdown of large droplets or particles, which occurs due to intense shear forces during the process. HPH serves as an important technology to perform pharmaceutical operations, including solid comminution and particle mixture and stabilization. Advancements in nanotechnology have led to significant improvements in the HPH technique, which has expanded its pharmaceutical applications domain through particle engineering processes. The deep knowledge of fluid dynamics has enabled scientists to create advanced models of high-pressure homogenizers, which show better processing performance and operational effectiveness (Luedtke et al., 2022). Research exists that demonstrates HPH generates nanosized liposomes using the method without organic solvents while maintaining good encapsulation efficiency (Ibišević et al., 2019; Wagner and Vorauer-Uhl, 2011).

14.3.5 Supercritical Fluid Technology (SCF)

Supercritical fluid technology produces nanoparticles by using supercritical fluids mainly carbon dioxide as solvents. The bioactive substance dissolves within a supercritical fluid before the solution is combined with a co-solvent or exposed to rapid pressure changes to produce nanoparticles. The process finds its use within pharmaceutical operations together with food production facilities to create drug, flavor, and nutraceutical capsules. The system produces nanoparticles with narrow size distribution while using no toxic substances, which makes it friendly toward the environment (Figure 14.4).

Supercritical carbon dioxide (CO_2) together with other supercritical fluid properties creates various process opportunities in pharmaceutical applications. The field of

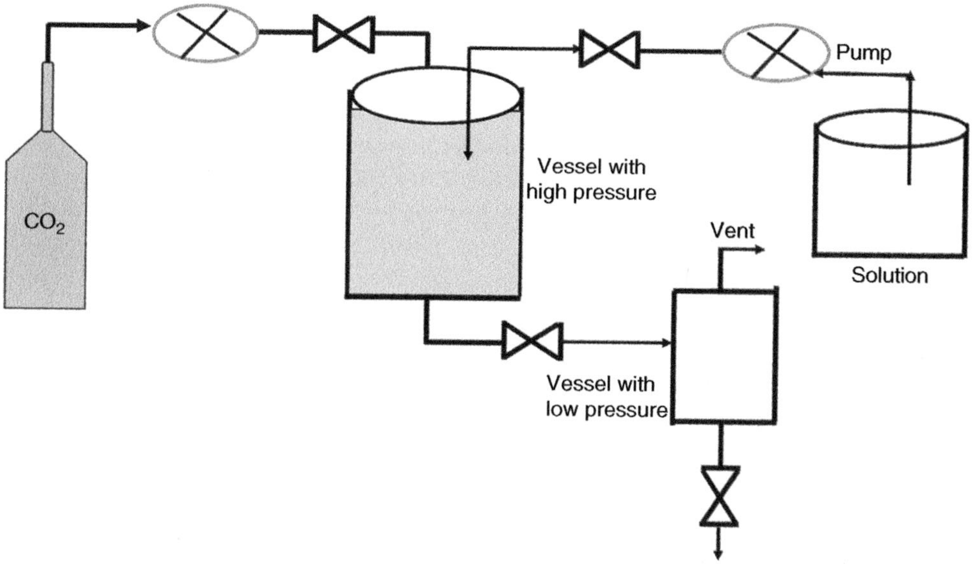

Figure 14.4 Depiction of the supercritical fluid (SCF) method for nanoencapsulation.

pharmaceuticals finds one of its potential uses in converting drugs into micro- and nanoencapsulation systems for delivery management purposes. Particle size design and drug loading become possible through supercritical CO_2 processes due to the adjustable properties of supercritical CO_2, which can be managed by altering operating conditions like flow ratio, temperature, and pressure parameters (Figure 14.4). Research presents an extensive breakdown of supercritical fluid processes using their distinctive properties by explaining the importance of supercritical carbon dioxide throughout processing together with formulation generation mechanisms for each technique evaluated (Soh and Lee, 2019).

14.4 Application of Nanoencapsulation in Food Processing: Nutrient Delivery, Flavor, and Aroma Retention

The emerging food industry technology called nanoencapsulation provides transformative solutions, which enhance nutrient delivery as well as flavor protection and elevated food quality standards. The encapsulation of bioactive compounds within nanoparticles allows improved optimization of stability alongside enhanced bioavailability as well as controlled substance release. The food processing industry benefits from several key applications of nanoencapsulation, as demonstrated below.

14.4.1 Nutrient Delivery and Bioavailability Enhancement

The most promising use of nanoencapsulation in food processing serves to improve nutritional delivery. Several bioactive compounds, including vitamins and antioxidants, together with essential fatty acids, demonstrate sensitivity to light exposure and heat and oxygen that leads to their degradation and reduced effectiveness. The protection of sensitive nutrients through nanoencapsulation creates a solution that safeguards their effectiveness from the damaging effects of oxidative conditions during processing and throughout storage and while being consumed. The body faces difficulties absorbing hydrophobic nutrients (including vitamin D and omega-3 fatty acids) because they do not dissolve well in water. The packaging of such nutrients inside nanocarriers helps boost their solubility properties and increases their absorption efficiency. The controlled release of nutrients occurs through nanoencapsulation within the digestive system, which directs these substances toward absorption-maximizing small intestine areas (Karunaratne et al., 2017; Nallamuthu et al., 2017).

Nanoencapsulation technology improves the availability of fat-soluble vitamins, including vitamin E and A, by encapsulating them in lipid-based nanocarriers. The encapsulation of vitamins within lipid-based nanocarriers, such as liposomes or solid lipid nanoparticles, enables better absorption by the body to enhance their health advantages (Pugazhendhi et al., 2024).

14.4.2 Flavor and Aroma Retention

Food products require pristine sensory elements for customer satisfaction because this preserves their flavor and aroma properties. Electrical energy pretreatment of food generates beneficial conditions for food processing and storage by protecting essential oil-derived flavors and aromatic substances, which are susceptible to heat, light, and oxygen. The volatile compounds remain intact through nanoencapsulation because this technology provides protective measures against degradation. Through nanoencapsulation manufacturers can manage how flavors and aromas become available in food products. Prepared food products experience a sustained intense flavor profile because nanocarrier technologies protect volatile compounds by controlled sequential release during food consumption. The controlled release emerges through nanoencapsulation and enables activated release based on defined conditions like temperature and pH along with mechanical forces (e.g., chewing) for enhancing dynamic flavor delivery (English et al., 2023).

The encapsulation process delivers flavors into nanoparticles to conceal unfavorable tastes of sometimes necessary active ingredients and nutrients, including fish oils or specific vitamins. Manufacturers use specific release timing of flavors from designed nanocarriers to develop more appealing products for consumers (Burgos et al., 2017). Nanoencapsulation protects food volatile aromatic compounds through its protective mechanism alongside maintaining flavor quality in food products. The processing and storage methods together with packaging processes lead to oxidation and evaporation of numerous basic food aromas. Nanocarriers based on cyclodextrins and lipid nanoparticles protect the enclosed volatile compounds by shielding them from air exposure and light rays during storage (Saifullah et al., 2019). Through encapsulation, the sensory aroma compounds survive the cooking process and heating treatments that matter for foods like sauces, soups, or beverages where aroma affects the sensory experience.

14.5 Application of Nanoencapsulation in Food Packaging

In the food industry, nanoencapsulation has been used for many years to improve the stability, bioavailability, and controlled release of active agents by enclosing bioactive compounds in nanometer-sized carriers, usually ranging from 1 to 100 nm. Natural or synthetic polymers are the most commonly used encapsulation materials (Chowdhury et al., 2024). Nanoencapsulation has developed as a disruptive method in food packaging, resulting in advanced systems that prolong shelf life, enhance food safety, and improve quality. The bioactive compounds exhibit enhanced stability and bioavailability, hence augmenting their physiological effects (Nyari et al., 2018). Vitamins, antioxidants, enzymes, bioactive peptides, dyes, and flavors, and other substances with insecticidal and antimicrobial properties, such as those found in essential oils and herbal extracts, are also being encapsulated (Timilsena et al., 2020). Biotechnology encounters significant challenges in managing volatile oils and extracts. The essential oil derived from savory leaves was integrated with natural polymers such as apple pectin and gum arabic. All polymers demonstrated an enhanced efficiency in encapsulation. The herbicidal effect of amaranth, tomato, and other plants was subsequently enhanced (Pavoni et al., 2019). In a long-term experiment, pepper oil was encapsulated, and its antibacterial activity was tested. Encapsulating this oil with malt dextrin had a repressive action against *Pseudomonas aeruginosa*, *Enterococcus faecalis*, and *Staphylococcus aureus*. Applications of nanoencapsulated compounds in food packaging also include.

14.5.1 Active Packaging System

The concepts of active packaging and intelligent packaging denote enhanced systems utilized for food and various other products. They contribute to prolonging shelf life, monitoring freshness, conveying quality information, enhancing safety, and increasing convenience (Dainelli et al., 2008). Active packaging engages with food items to preserve or enhance their state. Nanoencapsulation enables the integration of active agents such as antimicrobials, antioxidants, and oxygen scavengers into packaging materials. These nanocarriers safeguard the active agents from degradation and regulate their release, thereby improving the effectiveness and durability of the packaging (Anwar and Warsiki, 2018).

14.5.2 Antimicrobial Packaging and Its Substances

A major public health concern and financial burden on the food business, microbial contamination is a common occurrence in food items and is one of the primary causes of foodborne illnesses (Dhanasekar et al., 2018; Ju et al., 2019). Adding antimicrobial chemicals to packaging materials is known as antimicrobial packaging systems. These agents' primary goals are to prevent or lessen the growth of undesirable microbes on food surfaces, extending product shelf life and ensuring food safety. In an antimicrobial packaging system, a delivery mechanism is used, which inhibits the growth of microbes from all stages of transportation to ultimate consumption (Mangalassary, 2012; Giannakourou and Tsironi, 2021).

Table 14.1 Examples of antimicrobial packaging applicable to meat and cheese.

Type of cheese/ antimicrobial agent	Description	References
Saloio cheese	Whey protein isolate as a coating for lactic acid, natamycin, or chitooligosaccharides. A coating with natamycin and lactic acid was chosen as the optimal choice for cheese.	Nostro et al. (2013)
Cheddar cheese	Low-density polyethylene (LDPE) and cellulose films coated with peptides from *Bacillus licheniformis* Me1 demonstrate proven biopreservative effectiveness in curbing pathogen growth	Nithya, Murthy and Halami (2013)
Mozzarella cheese	Sachets made from microcellular foam starch infused with rosemary oil and thyme volatile oils also demonstrated suppressive effects on the growth of lactic acid bacteria (LAB) and total aerobic bacteria (TAB)	Han (2014)
Ultra-filtrated (UF) cheese	LDPE films integrated with silver (Ag), copper oxide (CuO), and zinc oxide (ZnO) nanoparticles show the best antibacterial effectiveness when Cu-ZnO is present without Ag nanoparticles	Beigmohammadi et al. (2016)
Nisin, Chitosan, Potassium sorbate	Assessing the efficacy of various antimicrobial agents on the microbiological traits of chicken drumsticks	Moreno et al. (2018)
Nanoencapsulated *Satureja khuzestanica* essential oil (SKEO)	Evaluation of chitosan coatings enhanced with SKEO in lamb meat	Pabast et al. (2018)
Gelatin/palm wax/ lemongrass essential oil (GPL)	Evaluate the antimicrobial properties in vacuum-packed sausages Assess the performance of the GPL-coated Kraft paper in preserving the quality of minced beef	Syahida et al. (2021)

The antimicrobial agent is directly incorporated into food particles or polymer films/ packaging to inhibit the activity of target microorganisms such as *Listeria monocytogenes*, *Mycobacterium smegmatis* (MTCC 943), *Pseudomonas aeruginosa* (MTCC 4676), *Escherichia coli* O157, *Salmonella*, *Staphylococcus aureus*, *Bacillus cereus*, *Campylobacter*, *Clostridium perfringens*, *Aspergillus niger*, *Saccharomyces cerevisiae*, etc (Zhao et al., 2019). A more comprehensive compilation of work on the incorporation of antimicrobial agents added in edible film and coating are given in Table 14.1.

14.5.3 Antioxidant Packaging

Antioxidants are increasingly important food ingredients in the food-processing sector. The traditional function of antioxidants, as suggested by their name, is to prevent the progression of oxidative rancidity in foods rich in fats, especially fried items, animal-derived foods, and dairy products. Synthetic antioxidants such as butylated hydroxytoluene (BHT), tertiary butyl hydroquinone (TBHQ), propyl gallate (PG), and butylated hydroxyanisole (BHA) are commonly employed to inhibit the oxidation of food (Xu et al., 2021).

Key natural antioxidants include tocopherols, essential oils, volatile phenols, flavanols, organic acids, and others. Incorporating antioxidants directly into food has specific constraints, as interactions between food components and antioxidants can diminish their effectiveness, potentially making the food unsafe. Delving into the process of lipid oxidation reactions, one might discover that most of these reactions begin at the surface of foods. Therefore, food scientists are working to encapsulate natural antioxidants and incorporate them into edible packaging to enhance shelf life by protecting food items from oxidation, oxidative rancidity, and color fading (Table 14.2). A broader collection of studies in which antioxidants and other nanoencapsulated compounds with nanomaterial are presented in Table 14.2.

14.5.4 Intelligent Packaging System

According to Vanderroost et al. (2014), intelligent packaging is a complete packaging solution that intelligently monitors changes in a product or its surroundings and actively responds to these changes. Intelligent packaging includes packaging technologies that specially designed features to monitor, communicate, or improve various aspects pertaining to the quality or safety packaged product. Intelligent packaging utilizes chemical sensors or biosensors to monitor the quality and safety of food throughout the supply chain from producers to consumers (Xu et al., 2021). Intelligent packaging utilizes several sensors to monitor the quality and safety of food by assessing attributes such as freshness, pH levels, carbon dioxide, pathogens, oxygen, and leaks, in addition to tracking time and temperature (Schaefer and Cheung, 2018).

The characteristics of intelligent packaging solutions vary based on the enclosed product (Kuswandi et al., 2013). Likewise, the particular situation that must be monitored, conveyed, or altered varies correspondingly. Intelligent packaging facilitates the monitoring and tracing of a product throughout its life cycle by enabling the analysis and management of the package's internal or external environment, thereby providing real-time information on the product's condition to manufacturers, retailers, or consumers (Schaefer and Cheung, 2018). Examples of gadgets or sensors frequently utilized in intelligent packaging to assess the quality and safety of food products are presented below.

14.5.5 Time–Temperature Indicators (TTIs)

These are devices employed to assess the temperature of items within a package (Liu et al., 2023). The temperature influences the duration for which a specific food item can be preserved (Realini and Marcos, 2014). Preserving the dietary value of food while protecting it from adverse temperature fluctuations is a significant concern and a formidable challenge for perishable food items. Inadequate or unregulated temperature control may lead to substantial food loss during procurement and throughout the supply chain (Drago et al., 2020). Research indicates that inadequate temperature regulation during the transportation and storage of perishable foods might lead to a 35% product loss (Göransson,

Table 14.2 Application of nanoencapsulated compounds in food industry.

Sr. No.	Nanoencapsulated compounds with nanomaterials	Function	Food material	References
1	Vitamin D3 with potato protein	Encouragement of human health	Beverage solutions	Meftah Kadmiri et al. (2021)
2	Vitamin E with edible mustard oil	Health supplement antioxidant	Mango juice	David and Livney (2016)
3	α-Tocopherol with canola oil	Antioxidant agent	Fish sausages	Dasgupta et al. (2016)
4	Quinoa peptide	Natural food preservative	Fresh beef burgers	Feng et al. (2020)
5	Pear peel fruit	Preservative	Guava juice	Yekta et al. (2020)
6	Olive leaf phenolics (oleuropein) with cholesterol	Functional food product	Yogurt	Mahmoud, Ali and Amin (2018)
7	Fish skin peptide	Preservative and vehicle for entrapping fishy smell	Pork patties	Lopes, Pinilla and Brandelli (2019)
8	Fish oils	Improvement of the nutritional value	Bread	Bai et al., (2017)
9	Prebiotics (insulin, lactulose)	Control of cholesterol, cancer prevention, constipation relief	Milk	Gómez et al. (2017)
11	Nisin and lysozyme	Inhibition of *Listeria monocytogens* and *Salmonella enteritidis*	Whole or skim UHT milk	Lopes, Pinilla and Brandelli, (2019)
12	*Thymus capitatus* EO	Improving physicochemical and microbial (*S. aureus*) quality of milk	Semi-skimmed UHT milk	Jemaa et al. (2017)
13	Xanthohumol	Prevention of microbial growth and lipid oxidation. Addition of nanoliposomes did not impair sensory properties	Beef sausage	Khatib et al. (2020)
14	Cumin seed essential oil encapsulated with nanoemulsions	Nanocarriers incorporated cellulosic papers had enhanced effects on beef hamburgers by significantly reducing thiobarbituric acid reactive substances (TBARS), total mesophilic and psychrophilic bacteria	Beef hamburger	Hemmatkhah, Zeynali and Almasi (2020)

Nilsson and Jevinger, 2018). Time–temperature indicators (TTIs) provide essential support in mitigating this issue. TTIs monitor the environment surrounding a package and utilize that data to identify fluctuations in temperature. It is also beneficial for identifying enzymatic and microbiological alterations in food products that are dependent on time and temperature (Wang et al., 2015). Temperature-sensitive products, particularly refrigerated and frozen commodities, are predominantly utilized for time–temperature indications. They are also employed to control pasteurization and sterilization procedures. Owing to its intrinsic characteristics, TTIs are considered user-friendly and easily accessible instruments (Drago et al., 2020).

14.5.6 Freshness Indicators

Freshness indicators are advanced instruments developed to assess and convey the freshness of food items throughout storage and transit by identifying microbial proliferation and chemical alterations (Table 14.2). These indicators, integrated into the packaging, respond to by-products of microbial activity, offering visual signals regarding the product's microbiological quality (Kuswandi et al., 2013). Freshness indicators, in contrast to temperature indicators, evaluate the onset of decomposition by tracking specific components such as amines, oxygen, and carbon dioxide. In contrast to temperature indicators, freshness indicators deliver immediate information regarding the item's state by analyzing the chemical processes associated with food degradation due to certain types of bacteria (Kalpana et al., 2019).

14.5.7 Biosensors and Gas Sensors

Biosensors detect alterations in food packaging resulting from biological processes. The primary components of biosensors are bioreceptors and transducers (Müller and Schmid, 2019). The bioreceptor identifies the bio-enzymatic component, which is subsequently converted into a quantifiable electrical output by the transducer. Biosensors are capable of detecting allergens, glucose, carbohydrates, amino acids, alcohols, fatty acids, pathogenic organisms, and various other analytes. They can also identify target metabolites generated during the degradation of food (Shi et al., 2022). A primary problem with these sensors is the immobilization of biological components within the receptor to avert denaturation and disintegration, typically accomplished using adhesive methods such as electrodeposition (Drago et al., 2020). It is imperative to prevent any detrimental effects from biological agents contaminating food (Vanderroost et al., 2014; Siracusa and Lotti, 2019). The transducer may be either electrochemical or optical (Kalakonda et al., 2023). Among the diverse biosensors examined in the literature, electrochemical sensors have been the most thoroughly investigated (Drago et al., 2020). The chemical composition of gases in the packaging atmosphere alters due to the food, packaging structure, and external influences upon exposure. This presents a challenge in preserving the nutritional integrity of food ingredients during the packaging process. Gas indicators have been developed to tackle this difficulty by monitoring gas fluctuations and evaluating the efficacy of active packaging components such as carbon dioxide and oxygen scavengers (Ghaani et al., 2016).

14.5.8 Improvement of Mechanical and Barrier Properties

The integration of nanoencapsulated fillers into packaging materials can markedly improve their mechanical strength and barrier characteristics. Nanocomposites, created by incorporating nanocarriers into polymer matrices, yield packaging that exhibits enhanced resistance to physical damage and improved barrier properties against gas and moisture transfer (Garcia, Shin and Kim, 2018). E.g., the incorporation of nanoencapsulated clay particles into biodegradable polymers has demonstrated enhanced durability and decreased permeability, consequently prolonging the shelf life of food goods. The enhanced food packaging incorporates nanoparticles to improve mechanical and physical attributes, including durability, strength, flexibility, biodegradability, thermal resistance, UV absorption, and impermeability to water vapor and oxygen. The integration of metal oxides into polymers enhances mechanical, barrier, and light transmittance qualities. (Arfat et al., 2017) developed bio composite films with fish skin gelatin and Ag–Cu nanoparticles. The films exhibited enhanced mechanical strength, thermal stability, UV barrier characteristics, and antibacterial efficacy against gram-positive (*L. monocytogenes*) and gram-negative (*S. typhimurium*) bacteria.

14.5.9 Edible Nanocoatings

Currently, nanocoating is employed in food packaging to create enhanced packaging solutions. Food coating can be applied as thin layers or films, produced from both edible and nonedible materials. The edible coating is a promising solution for extending the shelf life of food goods. Various biopolymers, including polysaccharides and lipid proteins, are utilized in the formulation of edible coatings (Jeevahan and Chandrasekaran, 2019). Single polymer coatings impart specific properties to food packaging, which may be inadequate for optimal food quality preservation. Consequently, the need for composites and nanocomposites has arisen to simultaneously deliver multiple properties and enhance preservation (Ashfaq et al., 2022). Nanofillers enhance the mechanical properties, barrier qualities, and coloration of edible films compared to standard edible films; nonetheless, further study is necessary to render them competitive with synthetic films. (Dash et al., 2019) created starch-pectin films infused with TiO_2. It was observed that TiO_2 nanoparticles enhance mechanical and oxygen barrier characteristics. It was determined that increasing the concentration of TiO_2 nanoparticles enhanced the UV impermeability of the starch–pectin blended films.

14.6 Benefits of Nanoencapsulation in Food Processing and Packaging

The application of nanotechnology to the surface of food is connected to packaging and ensuring food purity. Food can get polluted or deteriorate at various stages of the food chain; therefore, safe, harmless, affordable, high quality packaging material is key as nanoencapsulation in nanotechnology has appeared as an innovative technology in the food industry (Ameta et al., 2020). By utilizing nanotechnology, nanoencapsulation provides various advantages in food packaging and manufacturing, improving food quality, safety, and preservation time.

14.6.1 Enhanced Bioavailability and Nutrient Absorption

One of the main benefits of nanoencapsulation is the enhancement of bioavailability of nutrients and bioactive compounds. By decreasing the size of these compounds to the nanoscale, they possess a larger surface area relative to their volume, which helps improve absorption in the human digestive system. This improvement is mainly useful for delivering important nutrients, vitamins, and antioxidants that may otherwise have imperfect bioavailability. For example, nanoencapsulation has been used to increase the solubility and assimilation of flavonoids, compounds known for their health-supporting attributes but usually delayed by low water solubility and stability (Sekhon, 2010).

14.6.2 Keeping Our Food Safe – Nanomaterials Style

Food nanoparticles can be organic (carbon-based) or inorganic (non-carbon-based). Their performance in the human body depends on their composition, as well as their size and shape (McClements and Xiao, 2017). Silver, silicon dioxide, iron oxide, titanium dioxide, and zinc oxide are commonly used in the food industry. Silver nanoparticles are utilized as antimicrobial agents in various applications such as food wrapping, chopping boards, packaging containers, refrigerators, and dietary supplements. Zinc and zinc oxide nanoparticles are important nutrients for human health, and hence, they are used as preservatives in supplements and functional foods for nutrition. Zinc oxide nanoparticles serve as antibacterial agents in food packaging to evade bacterial contamination. Organic nanoparticles are composed of biochemical compounds such as carbohydrates, proteins, or lipids (Sirelkhatim et al., 2015). Nanopackaging decreases food contamination by improving packaging integrity and enabling real-time monitoring of food conditions. Nano has facilitated improving the heat resistance and mechanical features in food packaging constituents. It can also facilitate detect variations in food products, such as harmful microorganisms growing on the surface. Nanotechnology can also aid food industries to avoid contamination through substantiating and following the trace attributes of a food product utilizing nanobarcodes (Nam et al., 2003).

14.6.3 Extension of Shelf Life and Food Safety

Integrating nanoencapsulated antimicrobial agents into food packaging materials can actively prevent the growth of decomposition and pathogenic microorganisms, so prolonging the shelf life of foodstuffs. For example, packaging materials integrated with silver nanoparticles have revealed antimicrobial properties that support and sustain food integrity and freshness (Manohar et al., 2017).

14.6.4 Development of Functional and Smart Foods

The idea of smart foods for health, driven by nanoencapsulation technology, signifies a novel approach. It presents new avenues and confronts difficulties within the realms of both the food production and pharmaceutical sectors. Nanoencapsulation technology represents a revolutionary advancement, mostly in the domain of edible coatings. It allows the incorporation of a variety of smart nutrition and bioactive components, deliberately releasing them in specific locations. This accuracy significantly enhances the constancy, flavor, color, and quality of food products throughout their processing and storage stages. More, this technology reveals its capability by helping as a means to provide smart and bioactive substances in a customized and focused approach. This not only improves food safety but also expands the shelf life of foodstuffs (Singh et al., 2023).

14.7 Challenges and Limitations of Nanoencapsulation in Food Processing and Packaging

The food industry shows intense interest in nanoencapsulation because it offers improved methods for delivering bioactive substances and preservatives alongside nutrients while enhancing food storage period and product quality. The wide industry adoption of nanoencapsulation for food processing and packaging faces numerous challenges together with limitations which need resolution. Technical challenges together with regulatory hurdles as well as safety concerns and consumer acceptance factors hinder the practical application and successful integration of nanoencapsulation in food systems.

14.7.1 Safety Concerns and Toxicity

The main challenge with food-based nanoencapsulation emerges from the safety risks and toxic effects of nanomaterials. The small dimensions of nanoparticles compare to standard materials lead to modifications in their physical and chemical makeup and biological dynamics, which allow absorption through biological tissue such as skin and mucous membranes and the blood-brain barrier. Some nanomaterials demonstrate compatibility with biological tissue but other types prove harmful because they accumulate and lead to unplanned effects in human bodies. The complete understanding of long-term exposure hazards from nanomaterials still remains ambiguous because studies of the toxicological risks between food system nanoparticles (Biswas et al., 2022). Food migration of particles from packaging materials represents a primary concern because of their movement into edible substances. Nanomaterials running into contact with food elements or dissolving into food products creates potential issues for product quality and safety during the final stage. Nanomaterials require rigorous safety assessments and standardized protocols for evaluating their risks according to regulatory positions before receiving approvals for food applications (McClements and Xiao, 2017; Gupta, Guha, et al., 2024).

14.7.2 High Production Costs

The production of nanoencapsulated food products and packaging requires specialized equipment together with materials and expert knowledge, which leads to production expenses that exceed traditional methods. Research shows that nanoparticle synthesis uses advanced production methods including chemical vapor deposition, HPH, and solvent evaporation. The encapsulation method needs both skilled personnel and strict control of particle dimensions along with encapsulation performance along with surface properties during processing (Singh, 2016; Shamloo et al., 2024).

Nanomaterials used in food packaging applications through methods that include nanoparticle-based coatings and active packaging films result in cost elevation. Nanoencapsulation of food products along with related packaging becomes less affordable for businesses and customers primarily regarding basic food options and undeveloped marketplaces that prioritize price.

14.7.3 Stability and Storage Issues

Nanoencapsulation demonstrates effectiveness in maintaining the stability of various bioactive compounds such as vitamins alongside antioxidants and enzymes but nanocarriers tend to experience stability-related challenges. Nanosized materials tend to build aggregates during storage times while simultaneously undergoing oxidation processes and chemical breakdown mainly when stored under unfavorable environments including light, high temperatures, or moist conditions. Concerning nanocarrier degradation, the released encapsulated substances become ineffective and have a reduced shelf life (Pateiro et al., 2021). Food packaging stability depends heavily on the stability of nanomaterials within films or coatings since this determines food quality and maintains safety throughout storage periods. The storage stability of nanocarriers and active ingredients affects the benefits obtained from nanoencapsulation because decreased stability affects the performance and shelf life of the nanoencapsulated system.

14.8 Conclusion and Future Perspectives

Despite the many advantages that nanoencapsulation offers in food processing and packaging, such as enhanced stability, improved bioavailability, and extended shelf life, several challenges and limitations must be addressed. Safety concerns, regulatory hurdles, high production costs, stability issues, consumer resistance, environmental impact, and compatibility with existing processing methods are significant barriers that need to be overcome. Nanotechnology is being visualized as a quickly evolving field with prospective to revolutionize food systems. There have been tremendous progresses in nanotechnology applications in food science and research. From enhancing product shelf life to advanced isolation of microbes to better-quality food storage and safety to tracing-tracking of contaminants/pollutants to detection of toxins/pathogens to nutritional or functional supplements through food, nanotechnology has touched almost every spectrum of the food processing sector. The direct applications mostly include mixing of coloring agents, fragrances, nanopreservatives, antioxidants, and bioactive compounds like polyphenols, vitamins, and fatty acids. Indirect applications encompass the usage of nanosized substances in packaging material, nanosensors, and usage of catalysts in hydrogenation in case of fats. Recent advancements in the field of nanoscience-based applications have shown prospects to re-frame processing line, re-structure the production chain, promote natural resource conservation, and even re-define the food habits.

Such advancements have major implications for the agri-food systems and will ultimately have vital impacts on the quality, safety, storability, and security of foods benefiting both producers and consumers influencing the economic outlook.Continued research, innovation, and collaboration between researchers, regulators, and food manufacturers are essential to addressing these challenges and ensuring the safe, sustainable, and effective use of nanoencapsulation in the food industry. However, more research is needed based on precautionary principles, particularly in regards to NSM migration patterns in food matrix, human cytotoxicity of nanoparticles, and their potential effects on consumer health and safety, as well as the environment. Potential risks, toxicological issues, and environmental considerations must all be addressed. The concerned rules and regulations are only to be laid to overcome various prevailing safety challenges after which, it would rule the entire food industrial domain.

References

Alshawwa, S. Z., Kassem, A. A., Farid, R. M., Mostafa, S. K., & Labib, G. S. (2022). Nanocarrier drug delivery systems: Characterization, limitations, future perspectives and implementation of artificial intelligence. *Pharmaceutics, 14*(4), 883.

Ameta, S. K., Rai, A. K., Hiran, D., Ameta, R., & Ameta, S. C. (2020). Use of nanomaterials in food science. *Biogenic Nano-particles and their Use in Agro-ecosystems*, 457–488.

Anwar, R. W., & Warsiki, E. (2018). The comparison of antimicrobial packaging properties with different applications incorporation method of active material. In *IOP conference series: Earth and environmental science* (pp. 12002). IOP Publishing.

Arfat, Y. A., Ahmed, J., Hiremath, N., Auras, R., & Joseph, A. (2017). Thermo-mechanical, rheological, structural and antimicrobial properties of bionanocomposite films based on fish skin gelatin and silver-copper nanoparticles. *Food Hydrocolloids, 62*, 191–202.

Ashfaq, A., Khursheed, N., Fatima, S., Anjum, Z., & Younis, K. (2022). Application of nanotechnology in food packaging: Pros and Cons. *Journal of Agriculture and Food Research, 7*, 100270.

Bahrami, A., Delshadi, R., Assadpour, E., Jafari, S. M., & Williams, L. (2020). Antimicrobial-loaded nanocarriers for food packaging applications. *Advances in Colloid and Interface Science, 278*, 102140.

Bai, J.-J., Lee, J.-G., Lee, S.-Y., Kim, S., Choi, M.-J., & Cho, Y. (2017). Changes in quality characteristics of pork patties containing antioxidative fish skin peptide or fish skin peptide-loaded nanoliposomes during refrigerated storage. *Korean Journal for Food Science of Animal Resources, 37*(5), 752.

Bawa, R., Bawa, S. R., Maebius, S. B., Flynn, T., & Wei, C. (2005). Protecting new ideas and inventions in nanomedicine with patents. *Nanomedicine: Nanotechnology, Biology and Medicine, 1*(2), 150–158.

Beigmohammadi, F., Peighambardoust, S. H., Hesari, J., Azadmard-Damirchi, S., Peighambardoust, S. J., & Khosrowshahi, N. K. (2016). Antibacterial properties of LDPE nanocomposite films in packaging of UF cheese. *LWT, 65*, 106–111.

Biswas, R., Alam, M., Sarkar, A., Haque, M. I., Hasan, M. M., & Hoque, M. (2022). Application of nanotechnology in food: Processing, preservation, packaging and safety assessment. *Heliyon, 8*(11).

Burgos, N., Mellinas, A. C., García-Serna, E., & Jiménez, A. (2017). Nanoencapsulation of flavor and aromas in food packaging. In *Food packaging* (pp. 567–601). Elsevier.

Carmeliet, P., & Jain, R. K. (2000). Angiogenesis in cancer and other diseases. *Nature, 407*(6801), 249–257.

Chowdhury, S., Kar, K., & Mazumder, R. (2024). Exploration of different strategies of nanoencapsulation of bioactive compounds and their ensuing approaches. *Future Journal of Pharmaceutical Sciences, 10*(1), 72.

Clogston, J. D., & Patri, A. K. (2010). Zeta potential measurement. In *Characterization of Nanoparticles Intended for Drug Delivery* (pp. 63–70). Springer.

Cooley, M., Sarode, A., Hoore, M., Fedosov, D. A., Mitragotrib, S., & Gupta, A. S. (2018). Influence of particle size and shape on their margination and wall-adhesion: Implications in drug delivery vehicle design across nano-to-micro scale. *Nanoscale, 10*(32), 15350–15364.

Dainelli, D., Gontard, N., Spyropoulos, D., den Beuken, E. Z.-v., & Tobback, P. (2008). Active and intelligent food packaging: Legal aspects and safety concerns. *Trends in Food Science & Technology, 19*, S103–S112.

Dasgupta, N., Ranjan, S., Mundra, S., Ramalingam, C., & Kumar, A. (2016). Fabrication of food grade vitamin E nanoemulsion by low energy approach, characterization and its application. *International Journal of Food Properties, 19*(3), 700–708.

Dash, K. K., Ali, N. A., Das, D., & Mohanta, D. (2019). Thorough evaluation of sweet potato starch and lemon-waste pectin based-edible films with nano-titania inclusions for food packaging applications. *International Journal of Biological Macromolecules, 139*, 449–458.

David, S., & Livney, Y. D. (2016). Potato protein based nanovehicles for health promoting hydrophobic bioactives in clear beverages. *Food Hydrocolloids, 57*, 229–235.

Dhanasekar, M., Jenefer, V., Nambiar, R. B., Babu, S. G., Selvam, S. P., Neppolian, B., & Bhat, S. V. (2018). Ambient light antimicrobial activity of reduced graphene oxide supported metal doped TiO_2 nanoparticles and their PVA based polymer nanocomposite films. *Materials Research Bulletin, 97*, 238–243.

Dinh, L., Mahon, L., & Yan, B. (2024). Nano-encapsulation and conjugation applied in the development of lipid nanoparticles delivering nucleic acid materials to enable gene therapies. *Applied Nanoscience, 5*(3), 143–161.

Drago, E., Campardelli, R., Pettinato, M., & Perego, P. (2020). Innovations in smart packaging concepts for food: An extensive review. *Food, 9*(11), 1628.

Duan, X., & Li, Y. (2013). Physicochemical characteristics of nanoparticles affect circulation, biodistribution, cellular internalization, and trafficking. *Small, 9*(9-10), 1521–1532.

English, M., Okagu, O. D., Stephens, K., & Udenigwe, C. C. (2023). Flavour encapsulation: A comparative analysis of relevant techniques, physiochemical characterisation, stability, and food applications. *Frontiers in Nutrition, 10*, 1019211.

Feng, X., Tjia, J. Y. Y., Zhou, Y., Liu, Q., Fu, C., & Yang, H. (2020). Effects of tocopherol nanoemulsion addition on fish sausage properties and fatty acid oxidation. *LWT, 118*, 108737.

Fredenberg, S., Wahlgren, M., Reslow, M., & Axelsson, A. (2011). The mechanisms of drug release in poly (lactic-co-glycolic acid)-based drug delivery systems—A review. *International Journal of Pharmaceutics, 415*(1–2), 34–52.

Garcia, C. V., Shin, G. H., & Kim, J. T. (2018). Metal oxide-based nanocomposites in food packaging: Applications, migration, and regulations. *Trends in Food Science & Technology, 82*, 21–31.

Ghaani, M., Cozzolino, C. A., Castelli, G., & Farris, S. (2016). An overview of the intelligent packaging technologies in the food sector. *Trends in Food Science & Technology*, *51*, 1–11.

Ghorani, B., & Tucker, N. (2015). Fundamentals of electrospinning as a novel delivery vehicle for bioactive compounds in food nanotechnology. *Food Hydrocolloids*, *51*, 227–240.

Giannakourou, M. C., & Tsironi, T. N. (2021). Application of processing and packaging hurdles for fresh-cut fruits and vegetables preservation. *Food*, *10*(4), 830.

Gómez, B., Míguez, B., Yáñez, R., & Alonso, J. L. (2017). Extraction of oligosaccharides with prebiotic properties from agro-industrial wastes. In *Water extraction of bioactive compounds* (pp. 131–161). Elsevier.

Göransson, M., Nilsson, F., & Jevinger, Å. (2018). Temperature performance and food shelf-life accuracy in cold food supply chains–Insights from multiple field studies. *Food Control*, *86*, 332–341.

Gouin, S. (2004). Microencapsulation: industrial appraisal of existing technologies and trends. *Trends in Food Science & Technology*, *15*(7–8), 330–347.

Gupta, R. K., Guha, P., & Srivastav, P. P. (2024). Investigating the toxicological effects of nanomaterials in food packaging associated with human health and the environment. *Journal of Hazardous Materials Letters*, *5*, 100125.

Gupta, S., Khan, S., Muzafar, M., Kushwaha, M., Yadav, A. K., & Gupta, A. P. (2016). Encapsulation: Entrapping essential oil/flavors/aromas in food. In *Encapsulations* (pp. 229–268). Elsevier.

Han, J. H. (2014). A review of food packaging technologies and innovations. *Innovations in Food Packaging*, 3–12.

Hemmatkhah, F., Zeynali, F., & Almasi, H. (2020). Encapsulated cumin seed essential oil-loaded active papers: Characterization and evaluation of the effect on quality attributes of beef hamburger. *Food and Bioprocess Technology*, *13*, 533–547.

Honary, S., & Zahir, F. (2013). Effect of zeta potential on the properties of nano-drug delivery systems-a review (Part 2). *Tropical Journal of Pharmaceutical Research*, *12*(2), 265–273.

Hwisa, N. T., Katakam, P., Rao, C. B., & Adiki, S. K. U. M. A. R. I. (2013). Solvent evaporation techniques as promising advancement in microencapsulation. *Vedic Research International Biological Medicinal Chemistry*, *1*, 8–22.

Ibišević, M., Smajlović, A., & Arsić, I. (2019). Optimization of high pressure homogenization in the production of liposomal dispersions. *Technologica Acta: Scientific/Professional Journal of Chemistry and Technology*, *12*(2), 7–10.

Jahangirian, H., Lemraski, E. G., Webster, T. J., Rafiee-Moghaddam, R., & Abdollahi, Y. (2017). A review of drug delivery systems based on nanotechnology and green chemistry: Green nanomedicine. *International Journal of Nanomedicine*, 2957–2978.

Jahanshahi, M., & Babaei, Z. (2008). Protein nanoparticle: A unique system as drug delivery vehicles. *African Journal of Biotechnology*, *7*(25).

Jeevahan, J., & Chandrasekaran, M. (2019). Nanoedible films for food packaging: A review. *Journal of Materials Science*, *54*(19), 12290–12318.

Jemaa, M. B., Falleh, H., Neves, M. A., Isoda, H., Nakajima, M., & Ksouri, R. (2017). Quality preservation of deliberately contaminated milk using thyme free and nanoemulsified essential oils. *Food Chemistry*, *217*, 726–734.

Joye, I. J., & McClements, D. J. (2014). Biopolymer-based nanoparticles and microparticles: Fabrication, characterization, and application. *Current Opinion in Colloid & Interface Science*, *19*(5), 417–427.

Ju, J., Chen, X., Xie, Y., Yu, H., Guo, Y., Cheng, Y., Qian, H., & Yao, W. (2019). Application of essential oil as a sustained release preparation in food packaging. *Trends in Food Science & Technology*, *92*, 22–32.

Kalakonda, S. N., Bammidi, R., Edubilli, H., Medapati, S., Boyina, S. L. D., & Prasad, V. V. S. R. (2023). Biosensors-An Insight into the Electrochemical and Optical Biosensors. *International Journal of Pharmaceutical Investigation*, *13*(3).

Kalpana, S., Priyadarshini, S. R., Leena, M. M., Moses, J. A., & Anandharamakrishnan, C. (2019). Intelligent packaging: Trends and applications in food systems. *Trends in Food Science & Technology*, *93*, 145–157.

Kamaly, N., Xiao, Z., Valencia, P. M., Radovic-Moreno, A. F., & Farokhzad, O. C. (2012). Targeted polymeric therapeutic nanoparticles: Design, development and clinical translation. *Chemical Society Reviews*, *41*(7), 2971–3010.

Karunaratne, D. N., Siriwardhana, D. A. S., Ariyarathna, I. R., Rajakaruna, R. M. P. I., Banu, F. T., & Karunaratne, V. (2017). Nutrient delivery through nanoencapsulation. In *Nutrient delivery* (pp. 653–680). Elsevier.

Khatib, N., Varidi, M. J., Mohebbi, M., Varidi, M., & Hosseini, S. M. H. (2020). Replacement of nitrite with lupulon–xanthohumol loaded nanoliposome in cooked beef-sausage: Experimental and model based study. *Journal of Food Science and Technology*, *57*, 2629–2639.

Kuswandi, B., Maryska, C., Jayus, J., & heng, L. y. (2013). Real time on-package freshness indicator for guavas packaging. *Journal of Food Measurement and Characterization*, *7*, 29–39.

Kyriakoudi, A., Spanidi, E., Mourtzinos, I., & Gardikis, K. (2021). Innovative delivery systems loaded with plant bioactive ingredients: Formulation approaches and applications. *Plants*, *10*(6), 1238.

Lam, P.-L., Wong, W.-Y., Bian, Z., Chui, C.-H., & Gambari, R. (2017). Recent advances in green nanoparticulate systems for drug delivery: Efficient delivery and safety concern. *Nanomedicine*, *12*(4), 357–385.

Lengyel, M., Kállai-Szabó, N., Antal, V., Laki, A. J., & Antal, I. (2019). Microparticles, microspheres, and microcapsules for advanced drug delivery. *Scientia Pharmaceutica*, *87*(3), 20.

Li, M., Rouaud, O., & Poncelet, D. (2008). Microencapsulation by solvent evaporation: State of the art for process engineering approaches. *International Journal of Pharmaceutics*, *363*(1–2), 26–39.

Liu, H., Jian, R., Chen, H., Tian, X., Sun, C., Zhu, J., Yang, Z., Sun, J., & Wang, C. (2019). Application of biodegradable and biocompatible nanocomposites in electronics: Current status and future directions. *Nanomaterials*, *9*(7), 950.

Liu, Y., Li, L., Yu, Z., Ye, C., Pan, L., & Song, Y. (2023). Principle, development and application of time–temperature indicators for packaging. *Packaging Technology and Science*, *36*(10), 833–853.

Lopes, N. A., Pinilla, C. M. B., & Brandelli, A. (2019). Antimicrobial activity of lysozyme-nisin co-encapsulated in liposomes coated with polysaccharides. *Food Hydrocolloids*, *93*, 1–9.

Luedtke, F. L., Stahl, M. A., Grimaldi, R., Forte, M. B. S., Gigante, M. L., & Ribeiro, A. P. B. (2022). Optimization of high pressure homogenization conditions to produce nanostructured lipid carriers using natural and synthetic emulsifiers. *Food Research International*, *160*, 111746.

Mahmoud, K. F., Ali, H. S., & Amin, A. A. (2018). Nanoencapsulation of bioactive compounds extracted from Egyptian prickly pears peel fruit: Antioxidant and their application in Guava juice. *Journal of Asian Scientific Research, 11*(4), 574–586.

Mangalassary, S. (2012) 'Antimicrobial food packaging to enhance food safety: Current developments and future challenges'.

Manohar, C. M., Kundgar, S. D., & Doble, M. (2017). Betanin immobilized LDPE as antimicrobial food wrapper. *LWT, 80*, 131–135.

Maqbool, I., & Noreen, S. (2019). A review of novel techniques for nanoparticles preparation. *Global Drug Design & Development Review: GDDDR*, 41–50.

Masiero, J. F., Barbosa, E. J., de Oliveira Macedo, L., de Souza, A., Yukuyama, M. N., Arantes, G. J., & Bou-Chacra, N. A. (2021). Vegetable oils in pharmaceutical and cosmetic lipid-based nanocarriers preparations. *Industrial Crops and Products, 170*, 113838.

McClements, D. J. (2014). *Nanoparticle-and microparticle-based delivery systems: Encapsulation, protection and release of active compounds*. CRC press.

McClements, D. J., & Xiao, H. (2017). 'Is nano safe in foods? Establishing the factors impacting the gastrointestinal fate and toxicity of organic and inorganic food-grade nanoparticles. *npj Science of Food, 1*(1), 6.

Meftah Kadmiri, I., El Mernissi, N., Azaroual, S. E., Mekhzoum, M. E. M., Qaiss, A. E. K., & Bouhfid, R. (2021). Bioformulation of microbial fertilizer based on clay and alginate encapsulation. *Current Microbiology, 78*(1), 86–94.

Miladi, K., Fessi, H., Elaissari, A., & Sfar, S. (2016). Nanoprecipitation process: From particle preparation to in vivo applications. In *Polymer nanoparticles for nanomedicines: A guide for their design, preparation and development* (pp. 17–53).

Moreno, O., Atarés, L., Chiralt, A., Cruz-Romero, M. C., & Kerry, J. (2018). Starch-gelatin antimicrobial packaging materials to extend the shelf life of chicken breast fillets. *LWT, 97*, 483–490.

Müller, P., & Schmid, M. (2019). Intelligent packaging in the food sector: A brief overview. *Food, 8*(1), 16.

Nallamuthu, I., Khanum, F., Fathima, S. J., Patil, M. M., & Anand, T. (2017). Enhanced nutrient delivery through nanoencapsulation techniques: The current trend in food industry. In *Nutrient delivery* (pp. 619–651). Elsevier.

Nam, J.-M., Thaxton, C. S., & Mirkin, C. A. (2003). Nanoparticle-based bio-bar codes for the ultrasensitive detection of proteins. *Science, 301*(5641), 1884–1886.

Neethirajan, S., & Jayas, D. S. (2011). Nanotechnology for the food and bioprocessing industries. *Food and Bioprocess Technology, 4*, 39–47.

Nithya, V., Murthy, P. S. K., & Halami, P. M. (2013). Development and application of active films for food packaging using antibacterial peptide of Bacillus licheniformis Me1. *Journal of Applied Microbiology, 115*(2), 475–483.

Nostro, A., Scaffaro, R., D'Arrigo, M., Botta, L., Filocamo, A., Marino, A., & Bisignano, G. (2013). Development and characterization of essential oil component-based polymer films: A potential approach to reduce bacterial biofilm. *Applied Microbiology and Biotechnology, 97*, 9515–9523.

Nyari, N., Paulazzi, A., Zamadei, R., Steffens, C., Zabot, G. L., Tres, M. V., Zeni, J., Venquiaruto, L., & Dallago, R. M. (2018). Synthesis of isoamyl acetate by ultrasonic system using Candida antarctica lipase B immobilized in polyurethane. *Journal of Food Process Engineering, 41*(6), e12812.

Pabast, M., Shariatifar, N., Beikzadeh, S., & Jahed, G. (2018). Effects of chitosan coatings incorporating with free or nano-encapsulated Satureja plant essential oil on quality characteristics of lamb meat. *Food Control, 91*, 185–192.

Pateiro, M., Gómez, B., Munekata, P. E. S., Barba, F. J., Putnik, P., Kovačević, D. B., & Lorenzo, J. M. (2021). Nanoencapsulation of promising bioactive compounds to improve their absorption, stability, functionality and the appearance of the final food products. *Molecules, 26*(6), 1547.

Pavoni, L., Benelli, G., Maggi, F., & Bonacucina, G. (2019). Green nanoemulsion interventions for biopesticide formulations. In *Nano-biopesticides today and future perspectives* (pp. 133–160). Elsevier.

Perinelli, D. R., Palmieri, G. F., Cespi, M., & Bonacucina, G. (2020). Encapsulation of flavours and fragrances into polymeric capsules and cyclodextrins inclusion complexes: An update. *Molecules, 25*(24), 5878.

Powers, K. W., Palazuelos, M., Moudgil, B. M., & Roberts, S. M. (2007). Characterization of the size, shape, and state of dispersion of nanoparticles for toxicological studies. *Nanotoxicology, 1*(1), 42–51.

Pugazhendhi, A., Alshehri, M. A., Kandasamy, S., Sarangi, P. K., & Sharma, A. (2024). Deciphering the importance of nanoencapsulation to improve the availability of bioactive molecules in food sources to the human body. *Food Chemistry*, 141762.

Quintanilla-Carvajal, M. X., Camacho-Díaz, B. H., Meraz-Torres, L. S., Chanona-Pérez, J. J., Alamilla-Beltrán, L., Jimenéz-Aparicio, A., & Gutiérrez-López, G. F. (2010). Nanoencapsulation: A new trend in food engineering processing. *Food Engineering Reviews, 2*, 39–50.

Realini, C. E., & Marcos, B. (2014). Active and intelligent packaging systems for a modern society. *Meat Science, 98*(3), 404–419.

Rivas, C. J. M., Tarhini, M., Badri, W., Miladi, K., Greige-Gerges, H., Nazari, Q. A., Rodríguez, S. A. G., Román, R. Á., Fessi, H., & Elaissari, A. (2017). Nanoprecipitation process: From encapsulation to drug delivery. *International Journal of Pharmaceutics, 532*(1), 66–81.

Rostamabadi, H., Assadpour, E., Tabarestani, H. S., Falsafi, S. R., & Jafari, S. M. (2020). Electrospinning approach for nanoencapsulation of bioactive compounds; recent advances and innovations. *Trends in Food Science & Technology, 100*, 190–209.

Saifullah, M., Shishir, M. R. I., Ferdowsi, R., Rahman, R. T., & Van Vuong, Q. (2019). Micro and nano encapsulation, retention and controlled release of flavor and aroma compounds: A critical review. *Trends in Food Science & Technology, 86*, 230–251.

Sanguansri, P., & Augustin, M. A. (2006). Nanoscale materials development–A food industry perspective. *Trends in Food Science & Technology, 17*(10), 547–556.

Schaefer, D., & Cheung, W. M. (2018). Smart packaging: Opportunities and challenges. *Procedia CIRP, 72*, 1022–1027.

Scholes, P. D., Coombes, A. G. A., Illum, L., Davis, S. S., Watts, J. F., Ustariz, C., Vert, M., & Davies, M. C. (1999). Detection and determination of surface levels of poloxamer and PVA surfactant on biodegradable nanospheres using SSIMS and XPS. *Journal of Controlled Release, 59*(3), 261–278.

Sekhon, B. S. (2010). Food nanotechnology–An overview. *Nanotechnology, Science and Applications*, 1–15.

Shamloo, E., Nickfar, F., Mahmoudzadeh, M., Sarafraz, M., Salari, A., Darroudi, M., Abdi-Moghadam, Z., Amiryosefi, M. R., Rezagholizade-shirvan, A., & Rezaei, Z. (2024). Investigation of heavy metal release from variety cookware into food during cooking process. *International Journal of Environmental Analytical Chemistry*, *104*(19), 8085–8101.

Shetty, K., Bhandari, A., & Yadav, K. S. (2022). Nanoparticles incorporated in nanofibers using electrospinning: A novel nano-in-nano delivery system. *Journal of Controlled Release*, *350*, 421–434.

Shi, J., Xiao, Z., Kamaly, N., & Farokhzad, O. C. (2011). Self-assembled targeted nanoparticles: Evolution of technologies and bench to bedside translation. *Accounts of Chemical Research*, *44*(10), 1123–1134.

Shi, S., Xie, Y., Wang, G., & Luo, Y. (2022). Metabolite-based biosensors for natural product discovery and overproduction. *Current Opinion in Biotechnology*, *75*, 102699.

da Silva Barbosa, R. F., de Souza, A. G., Rangari, V., & dos Santos Rosa, D. (2021). The influence of PBAT content in the nanocapsules preparation and its effect in essential oils release. *Food Chemistry*, *344*, 128611.

Singh, A. K., Pal, P., Pandey, B., Goksen, G., Sahoo, U. K., Lorenzo, J. M., & Sarangi, P. K. (2023). Development of "Smart Foods" for health by nanoencapsulation: Novel technologies and challenges. *Food Chemistry: X*, *20*, 100910.

Singh, H. (2016). Nanotechnology applications in functional foods; opportunities and challenges. *Preventive Nutrition and Food Science*, *21*(1), 1.

Singh, R., & Lillard, J. W., Jr. (2009). Nanoparticle-based targeted drug delivery. *Experimental and Molecular Pathology*, *86*(3), 215–223.

Siracusa, V., & Lotti, N. (2019). Intelligent packaging to improve shelf life. In *Food quality and shelf life* (pp. 261–279). Elsevier.

Sirelkhatim, A., Mahmud, S., Seeni, A., Kaus, N. H. M., Ann, L. C., Bakhori, S. K. M., Hasan, H., & Mohamad, D. (2015). Review on zinc oxide nanoparticles: Antibacterial activity and toxicity mechanism. *Nano-Micro Letters*, *7*, 219–242.

Soh, S. H., & Lee, L. Y. (2019). Microencapsulation and nanoencapsulation using supercritical fluid (SCF) techniques. *Pharmaceutics*, *11*(1), 21.

Staff, R. H., Landfester, K., & Crespy, D. (2013). Recent advances in the emulsion solvent evaporation technique for the preparation of nanoparticles and nanocapsules. In *Hierarchical macromolecular structures: 60 years after the staudinger nobel prize II* (pp. 329–344). Springer.

Syahida, S. N., Ismail-Fitry, M. R., Ainun, Z. M. A., & Hanani, Z. A. N. (2021). Effects of gelatin/palm wax/lemongrass essential oil (GPL)-coated Kraft paper on the quality and shelf life of ground beef stored at 4°C. *Food Packaging and Shelf Life*, *28*, 100640.

Timilsena, Y. P., Haque, M. A., & Adhikari, B. (2020). Encapsulation in the food industry: A brief historical overview to recent developments. *Food and Nutrition Sciences*, *11*(6), 481–508.

Vanderroost, M., Ragaert, P., Devliegherea, F., & De Meulenaer, B. (2014). Intelligent food packaging: The next generation. *Trends in Food Science & Technology*, *39*(1), 47–62.

Wagner, A., & Vorauer-Uhl, K. (2011). Liposome technology for industrial purposes. *Journal of Drug Delivery*, 2011 *1*, 591325.

Wang, S., Liu, X., Yang, M., Zhang, Y., Xiang, K., & Tang, R. (2015). Review of time temperature indicators as quality monitors in food packaging. *Packaging Technology and Science, 28*(10), 839–867.

Wang, S., Liu, R., Fu, Y., & Kao, W. J. (2020). Release mechanisms and applications of drug delivery systems for extended-release. *Expert Opinion on Drug Delivery, 17*(9), 1289–1304.

Weiss, J., Takhistov, P., & McClements, D. J. (2006). Functional materials in food nanotechnology. *Journal of Food Science, 71*(9), R107–R116.

Wischke, C., & Schwendeman, S. P. (2011). Degradable polymeric carriers for parenteral controlled drug delivery. *Fundamentals and Applications of Controlled Release Drug Delivery*, 171–228.

Witika, B. A., Makoni, P. A., Matafwali, S. K., Chabalenge, B., Mwila, C., Kalungia, A. C., Christian, N., I, Bapolisi, A. M., & Walker, R. B. (2020). Biocompatibility of biomaterials for nanoencapsulation: Current approaches. *Nanomaterials, 10*(9), 1649.

Xu, X., Liu, A., Hu, S., Ares, I., Martínez-Larrañaga, M.-R., Wang, X., Martínez, M., Anadón, A., & Martínez, M.-A. (2021). Synthetic phenolic antioxidants: Metabolism, hazards and mechanism of action. *Food Chemistry, 353*, 129488.

Yekta, M. M., Rezaei, M., Nouri, L., Azizi, M. H., Jabbari, M., Eş, I., & Khaneghah, A. M. (2020). Antimicrobial and antioxidant properties of burgers with quinoa peptide-loaded nanoliposomes. *Journal of Food Safety, 40*(2), e12753.

Yu, H., Park, J.-Y., Kwon, C. W., Hong, S.-C., Park, K.-M., & Chang, P.-S. (2018). An overview of nanotechnology in food science: Preparative methods, practical applications, and safety'. *Journal of Chemistry, 1*, 5427978.

Zhao, Y., Teixeira, J. S., Saldaña, M. D. A., & Gänzle, M. G. (2019). Antimicrobial activity of bioactive starch packaging films against Listeria monocytogenes and reconstituted meat microbiota on ham. *International Journal of Food Microbiology, 305*, 108253.

15

Potential Health Risks Associated with Nanoparticles

Zakir Ullah[1,2], Javed Iqbal[3], Banzeer Ahsan Abbasi[4], Shumaila Ijaz[5], Rooma Waqar[1], Akhtar Munir[6], Muhammad Nasir Hussain[6], Tariq Mahmood[1], Shah Fahad[7], and Naila Ijaz[8]

[1] Department of Plant Sciences, Faculty of Biological Sciences, Quaid-i-Azam University, Islamabad, Pakistan
[2] Center for Pharmaceutical Engineering Science and Institute of Cancer Therapeutics, School of Pharmacy and Medical Sciences, Faculty of Life Sciences, University of Bradford, Bradford, United Kingdom
[3] Department of Botany, Bacha Khan University, Charsadda, Khyber Pakhtunkhwa, Pakistan
[4] Department of Botany, Rawalpindi Women University, Rawalpindi, Pakistan
[5] Department of Burn and Plastic Surgery, Shenzhen Institute of Translational Medicine, The First Affiliated Hospital of Shenzhen University, School of Biomedical Engineering, Shenzhen University Medical School, Shenzhen University, Shenzhen, China
[6] Department of Chemistry, Quaid-i-Azam University, Islamabad, Pakistan
[7] Department of Agronomy, Abdul Wali Khan University Mardan, Khyber Pakhtunkhwa, Pakistan
[8] Nano composite lab, Department of Physics, University of agriculture Faisalabad, Faisalabad, Pakistan

15.1 Introduction

Nanotechnology research has continued as an established field throughout the past hundred years (Abbasi et al., 2023). Nanotechnology came into use through Nobel laureate Richard P. Feynman, who first presented it in his famous 1959 lecture "There's Plenty of Room at the Bottom" (Abbasi et al., 2023; Abdelnour et al., 2021). Multiple groundbreaking developments have taken place in this field since its foundation. Nanotechnology provides methods to create numerous materials on the nanometer scale (Abdelnour et al., 2021). NPs comprise a wide range of materials whose smallest dimension measures less than 100 nm as per (Abbasi et al., 2023). The morphology of NPs allows researchers to classify them into zero-dimensional (0D), one-dimensional (1D), two-dimensional (2D), and three-dimensional (3D) structures (Aboyewa et al., 2021). Scientists observed that both the size and shape of these materials determine their physicochemical behavior, including their optical properties. The optical properties of gold (Au), platinum (Pt), silver (Ag), and palladium (Pd) NPs with a 20-nm diameter result in specific color appearances (wine-red, yellowish-gray, black, and dark black), respectively (Ali et al., 2024). This explains that optical changes result from nanoparticle dimensions and structure that lead to their success in bioimaging applications (Ali et al., 2024). Absorption characteristics and color expressions stem from the three

Nanotechnology Innovations for Food Security and Sustainable Agriculture, First Edition. Edited by Zia Ur Rehman Mashwani, Shah Fahad, Ilyas Ahmad, Shah Saud, Sajid Fiaz, and Taufiq Nawaz.

main variables—aspect ratio, nanoshell thickness, and gold content (Almanza-Reyes et al., 2021).

The composition of NPs consists of a surface layer with metal ions and small molecules, surfactants, or polymers, along with functionalization possibilities, a shell layer with a chemical nature different from the core material, and the core that involves the central section of the NPs. Various scientists from diverse expert fields have demonstrated substantial interest in these unique structural properties (Almanza-Reyes et al., 2021).

The arranged structure of mesoporous architectures enhances the functional properties of NPs. Scientists have intensely studied NPs for various applications because of their exceptional qualities, including drug delivery (Aboyewa et al., 2021), gas sensing (Ali et al., 2024), chemical and biological sensing (Almanza-Reyes et al., 2021), carbon dioxide capture (Almanza-Reyes et al., 2021), and other technological developments (Paras et al., 2022).

15.2 Classification of Nanomaterials

NMs can be systematically classified based on their key characteristics, including dimensionality, morphology, physical state, and chemical composition (Asghar et al., 2024). To qualify as an NM, at least one dimension must fall within the nanoscale range of 1–100 nm (Paras et al., 2022). Based on their structural dimensions and overall geometry, NMs are broadly categorized into four distinct classes, providing a comprehensive framework for their classification and application (Almanza-Reyes et al., 2021).

a) The nanoscale dimensions comprise the whole area of 0D NMs, which measure less than 100 nm. The nanomaterial categories include spherical NPs along with nanocubes and nanorods, as well as hollow spheres, quantum dots (QDs) and core–shell structures (Paras et al., 2022).
b) The classification of 1D NMs requires nanoscale dimensions to exist across two parameters and an extended dimension that exists at the macro level. The family of nanomaterials includes nanowires and nanotubes, together with nanorods and nanofibers with metal, polymer, and ceramic material composition (Asghar et al., 2024).
c) NMs that belong to the 3D category have one dimension falling within nanoscale boundaries. Righteous nanofilms join nanoplates alongside nanosheets, while coatings of both crystalline and amorphous types belong to this category (Paras et al., 2022.
d) 3D NMs possess dimensions that exceed 100 nm by incorporating assembled nanocrystals or nanostructures. Various carbon nanobuds, nanoflowers, polycrystals, foams, and nanostructured frameworks fall under the category of 3D NMs (Asghar et al., 2024).

A key factor in the characteristics of NPs is their morphology, which includes flatness, aspect ratio, and sphericity. NPs can be either isometric or anisometric based on homogeneity, and their magnetic characteristics and surface charges can determine whether they are scattered, agglomerated, or clustered (Asmatulu et al., 2024). Functionalization and surface alterations affect whether they are hydrophilic or hydrophobic (Azizi-Lalabadi et al., 2021). The following are examples of morphological variations: nanostars, nanozigzags, nanocubes, nanohooks, nanohelices, and nanoplates.

15.2.1 Classification

15.2.1.1 Metal Nanomaterials

In contrast to their bulk counterparts, metallic NPs such as Au, Ag, and Fe exhibit distinct chemical, optical, and electrical characteristics (Bibi et al., 2023). Phonetic and sonoelectrochemical synthesis methods generate these characteristics according to (Bruna et al., 2021). The high surface area-to-volume ratio present in these NPs improves their reactivity, thereby making them valuable for applications in catalysis and photonics. Platinum, palladium, and nickel NPs exhibit inter-band transition absorption features, whereas the intra-band transition behavior occurs in Au, Ag, and Cu systems. When subjected to light radiation, the NPs show enhanced catalytic activity according to (Bibi et al., 2023).

15.2.2 Metal Oxide Nanomaterials

Different metal oxide NPs, such as TiO_2, ZnO, Fe_2O_3, Al_2O_3, and SiO_2, can be synthesized using hydrothermal or sol–gel methods. These nanoparticles exhibit excellent chemical detection capabilities, catalytic properties, and semiconducting behavior (Bibi et al., 2023; Ullah et al., 2024). Their surface characteristics, including hydrophobicity, target specificity, and binding affinity, can be finely tuned by incorporating polymers, fatty acids, or organic substances such as epoxies, amines, and thiols (Bruna et al., 2021).

15.2.3 Bimetallic Nanomaterials

Such materials consist of two different metals that combine to enhance stability alongside reactive properties and optical and magnetic characteristics. The structural behavior of metals depends on three critical elements, which are composition, form, and miscibility properties. Various systems used for catalysis comprise Au/Pd alongside Fe–Cu as well as Ag–Au and Pd/Cu according to de Dicastillo and Correa (2021).

15.2.4 Composite Nanomaterials

Nanocomposites improve characteristics through matrix material and nanoparticle combination to attain conductivity, strength, and water resistance (Dhaka et al., 2023). Nanoclays blended with polymers and titanium dioxide connected to chitosan represent common types of composites. Such materials establish interconnected platforms by hydrogen bonding, π–π stacking, and electrostatic forces, which enhance drug delivery effectiveness and catalysis and adsorption capabilities (Ding et al., 2022).

15.2.5 Carbon-Based Nanomaterials

Numerous exceptional properties define the mechanical features, together with thermal resistance and electrical capability of carbon nanomaterials. Types consist of: The carbon molecule known as C60 in the form of Buckyballs functions as an antioxidant by scavenging radicals (0D). Moreover, nanotubes serve as the fundamental element in electronics and sensors because they exist as hexagonally arranged, cylindrical, hollow carbon structures (1D). The one-atom-thin structure of sp^2-linked carbon elements constitutes graphene, which exhibits exceptional conductivity features along with extended surface area properties (2D). Multilayered carbon structures that use optical, abrasive, and semantic properties are known as graphite/nanodiamonds (3D) (Ding et al., 2022).

15.2.6 Zeolite and Silica-Based Nanomaterials

The adjustable pore structures of zeolites and mesoporous silica NPs match with their spacious surface areas. Hydrolysis rate, together with template selection and pH value, determines the synthesis process of these materials (Egbuna et al., 2021). The addition of metal complexes to nanoclays enables makers to use these materials for catalytic purposes. The combination of carbon nanotubes (CNTs) with silica NPs shows promise for environmental and catalytic applications, according to Ettlinger et al. (2022).

15.2.7 Ceramic Nanomaterials

Ceramic NMs work as heat-resistant inorganic materials that find use in mechanical applications alongside high-temperature operations. Nanostructured oxides, carbides, and nitrides function together for their high hardness along with corrosion resistance and thermal stability properties.

15.3 Synthesis of NPs

The synthesis of NMs uses two fundamental methods known as bottom-up and top-down (Abbasi et al., 2023; Abdelnour et al., 2021; Egbuna et al., 2021). NPs produced at the atomic level through the bottom-up approach eventually get integrated into the desired material (Abdelnour et al., 2021). The synthesis utilizes colloidal dispersion to form NPs before proceeding to powder synthesis through sol–gel methods and then integrates them (Ding et al., 2022) describe the many techniques, such as atomic layer deposition, molecular self-assembly, vapor phase deposition of NMs, spinning, biological synthesis, sedimentation, reduction, green synthesis, and deoxyribonucleic acid scaffolding of nanoelectronics. The top-down approach begins with a bulk material at the macroscopic level and uses a variety of reduction techniques to transform it into the required NPs (Ettlinger et al., 2022). The bottom-up fabrication approach consists of various techniques, including etching, ball milling, grinding/milling, CVD, physical vapor deposition (PVD), and optical lithography, in addition to e-beam lithography, soft lithography, nanoimprint lithography, block copolymer lithography, and scanning probe lithography (Table 15.1) (Ettlinger et al., 2022).

15.3.1 Functionalization of Nanomaterials

Applications of NMs call for their modification. Because of intermolecular interactions, including dipole–dipole and van der Waals forces, the stable agglomerates of unmodified NPs are formed. By functionalizing with other chemical species and different functional groups, NMs can be modified to develop specific active sites on their surface or

Table 15.1 A general overview of synthesis methods, types, and applications of various nanoparticles.

NPs	Synthesis methods	Types	Applications	References
Metal NPs (e.g., Au, Ag, Pt, Pd)	Chemical reduction, green synthesis, electrochemical methods	Spherical, rod-shaped, cubic	Cancer therapy, antibacterial agents, catalysis, imaging	Abbasi et al., (2023; Ullah et al., 2025)
Metal Oxide NPs (e.g., Fe_3O_4, ZnO, TiO_2)	Sol-gel, co-precipitation, hydrothermal, green	Hollow spheres, nanosheets, nanorods	MRI contrast agents, drug delivery	Abdelnour et al., (2021)
Carbon-based NPs (e.g., graphene, CNTs, carbon dots)	Chemical vapor deposition (CVD), arc discharge, laser ablation	Tubes, sheets, dots	Electronics, biosensors, drug delivery, energy storage	Aboyewa et al., (2021)
Polymeric NPs (e.g., PLGA, chitosan, PEG-based)	Emulsion–solvent evaporation, nanoprecipitation, ionic gelation	Nanospheres, nanocapsules	Controlled drug delivery, gene delivery, vaccine delivery	Ali et al., (2024)
Lipid-based NPs (e.g., liposomes, solid lipid NPs)	Thin-film hydration, high-pressure homogenization	Vesicles, nanocapsules	Cancer therapy, mRNA vaccine delivery, antiviral therapies	Almanza-Reyes et al., (2021)
Dendrimers (e.g., PAMAM, PPI)	Divergent and convergent synthesis methods	Tree-like branched structures	Drug delivery, gene therapy, diagnostic imaging	Paras et al. (2022)
Silica NPs (e.g., mesoporous silica)	Stöber process, sol–gel synthesis	Spheres, rods	Targeted drug delivery, catalysis, biosensing	Asghar et al., (2024)
Quantum Dots (QDs) (e.g., CdSe, ZnS)	Hot injection, colloidal synthesis	Spherical or core-shell structures	Bioimaging, photovoltaics, LEDs	Asmatulu et al., (2024)
Ceramic NPs (e.g., ZrO_2, Al_2O_3, SiO_2)	Sol–gel, spray pyrolysis	Hollow particles, nanoparticles, nanoplatelets	Structural materials, sensors, drug delivery	Azizi-Lalabadi et al., (2021)
Magnetic NPs (e.g., Fe_3O_4, $CoFe_2O_4$)	Co-precipitation, thermal decomposition	Spherical, cubic, octahedral	Hyperthermia therapy, targeted drug delivery, MRI	Bibi et al., (2023)

alter their surface characteristics. Functionalized NMs have been used in several applications and fields because of these unique surface interactions (Ullah et al., 2025). Noncovalent and covalent functionalities can be used for functionalization via electrostatic, van der Waals, and hydrogen bond forces. Conditions like reaction temperature, solvent, and surfactant can be selectively changed to control functionalities as well as the select configurations (Fatima et al., 2021). Chemical bonds may be formed between atoms present on the surface of NMs through chemical reaction. Noncovalent functionalizations of NM surfaces are achieved through hydrophobic interaction, and thus, on NM surfaces, the electronic configurations and properties of the atoms on the NM surface remain mostly the same.

This is also in addition to ionic bonding, stacking, and hydrogen bonding (Fatima et al., 2021). There are several ways to carry out NM functionalization, including the following:

i) Chemical methods, which include acid, alkali, O_2, heat, and modification with appropriate silane coupling agents and metal ethoxides and alkoxides (Gao et al., 2023);
ii) Ligand exchange processes using ascorbic acid, oleic acid, oleylamine, trioctylphosphine, polyvinyl alcohol, polyvinylpyrrolidone, and cetyl trimethyl ammonium bromide (Ullah et al., 2025);
iii) In situ surface modifications, which are accomplished during the NP synthesis phase by the thermal breakdown of organometallic chemicals, reverse micelle approaches, and polyol methods (Fatima et al., 2021) to prevent aggregations, surfactants, and capping agents, such as oleic acid, amines, and trioctylphosphine oxides that dissolve in the reaction media. NPs (such as Al_2O_3, TiO_2, and Fe_2O_3) are dispersed using anionic surfactants based on a variety of polycarboxylic acids and their salts (Ullah et al., 2025) and
iv) Grafted synthetic polymers using grafting-to, grafting-from, and grafting-through techniques.

NM surfaces are often directly connected to end-functionalized polymers in the grafting-to process (Ettlinger et al., 2022). Grafting via process involves the co-polymerization of a low molecular weight monomer with another monomer on the surface of the NMs, whereas grafting from procedure creates polymer chains from the starting sites on the surface (Fatima et al., 2021). Depending on the necessary uses and the physicochemical characteristics of the intended NP applications, several pathways for modifying NPs with polymers can be developed (Gao et al., 2023). However, selecting the right synthesis method is essential to producing NMs with the right characteristics. Many synthesis processes may be divided into two categories: (ii) using premade polymers and (iii) polymerizing monomers (Fatima et al., 2021). A good strategy should be adopted to obtain certain desirable qualities and to avoid some of the potential downsides.

15.3.2 Characterization of NPs

The use of NPs in many areas, such as medicine, agriculture, environmental science, and materials engineering, necessitates the search for a general method of producing NPs by stabilizing them in water. Therefore, comprehensive characterization of NPs is necessary to ensure the safe and effective application. Evaluation of their chemical, physical, and biological properties is made using a broad spectrum of advanced analytical techniques (Gao et al., 2023).

Evaluation of the most important characteristics of NPs includes their composition, size, shape, surface area, distribution, stability, surface chemistry, roughness, and topography (García-Quintero & Palencia, 2021; Ullah et al., 2023).

15.3.2.1 Spectroscopic Techniques

Raman and infrared (IR) spectroscopy and surface-enhanced Raman scattering (SERS) are used to investigate the structural characteristics and functional groups of NPs, and especially bioconjugates, including their effect on particle function. Nuclear magnetic resonance (NMR) is a technique that reveals sequence structure, chemical composition, purity, and conformational change. Often, optical characterization achieved by surface plasmon resonance (SPR) using UV–Vis spectroscopy, as well as the creation of NPs, is used to validate process methodology (Abbasi et al., 2023; Abdelnour et al., 2021; Aboyewa et al., 2021). Fourier transform infrared spectroscopy (FTIR) and attenuated total reflection (ATR) help identification of surface functional groups and bonding properties (Almanza-Reyes et al., 2021; Asmatulu et al., 2024; Bruna et al., 2021). XPS is used to determine the elemental and chemical composition of the NP surface (de Dicastillo & Correa, 2021). Raster or mapping images are often obtained, as elemental maps and composition investigation by electron microscopy and energy dispersive X-ray spectroscopy (EDS or EDX) are commonly performed (Dhaka et al., 2023; Ding et al., 2022). Features such as size, shape, and surface texture of NPs are studied using SEM, FESEM, and STM. Transmission electron microscopy (TEM) provides high-resolution images of the interior structure, size, and form of a wide range of heterogeneity and aggregation (Egbuna et al., 2021; Ettlinger et al., 2022). Atomic force microscopy (AFM) evaluates the mechanical characteristics, surface roughness, and topographical information of surfaces at the nanoscale (Ullah et al., 2025).

15.3.2.2 Size and Surface Area Analysis

Dynamic light scattering (DLS) is used to measure both the hydrodynamic size distribution of NPs in suspension and the polydispersity index. Brunauer Emmett Teller (BET) analysis is used to calculate porosity and specific surface area. X-ray diffraction (XRD) as a tool for peak broadening analysis can be used to determine average particle size, phase identification, and crystallinity (Gao et al., 2023). Measurement of surface charge and zeta potential analysis helps in determining the surface charge, zeta potential, and stability of colloidal NPs. The electrokinetic and cyclic voltammetry studies are used to examine the electrical characteristics of nanomaterials in electrochemical applications (García-Quintero & Palencia, 2021). Differential scanning calorimetry (DSC) is performed to investigate the physicochemical condition of drug–nanoparticle complexes and their thermal behavior, including interactions between drug and carrier material (Ullah et al., 2025).

15.3.2.3 Optical and Molecular Characterization

Using optical microscopy (NSOM), the NP size and shape can be searched at an extremely fine resolution in the near field. Methods such as secondary ion mass spectrometry (SIMS), some of which are available on instruments with high-performance mass spectrometry, are used to ascertain molecular weight, composition, and surface characteristics. Surface changes or active component concentration are measured by high-performance liquid chromatography (HPLC) (García-Quintero & Palencia, 2021).

15.4 Properties

15.4.1 Electronic Properties

All the electrical characteristics of NPs depend on their size, surface area, chemical makeup, and surface modifications. For instance, the use of organic ligands to change the surface properties and electrical characteristics of AlO_3 (aggregation and particle size control). The addition of inorganic substances such as barium titanate (BT) to polymer matrices because of their piezoelectric, perovskite properties, serves to improve electrical conductivity and dielectric constants (García-Quintero & Palencia, 2021).

15.4.2 Optical Properties

Optical characteristics are required for uses such as photovoltaics and photocatalysis. Among the variables affecting these characteristics are size, shape, and doping. Semiconductor NPs (NPs) like Nd-doped NiO have variable band gaps due to electron exchange interaction. UV–Vis spectroscopy is often used to examine such alterations. Scattering behavior and reflectance, as well as spectral patterns and reflectivity, are dependent on the particle size; larger particles shift spectral patterns and increase reflectivity (Sasikala et al., 2022).

15.4.3 Magnetic Properties

Magnetic NPs are widely used in drug delivery, environmental applications, and medical diagnostics. What makes them magnetic is their size, content, and production techniques. Particulates smaller than 35 nm are frequently at their optimal performance (Gigault et al., 2021). Magnetic characteristics in a multicomponent system depend on the quantity of lone pair electrons and lattice mismatches that can cause surface stress and change magnetism (Gudkov et al., 2021).

15.4.4 Mechanical Properties

However, NPs offer better mechanical characteristics due to their large surface area and ability to improve matrices. When incorporated in acrylic polyurethane, together with inorganic NPs like SnO_2, the adhesion, hardness, and impact resistance are significantly increased. Thorough mixing and agglomeration due to weakening interactions amongst components are problems, but over-addition may lead to all of the above. Mechanical characteristics are also affected by the particle size (Gul et al., 2025).

15.4.5 Thermal Properties

This is since NPs have an enormous surface area, which leads to increased thermal conductivity as compared to that of the bulk or fluid. It is seen that the addition of metal oxides, like SiO_2, to polymer matrices helps in limiting the mobility of polymer chains and thereby improves thermal stability. Thermal behavior is affected by, among other things, surface area, NP concentration, and the inherent thermal conductivity of other fillers (Gul et al., 2025).

15.4.6 Applications of NPs

Considering their unique properties, NPs can be used in a variety of applications. Some of these are given below (Figure 15.1).

15.4.6.1 Applications in Drugs and Medications

Metallic and semiconductor NPs possess significant potential for applications in cancer diagnosis and therapy due to their strong light scattering and absorption properties, which arise from the localized SPR (LSPR) effect (Gupta et al., 2022). Among them, gold (Au) NPs demonstrate efficient light absorption and can convert light energy into localized heat, enabling photothermal therapy (PTT) to selectively induce cancer cell death within tumor tissues (Haleem et al., 2023). Additionally, their unique optical properties make AuNPs effective agents for photodynamic therapy (PDT), where they facilitate the activation of light-sensitive drugs to target and destroy cancer cells (Holmannova et al., 2022). Nanoparticles based on gadolinium (Gd) have also shown promising results, demonstrating the ability to inhibit tumor growth (Hong et al., 2021), prevent metastasis (Gigault et al., 2021), and enhance tumor-specific magnetic resonance imaging (MRI) contrast (Hong et al., 2021). Beyond therapeutic applications, targeted drug delivery represents another critical role of nanoparticles. ZnO and Fe_3O_4 nanoparticles have been successfully utilized in targeted drug delivery systems to selectively eliminate tumor cells (Horie & Tabei, 2021). Moreover, NPs are extensively applied in various detection platforms, including cellular imaging (Horie & Tabei, 2021), DNA detection (Ijaz et al., 2023), carbohydrate detection (Ijaz et al., 2023), protein sensing (Horie & Tabei, 2021), heavy metal ion detection (Hong et al., 2021), blood glucose monitoring (Ijaz et al., 2024), and the detection of bacteria (Hong et al., 2021), and viruses (Ijaz et al., 2023). Of note, AuNPs conjugated with SARS-CoV-2 antigens have been employed in rapid antibody detection assays from blood samples, achieving results within 10–15 minutes (Ijaz et al., 2024). Given their inherent antimicrobial and antibacterial properties, nanoparticles are also

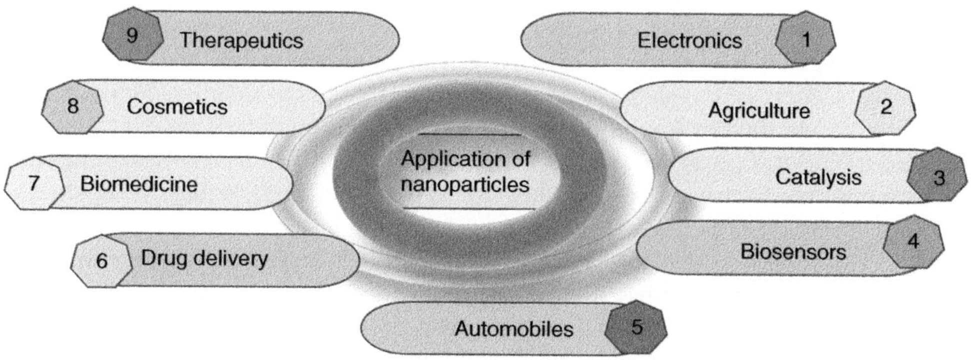

Figure 15.1 The generalized view of nanomaterials and their applications in various sectors.

incorporated into medical devices, such as catheters, to confer antimicrobial and antibacterial functions (Horie & Tabei, 2021).

15.4.6.1.1 Applications in Electronics NPs show great potential for imaging and in electronics as well. Gd-based NPs, however, can be used to ameliorate the imaging quality without increasing the dose of the contrast agent. However, Gd_2O_3 NPs are more effective than conventional Gadolinium–tetraazacyclododecanetetraacetic acid (Gd-DOTA) agents at the same level of concentration (Ijaz et al., 2025), and $GdPO_4$ NPs can detect tumors with one-tenth of the normal dose of Gd-DTPA (Ijaz et al., 2024). Single-molecule tracking and information about the organization or interaction with Eu^3-doped oxide NPs have also been reported with NPs, for example, the 30 nm localization precision (Horie & Tabei, 2021).

NPs enhance separator efficiency as well as ion transport in battery technology. ZnO NPs also form an aerogel-like structure, and then, the energy storage in battery separators can be improved greatly (Ijaz et al., 2025). Pt–Au bimetallic NPs conduct oxygen reduction and evolution reactions better when used in lithium–air batteries (Ijaz et al., 2024). The charging requirements of the batteries of Ni and metal-hydride technical systems are reduced, and their lifespans are longer (Ijaz et al., 2025). CdS and ZnSe NPs are also used in display technologies to increase brightness and to extend the life of the screen of LEDs (Ijaz et al., 2025). CdTe NPs are used to improve the efficiency of liquid crystal displays (LCDs) (Ijaz et al., 2025) in this way.

15.4.6.1.2 Applications in Agriculture NPs offer revolutionary solutions for modern agriculture, primarily as nanofertilizers and nanopesticides. Unlike conventional fertilizers, nanofertilizers do not leach and volatilize like conventional fertilizers, and they supply the necessary nutrients in a controlled and efficient fashion, at a much lower application rate (Ijaz et al., 2025; Ullah et al., 2025). NPs such as SiO_2, ZnO, CuO, Fe, and Mg can be used to enhance N fixation (Ijaz et al., 2023) to seed germination, drought tolerance, seed weight, and photosynthetic efficiency. Since they have a large surface area and nanoscale size (Ijaz et al., 2025). Besides this, the NPs have antimicrobial, insecticidal, and nematicidal properties that can serve as eco-friendly pesticide alternatives. For example, TiO_2 NPs inhibit the growth of *Xanthomonas perforans* (tomato spot disease) (Ijaz et al., 2025; Ullah et al., 2025), CuO NPs inhibit the growth of *Spodoptera littoralis* (cotton leafworm) (Ijaz et al., 2025), and the growth of *Meloidogyne* spp. (root knot nematodes) (Bruna et al., 2021) is suppressed by AgNPs.

Metallic and semiconductor NPs also hold promise for cancer therapy, e.g., $[Gd@C_{82}(OH)_{22}]_n$ NPs show antineoplastic activity with minimal toxicity (de Dicastillo & Correa, 2021). Because of high broad spectrum of effectiveness as an antimicrobial agent, AgNPs are widely used in wound dressings, catheters, and household products (Dhaka et al., 2023; Ding et al., 2022). In this case, TiO_2, ZnO, $BiVO_4$, Cu, and Ni NPs are functionalized to enhance selective antimicrobial activity (Egbuna et al., 2021; Ettlinger et al., 2022).

15.4.6.1.3 Applications in the Food Industry Vast potential in imaging and electronics is also shown by NPs. However, Gd-based NPs can be used to improve imaging quality while minimizing the dose of the contrast agent. For instance, Gd_2O_3 NPs are more effective than conventional Gd-DOTA agents at the same levels of concentration (Ullah et al., 2025), whereas $GdPO_4$ NPs detect tumors with one-tenth of the normal Gd-DTPA dose (Ijaz et al., 2025). NPs have also been applied to single-molecule tracking and can provide information about the organization or interactions with Eu^3-doped oxide NPs, for example, achieving 30 nm localization precision (Ijaz et al., 2024).

For battery technology, NPs increase separator efficiency and ion transport. Since ZnO NPs form into aerogel-like structure, energy storage in battery separators can be significantly improved (Ijaz et al., 2023). In lithium–air batteries, Pt–Au bimetallic NPs improve oxygen reduction and evolution reactions (Ijaz et al., 2024). Batteries of nanocrystalline nickel (Ni) and metal-hydride technical systems show lower charging needs and longer lifespans (Ikram et al., 2021). NPs such as CdS and ZnSe are used in display technologies for both increasing brightness and elongating screen life in LEDs (Jaswal & Gupta, 2023). In this way, CdTe NPs are used in improving the efficiency of LCDs (Joseph et al., 2023).

15.4.6.1.4 Applications in Manufacturing and Materials Highly size-dependent, yet unique properties of nanocrystalline materials are highly valued for materials science because they significantly differ from the bulk material. Discrete electrical, mechanical, optical, and imaging properties of manufactured NPs make them essential in medical, commercial, and environmental sectors (Gigault et al., 2021; Khan et al., 2024). Describing biological or non-biological structures <100 nm, nanotechnology enables the design of new functional properties. NPs are increasingly found in commercial products, such as in microelectronics, aerospace, and pharmaceuticals (Gudkov et al., 2021). One of these applications is in the health and fitness product segment, which is the largest consumer segment. Resonance energy transfer (RET) systems that make use of organic dyes and noble metal NPs have been widely applied in both biophotonics and materials science (Gul et al., 2025). Noble metal NPs, such as Au and Ag, have vivid visible colors because of SPR, i.e., electron oscillations. Particle size, shape, spacing, and surrounding dielectric properties influence the bands of resonance wavelength. For these same reasons, they can be used as chemical and biosensors (Gul et al., 2025).

15.4.6.1.5 Applications in the Environment The extensive use of engineered NPs in industrial and household products has raised concerns regarding their environmental impact. The mobility, reactivity, ecotoxicity, and persistence of NPs in natural ecosystems, therefore, must be assessed (Gupta et al., 2022; Haleem et al., 2023). The primary environmental reservoirs of NPs are groundwater and soil, where their high surface to mass ratio can influence the transport and fate of contaminants (Holmannova et al., 2022). During the NPs formation and environmental transformations, they can adsorb the contaminants, become co-precipitated, or trap the pollutants (Figure 15.2).

15.5 Toxicity of NP

Nanomaterials (NMs) have numerous industrial and medical applications but are also associated with various toxicities (Ijaz et al., 2024). Proper management of these toxic effects is essential, requiring a fundamental understanding of their mechanisms. Nanoparticles (NPs) can unintentionally infiltrate the environment through air, water, and soil as a result of human activities. Additionally, there is growing interest among

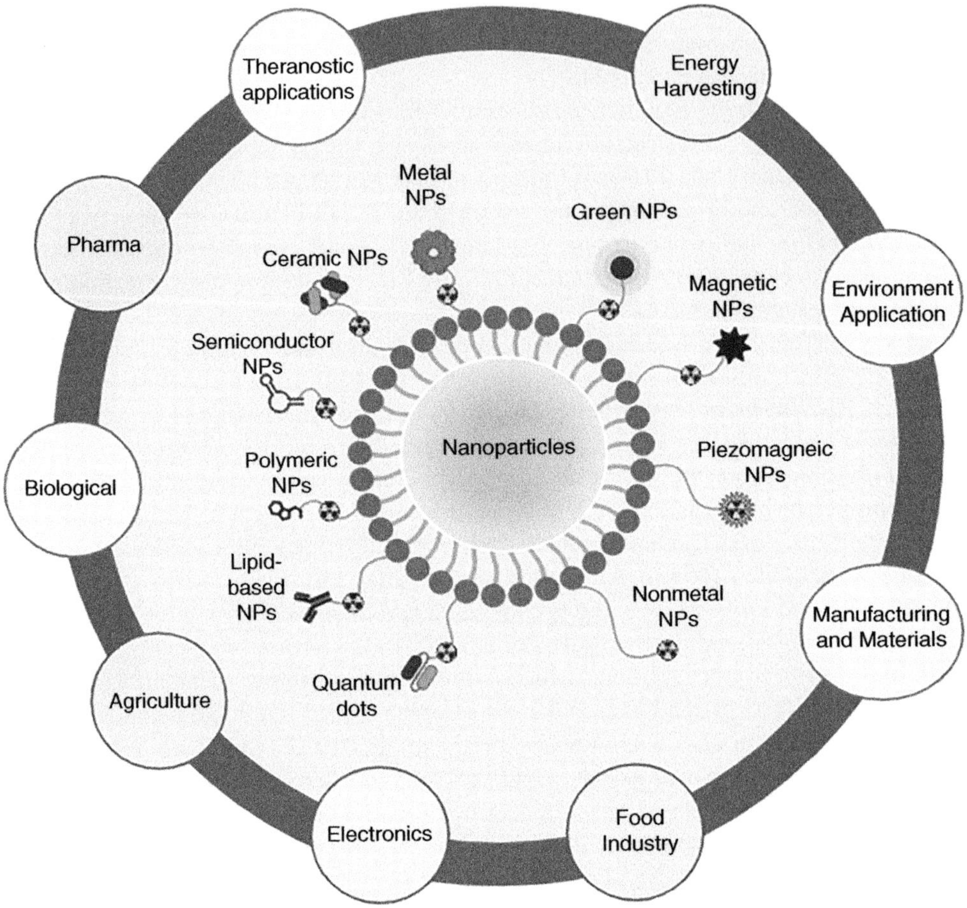

Figure 15.2 Different types of NPs and their roles in various industries.

stakeholders in utilizing NPs deliberately for environmental remediation, such as by injecting or releasing engineered NPs into soil or aquatic systems. While magnetic NPs offer advantages due to their small size, high reactivity, and large surface area, these same properties may pose a risk by inducing toxic cellular effects not typically observed with larger, micron-sized particles. Studies have demonstrated that NPs can enter organisms via ingestion, inhalation, and dermal contact, and subsequently be transported to various organs and tissues, where their reactivity may elicit toxicological effects (Jaswal & Gupta, 2023).

Although some research has addressed the toxic effects of NPs on animal and plant cells, the toxicological impact of magnetic NPs on plants remains insufficiently explored. Silver (Ag) NPs, widely used in various consumer products, are often released into aquatic environments where they can dissolve and exert toxic effects on aquatic organisms, including bacteria, algae, fish, and daphnia (Joseph et al., 2023). A unique target of NP toxicity is the respiratory system, which is exposed to inhaled particles and also receives the entire cardiac output, facilitating particle deposition (Khan et al., 2021). Despite rapid advances in nanobiotechnology, the potential adverse health effects of prolonged NP exposure at various concentrations on the human body and the environment remain largely undetermined. The anticipated large-scale future use of NPs is expected to have significant environmental impacts. Another important aspect of NP toxicity involves their capacity to bind with proteins, a process influenced by particle size, shape, curvature, surface charge, functional groups, and free energy (Kumah et al., 2023). Such interactions can lead to adverse biological effects, including protein unfolding, fibrillation, thiol crosslinking, and loss of enzymatic activity.

A secondary mechanism of toxicity arises from the release of harmful ions, particularly when the properties of the material near thermodynamic equilibrium favor the dissolution of particles into the surrounding medium or biological environment (Kumar et al., 2022). Additionally, aggregation of NPs tends to occur in seawater and hard water, while the presence of organic matter and natural colloids in freshwater systems significantly affects NP behavior. Several abiotic factors, including pH, salinity, organic materials, and the dispersion state of NPs, influence their ecotoxicity and require systematic investigation (Kus-Liśkiewicz et al., 2021).

Although NPs and other nanomaterials are widely used in industry and, increasingly, in medical practice, there is no shortage of questions about their toxicity (Landa, 2021). It is necessary to control them and have a fundamental grasp of their harmful impacts. NPs may unintentionally infiltrate the environment through the air, soil, or water during a variety of human activities. The intentional use of NPs in environmental remediation, such as releasing or dumping designed NPs into soil or the aquatic system, further stoked concerns about NPs (Malik et al., 2023). These same benefits and drawbacks (small size, high reactivity, and capacity) of magnetic nanoparticles, however, may have lethal consequences in these uncommon cells that would not always be seen in larger, micron-sized particles. It has been shown that NPs may enter an organism by ingestion or inhalation and then enter several organs or tissues, where their reactivity might produce toxicological consequences (Manuja et al., 2021). Although some studies has looked at the harmful effects of NPs on animal and plant cells, there are still few investigations on the toxicity of magnetic NPs in plants. The extensive usage of silver (Ag) nanoparticles in mass-produced consumer goods eventually results in their release into the aquatic environment, where they dissolve and have been demonstrated to have harmful effects on fish, bacteria, algae, and daphnia (Kumah et al., 2023). The respiratory system is considered a special target for NP toxicity because it serves as an entrance site for inhaled particles and, of course, as the recipient of the whole cardiac output in addition (Jaswal & Gupta, 2023). Even while nanobiotechnology has advanced quickly and gained acceptance in many scientific settings, it is still unclear how long-term exposure to NPs at different concentrations may affect people and the environment.

In the future, however, the environmental impacts of NPs are expected to increase (Kumah et al., 2023). A known aspect of the NP toxicity is their ability to bind with proteins, where the binding depends on size, shape, curvature, the surface charge, the functional groups, and the free energy of the particles. This interaction can induce adverse biological effects, including protein alteration as unfolding, fibrillation, thiol cross-linking, and the loss of enzymatic functions (Jaswal & Gupta, 2023). Another toxicity mechanism is the release of harmful ions because the thermodynamic properties of the material favor particle dissolution in the surrounding medium or biological environment (Joseph et al., 2023). In hard water and seawater, NPs usually aggregate, and their behavior is strongly affected by the nature of organic matter or other natural colloids in freshwater. The ecotoxicity of NPs also depends on their dispersion state, but many abiotic factors such as pH, salinity, and the presence of organic material, are yet to be thoroughly explored in the ecotoxicological research (Kus-Liśkiewicz et al., 2021).

NPs are generated through processes such as photochemical processes, where the NPs are naturally formed, or they are manufactured from various processes like volcanic eruptions, forest fires, and simple erosion. Pollution is normally connected to human activities, including industry, for instance, fuel burning and charcoal burning, and natural processes, for instance, dust storms (Manuja et al., 2021). The toxicity of NPs also depends on their origin, dose, dimension, and durability after exposure or intake through several routes (Miyazawa et al., 2021). The toxicity of NPs may depend on different parameters, for example, mass, number, surface area, size, bulk or surface chemistry, and aggregation of NPs (Mubeen et al., 2021). Other factors that influence the toxicity of NPs include crystalline structure, aspect ratio, surface coating, and functionalization. Inhalation, injection, dermal uptake of NPs, nervous system uptake of NPs, and translocation to the lymphatic and the circulatory systems are the ways of occurrence of the toxicology of NPs (Musielak et al., 2022). A comparison can be made to the toxic effects of NPs with those of different bulk forms to evaluate the toxicity of NMs. Several have come to pioneer this initiative and have been analyzed in other reviews (Kus-Liśkiewicz et al., 2021). We reported that inhaled 20 nm NPs TiO_2 low dose of 10 mg/m^3 developed a greater lung tumor incidence than the high dose of 250 mg/m^3 exposure to TiO_2 300 nm (Ijaz et al., 2023). The exposure or concentration of NPs in air, water, food, and other sources, multiplied by the contact time, determines the dose of NPs reaching a biological system, which is directly correlated with the number of NPs that enter the system (Kus-Liśkiewicz et al., 2021).. Ag, TiO_2, Fe_2O_3, Al_2O_3, ZrO_2, Si_3N_4, and CNTs were reported to be more toxic than silica and carbon black (Jaswal & Gupta, 2023). The toxicity test, however, depends on the CNT types (single-walled CNT or multiple-walled CNT) and diameters, and functionalities. It was reported that silver NP aggregates are more toxic than asbestos, titania, alumina, iron oxide, and zirconia (Kumar et al., 2022). When water met NPs of ZnS, 3 nm in size, the crystals changed their structure to that of bulkier ZnS because of surface effects (Miyazawa et al., 2021).

Moreover, the NMs are small and can trigger problems and be harmful to the environment. Malik et al. (2023) studied Fe NPs toxicity in the environment where NPs can be accumulated in organisms and have several effects, such as oxidative stress, apoptosis, lipid peroxidation estimation, and reactive oxygen species (ROS) production. The accumulation mechanism of NMs in organisms in the body occurs by inhalation and ingestion and, through spreading in different tissues and organs, can lead to toxicological effects on the body (Kumar et al., 2022). Despite several works reported regarding the toxicological effects of NMs on plant and animal cells, toxicological studies on magnetic NPs on plants are still limited. Khan et al., (2021) has shown that CuO NPs change as they interact with *Nicotiana tabacum* plant cells (Ijaz et al., 2023). The study's findings demonstrated that the toxicity of NPs is dependent on the degree of transformation, which is induced by several plant cell components. For one, in the presence of galacturonic acid in the protoplast, CuO can be reduced to Cu_2O. The Cu_2O form will generate more toxicity than the CuO form, and more transformation into other metal oxides, Cu_2O, will result from the transformation into oxidized materials (Joseph et al., 2023).

NP toxicity in biotechnology is therefore directly dependent on particle size, active surface area, shape, and attached functional groups on the surface of materials, because they may regulate protein binding and processes (Ijaz et al., 2025). This binding process may lead to the loss of protein function (enzyme), protein breakdown, and production of additional toxic biological chemicals (Ijaz et al., 2025). Concerning environmental factors, the type of water, whether it be hard or saltwater, along with any organic elements of the compounds can influence aggregation of the material. The process is further affected by pH, salinity, and colloidal forming chemicals. However, more kinetic and biochemical interaction investigations are needed to advance research on NP accumulation (Ijaz et al., 2023).

15.5.1 Mechanisms of Toxicity

The toxicity of NPs is controlled by certain processes during which NPs interact with biological systems. These pathways can differ depending on the type of nanoparticle and the cell or organ it may be targeting (Ijaz et al., 2024). There is also an emerging method of direct cellular interactions, in which, when nanomaterials break down, metal ions are released that are toxic to cells and can further shatter enzymes important for cells to operate correctly (Table 15.2) (Ikram et al., 2021).

The second important process is the production of ROS. NPs can cause oxidative stress when they contact cells by producing harmful oxygen radicals in a vicious cycle that upsets cellular redox equilibrium. This imbalance may hamper several biological functions and prove to be harmful to cells (Jaswal & Gupta, 2023). Also, NPs have the potential to bind to noncancerous cells, triggering the release of harmful ions and increasing cellular toxicity (Kumar et al., 2022). Besides the direct interaction between the NPs and the cells, NPs can interact with the cytoskeleton, the cell membrane, or organelles and cause their malfunctioning (Landa, 2021). They also have the capacity for exerting epigenetic impacts that can change cellular activity and gene expression patterns. These mechanisms—cellular interactions, ROS generation, internalization, and functional impairment—collectively contribute to nanoparticle-induced toxicity (Kus-Liśkiewicz et al., 2021).

In Figure 15.4, nanotoxicology, a critical topic in nanoscience, attempts to understand the potential harms of engineered NPs, especially in biomedical applications. Although they have been widely utilized in nanomedicine, questions regarding nanoparticle safety are still largely present because of their small size and novel physicochemical properties (Hong et al., 2021). Particle size, shape, surface functionalization, and dosage, on the other hand, are key factors that affect the toxicity (Horie & Tabei, 2021). Residual synthetic chemicals in chemically synthesized NPs may pose a higher toxicity, while that is lower for biosynthesized NPs having surface groups that are biocompatible. Nanoparticle biocompatibility is strongly dependent on the choice of synthesis method and surface functionalization (Horie & Tabei, 2021).

Larger MNPs accumulate in various anatomical organs, for instance, in the kidneys to concentrate in Bowman's space (Ijaz et al., 2023). The kidneys provide high reabsorption capacity for tubular epithelial cells, resulting in an accumulation of MNPs that leads to kidney damage and disruption of function (Ijaz et al., 2025). Ijaz et al. (2025) also reported

Table 15.2 Mechanisms of nanoparticle toxicity.

Mechanism	Pathways involved	Target organelles/cells	Cellular effects	Biological outcomes	Example nanoparticles	References
Oxidative Stress	ROS and RNS generation, antioxidant depletion	Mitochondria, nucleus, cytoplasm	Lipid peroxidation, protein carbonylation, and DNA oxidation	Mitochondrial dysfunction, apoptosis, necrosis	TiO_2 NPs, ZnO NPs, Ag NPs	Abbasi et al. (2023), Abdelnour et al. (2021), and Aboyewa et al. (2021)
Cell Membrane Damage	Lipid bilayer disruption, nanoparticle insertion	Plasma membrane, endosomal membranes	Loss of integrity, ion imbalance, and membrane fluidity changes	Cell lysis, endocytosis alteration	SiO_2 NPs, graphene oxide NPs	Ali et al. 2024, Almanza-Reyes et al. (2021) and Paras et al., 2022)
Inflammation	Activation of NF-κB pathway, inflammasome activation (NLRP3)	Macrophages, epithelial cells	Cytokine release (IL-6, TNF-α), leukocyte recruitment	Chronic inflammation, fibrosis, and immunotoxicity	Carbon nanotubes (CNTs), Fe_3O_4 NPs	Asghar et al., (2024), Asmatulu et al. (2024), and Azizi-Lalabadi et al. (2021)
Genotoxicity	DNA strand breaks, oxidative DNA lesions, chromosomal aberrations	Nucleus	DNA mutations, micronucleus formation, chromosome loss	Carcinogenesis, heritable mutations	Quantum dots (CdSe/ZnS), metal oxide NPs	Bibi et al. (2023), Bruna et al. (2021), and de Dicastillo & Correa (2021)
Mitochondrial Dysfunction	Electron transport chain disruption, mPTP opening	Mitochondria	ATP depletion, Cytochrome c release	Apoptosis, Energy failure	Ag NPs, Au NPs	Dhaka et al. (2023) and Ding et al. (2022)
Lysosomal Dysfunction	Lysosomal membrane permeabilization (LMP)	Lysosomes	Enzyme leakage, pH imbalance	Cell death, autophagy dysregulation	SiO_2 NPs, TiO_2 NPs	Egbuna et al. (2021) and Ettlinger et al. (2022)
Endoplasmic Reticulum (ER) Stress	Unfolded protein response (UPR) activation	ER	Protein misfolding, calcium dysregulation	Apoptosis, impaired protein synthesis	ZnO NPs, Au NPs	Ettlinger et al. (2022) and Ullah et al. (2025)
Autophagy Dysregulation	Autophagosome accumulation, impaired lysosome fusion	Cytoplasm	Impaired cell clearance, cell stress	Cell survival alteration, cell death	Carbon dots, graphene oxide NPs	Fatima et al. (2021) and Gao et al. (2023)
Epigenetic Modifications	DNA methylation, histone modification, miRNA deregulation	Genome-wide	Gene expression alteration	Cancer, developmental toxicity	Ag NPs, Fe_3O_4 NPs	García-Quintero & Palencia (2021) and Sasikala et al. (2022)

that exposure to NPs may cause glomerular swelling, thickened basement membrane, and degeneration and necrosis of the renal tubular cells (Ijaz et al., 2024). All these adverse effects are mainly due to the oxidative stress, inflammation, autophagy, DNA damage, and the endoplasmic reticulum (ER) stress caused by NPs (Ijaz et al., 2025). Are previous studies about the damage of engineered nanomaterials (ENMs) and nephrotoxicity helpful? Yes, they provide molecular mechanisms in which NPs can cause damage to the kidneys during processes that lead to the safer design and use of NPs (Ijaz et al., 2025). The main reason for this is that they are nanoscale, which allows them to penetrate cells and other organs quickly. NPs exposure may produce effects that include oxidative stress, inflammation, DNA damage, and eventually, cell death (Jaswal & Gupta, 2023), even though NPs could be safe or even beneficial at lower concentrations (Figure 15.3).

15.5.2 Toxic Effects of NPs on the Respiratory System

The respiratory system is composed of organs that are responsible for gas exchange between the human body and the outside world (Khan et al., 2021). The nose, pharynx, larynx, trachea, bronchi, and lungs are composed of all of the other tissues, including alveoli, blood vessels, lymphatic vessels, nerves, and pleura. Factors affecting the deposition of NPs in the respiratory system are size, shape, and surface chemistry of NPs. When NPs are on smaller scales, there is a higher chance of them entering the lower airways, and thus they can interact with lung cells by penetrating the alveolar area. The surface chemistry of NPs also plays a role in deposition; hydrophilic particles are deposited to a lesser extent than hydrophobic ones in the respiratory tract (Joseph et al., 2023). Research has shown, for example, that titanium dioxide (TiO_2) NPs cause lung effects such as inflammation and fibrosis in animal models. In addition, exposure to NPs can aggravate the effect that an underlying respiratory

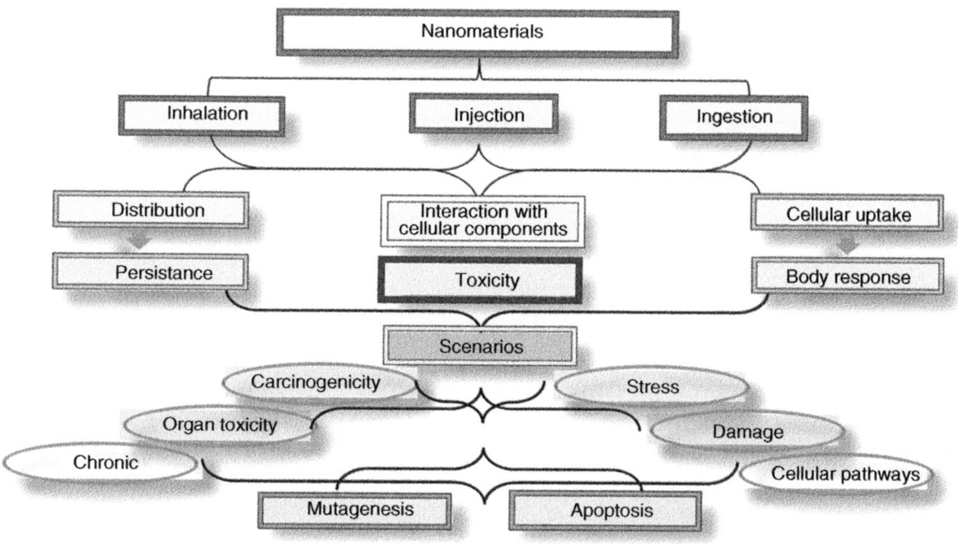

Figure 15.3 Nanomaterials and their interactions with living cells cause serious problems.

problem will have in people, such as asthma or chronic obstructive pulmonary disease (COPD) (Jaswal & Gupta, 2023).

In recent years, biomarkers have proved useful in investigating the health effects of NPs in the relationship between NPs exposure and the environment, using poor respiratory outcomes. Negative consequences of NPs on the respiratory system are mainly oxidative stress, inflammation, epithelium damage, fibrosis, and genetic changes (Jaswal & Gupta, 2023). This is illustrated with the example of titanium dioxide (TiO_2) NPs in animal models, where it was shown to induce lung inflammation and fibrosis. In addition, breathing may become even more troublesome if NPs are inhaled by those who already suffer from respiratory disorders such as asthma or COPD (Musielak et al., 2022).

Recently, biomarkers have been an important tool to investigate the relationship between NPs and their health consequences. They shed light on the relation of exposure to the environment and poor respiratory outcomes (Mubeen et al., 2021). The fibrosis, inflammation, oxidative stress, epithelium damage and genetic changes appeared to be the main negative consequences of NPs on the pulmonary system. Studies have proven that certain NPs, including carbon-based NPs, silica (SiO_2) NPs, silver (Ag) NPs, and zinc oxide (ZnO) NPs, can result in an inflammatory reaction and respiratory immune toxicity. Research by Jaswal & Gupta (2023) indicated that human lung epithelial cells engulfed a variety of sized nanoplastics, resulting in death and cell cycle arrest. Similarly, studies have shown that SiO_2 nanoparticles (NPs) can enter systemic circulation, cross the air–blood barrier, and cause cardiotoxic effects (Lozano 2020) Ag NP exposure has also resulted in Drosophila respiratory distress and elevated Drosophila heat shock protein 70 (Hsp70). Moreover, it has been shown that rats that are provided with intratracheal copper oxide (CuO) NPs may develop tumors, inflammation, and oxidative stress (Sasikala et al., 2022).

In addition to being toxic to the respiratory system, NPs can disturb the equilibrium of the immune system leading to either immunosuppression or immunity overactivation and thus compromise the body's defence against respiratory infections. In addition to NPs, CNTs, among other nanomaterials, are also considered to impact the respiratory system. It has been shown in animal models that CNTs promote airway and lung remodeling, recruiting cytokines, including tumor necrosis factor (TNF)-α, interleukin (IL)-1 beta (IL-1β), monocyte chemoattractant protein1, and IL-13, as well as inflammatory blood cells. Transforming growth factor-beta (TGF-β), TNF-α, and osteopontin are CNT-induced extracellular remodeling pathways.

Gao et al. (2023) further researched that airborne nanoplastics readily accumulate in human and animal alveoli and can cause epithelial–mesenchymal transition (EMT) in alveolar epithelial A549 cells via NADPH oxidase 4 (NOX4) and ROS, resulting in pulmonary fibrosis. It was also reported by Jaswal & Gupta (2023) that nanoplastics could cause mitochondrial dysfunction in lung cells, thus leading to metabolic toxicity. Prior comparative studies on rats have shown that rats exposed to TiO_2 NPs have a more severe pulmonary alveolar proteinosis and an increase in inflammatory biomarkers, whereas those exposed to SiO_2 NPs had severe fibrosis and granuloma formation (Gupta et al., 2023; Ullah et al., 2025).

15.5.3 Toxic Effects of NPs on the Nervous System

One of the most remarkable, if not the most remarkable, and complex attributes of humans is that of the nervous system, particularly the brain and its cognitive functions. Its primary role in your body is to control temperature, regulate physiological functions, and coordinate the activities of the body (Ullah et al., 2025). The nervous system is composed principally of nervous tissue and is divided into two main parts that include the central nervous system (CNS), comprising brain and spinal cord, and peripheral nervous system (PNS), comprising cranial and spinal nerves, which are increasingly associated with various systemic comorbidities, including immunological and bioenergetic imbalance, and disruption in gut microbioma. To date, most of the research on the interaction between NPs and the nervous system has been directed toward designing NPs for drug delivery. It has been shown in several studies that NPs could cross the blood–brain barrier (BBB) and as a result, enter the brain and modulate neural function. Specific NPs designed to evade phagocytosis by the reticuloendothelial system (RES) can therefore increase the drug concentration in the brain significantly (Bruna et al., 2021). For instance, it has been proven that liposomes circulate in the bloodstream for longer duration when they are modified with polyethylene glycol (PEG) (Landa, 2021).

The use of NPs for drug delivery is rapidly expanding, but their potential neurotoxic effects should not be overlooked. Several mechanisms have been identified through which NPs can induce neurotoxicity.Alterations to cellular morphology, cell death, genotoxicity, oxidative stress, and proinflammatory reactions have most recently been reported in vitro in responses to NP exposure (Kus-Liśkiewicz et al., 2021). We show that polystyrene NPs (PS-NPs) damaged the intestinal barrier of zebrafish and triggered gut inflammation and developmental impairments, with the involvement of zebrafish brain and gut in the toxin detection, which in turn altered microbiota composition, resulted in growth inhibition, impaired gut–brain communication, and triggered intestinal damage. Additionally in rats, SiO_2 NPs, when instilled intranasally, also accumulated in the striatum, which caused oxidative damage and an inflammatory response (Ettlinger et al., 2022). Furthermore, in vitro studies showed that SiO_2 NP exposure decreased cell viability, enhanced lactate dehydrogenase (LDH) levels, stimulated oxidative stress, disrupted cell cycle, caused cell apoptosis, and activated p53-mediated signaling pathways. In addition, studies of (Bruna et al., 2021) showed that Fe_2O_3 NP exposure resulted in oxidative damage and neurotoxicity in the mouse brain. Apart from these, Ag NPs, PbO NPs, and ZnO NPs have also been found to result in different levels of neural damage. Overall, NPs have the capability of entering the body through inhalation and crossing the BBB and therefore can cause serious harm to the neurological systems (Bibi et al., 2023).

15.5.4 Toxic Effects of NPs on the Endocrine System

Endocrine tissues are made up of several endocrine tissues including traditional endocrine glands and neuroendocrine cell groups in organs such as the lungs, gastrointestinal tract, thymus, mammary gland, prostate, adrenal glands, and the paraganglia (Sasikala et al., 2022). Hormones are produced and released directly into the bloodstream by these tissues, and these tissues are essential for controlling bodily functions. The endocrine system serves as a working partner with the nervous and immune systems, and aids in maintaining homeostasis, energy balance, and controlling processes such as growth, development, and reproduction (Gao et al., 2023).

In mammals and other species, NPs can perturb the endocrine system, resulting in both detrimental (adverse) and beneficial effects (Fatima et al., 2021). They interact with the endocrine system in many ways. They are able, importantly, to mimic hormones. It is possible that upon entering cells, NPs could bind to hormone receptors and either activate or inhibit associated signaling pathways. For example, some metal-based NPs, e.g., silver (Ag) NPs and zinc oxide (ZnO) NPs, have been found to have the ability to interact with estrogen receptors, to act like estrogen. Especially dangerous for pregnant women and children, the effect of this hormonal disruption can be concerning (Ullah et al., 2025).

Additionally, there is an induction of oxidative stress. ROS generation by NPs at the cellular level can cause electron donation to molecular oxygen and lead to its condition of cellular damage and disruption of hormone regulatory pathways (Ettlinger et al., 2022). Furthermore, the inhibited enzymes and transporters involved in the endocrine function may be impaired by NPs. For example, carbon-based NPs have been shown to suppress aromatase activity, which is the enzyme crucial for estrogen generation, and these effects can cause hormonal imbalance (Egbuna et al., 2021). Furthermore, many studies have shown such effects. Lei et al. (2020) discovered that titanium dioxide (TiO_2) NPs increased the thyroid-disrupting effects of PCP exposure in zebrafish. Lozano (2020), reported that, at non-toxic concentrations, SiO_2 nanoparticles (NPs) increased thyroid disruption in juvenile zebrafish by enhancing the bioavailability and tissue accumulation of co-exposed compounds. Besides, Miao et al. showed that increased bioconcentration of lead in larval zebrafish due to TiO_2 NPs resulted in thyroid and neuronal dysfunction (de Dicastillo & Correa, 2021).

Additionally, NPs have been associated with disturbances of the male reproductive endocrine function. Ag NPs can disrupt the hypothalamic–pituitary–gonadal axis or damage the testicular cells directly as suggested by the studies (Ijaz et al., 2025). ZnO NPs were reported to inhibit steroid production in Leydig cells in vivo but paradoxically augment steroidogenesis in TM-3 cells in vitro. Overall, the effect of NPs on the endocrine system is crucial and may pose serious health risks, such as reproductive and metabolic dysfunction. There is, however, a growing body of evidence and a poorly understood mechanism as to how NPs cause endocrine disruption. Thus, more research is required to completely estimate the potential risk of different kinds of NPs in pathogenic endocrine disorders (Joseph et al., 2023).

15.5.5 Toxic Effects of NPs on the Immune System

The immune system could be divided roughly into innate and acquired immune responses. The first line of defense against foreign substances, including antigenic invaders, is innate immunity, which is a nonspecific mechanism to recognize foreign substances and expel them (Kumar et al., 2022). The response to this involves the utilization of neutrophils, monocytes, macrophages, cytokines, complement proteins, and acute phase proteins. Innate immunity is highly conserved in evolution even in the simplest organisms, and its importance in survival is emphasized. In higher animals, acquired (specific) immunity is characterized by targeted responses mediated by T and B lymphocytes. Specific immunity is precise, but it takes several days or weeks to develop and thus is different from innate immunity (Kus-Liśkiewicz et al., 2021).

NPs can interact with immune cells and influence immune responses. However, most studies have focused on their impacts on innate immunity (Malik et al., 2023). There is research that NPs of cadmium (Cd-NPs) induce the suppression of immune functions. Exposure to

Cd-NPs decreases hemophagocytosis and immunosuppression in oysters (*Crassostrea gigas*) as shown by Auffret et al. Like Bruneau et al., they reported that Cd-NPs caused Cd 158 monocyte viability in mice, resulting in lymphocyte alteration and immunodeficiency (Kus-Liśkiewicz et al., 2021).

It was found that some NPs alter immune responses by the regulation of cytokine production and phagocytic activity. For instance, Manuja et al. (2021) observed that exposure of macrophages to carbon black NPs increased proinflammatory cytokine expression (TNF-α), while decreasing the phagocytic capacity. Besides these, NPs are known to activate innate immunity, for example, polystyrene (PS) NPs, ZnO NPs, Ag NPs, and Au NPs also cause the transcription and translation of inflammatory cytokines, chemokines, and antimicrobial peptides. Benmerzoug et al. further found that SiO_2 NPs induce sterile lung inflammation and exacerbate *Mycobacterium tuberculosis* infection through a STING-dependent type 2 immune response (Manuja et al., 2021). Interestingly, although gold nanoparticles (Au NPs) may have identical surface chemistry, differences in morphology can elicit distinct responses from the innate immune system. In other words, interactions between NPs and living systems are complex, as nanoparticles with the same surface chemistry can provoke different biological responses depending on their shape and structure (Miyazawa et al., 2021).

15.5.6 Toxic Effects of NPs on the Reproductive System

The reproductive system carries out this function through many physiological processes, including producing and containing the sperm, while the female reproductive system is responsible for producing eggs and helps in conception (Mubeen et al., 2021). NPs can interact with many cells both in the tissues and in the blood; they can enter the circulatory system through different routes and, finally, penetrate through the Sertoli cell barrier and get to the reproductive organs, where they can be damaging. While the exact processes are still unclear, there is increasing evidence that NP exposure may have adverse effects on the reproductive system. NP-induced DNA damage can cause infertility and other reproductive diseases or lead to cell death and genetic abnormalities (Musielak et al., 2022).

In men, NPs can also cross the blood–testis barrier and enter the testes, thereby protecting them from hazardous chemicals. Once inside, they can cause oxidative stress and inflammation, the latter of which damages cells and causes malfunction. A decreased sperm motility and count may then increase the risk of infertility (Nie et al., 2023).

NPs can also pass through the placental barrier and into the fetal circulation, which, in turn, may influence the reproductive health of a female and have subsequent effects on the ability of her offspring to reproduce. NPs can hurt ovarian tissue, which could result in a decrease in the number and quality of eggs produced and lead to infertility or other reproductive problems (Noureen et al., 2025).

Due to their antibacterial qualities, silver NPs (Ag NPs), which are frequently found in consumer goods like apparel and cosmetics have been demonstrated to have detrimental effects on reproductive health (Ozdal & Gurkok, 2022). According to studies, a variety of NPs, including CNTs, silicon dioxide (SiO_2-NPs), zinc oxide (ZnO NPs), cerium oxide (CeO_2 NPs), gold (Au NPs), and silver (Ag NPs), can build up in the testes and cause oxidative stress, hormone imbalances, inflammation, and damage to germ cells. Several of these

NPs, such as Au NPs, Ag NPs, TiO_2 NPs, and CeO_2 NPs, can remain in the testes where they can cause long-term toxicological consequences. Curiously, the reproductive toxicity of CeO_2 NPs has been observed to be associated with interference with spermatogenesis in mice, while reproductive toxicity of other NPs is often linked to inflammation and tissue damage (Perut et al., 2021; Pogribna & Hammons, 2021).

Remember, NPs first need to penetrate the Sertoli cell barrier and may require interaction with stromal cells to arrive at the vas deferens. NPs are widely harmful to reproduction because of their unique physicochemical characteristics to evade biological barriers, to harm reproductive tissue, and to induce chronic inflammatory reactions (Perut et al., 2021). Still, no preventative measures or defenses against NP reproductive damage have been proven successful to date. But more research is needed to fully understand their effect on reproductive health and to formulate safety measures that could minimize or eliminate any hazards (Pogribna & Hammons, 2021).

15.6 Specific Health Risks Associated with Different Types of NPs

Based on the discovery of the CNTs, studies on the possible toxicity of NPs to humans were begun in the early 2000s. A great deal of research has since assessed the harmful consequences of NP exposure on the brain, liver, kidneys, lungs, and other organs. Research suggests that when NPs get into the body through the mouth, nose, or skin, they damage cells, tissues, and organs (Pokrajac et al., 2021).

Recent research has been focused on explaining the mechanisms of NP toxicity and identifying key components in their detrimental effect. A first conclusion is that all of these characteristics—size, shape, surface area, and chemical makeup—affect NP toxicity (Rai et al., 2021). In addition, the hazardous potential of NPs can be further controlled by physiological interaction with biological elements such as proteins, enzymes, or DNA (Said et al., 2023).

The main objective of the following stage of NP toxicity research is to develop methods to mitigate the associated dangers of NP exposure. This includes creating safer and more efficient synthetic NP methods and determining biomarkers for NP toxicity as well as supplying safety measures in enterprises where NP is used. Further thorough study is also required to fully understand the long-term impact of exposure to NP on the ecosystem and human health (Silva et al., 2022).

15.6.1 Carcinogenicity of NPs

Cancer is still among the leading causes of mortality and continues to be one of the most complex diseases that affects millions of people around the globe, despite all the great progress that has been made in terms of both prevention and treatment (Kus-Liśkiewicz et al., 2021). Carcinogenesis (the process by which healthy cells are turned into malignant ones) is a progression in three steps, namely initiation, promotion, and progression. In the start phase, genetic or epigenetic changes in healthy cells lead to the condition of precancerous lesions (Silva et al., 2022). During the promotion period, these lesions increase, develop, and become benign tumors. During this phase, additional genetic or epigenetic alterations that lead to malignancy occur (Sim & Wong, 2021).

Some NPs are carcinogenic because they can cause DNA damage by the formation of ROS, mutations, apoptosis, cell cycle arrest and increased secretion of cytokines and chemokines. These consequences, particularly in cells, that are involved in the immune system, lead to reduced cell viability, immunosuppression, and inflammatory reactions. This study reports on the metal-based NPs because of their implications on the carcinogenicity of NPs based on their intrinsic chemical characteristics (Sonwani et al., 2021). Unlike metal NPs such as nickel NPs, hexavalent chromium NPs, and zinc oxide (ZnO NPs), whether other forms of NPs can cause cancer also needs more study. Since cancer formation is a protracted process, future research needs to be focused on understanding the relationship between NP carcinogenicity, properties of NP materials, and particle size (Sonwani et al., 2021).

In recent years, the harmful effects of NPs on human health have been well studied, and many adverse results, such as cancer, neurological disorders, cardiovascular diseases, and respiratory diseases, have been reported (Stater et al., 2021). The NP toxicity is a complex process in which the size, shape, surface chemistry, and physicochemical characteristics of NPs play a significant role. Despite the lack of knowledge about precise toxicological processes, it is believed that the main source of NP toxicity lies in the production of ROS. Oxidative stress can lead to damage to cellular components, like proteins, lipids, and DNA; it can promote cell death or malfunction (Uddin et al., 2025). Additionally, NPs may stimulate the immune system and generate proinflammatory cytokines, leading to inflammatory response (Figure 15.4).

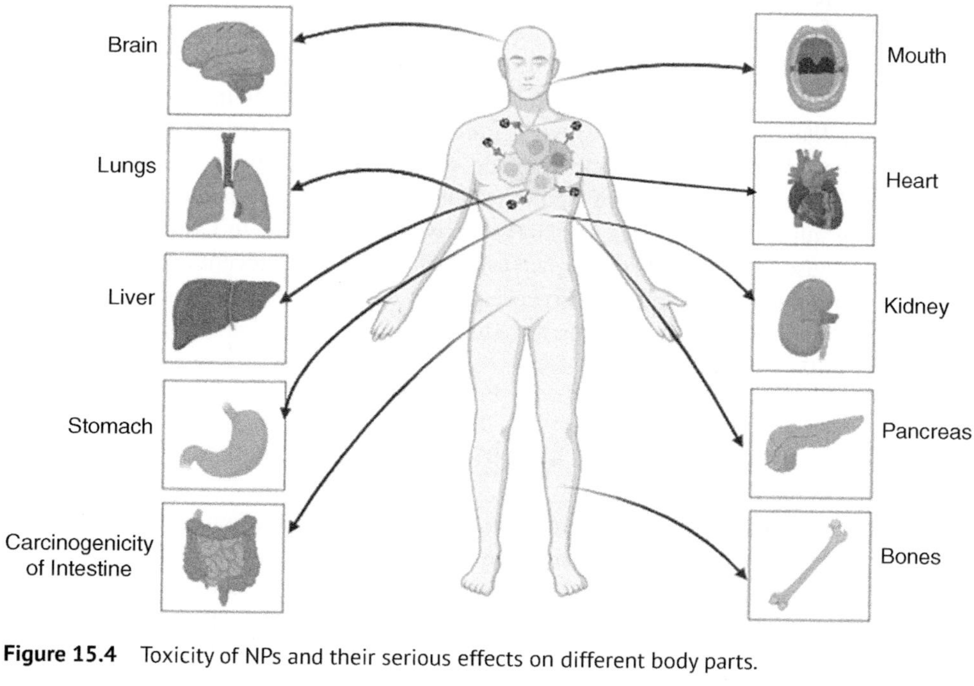

Figure 15.4 Toxicity of NPs and their serious effects on different body parts.

15.7 Occupational Health Risks

Health and safety associated with the use of nanomaterials include potential toxicity from different kinds of NPs and fire and dust explosion possibilities (Ullah et al., 2023). Nanotechnology is still a young science, and the health consequences of exposure to nanomaterials and appropriate degrees of exposure to nanomaterials are currently being studied (Suliasih et al., 2024). There is concern, given that some nanomaterials can cause lung problems such as inflammation, fibrosis, and carcinogenicity in animal studies. Inhalation exposure ranks as the most concerning of several risks. Other issues of note include ingestion, skin contact, and dust explosions (Szczyglewska et al., 2023).

Hazard controls and exposure reduction levels have been defined by guidelines. The use of safer forms of nanomaterials, engineering controls such as proper ventilation, and personal protective equipment as a final line of defense are also recommended (Ullah et al., 2024). Occupational exposure limits have been established for some materials and indicate the highest safe concentration of nanomaterials in the air, and exposure assessments with standard industrial hygiene sampling methods are possible. It is also important for worker protection to have an ongoing occupational health surveillance program. Furthermore, microplastics and NPs leached from plastic containers are becoming a concern (Sani et al., 2021). Occupational health surveillance is the process of systematically collecting, analyzing, and disseminating data on exposure and health outcomes in workers to prevent or diagnose disease or evaluate the outcome of intervention programs. It incorporates both medical and hazard surveillance. An initial medical assessment, recurring follow-up exams, post-event assessments, employee education, and tracking patterns in medical screening data are all components of basic medical surveillance (Shankar et al., 2021).

The goal of medical screening is to identify health issues in individual employees early on so that interventions may be made before disease processes start. A review of work history, a physical examination, and medical tests may be part of this. Currently, no testing or health assessments exist for health impacts resulting from exposure to manmade nanomaterials alone (Ullah et al., 2025). However, guidelines for medical screening of the bulk substance from which a nanoparticle is made remain relevant. In 2013, the National Institute for Occupational Safety and Health (NIOSH) issued special medical surveillance and screening guidelines (Sahu et al., 2021) for workers exposed to CNTs and carbon nanofibers because of growing toxicological evidence. Secondary prevention also involves medical screening and interventions; however, these do not replace primary prevention activities involving controls on hazards to reduce exposure of workers to nanomaterials (Ullah et al., 2025).

15.8 Environmental Impact and Indirect Health Risks

Scientists agree that nanotechnology stands as the upcoming major scientific breakthrough through the unification of engineering practice with biology along with chemistry, physics, and medicine. Materials exhibit significant changes in physical properties together with chemical properties when scientists reduce them down to nanometer dimensions (Sibuyi et al., 2021). The present nanotechnology divides its resources to build devices starting from microscopic objects all the way down to molecular dimensions with potential applications in medicine alongside environmental protection, energy generation, and space exploration (Ullah et al., 2023). Nanotechnology advances continue to expand human ability to engineer new products, thus creating the potential for future portable history storage devices and living organism-based molecular enhancement systems (Ullah et al., 2025).

During recent times, nanotechnology gained popularity as an innovative technology with tremendous potential. Nanotechnology enables functional system development through atomic-level manipulation, which reveals previously unattainable properties and produces evolutionary phenomena. Sensor design shows improved features including compact size, decreased weight and power usage, higher sensitivity, and better specificity ability (Ullah et al., 2024). Transistor miniaturization has acted as an essential industrial force since many years, yet nanolithography approaches attainability because transistors reach nanometric scales (Ullah et al., 2025). Nanotechnology sets one main objective to decrease computer costs and length and boost their processing speed. Future nanocomputers based on molecular switches and logic rods will enable many applications including substance delivery for medical purposes, bacterial and virus disinfection, and cancer cell destruction. The use of microscopic robots will enable the construction of intricate devices, as well as the removal of pollutants from environmental spaces (Ullah, 2017; Hou et al., 2024).

The rapid development of nanoscience and nanotechnology attracts significant public attention because nanomaterials have multiple industrial applications across agriculture, business, medicine, and public health sectors. From 1997 to 2005, global nanotechnology research investments grew from $432 million to reach $4.1 billion while the industry took over government funding during 2005 (Ullah et al., 2023). In 2015, the global economy was projected to receive $1 trillion in revenue from nanotechnology-based products supported by 2 million direct nanotechnology industry workers, as well as a second group of three million supporting personnel (Ullah et al., 2023; Aydoğdu et al., 2018).

Many ENMs exist in everyday products, including cosmetics, food packaging, drug delivery systems, therapeutics, and biosensors. Nanomaterials find extensive applications in wound dressings, detergents, and antimicrobial coatings because of their favorable dimensions matching biological macromolecules, as well as their antibacterial and odor-fighting properties. Weekly market entries of new nanotechnology products increase at a fast rate according to the Project on Emerging Nanotechnologies' nanotechnology consumer product inventory, which reports three to four new launches per week (Ullah et al., 2023; Ullah et al., 2025).

The rising usage of nanomaterials coincides with their broader human contact opportunities. People remain concerned about how NPs used in daily items affect the environment and display toxic properties (Wang et al., 2023). Due to their many benefits, small materials remain under investigation for their unknown environmental impacts (Figure 15.5). Scientists need to understand that studying nanomaterial breakdown patterns within the body remains crucial because analyzing resulting biological effects, such as genetic alterations and cellular organelle damage, is essential to their safety assessment. Several essential questions require clarification concerning whether nanoparticle toxicity eclipses or is equivalent to their standard-sized counterparts (Gul et al., 2025). The environment poses a risk of transforming NPs into more dangerous chemical compositions (Ullah et al., 2024; Waqar et al., 2024).

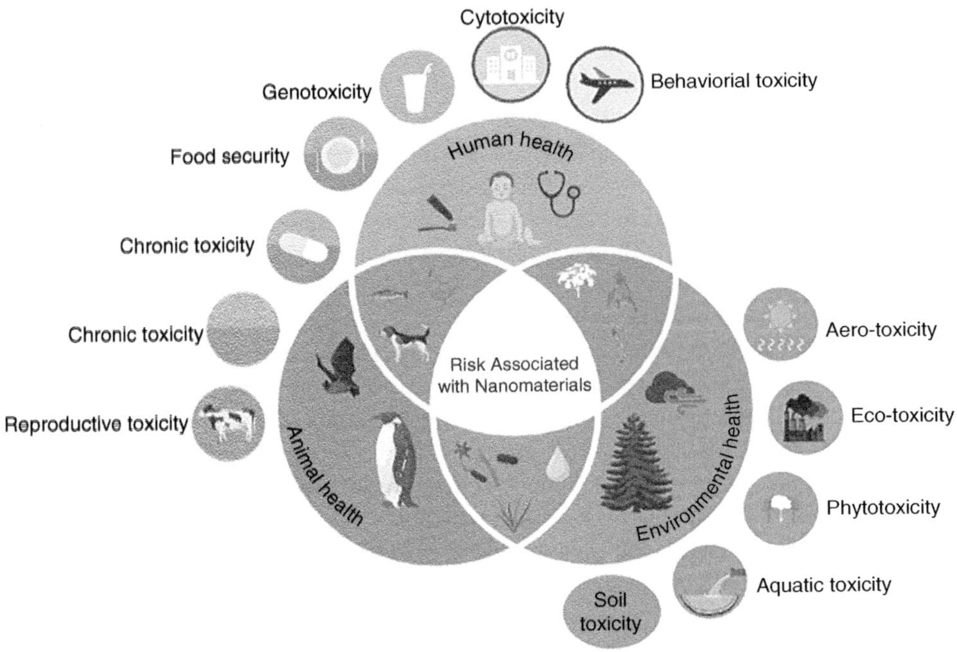

Figure 15.5 Illustration depicting pathways and risks linked to nanoparticle exposure.

15.9 Conclusion

Considering that NPs can result in great improvements in various industries, it is important, however, that the toxicological effects of these distinguishing physicochemical properties be adequately understood. Toxicity profiles of NPs depend heavily on their size, surface charge, functionalization, and aggregation behavior, which should all be analyzed in the context of intricate interactions that NPs have with biological systems. After ingestion, inhalation, skin absorption, or systemic administration, NPs accumulate in organs of vital importance and are toxic through various harmful mechanisms leading to oxidative stress, inflammation, genotoxicity, neurotoxicity, and immunotoxicity. More research is needed to adequately test the long-term effects of NPs in the body, including bioaccumulation and their relationship to gut flora. There are further worries for neurotoxicological consequences when manufactured NPs, and particularly those designed for medicinal uses, could break barriers, for example, the BBB. As the use of NPs expands, there is a desperate need for standardized toxicity assessment procedures and extensive long-term in vivo research to determine the effects of NPs in the body and ensure their safe use. Reducing possible danger related to NPs is crucial to achieve this and requires a multidisciplinary strategy bridging clinical sciences, bioengineering, and nanotoxicology. With this strategy, it will be simpler to manufacture safer nanomaterials that will have less harmful biological effects and could be responsibly used in many areas. Moreover, the balance between the benefits of nanotechnology on the one hand and keeping the environment and human health on the other hand will require improvement of a further regulatory framework.

References

Abbasi, R., Shineh, G., Mobaraki, M., Doughty, S., & Tayebi, L. (2023). Structural parameters of nanoparticles affecting their toxicity for biomedical applications: A review. *Journal of Nanoparticle Research*, *25*(3), 43.

Abdelnour, S. A., Alagawany, M., Hashem, N. M., Farag, M. R., Alghamdi, E. S., Hassan, F. U., & Attia, Y. A. (2021). Nanominerals: Fabrication methods, benefits and hazards, and their applications in ruminants with special reference to selenium and zinc nanoparticles. *Animals*, *11*(7), 1916.

Aboyewa, J. A., Sibuyi, N. R., Meyer, M., & Oguntibeju, O. O. (2021). Green synthesis of metallic nanoparticles using some selected medicinal plants from Southern Africa and their biological applications. *Plants*, *10*(9), 1929.

Ali, N., Katsouli, J., Marczylo, E. L., Gant, T. W., Wright, S., & De La Serna, J. B. (2024). The potential impacts of micro-and-nano plastics on various organ systems in humans. *eBioMedicine*, *99*.

Almanza-Reyes, H., Moreno, S., Plascencia-López, I., Alvarado-Vera, M., Patrón-Romero, L., Borrego, B., Reyes-Escamilla, A., Valencia-Manzo, D., Brun, A., Pestryakov, A., & Bogdanchikova, N. (2021). Evaluation of silver nanoparticles for the prevention of SARS-CoV-2 infection in health workers: In vitro and in vivo. *PLoS One*, *16*(8), e0256401.

Asghar, A., Ullah, Z., Ali, T., Iqbal, J., Abbasi, B. A., Ijaz, S., Yaseen, T., Iqbal, R., Murtaza, G., Kanwal, S., & Mahmood, T. (2024). Introduction to nc-RNA and their crucial role in biological regulation under environmental stress. In *ncRNAs: Mediated regulation: The unarguable biological function to environmental constrains* (pp. 1–46). Cham: Springer Nature Switzerland.

Asmatulu, R., Khan, W. S., & Asmatulu, E. (2024). *Nanotechnology safety*. Elsevier.

Azizi-Lalabadi, M., Garavand, F., & Jafari, S. M. (2021). Incorporation of silver nanoparticles into active antimicrobial nanocomposites: Release behavior, analyzing techniques, applications and safety issues. *Advances in Colloid and Interface Science*, *293*, 102440.

Bibi, S., Ullah, R., Burni, T., Ullah, Z., & Kazi, M. (2023). Impact of resorcinol and biochar application on the growth attributes, metabolite contents, and antioxidant systems of tomato (*Lycopersicon esculentum* Mill.). *ACS Omega*, *8*(48), 45750–45762.

Bruna, T., Maldonado-Bravo, F., Jara, P., & Caro, N. (2021). Silver nanoparticles and their antibacterial applications. *International Journal of Molecular Sciences*, *22*(13), 7202.

Dhaka, A., Mali, S. C., Sharma, S., & Trivedi, R. (2023). A review on biological synthesis of silver nanoparticles and their potential applications. *Results in Chemistry*, *6*, 101108.

de Dicastillo, C. L., & Correa, M. G. (2021). *Dioxide nanoparticles. Antimicrobial resistance: A one health perspective*, IntechOpen 95.Editer Mihai Mares

Ding, X., Pu, Y., Tang, M., & Zhang, T. (2022). Environmental and health effects of graphene-family nanomaterials: Potential release pathways, transformation, environmental fate and health risks. *Nano Today*, *42*, 101379.

Egbuna, C., Parmar, V. K., Jeevanandam, J., Ezzat, S. M., Patrick-Iwuanyanwu, K. C., Adetunji, C. O., Khan, J., Onyeike, E. N., Uche, C. Z., Akram, M., Ibrahim, M. S., El Mahdy, N. M., Awuchi, C. G., Saravanan, K., Tijjani, H., Odoh, U. E., Messaoudi, M., Ifemeje, J. C., Olisah, M. C., Ezeofor, N. J., Chikwendu, C. J., & Ibeabuchi, C. G. (2021). Toxicity of nanoparticles in biomedical application: Nanotoxicology. *Journal of Toxicology*, *2021*(1), 9954443.

Ettlinger, R., Lächelt, U., Gref, R., Horcajada, P., Lammers, T., Serre, C., Couvreur, P., Morris, R. E., & Wuttke, S. (2022). Toxicity of metal–organic framework nanoparticles: From essential analyses to potential applications. *Chemical Society Reviews*, *51*(2), 464–484.

Fatima, F., Hashim, A., & Anees, S. (2021). Efficacy of nanoparticles as nanofertilizer production: A review. *Environmental Science and Pollution Research*, *28*(2), 1292–1303.

Gao, M., Chang, J., Wang, Z., Zhang, H., & Wang, T. (2023). Advances in transport and toxicity of nanoparticles in plants. *Journal of Nanobiotechnology*, *21*(1), 75.

García-Quintero, A., & Palencia, M. (2021). A critical analysis of environmental sustainability metrics applied to green synthesis of nanomaterials and the assessment of environmental risks associated with nanotechnology. *Science of the Total Environment*, *793*, 148524.

Gigault, J., El Hadri, H., Nguyen, B., Grassl, B., Rowenczyk, L., Tufenkji, N., Feng, S., & Wiesner, M. (2021). Nanoplastics are neither microplastics nor engineered nanoparticles. *Nature Nanotechnology*, *16*(5), 501–507.

Gudkov, S. V., Burmistrov, D. E., Serov, D. A., Rebezov, M. B., Semenova, A. A., & Lisitsyn, A. B. (2021). A mini review of antibacterial properties of ZnO nanoparticles. *Frontiers in Physics*, *9*, 641481.

Gul, F., Ullah, Z., Iqbal, J., Abbasi, B. A., Ali, S., Kanwal, S., Uddin, J., & Mahmood, T. (2025). Ecofriendly synthesis characterization and biological activities of *Eruca sativa* mediated silver oxide nanoparticles. *Scientific Reports*, *15*(1), 13466.

Gul, F., Ullah, Z., Iqbal, J., Abbasi, B. A., Ijaz, S., Uddin, S., Yaseen, T., Ali, H., Jamil, M., Kanwal, S., & Mahmood, T. (2025). Bioactive phytochemicals from garlic (*Allium sativum* L.) by-products. In *Bioactive phytochemicals in by-products from bulb, flower and fruit vegetables* (pp. 23–83). Cham: Springer Nature Switzerland.

Gupta, V., Mohapatra, S., Mishra, H., Farooq, U., Kumar, K., Ansari, M. J., Aldawsari, M. F., Alalaiwe, A. S., Mirza, M. A., & Iqbal, Z. (2022). Nanotechnology in cosmetics and cosmeceuticals—A review of latest advancements. *Gels*, *8*(3), 173.

Haleem, A., Javaid, M., Singh, R. P., Rab, S., & Suman, R. (2023). Applications of nanotechnology in medical field: A brief review. *Global Health Journal*, *7*(2), 70–77.

Holmannova, D., Borsky, P., Svadlakova, T., Borska, L., & Fiala, Z. (2022). Carbon nanoparticles and their biomedical applications. *Applied Sciences*, *12*(15), 7865.

Hong, J., Wang, C., Wagner, D. C., Gardea-Torresdey, J. L., He, F., & Rico, C. M. (2021). Foliar application of nanoparticles: Mechanisms of absorption, transfer, and multiple impacts. *Environmental Science: Nano*, *8*(5), 1196–1210.

Horie, M., & Tabei, Y. (2021). Role of oxidative stress in nanoparticles toxicity. *Free Radical Research*, *55*(4), 331–342.

Hou, J. Y., Liu, H. T., Huang, L. T., Wu, S. B., & Zhang, Z. L. (2024). Recent advances in micro/nano-robots for environmental pollutant removal: Mechanism, application, and prospect. *Chemical Engineering Journal*, 498(155135.

Ijaz, S., Iqbal, J., Abbasi, B. A., Tufail, A., Ullah, Z., Sharifi-Rad, J., Yaseen, T., & Uddin, S. (2023). Biofortification: Lessons from the past and strategies for future food security. In *Legumes biofortification* (pp. 521–545). Cham: Springer International Publishing.

Ijaz, S., Iqbal, J., Abbasi, B. A., Tufail, A., Ullah, Z., Yaseen, T., Ali, I., Uddin, S., & Iqbal, R. (2024). Heavy metal–polluted arable land and its consequences: A global scenario. In *Biochar-assisted remediation of contaminated soils under changing climate* (pp. 71–99). Elsevier.

Ijaz, S., Iqbal, J., Abbasi, B. A., Ullah, Z., Ijaz, N., Yaseen, T., Iqbal, R., Murtaza, G., Usman, M., Sampath, S., Hafeez, M. B., Kanwal, S., & Mahmood, T. (2024). Regulatory and ethical concerns of nanotechnology in agriculture. In *Revolutionizing agriculture: A comprehensive exploration of agri-nanotechnology* (pp. 395–427). Cham: Springer Nature Switzerland.

Ijaz, S., Iqbal, J., Abbasi, B. A., Ullah, Z., Uddin, S., Yaseen, T., Zarshan, F., Imtiaz, M., Ali, H., Jamil, M., Kanwal, S., & Mahmood, T. (2025). Bioactive phytochemicals from beetroot (*Beta vulgaris*) by-products. In *Bioactive phytochemicals in by-products from leaf, stem, root and tuber vegetables* (pp. 131–180). Cham: Springer Nature Switzerland.

Ijaz, S., Iqbal, J., Abbasi, B. A., Ullah, Z., Uddin, S., Yaseen, T., Zarshan, F., Ali, H., Sher, H., Ullah, Z., Ali, A., Kanwal, S., & Mahmood, T. (2025). Bioactive phytochemicals from ginger (*Zingiber officinale*) by-products. In *Bioactive phytochemicals in by-products from Leaf, stem, root and tuber vegetables* (pp. 281–303). Cham: Springer Nature Switzerland.

Ijaz, S., Iqbal, J., Abbasi, B. A., Ullah, Z., Yaseen, T., Munir, F., Hyder, S., Kanwal, S., Sahito, Z. A., & Mahmood, T. (2024). From single nc-RNAs to networks: Understanding the complexity of environmental adaption. In *ncRNAs: Mediated regulation: The unarguable biological function to environmental constrains* (pp. 283–310). Cham: Springer Nature Switzerland.

Ijaz, S., Iqbal, J., Abbasi, B. A., Ullah, Z., Yaseen, T., Murtaza, G., Iqbal, R., Zahid, S. M. H., Mahmood, T., & Ijaz, N. (2025). Iron and soil chemistry: Beneficial elements and pollution mitigation. In *Beneficial elements for remediation of heavy metals in polluted soil* (pp. 197–220). Elsevier.

Ijaz, S., Iqbal, J., Abbasi, B. A., Ullah, Z., Yaseen, T., Murtaza, G., & Mahmood, T. (2025). Biofortification of cereals and pulses using new breeding techniques. In *Crop biofortification: Biotechnological approaches for achieving nutritional security under changing climate* (pp. 33–50).

Ijaz, S., Iqbal, J., Abbasi, B. A., Ullah, Z., Yaseen, T., Kanwal, S., Mahmood, T., Sydykbayeva, S., Ydyrys, A., Almarhoon, Z. M., Sharifi-Rad, J., Hano, C., Calina, D., & Cho, W. C. (2023). Rosmarinic acid and its derivatives: Current insights on anticancer potential and other biomedical applications. *Biomedicine & Pharmacotherapy, 162*, 114687.

Ijaz, S., Iqbal, J., Abbasi, B. A., Yaseen, T., Ullah, Z., Alqahtani, T., Iqbal, R., Murtaza, D. G., Kanwal, S., & Mahmood, T. (2024). nc-RNA: A key player in stress response pathways, a molecular perspective. In *ncRNAs: Mediated regulation: The unarguable biological function to environmental constrains* (pp. 139–163). Cham: Springer Nature Switzerland.

Ikram, M., Javed, B., Raja, N. I., & Mashwani, Z. U. R. (2021). Biomedical potential of plant-based selenium nanoparticles: A comprehensive review on therapeutic and mechanistic aspects. *International Journal of Nanomedicine*, 249–268.

Jaswal, T., & Gupta, J. (2023). A review on the toxicity of silver nanoparticles on human health. *Materials Today: Proceedings, 81*, 859–863.

Joseph, T. M., Kar Mahapatra, D., Esmaeili, A., Piszczyk, Ł., Hasanin, M. S., Kattali, M., Haponiuk, J., & Thomas, S. (2023). Nanoparticles: Taking a unique position in medicine. *Nanomaterials, 13*(3), 574.

Khan, A., Ahmad, M., Sultan, A., Ullah, Z., Majeed, K., Ullah, T., & Zafar, M. (2024). Contribution to the pollen morphology of *Astragalus* L. section *Aegacantha* Bunge (Galegeae-Fabaceae) and its systematic significance. *Palynology, 48*(1), 2264352.

Khan, S., Naushad, M., Al-Gheethi, A., & Iqbal, J. (2021). Engineered nanoparticles for removal of pollutants from wastewater: Current status and future prospects of nanotechnology for remediation strategies. *Journal of Environmental Chemical Engineering, 9*(5), 106160.

Kumah, E. A., Fopa, R. D., Harati, S., Boadu, P., Zohoori, F. V., & Pak, T. (2023). Human and environmental impacts of nanoparticles: A scoping review of the current literature. *BMC Public Health, 23*(1), 1059.

Kumar, A., Choudhary, A., Kaur, H., Guha, S., Mehta, S., & Husen, A. (2022). Potential applications of engineered nanoparticles in plant disease management: A critical update. *Chemosphere, 295*, 133798.

Kus-Liśkiewicz, M., Fickers, P., & Ben Tahar, I. (2021). Biocompatibility and cytotoxicity of gold nanoparticles: Recent advances in methodologies and regulations. *International Journal of Molecular Sciences, 22*(20), 10952.

Lei, L., Qiao, K., Guo, Y., Han, J., & Zhou, B. (2020). Titanium dioxide nanoparticles enhanced thyroid endocrine disruption of pentachlorophenol rather than neurobehavioral defects in zebrafish larvae. *Chemosphere, 249*, 126536.

Lozano, O. (2020). Amorphous SiO2 nanoparticles promote cardiac dysfunction via the opening of the mitochondrial permeability transition pore in rat heart and human cardiomyocytes. *Particle and Fibre Toxicology, 17*(1), 15.Landa, P. (2021). Positive effects of metallic nanoparticles on plants: Overview of involved mechanisms. *Plant Physiology and Biochemistry, 161*, 12–24.

Malik, S., Muhammad, K., & Waheed, Y. (2023). Emerging applications of nanotechnology in healthcare and medicine. *Molecules, 28*(18), 6624.

Manuja, A., Kumar, B., Kumar, R., Chhabra, D., Ghosh, M., Manuja, M., Brar, B., Pal, Y., Tripathi, B. N., & Prasad, M. (2021). Metal/metal oxide nanoparticles: Toxicity concerns associated with their physical state and remediation for biomedical applications. *Toxicology Reports, 8*, 1970–1978.

Miyazawa, T., Itaya, M., Burdeos, G. C., Nakagawa, K., & Miyazawa, T. (2021). A critical review of the use of surfactant-coated nanoparticles in nanomedicine and food nanotechnology. *International Journal of Nanomedicine*, 3937–3999.

Mubeen, B., Ansar, A. N., Rasool, R., Ullah, I., Imam, S. S., Alshehri, S., Ghoneim, M. M., Sami, A., I, Nadeem, M. S., & Kazmi, I. (2021). Nanotechnology as a novel approach in combating microbes providing an alternative to antibiotics. *Antibiotics, 10*(12), 1473.

Musielak, E., Feliczak-Guzik, A., & Nowak, I. (2022). Synthesis and potential applications of lipid nanoparticles in medicine. *Materials, 15*(2), 682.

Nie, P., Zhao, Y., & Xu, H. (2023). Synthesis, applications, toxicity and toxicity mechanisms of silver nanoparticles: A review. *Ecotoxicology and Environmental Safety, 253*, 114636.

Noureen, S., Iqbal, J., Abbasi, B. A., Yaseen, T., Ijaz, S., Ullah, Z., Ullah, Z., Ali, A., Murtaza, G., Iqbal, R., Waqar, R., & Harsonowati, W. (2025). Exploring beneficial elements in the soil environment: Fluorine and iodine perspectives. In *Beneficial elements for remediation of heavy metals in polluted soil* (pp. 161–196). Elsevier.

Ozdal, M., & Gurkok, S. (2022). Recent advances in nanoparticles as antibacterial agent. *ADMET and DMPK, 10*(2), 115–129.

Paras, Yadav, K., Kumar, P., Teja, D. R., Chakraborty, S., Chakraborty, M., Mohapatra, S. S., Sahoo, A., Chou, M. M. C., Liang, C.-T., & Hang, D.-R. (2022). A Review on Low-Dimensional Nanomaterials: Nanofabrication, Characterization and Applications. *Nanomaterials*, 13(1), 160. https://doi.org/10.3390/nano13010160.

Perut, F., Roncuzzi, L., Avnet, S., Massa, A., Zini, N., Sabbadini, S., Giampieri, F., Mezzetti, B., & Baldini, N. (2021). Strawberry-derived exosome-like nanoparticles prevent oxidative stress in human mesenchymal stromal cells. *Biomolecules, 11*(1), 87.

Pogribna, M., & Hammons, G. (2021). Epigenetic effects of nanomaterials and nanoparticles. *Journal of Nanobiotechnology, 19*, 1–18.

Pokrajac, L., Abbas, A., Chrzanowski, W., Dias, G. M., Eggleton, B. J., Maguire, S., Maine, E., Malloy, T., Nathwani, J., Nazar, L., Sips, A., Sone, J.'i., van den Berg, A., Weiss, P. S., & Mitra, S. (2021). Nanotechnology for a sustainable future: Addressing global challenges with the international network4sustainable nanotechnology. *ACS Nano, 15*(12), 18608–18623.

Rai, P. K., Lee, J., Brown, R. J., & Kim, K. H. (2021). Environmental fate, ecotoxicity biomarkers, and potential health effects of micro-and nano-scale plastic contamination. *Journal of Hazardous Materials, 403*, 123910.

Sahu, T., Ratre, Y. K., Chauhan, S., Bhaskar, L. V. K. S., Nair, M. P., & Verma, H. K. (2021). Nanotechnology based drug delivery system: Current strategies and emerging therapeutic potential for medical science. *Journal of Drug Delivery Science and Technology, 63*, 102487.

Said, Z., Sohail, M. A., Pandey, A. K., Sharma, P., Waqas, A., Chen, W. H., Nguyen, P. Q. P., Nguyen, V. N., Pham, N. D. K., & Nguyen, X. P. (2023). Nanotechnology-integrated phase change material and nanofluids for solar applications as a potential approach for clean energy strategies: Progress, challenges, and opportunities. *Journal of Cleaner Production, 416*, 137736.

Sani, A., Cao, C., & Cui, D. (2021). Toxicity of gold nanoparticles (AuNPs): A review. *Biochemistry and Biophysics Reports, 26*, 100991.

Sasikala, A. R. K., Kaliannagounder, V. K., Alluri, N. R., Shrestha, B. K., Kim, S. J., Ali-Boucetta, H., Park, C. H., & Unnithan, A. R. (2022). Development of self-powered multifunctional piezomagnetic nanoparticles for non-invasive post-surgical osteosarcoma theranogeneration. *Nano Energy, 96*, 107134.

Shankar, S., Jaiswal, L., & Rhim, J. W. (2021). New insight into sulfur nanoparticles: Synthesis and applications. *Critical Reviews in Environmental Science and Technology, 51*(20), 2329–2356.

Sibuyi, N. R. S., Moabelo, K. L., Fadaka, A. O., Meyer, S., Onani, M. O., Madiehe, A. M., & Meyer, M. (2021). Multifunctional gold nanoparticles for improved diagnostic and therapeutic applications: A review. *Nanoscale Research Letters, 16*, 1–27.

Silva, L. F., Oliveira, M. L., Crissien, T. J., Santosh, M., Bolivar, J., Shao, L., Dotto, G. L., Gasparotto, J., & Schindler, M. (2022). A review on the environmental impact of phosphogypsum and potential health impacts through the release of nanoparticles. *Chemosphere, 286*, 131513.

Sim, S., & Wong, N. K. (2021). Nanotechnology and its use in imaging and drug delivery. *Biomedical Reports, 14*(5), 42.

Sonwani, S., Madaan, S., Arora, J., Suryanarayan, S., Rangra, D., Mongia, N., Vats, T., & Saxena, P. (2021). Inhalation exposure to atmospheric nanoparticles and its associated impacts on human health: A review. *Frontiers in Sustainable Cities, 3,* 690444.

Stater, E. P., Sonay, A. Y., Hart, C., & Grimm, J. (2021). The ancillary effects of nanoparticles and their implications for nanomedicine. *Nature Nanotechnology, 16*(11), 1180–1194.

Suliasih, B. A., Budi, S., & Katas, H. (2024). Synthesis and application of gold nanoparticles as antioxidants. *Pharmacia, 71,* 1–19.

Szczyglewska, P., Feliczak-Guzik, A., & Nowak, I. (2023). Nanotechnology–general aspects: A chemical reduction approach to the synthesis of nanoparticles. *Molecules, 28*(13), 4932.

Uddin, S., Iqbal, J., Abbasi, B. A., Tareen, A. R., Ijaz, S., Ullah, Z., Ali, A., & Mahmood, T. (2025). Bioactive phytochemicals from cabbage (*Brassica oleracea*) by-products. In *Bioactive phytochemicals in by-products from leaf, stem, root and tuber vegetables* (pp. 31–57). Cham: Springer Nature Switzerland.

Ullah, R., Ullah, Z., Iqbal, J., Chalgham, W., & Ahmad, A. (2023). Aspartic acid-based nano-copper induces resilience in *Zea mays* to applied lead stress via conserving photosynthetic pigments and triggering the antioxidant biosystem. *Sustainability, 15*(16), 12186.

Ullah, Z. (2017). Estimation of genetic diversity in the germplasm of *Brassica Campestris* (L.) Using morphological and biochemical markers. *Journal of Botany, 1*(1), 45–59.

Ullah, Z., Gul, F., Iqbal, J., Abbasi, B. A., Kanwal, S., Chalgham, W., El-Sheikh, M. A., Diltemiz, S. E., & Mahmood, T. (2023). Biogenic synthesis of multifunctional silver oxide nanoparticles (Ag_2ONPs) using *Parieteria alsinaefolia* delile aqueous extract and assessment of their diverse biological applications. *Microorganisms, 11*(4), 1069. PMID: 37110492 PMCID: PMC10142072 DOI: 10.3390/microorganisms11041069

Ullah, Z., Iqbal, J., Abbasi, B. A., Akhtar, W., Kanwal, S., Ali, I., Chalgham, W., El-Sheikh, M. A., & Mahmood, T. (2023). Assessment of gus expression induced by anti-sense Os PPO gene promoter and antioxidant enzymatic assays in response to drought and heavy metal stress in transgenic *Arabidopsis thaliana*. *Sustainability, 15*(17), 12783.

Ullah, Z., Iqbal, J., Abbasi, B. A., Gul, F., Ali, S., Kanwal, S., Elsadek, M. F., Ali, M. A., Iqbal, R., Elsalahy, H. H., & Mahmood, T. (2025). Eco-friendly synthesis of iron oxide nanoparticles using *Parietaria alsinifolia* extracts and evaluation of biological applications. *Applied Biochemistry and Biotechnology*, 1–28. PMID: 39752127 DOI: 10.1007/s12010-024-05151-7

Ullah, Z., Iqbal, J., Abbasi, B. A., Ijaz, S., Ahmad, S., Khan, S., Sampath, S., Iqbal, R., Murtaza, G., Mehmood, Y., Kanwal, S., & Mahmood, T. (2024). Nano-formulation for agriculture applicability. In *Revolutionizing agriculture: A comprehensive exploration of agri-nanotechnology* (pp. 325–367). Cham: Springer Nature Switzerland.

Ullah, Z., Iqbal, J., Abbasi, B. A., Ijaz, S., Ahmad, S., Yaseen, T., Iqbal, R., Murtaza, D. G., Saleem, A., Kanwal, S., & Mahmood, T. (2025). Future horizons: Emerging "omics" technologies and challenges in tomato. In *Omics Approaches for Tomato Yield and Quality Trait Improvement* (Vol. 347). Springer.

Ullah, Z., Iqbal, J., Abbasi, B. A., Ijaz, S., Anjum, A., Yaseen, T., Murtaza, G., Iqbal, R., Hyder, S., Kanwal, S., & Mahmood, T. (2025). Current challenges and recent advancements in the adoption of omics to enhance biofortification. In *Crop biofortification: Biotechnological approaches for achieving nutritional security under changing climate* (pp. 61–88). Wiley.

Ullah, Z., Iqbal, J., Abbasi, B. A., Ijaz, S., Munir, M., Yaseen, T., Sampath, S., Kanwal, S., Sher, H., Ullah, Z., Ali, A., & Mahmood, T. (2025). Applications of nanoparticles in biofortification of crops: Amplifying nutritional quality. In *Crop biofortification: Biotechnological approaches for achieving nutritional security under changing climate* (pp. 321–350). Wiley.

Ullah, Z., Iqbal, J., Abbasi, B. A., Ijaz, S., Sharifi-Rad, J., Yaseen, T., Uddin, S., & Mahmood, T. (2023). How genome editing can be helpful in the biofortification of legumes. In *Legumes biofortification* (pp. 207–232). Cham: Springer International Publishing.

Ullah, Z., Iqbal, J., Abbasi, B. A., Ijaz, S., Yaseen, T., Murtaza, G., Iqbal, R., Noreen, S., Waqar, R., Kanwal, S., & Mahmood, T. (2025). Arsenic pollution: Sources, types, and effects on soil chemistry. In *Beneficial elements for remediation of heavy metals in polluted soil* (pp. 55–91). Elsevier.

Ullah, Z., Iqbal, J., Abbasi, B. A., Ijaz, S., Yaseen, T., Waqar, R., Kanwal, S., Sher, H., Ullah, Z., Ali, A., & Mahmood, T. (2024). The green revolution: Promoting environmental stewardship and plant growth. In *Environment, climate, plant and vegetation growth* (pp. 425–469). Cham: Springer Nature Switzerland.

Ullah, Z., Iqbal, J., Gul, F., Abbasi, B. A., Kanwal, S., Elsadek, M. F., Ali, M. A., Iqbal, R., Elsalahy, H. H., & Mahmood, T. (2024). Biogenic synthesis, characterization, and in vitro biological investigation of silver oxide nanoparticles (AgONPs) using *Rhynchosia capitata*. *Scientific Reports, 14*, 10484. PMID: 38714767 PMCID: PMC11076632 DOI: 10.1038/s41598-024-60694-3

Wang, M., Li, Q., Shi, C., Lv, J., Xu, Y., Yang, J., Chua, S. L., Jia, L., Chen, H., Liu, Q., Huang, C., Huang, Y., Chen, J., & Fang, M. (2023). Oligomer nanoparticle release from polylactic acid plastics catalysed by gut enzymes triggers acute inflammation. *Nature Nanotechnology, 18*(4), 403–411.

Waqar, R., Iqbal, J., Abbasi, B. A., Mumtaz, A. S., Ijaz, S., Ullah, Z., Murtaza, G., Iqbal, R., Hussain, M., & Butt, N. I. (2024). Climate resilience: Strategies for enhancing plant and vegetation growth. In *Environment, climate, plant and vegetation growth* (pp. 471–501). Cham: Springer Nature Switzerland.

16

Cost–Benefit Analysis of Nanoagricultural Technologies

Amir Abdullah Khan[1], Qamar Abbas[2], Rasheed Akbar[1], and Muhammad Ramzan[3]

[1] *School of Environmental sciences and Engineering, Jiangsu University, Zhenjiang, China*
[2] *School of Business, Hengxing University Qingdao, Shandong, China*
[3] *School of Finance and Economics, Jiangsu University, Zhenjiang, China*

16.1 Nanoagricultural Technologies

By using the unique physical and chemical properties of materials made at the nanoscale (1–100 nm), nanoagricultural technologies increase crop production by improving processes at the molecular and cellular levels (Zaman et al., 2025). Metallic and metal oxide nanoparticles, polymeric nanocapsules, nanoemulsions, layered double hydroxides, and carbon-based nanostructures are just a few examples of the nanomaterials that allow for controlled release, high specific surface area, and agrochemical and biological agent delivery and monitoring with previously unseen levels of accuracy (Adetutu et al., 2024). These systems can improve sustainability and yield stability under varying environmental conditions by modifying interactions at the plant–soil and plant–pathogen interfaces (Fathi Qarachal & Alizadeh, 2025; Khan et al., 2025). They can also reduce doses, improve absorption efficiencies, and lessen off-target impacts. Essential macro- and micronutrients are complexed or encapsulated within nanoscale carriers in nanofertilizers, a revolutionary class of inputs in nutrient management (Tomar et al., 2024). These products, which contain zinc oxide quantum dots, iron hydroxide nanoparticles, and urea covered with mesoporous silica nanoparticles, reduce nutrient loss from evaporation and washing away while keeping them available for plants in the soil (Yadav et al., 2023). Because of better interactions with root surfaces and longer-lasting availability of nutrients, experiments show that wheat and rice seedlings treated with nanoencapsulated nitrogen, phosphorus, and potassium take up to 30% more nutrients and grow better compared to traditional fertilization methods (Mehara et al., 2025).

The application of nanotechnology in agriculture can be done on the base of different factors such as to enhance the growth of plants, to ameliorate the negative effects of stressors, and to replace the chemical fertilizers, which have detrimental effects on environment

Nanotechnology Innovations for Food Security and Sustainable Agriculture, First Edition. Edited by Zia Ur Rehman Mashwani, Shah Fahad, Ilyas Ahmad, Shah Saud, Sajid Fiaz, and Taufiq Nawaz.

as shown in Figure 16.1. The integration of insecticides, fungicides, and herbicides into stimuli-sensitive delivery systems is known as nano-enabled crop protection (Camara et al., 2019). To ensure that active substances are released on-site only when a pathogen touches them, biodegradable polymeric nanoparticles and lipid-based nanoemulsions can be functionalized to react to pH, redox potential, or enzymatic triggers found in target organisms (Yeh et al., 2020). In addition to reducing residues in soil and water, this focused strategy can reduce the necessary dosage of active substances by up to an order of magnitude (Garcia et al., 2022). At the same time, developing tiny carriers that hold defense-boosting peptides and plant hormones has allowed us to prepare the plant's natural defense systems, making them better at fighting off viral, bacterial, and fungal infections (Arunachalam et al., 2023). In nanoagriculture, real-time diagnostics use biosensors made of fluorescent quantum dots, carbon nanotubes, and gold nanoparticles coupled with aptamers, antibodies, or molecularly imprinted polymers (Thirugnanasambandan, 2021). These sensors detect tiny amounts of specific proteins from pathogens, harmful substances such as mycotoxins, or stress indicators in plants (like malondialdehyde or H_2O_2) (Chugh et al., 2022). Crop losses can be decreased, and precise treatments can be informed by in-field monitoring of soil characteristics and crop health with high temporal resolution through the integration of portable electrochemical or optical instruments (Pal et al., 2024). Furthermore, with the use of sophisticated imaging techniques like confocal Raman microscopy and synchrotron-based X-ray fluorescence, nanotracers have proven crucial in

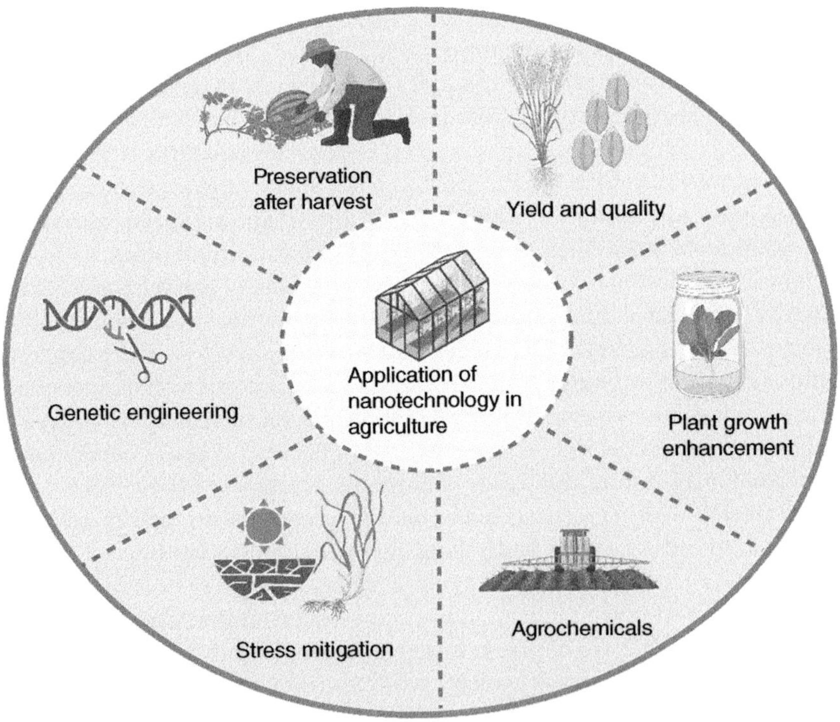

Figure 16.1 Application of nanoagricultural technologies in agriculture sector.

clarifying the water and nutrient transport channels within plant tissues (Kopittke et al., 2018). Under less-than-ideal circumstances, seed treatments that include nanocarriers have significantly increased seed vigor and germination (Shelar et al., 2023). During soaking, tiny plastic particles with helpful microbes, growth boosters, or nutrients can attach to seed surfaces and release their contents as the seeds absorb water. Studies on soybeans and corn have shown that seeds treated with special bacterial nanoformulations or silica-coated zinc nanocomposites can germinate up to 25% better when facing drought or salt stress, suggesting that these treatments work together to help protect the seeds (Maswada et al., 2018).

There is hope that new nano-enabled gene delivery devices will overcome the drawbacks of conventional transformation methods (Niazian et al., 2021). To achieve transient or sustained expression without the use of genomic integration vectors, functionalized mesoporous silica and layered double hydroxide nanosheets can transport nucleic acids, ribonucleoprotein complexes (like CRISPR-Cas9), or secondary metabolites into plant cells via endocytosis (Miyamoto & Numata, 2023; Zhang et al., 2024). Early results in rice and tobacco show that editing success rates are similar to those achieved with Agrobacterium methods but require less tissue culture. Even with these developments, thorough toxicological and environmental fate analyses are still necessary. Studies must measure the dissolving kinetics of nanoparticles, their potential for trophic transmission, and their effects on the function of soil microbiota, including nitrogen cycling and mycorrhizal symbioses. For safe field-scale deployment, standardized regulatory guidelines should incorporate life-cycle studies, ecotoxicity tests, and physicochemical characterization.

16.2 Economic Costs of Nanoagricultural Technologies

The economic feasibility of nanotechnology is still debatable, despite the fact that it has the potential to revolutionize agriculture, particularly through improved productivity and environmental sustainability as shown in Table 16.1 and Figure 16.2. Its practical application is hampered by high production costs, unclear regulations, and limited scalability, especially for smallholder farmers. Economic costs significantly influence the adoption and widespread usage of nanoagricultural technologies, especially in developing countries (Preethi et al., 2024). A fundamental evidence-based evaluation entails weighing the possible financial advantages and hazards of these technologies in addition to their direct and indirect costs (Liberty et al., 2024). Because it requires specialized equipment and skilled labor, the synthesis and manufacture of nanomaterials can be costly (Xin et al., 2020). A thorough assessment of the cost-effectiveness of increasing output for agricultural applications is necessary (Xin et al., 2020). Depending on how you apply the nanomaterials to the soil or crops, the cost may increase. The equipment and labour needed to administer nanopesticides, nanofertilizers, or nanosensors are included in this (Sembada & Lenggoro, 2024). Observing safety and environmental rules may result in extra expenses. It is crucial to evaluate the possible effects of nanomaterials on the environment and human health and to take into account the expenses related to obtaining regulatory clearances (Iavicoli et al., 2014, 2017). Infrastructure investments, such as specialized storage facilities or application equipment, may be necessary for the successful implementation of nanoagricultural

Table 16.1 Economic implications of nanoagricultural technologies at large scale.

Technology assessed	Key findings	Economic implications	References
Nanopesticides	Use of active ingredients goes down by 20–30%, but production costs go up by 30–50%.	Higher initial expenses may inhibit adoption despite long-term gains.	Kah et al. (2019)
Nanofertilizers	20–40% better nutrient absorption, but 2–3 times higher costs of synthesis than normal.	Cost-effectiveness hinges on getting more yield and losing less in the environment.	Liu and Lal (2015)
General nanoagri inputs	No large-scale application	Hard to scale the product.	Fraceto et al. (2016)
Nanofertilizers	Hard to commercialize the product.	Huge investment needed for commercialization.	Mukhopadhyay et al. (2022)
Nanofertilizers (wheat)	After two to three seasons, the return on investment (ROI) is higher than with regular fertilizer.	Even if there are long-term benefits, smallholders have trouble getting cash.	Singh et al. (2023)

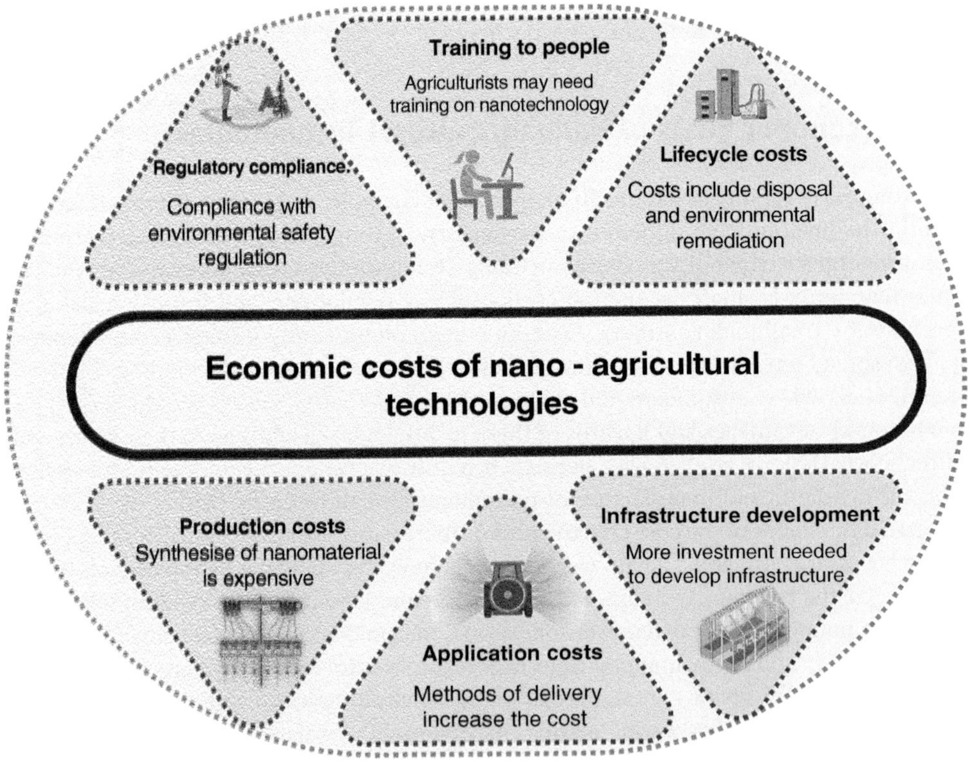

Figure 16.2 Shows the economic costs that make nanoagricultural technologies expensive.

technologies. For farmers and agricultural workers to use these technologies efficiently, they could also require training (Garg et al., 2024). According to Yudhistira et al. (2023), a thorough economic analysis should take into account the full lifespan of nanoagricultural products, including disposal and any environmental remediation expenses. When weighing the possible advantages of nanoagricultural technology, such as higher crop yields, less pesticide usage, or better soil health, a thorough cost–benefit analysis (CBA) should be carried out (Flores et al., 2024). The analysis ought to take into account both immediate and long-term effects. Engaging with all pertinent stakeholders, such as farmers, researchers, policymakers, and the general public, is also essential in order to collect data, verify assumptions, and guarantee that the evaluation is thorough and pertinent (Medani et al., 2024). In contrast to traditional agricultural methods and other alternative technologies, evaluate the advantages and disadvantages of nanoagricultural technology (Priya et al., 2023). This comparative analysis can assist in determining the best economical and environmentally friendly alternatives (Singh, 2023).

16.3 Potential Benefits of Nanoagricultural Technologies

Although they are expensive, nanoagricultural technologies have a number of potential financial advantages (Medani et al., 2024). For example, nanopesticides and fertilizers can enhance pest management and nutrient uptake, increasing crop yields and farmers' profits (Otari et al., 2024). Likewise, precision farming methods made possible by nanosensors can maximize the use of herbicides, fertilizers, and water, lowering input costs and minimizing environmental effects (Medani et al., 2024). Additionally, according to Sun et al. (2022), nanomaterials can be used to enhance the nutritional value and shelf life of agricultural products, boosting their market value and lowering food waste. According to Kamyab et al. (2024), nanoagricultural technologies can also support sustainable farming methods including integrated pest control, cover crops, and less tillage, which can improve soil health and lower long-term environmental costs. Additionally, nanotechnology can help turn agricultural waste into useful goods like bioenergy, which lowers waste management expenses and opens up new sources of income (Preethi et al., 2024).

16.4 Social Impacts of Nanotechnological Technologies

Nanotechnology, which is the ability to change matter at the nanoscale (1–100 nm), has a lot of potential in many fields, such as health, medicine, farming, and manufacturing (Singla, 2021). However, the widespread use of nanotechnology also presents serious social, moral, and health issues that need to be thought about carefully (Charitidis, 2020). Nanotechnology has a wide range of effects on society, including the economy, the workforce, and how people see it. Nanotechnologies in farming have the potential to change civilization by increasing crop yields, lowering the need for chemicals, and making food more secure. By making farmers more productive and giving them more money, these improvements can help reduce poverty and revitalize rural areas (Barzinjy & Hamadamen, 2022). But if not everyone has access to nanotech advancements, it could worsen societal inequities between regions that are technologically advanced and those that are resource-poor. Concerns about health and environmental issues can also influence public perception of a product and the functioning of laws related to it. Furthermore, the move toward high-tech farming could make traditional farming knowledge and practices less important, which raises moral and cultural problems concerning the future of farming communities and sustainable agricultural growth.

16.4.1 Economic Implications

Kalita and Baruah (2019) predicts that nanotechnology will boost the economy by creating new sectors, changing old ones, and creating jobs that require a high level of skill. Investing in nanotechnology research and development can lead to new goods, better materials, and more efficient ways to do things. Such developments can make the economy more productive and competitive (Shastri & Gandhi, 2024).

16.4.2 Workforce Transformations

Nanotechnology could change the needs of the workforce, which would mean that workers would need to learn new skills and be able to adapt to new employment responsibilities (Weh, 2024).

Nanotechnology-related industries may produce new jobs, but they may also lead to job loss in industries that are affected by these new technologies (Van de Voorde & Vlerick, 2021).

16.4.3 Social Equity

Nanotechnology may not benefit everyone in society equally, which could make socioeconomic disparities worse (Van de Voorde & Vlerick, 2021). Some groups may not be able to use nanotechnology-based products and technologies, which could create a "nano-divide" where some groups get greater benefits than others (Landers et al., 2024).

16.4.4 Public Perception and Acceptance

The way people see nanotechnology has a big effect on how widely it is used and accepted in agriculture and other fields (Kamarulzaman et al., 2020). When people don't have clear, accurate information or are exposed to false information and inflated claims, they may not comprehend, be afraid of, or resist nanotechnology-based products and new ideas (Singh et al., 2018). This is especially true in places where people are still very worried about possible health and environmental threats. To deal with these problems, clear and fact-based communication is essential. Active public engagement—through education, conversation, and developing policies that include everyone—along with strong ethical principles, helps people trust the responsible development and use of nanotechnolgy (Khan et al., 2025a,b; Patil et al., 2020).

16.4.5 Risk Governance and Regulations

It is important to have good risk management frameworks in place for nanotechnology since it has a lot of complicated and sometimes unforeseen dangers. These frameworks should also encourage innovation and maximize the benefits to society (Malakar et al., 2022). Nanomaterials are very different from regular materials because they are very small, react more quickly, and interact with living and nonliving systems in ways that are not fully understood, which might have unexpected hazardous effects. This level of complexity makes it challenging for existing risk assessment models to work and calls for the creation of new, customized regulatory approaches. Because of this, regulatory bodies need to change by making detailed standards, strong testing procedures, and explicit safety rules that are made just for nanomaterials and how they act. This includes making sure that there are standard ways to check for exposure, test for toxicity, and look at the consequences on the environment that take into account long-term and chronic effects. Also, risk governance needs to include openness, involvement of stakeholders, and ongoing monitoring to adapt properly to new scientific evidence and technological advances. This kind of flexible, interdisciplinary regulatory framework is necessary not just to protect people's health and the environment, but also to gain the public's trust and support for long-term progress in nano-enabled agriculture and other fields (Khan et al., 2023, 2024; Vogel et al., 2023). Without these steps, there is a chance that poor regulation might either slow down innovation or put society in danger of harm that was not planned for. This shows how important it is for good risk governance to find the right balance.

16.5 Health Impacts of Nanotechnology

Nanotechnology has the ability to help us find, cure, and prevent diseases, but it also raises concerns about the health dangers that can come from being around nanomaterials (Hull, 2019).

16.5.1 Toxicity

Because of their small size, large surface area, and ability to interact with living things, nanomaterials can have unique toxicological features (Elmanakhly et al., 2021). Some nanomaterials have been proven to cause oxidative stress, inflammation, DNA damage, and cell death in both lab and real-world settings (Rai & Biswas, 2018). Depending on their size, shape, chemical makeup, surface charge, and grouping, nanomaterials can cause harm in different ways.

16.5.2 Exposure Pathways

People can come into contact with nanomaterials through inhalation, ingestion, skin contact, or injection (Forbe et al., 2011). Workers who make, handle, and process nanomaterials are most worried about occupational exposure (Al-Bayati & Al-Zubaidi, 2018). Nano-enabled items, including cosmetics, food packaging, and textiles, can put people at risk (Maria et al., 2019; Sridhar et al., 2021). Nanomaterials can get into the air, water, and soil when they are released into the environment as a result of industrial activities and waste disposal (Rai & Biswas, 2018).

16.5.3 Target Organs and Systems

Nanomaterials can build up in many organs and tissues, like the lungs, liver, brain, and cardiovascular system. This could be harmful for your health (Sinitsyna et al., 2018).

Inhalation of nanoparticles that are in the air can induce lung cancer, fibrosis, and inflammation of the lungs (Hull, 2019). Nanomaterials can get beyond the blood–brain barrier and cause neurotoxicity, which can change how people think and act (Beura et al., 2024). Increased blood pressure, endothelial dysfunction, and thrombosis are some of the impacts on the cardiovascular system (Syahputra et al., 2024).

16.5.4 Risk Assessment and Management

It is critical to conduct a full risk assessment to find and describe the possible health risks that nanomaterials pose (Omari Shekaftik et al., 2023). This process includes figuring out how much exposure there is, testing for toxicity, and creating dose–response connections. Risk management techniques should focus on reducing exposure through engineering controls, personal protective equipment, and safe handling procedures (Miyamoto & Numata, 2023).

16.5.5 Nanomedicines and Targeted Therapies

Nanotechnology opens up new ways to deliver drugs to specific areas, perform gene therapy, and use regenerative medicine (Narayanan & Wakaskar, 2018). Nanomaterials can be designed to only attack cancer cells, get drugs across biological barriers, and help tissues (Kader, 2025). But we need to conduct many clinical trials to make sure that the benefits of nanomedicine applications outweigh the risks. Nanotechnology is moving forward quickly, which brings up a number of ethical issues that need to be dealt with to make sure it is developed and used responsibly (Berne, 2004). There are worries about privacy and data security when nanosensors and nanodevices are used for health monitoring and diagnosis (Kader, 2025). There need to be protections in place to keep sensitive health information safe from being accessed, used, or shared without permission. Patients should be able to choose how their data is used and shared and have control over it. Nanotechnology may have more negative effects on the environment and the health of vulnerable groups, like low-income communities and developing countries (Babatunde et al., 2019). We need to work to make sure that the advantages of nanotechnology are shared fairly and that the hazards are not unfairly spread out. Nanotechnology could be utilized to improve human skills, which raises moral problems about what it means to be "normal" and the possibility of making a "superhuman" split (Schummer, 2006). It is important to think carefully about the moral implications of employing nanotechnology to improve mental, physical, and emotional health.

16.6 Conclusion

Nanoagricultural technologies are a revolutionary way of farming that uses the unique physical and chemical features of nanoscale materials to improve crop yield, fertilizer usage efficiency, stress resistance, and environmental sustainability. There are several uses for these technologies, including nanofertilizers, nanopesticides, biosensors, and gene delivery systems. They all allow for regulated and targeted interventions at the molecular level. These new ideas make it easier for plants to take in nutrients, cut down on the loss of agrochemicals, make plant defense systems stronger, and allow for real-time diagnostics for precision farming.

Although these developments are intriguing, their large-scale implementation is challenging due to significant issues related to the economy, regulations, and growth. High production costs, a lack of standard norms, the requirement for qualified workers and specialized infrastructure, and limited access for smallholder farmers are all big problems. We also need to carefully examine concerns about environmental safety, long-term toxicity, and the fate of nanoparticles in agroecosystems.

16.7 Future Recommendations

1) Full Toxicological and Environmental Assessments: Future research should focus on figuring out how nanomaterials affect soil microbiota, beneficial creatures, and ecosystem functions, as well as how they get into the environment and how they build up in living things.
2) Creating Technologies That Are Affordable and Expandable: The focus should be on creating low-cost synthesis processes and scalable nano-formulations that can be used on small and medium-sized farms, especially in developing areas.
3) Standardization and Regulatory Frameworks: To make sure that nanoagricultural products can be used safely, we need to set up rules and safety standards that are the same all over the world. Life-cycle analysis, ecotoxicity testing, and risk–benefit assessments can accomplish this.
4) Engage farmers, agronomists, politicians, and academics in developing user-friendly nanoagricultural solutions by using participatory methods that raise awareness, provide training, and collect feedback.
5) Funding and Public–Private Partnerships: Encourage governments, universities, and businesses to work together and invest in new ideas to help them get started, lower their costs, and make them easier to sell.
6) Combining nanoagriculture with conservation methods like organic farming, integrated pest management, and reduced tillage can help the environment and make farming more sustainable in the long run.
7) Socioeconomic and Ethical Analysis: Conduct research that includes everyone to look at the socioeconomic effects, moral issues, and public opinions about nanotechnology in agriculture. This will make sure that everyone has fair access and that innovation is done responsibly.

References

Adetutu, B. O., Buba, A. B., Akinhanmi, F., Gana, M., Utazi, E. B., Oyewole, O. A., Adetunji, C. O., Eniola, K., & Yerima, M. B. (2024). Relevance of nanotechnology in agriculture. In C. O. Adetunji, C. Egbuna, A. Ficai, & O. A. Ijabadeniyi (Eds.), *Handbook of agricultural biotechnology* (pp. 413–462). Wiley.

Al-Bayati, A. J., & Al-Zubaidi, H. A. (2018). Inventory of nanomaterials in construction products for safety and health. *Journal of Construction Engineering and Management, 144*(9), 06018004.

Arunachalam, K., Krishnan, G. P., Sethuraman, S., Abraham, S. V. P. I., Krishnan, S. T., Venkateswar, A., Arunkumar, J., Shi, C., & MubarakAli, D. (2023). Exploring possible ways to enhance the potential and use of natural products through nanotechnology in the battle against biofilms of foodborne bacterial pathogens. *Pathogens, 12*(2), 270.

Babatunde, D. E., Denwigwe, I. H., Babatunde, O. M., Gbadamosi, S. L., Babalola, I. P., & Agboola, O. (2019). Environmental and societal impact of nanotechnology. *IEEE Access, 8*, 4640–4667.

Barzinjy, A. A., & Hamadamen, V. N. (2022). Investigating physical properties and formation mechanism of biosynthesized zinc oxide nanoparticles using dill (*Anethum graveolens*) leaf extract. *Nanoscience & Nanotechnology-Asia, 12*(6), 27–38.

Berne, R. W. (2004). Towards the conscientious development of ethical nanotechnology. *Science and Engineering Ethics, 10*, 627–638.

Beura, S. K., Panigrahi, A. R., Yadav, P., Palacio, I., Casero, E., Quintana, C., Singh, J., Singh, M. K., Gago, J. A. M., & Singh, S. K. (2024). Harnessing two-dimensional nanomaterials for diagnosis and therapy in neurodegenerative diseases: advances, challenges and prospects. *Ageing Research Reviews, 94*, 102205.

Camara, M. C., Campos, E. V. R., Monteiro, R. A., do Espirito Santo Pereira, A., de Freitas Proença, P. L., & Fraceto, L. F. (2019). Development of stimuli-responsive nano-based pesticides: emerging opportunities for agriculture. *Journal of Nanobiotechnology, 17*, 1–19.

Charitidis, C. A. (2020). Quest for ethical and socially responsible nanoscience and nanotechnology. In B. Bhushan (Ed.), *Handbook of research ethics and scientific integrity* (pp. 983–997). Cham: Springer International Publishing.

Chugh, B., Poddar, D., Singh, A., Yadav, P., Thakur, S., Nguyen, T. A., & Rajendran, S. (2022). Nanoparticles-based sensors for agricultural application. In A. Denizli, S. Rajendran, & A. K. Nadda (Eds.), *Nanosensors for Smart Agriculture* (pp. 117–146). Elsevier.

Elmanakhly, M. E., Rizk, M. G., & Eslam, S. M. G. O. (2021). Nanotechnology risk, toxicity and its effect on safety and human health. *Врач скорой помощи, 2*, 33–37.

Fathi Qarachal, J., & Alizadeh, M. (2025). Harnessing precision agriculture and nanotechnology for sustainable farming: a review. *Nanotechnology For Environmental Engineering, 10*(2), 1–17.

Flores, L. A., González-Hernández, I. J., Porras-Loaiza, A. P., & Watters, C. (2024). Advancements in inventory management within the agricultural supply chain: implications for waste reduction and sustainability. *Management Review Quarterly, September*, 1–26.

Forbe, T., García, M., & Gonzalez, E. (2011). Potencial risks of nanoparticles. *Food Science and Technology, 31*, 835–842.

Fraceto, L. F., Grillo, R., de Medeiros, G. A., Scognamiglio, V., Rea, G., & Bartolucci, C. (2016). Nanotechnology in agriculture: which innovation potential does it have? *Frontiers in Environmental Science, 4*, 20.

Garcia, C. R., Malik, M. H., Biswas, S., Tam, V. H., Rumbaugh, K. P., Li, W., & Liu, X. (2022). Nanoemulsion delivery systems for enhanced efficacy of antimicrobials and essential oils. *Biomaterials Science, 10*(3), 633–653.

Garg, M. C., Kumari, S., & Malik, N. (2024). Role of nanomaterials in advanced membrane technologies for groundwater purification. *Environmental Science: Water Research & Technology, 10*, 2628–2645.

Hull, M. S. (2019). Multidimensional impacts of nanotechnology on public health. In B. Roig, K. Weiss, & V. Thireau (Eds.), *Management of emerging public health issues and risks* (pp. 65–85). Elsevier.

Iavicoli, I., Leso, V., Ricciardi, W., Hodson, L. L., & Hoover, M. D. (2014). Opportunities and challenges of nanotechnology in the green economy. *Environmental Health, 13*, 1–11.

Iavicoli, I., Leso, V., Beezhold, D. H., & Shvedova, A. A. (2017). Nanotechnology in agriculture: opportunities, toxicological implications, and occupational risks. *Toxicology and Applied Pharmacology, 329*, 96–111.

Kader, N. (2025). Nanomaterials in medicine: advancing drug delivery, diagnostics, and therapeutics. *Journal of Basic and Applied Research in Biomedicine, 10*(1), 29–49.

Kah, M., Tufenkji, N., & White, J. C. (2019). Nano-enabled strategies to enhance crop nutrition and protection. *Nature Nanotechnology, 14*(6), 532–540.

Kalita, D., & Baruah, S. (2019). The impact of nanotechnology on food. In R. F. do Nascimento & A. J. De Paula (Eds.), *Nanomaterials applications for environmental matrices* (pp. 369–379). Elsevier.

Kamarulzaman, N. A., Lee, K. E., Siow, K. S., & Mokhtar, M. (2020). Public benefit and risk perceptions of nanotechnology development: psychological and sociological aspects. *Technology in Society, 62*, 101329.

Kamyab, H., Saberi Kamarposhti, M., Hashim, H., & Yusuf, M. (2024). Carbon dynamics in agricultural greenhouse gas emissions and removals: a comprehensive review. *Carbon Letters, 34*(1), 265–289.

Khan, A. A., Alabbosh, K. F., Kashif, Iqbal, B., Manan, S., Alhoqail, W. A., ... & Wang, Y. F. (2025). Ion homeostasis, osmotic adjustment, and ROS detoxification underlie pea salinity tolerance induced by *pseudomonas putida RT12. Microbiology Research, 16*(11), 227.

Khan, A. A., AlHaithloul, H. A. S., Alghanem, S. M., Senousy, H. H., Alhoqail, W. A., Shi, F., Ali, I., & Soliman, M. H. (2023). *Petrosolinum crispum* under stressful conditions. In A. A. H. A. Latif (Ed.), *Medicinal plant responses to stressful conditions* (pp. 299–310). CRC Press.

Khan, A. A., Hayat, K., Wang, Y.-F., ALHaithloul, H. A. S., Alghanem, S. M. S., Alhoqail, W. A., & Soliman, M. H. (2024). Biochar versus heavy metals under changing environmental conditions. In S. Fahad, S. Saud, & R. Zhou (Eds.), *Biochar-assisted remediation of contaminated soils under changing climate* (pp. 167–200). Elsevier.

Khan, A. A., Wang, Y.-F., Akbar, R., & Alhoqail, W. A. (2025a). Mechanistic insights and future perspectives of drought stress management in staple crops. *Frontiers in Plant Science, 16*, 1547452.

Khan, A. A., Wang, Y.-F., Zeb, A., Hayat, K., Alhoqail, W. A., & Soliman, M. H. (2025b). Beneficial elements and their roles against soil pollution. In S. Saud & D. Wang (Eds.), *Beneficial elements for remediation of heavy metals in polluted soil* (pp. 1–32). Elsevier.

Kopittke, P. M., Punshon, T., Paterson, D. J., Tappero, R. V., Wang, P., Blamey, F. P. C., Van Der Ent, A., & Lombi, E. (2018). Synchrotron-based X-ray fluorescence microscopy as a technique for imaging of elements in plants. *Plant Physiology, 178*(2), 507–523.

Landers, C., Blasimme, A., & Vayena, E. (2024). Sync fast and solve things-best practices for responsible digital health. *npj Digital Medicine, 7*(1), 113.

Liberty, J. T., Anil, A., Ijimdiya, S. J., Kwaji, M. J., & Ijimdiya, R. U. (2024). Harnessing the potential of nanostructured materials for sustainable development. *Nano-Structures & Nano-Objects, 38*, 101216.

Liu, R., & Lal, R. (2015). Potentials of engineered nanoparticles as fertilizers for increasing agronomic productions. *Science of the Total Environment, 514*, 131–139.

Malakar, Y., Lacey, J., & Bertsch, P. M. (2022). Towards responsible science and technology: how nanotechnology research and development is shaping risk governance practices in Australia. *Humanities and Social Sciences Communications, 9*(1), 1–14.

Maria, L., Moses, J., & Anandharamakrishnan, C. (2019). Ethical and regulatory issues in applications of nanotechnology in food. In C. Anandharamakrishnan & S. Parthasarathi (Eds.), *Food nanotechnology* (pp. 67–92). CRC Press.

Maswada, H., Djanaguiraman, M., & Prasad, P. (2018). Seed treatment with nano-iron (III) oxide enhances germination, seeding growth and salinity tolerance of sorghum. *Journal of Agronomy and Crop Science, 204*(6), 577–587.

Medani, K. R., Neill, A., Garrod, G., Ojo, M., & Hubbard, C. (2024). Societal perceptions and attitudes towards genetically modified (GM) crops, feed, and food products in the Middle East, North Africa, and Turkey (MENAT) region: a systematic literature review. *Food Quality and Preference, 117*, 105148.

Mehara, A. K., Kumari, A., Verma, N. K., Marwaha, P., Rai, A., Singh, M. K., & Singh, A. K. (2025). Fabrication of nanomaterials and their potential advantage for sustainable agriculture. In P. Singh, A. K. Singh, V. K. Singh, & V. Kumar (Eds.), *Nanotechnology-based sustainable agriculture* (pp. 1–46).

Miyamoto, T., & Numata, K. (2023). Advancing biomolecule delivery in plants: harnessing synthetic nanocarriers to overcome multiscale barriers for cutting-edge plant bioengineering. *Bulletin of the Chemical Society of Japan, 96*(9), 1026–1044.

Mukhopadhyay, R., Sarkar, B., Khan, E., Alessi, D. S., Biswas, J. K., Manjaiah, K., Eguchi, M., Wu, K. C., Yamauchi, Y., & Ok, Y. S. (2022). Nanomaterials for sustainable remediation of chemical contaminants in water and soil. *Critical Reviews in Environmental Science and Technology, 52*(15), 2611–2660.

Narayanan, E., & Wakaskar, R. (2018). Utilization of nanoparticulate therapy in cancer targeting. *Cogent Medicine, 5*(1), 1504504.

Niazian, M., Molaahmad Nalousi, A., Azadi, P., Ma'mani, L., & Chandler, S. F. (2021). Perspectives on new opportunities for nano-enabled strategies for gene delivery to plants using nanoporous materials. *Planta, 254*(4), 83.

Omari Shekaftik, S., Jadid, H., Dehdashti, M. J., Mohammadian, Y., & Ebrahimi, W. (2023). Toward a minimum data set for assessing the risks of activities involved with nanomaterials. *Journal of Nanoparticle Research, 25*(2), 25.

Otari, S., Bapat, V. A., Lakkakula, J., Kadam, U. S., & Suprasanna, P. (2024). Advancements in bionanotechnological applications for climate-smart agriculture and food production. *Biocatalysis and Agricultural Biotechnology, 57*, 103117.

Pal, A., Dubey, S. K., Goel, S., & Kalita, P. K. (2024). Portable sensors in precision agriculture: assessing advances and challenges in soil nutrient determination. *TrAC Trends in Analytical Chemistry, 180*, 117981.

Patil, A. S., Parishwad, G. V., Kale, S. K., & Husainy, A. S. (2020). Modern ethical and societal implications of nanotechnology. *Journal of Advanced Research in Manufacturing, Material Science and Metallurgical Engineering, 7*, 1–8.

Preethi, B., Karmegam, N., Manikandan, S., Vickram, S., Subbaiya, R., Rajeshkumar, S., Gomadurai, C., & Govarthanan, M. (2024). Nanotechnology-powered innovations for agricultural and food waste valorization: a critical appraisal in the context of circular economy implementation in developing nations. *Process Safety and Environmental Protection, 184*, 477–491.

Priya, S., Kumar, A., Upadhayay, V. K., Chaudhary, A., Parveen, H., & Kumar, G. (2023). 10 Impact of Nanoparticles. In *advances in nanotechnology for smart agriculture: Techniques and applications* (pp. 197). Taylor & Francis.

Rai, M., & Biswas, J. K. (2018). *Nanomaterials: ecotoxicity, safety, and public perception*. Springer.

Schummer, J. (2006). 'Societal and ethical implications of nanotechnology': Meanings, interest groups, and social dynamics. In *Nanotechnology challenges: Implications for philosophy, ethics and society* (pp. 413–449).

Sembada, A. A., & Lenggoro, I. W. (2024). Transport of nanoparticles into plants and their detection methods. *Nanomaterials, 14*(2), 131.

Shastri, D. H., & Gandhi, S. (2024). Nanorevolution unleashing the power of nanotechnology. *Current Nanomedicine (Formerly: Recent Patents on Nanomedicine), 14*(3), 227–246.

Shelar, A., Nile, S. H., Singh, A. V., Rothenstein, D., Bill, J., Xiao, J., Chaskar, M., Kai, G., & Patil, R. (2023). Recent advances in nano-enabled seed treatment strategies for sustainable agriculture: challenges, risk assessment, and future perspectives. *Nano-Micro Letters, 15*(1), 54.

Singh, S. (2023). Assessing and comparing the sustainability of organic and conventional wheat farming in India: an indicator-based approach. *Journal of Cleaner Production, 423*, 138652.

Singh, P., Kahlon, K. S., Sawhney, R. S., Vohra, R., & Kaur, S. (2018). Social media buzz created by# nanotechnology: insights from Twitter analytics. *Nanotechnology Reviews, 7*(6), 521–528.

Singh, A., Rawat, S., Rajput, V. D., Ghazaryan, K., Minkina, T., Al Tawaha, A. R. M., & Varshney, A. (2023). Impact of nanofertilizers for the mitigation of multiple environmental stresses. In K. A. Abd-Elsalam & M. A. Alghuthaymi (Eds.), *Nanofertilizers for sustainable agroecosystems: recent advances and future trends* (pp. 431–454). Springer.

Singla, V. K. (2021). Impact of nanotechnology on human life. In C. Bhargava, P. K. Sharma, & A. Sachdeva (Eds.), *The era of nanotechnology* (pp. 51–62). Apple Academic Press.

Sinitsyna, O., Meshkov, G., & Yaminsky, I. (2018). Carbon nanomaterials: potential risks to human health and the environment. In Mahendra Rai, Jayanta Kumar Biswas *Nanomaterials: ecotoxicity, safety, and public perception* (pp. 237–252). Springer Internation Publishing.

Sridhar, A., Ponnuchamy, M., Kumar, P. S., & Kapoor, A. (2021). Food preservation techniques and nanotechnology for increased shelf life of fruits, vegetables, beverages and spices: a review. *Environmental Chemistry Letters, 19*, 1715–1735.

Sun, Y., Zhu, G., Zhao, W., Jiang, Y., Wang, Q., Wang, Q., Rui, Y., Zhang, P., & Gao, L. (2022). Engineered nanomaterials for improving the nutritional quality of agricultural products: a review. *Nanomaterials, 12*(23), 4219.

Syahputra, R. A., Dalimunthe, A., Utari, Z. D., Halim, P., Sukarno, M. A., Zainalabidin, S., Salim, E., Gunawan, M., Nurkolis, F., & Park, M. N. (2024). Nanotechnology and flavonoids: current research and future perspectives on cardiovascular health. *Journal of Functional Foods, 120*, 106355.

Thirugnanasambandan, T. (2021). Advances in nanotechnology for bio-sensing in agriculture and food. In M. Rai & J. K. Biswas (Eds.), *Biosensors in agriculture: recent trends and future perspectives* (pp. 27–49).

Tomar, B., Patle, T., Parihar, S. S., Singh, P., & Tomar, S. S. (2024). Enhancing nutrient uptake with nano fertilizers and soil amendments. In *Harnessing nanoomics and nanozymes for sustainable agriculture* (pp. 43–59). IGI Global Scientific Publishing.

Van de Voorde, M., & Vlerick, M. (2021). 3 Nanotechnology: ethical guidelines for a disruptive technology. In *Handbook of nanoethics* (pp. 43). De Gruyter.

Vogel, U., Jensen, K. A., Saber, A. T., Hadrup, N., Andersen, M. H. G., Hougaard, K. S., Jacobsen, N. R., & Sørensen, O. H. (2023). Lessons from nanosafety research for effective regulation and safer handling of nanomaterials at the workplace. *Societal Impacts*, *1*(1–2), 100016.

Weh, L. (2024). An integrated embodiment concept combines neuroethics and AI ethics–relational perspectives on artificial intelligence, emerging neurotechnologies and the future of work. *NanoEthics*, *18*(2), 8.

Xin, X., Judy, J. D., Sumerlin, B. B., & He, Z. (2020). Nano-enabled agriculture: from nanoparticles to smart nanodelivery systems. *Environmental Chemistry*, *17*(6), 413–425.

Yadav, A., Yadav, K., & Abd-Elsalam, K. A. (2023). Nanofertilizers: types, delivery and advantages in agricultural sustainability. *Agrochemicals*, *2*(2), 296–336.

Yeh, Y.-C., Huang, T.-H., Yang, S.-C., Chen, C.-C., & Fang, J.-Y. (2020). Nano-based drug delivery or targeting to eradicate bacteria for infection mitigation: a review of recent advances. *Frontiers in Chemistry*, *8*, 286.

Yudhistira, B., Punthi, F., Gavahian, M., Chang, C.-K., Hazeena, S. H., Hou, C.-Y., & Hsieh, C.-W. (2023). Nonthermal technologies to maintain food quality and carbon footprint minimization in food processing: a review. *Trends in Food Science & Technology*, *141*, 104205.

Zaman, W., Ayaz, A., & Park, S. (2025). Nanomaterials in agriculture: a pathway to enhanced plant growth and abiotic stress resistance. *Plants*, *14*(5), 716.

Zhang, Y., Huang, C., & Xiong, R. (2024). Advanced materials for intracellular delivery of plant cells: strategies, mechanisms and applications. *Materials Science and Engineering: R: Reports*, *160*, 100821.

17

Ethical Safety and Regulatory Considerations in Nanotechnology for Agriculture

Sameel Hassan[1], Nizakat Bibi[2], Waseem Ahmed Khattak[3], Amjid Khan[3,4], Sadia Riaz[3], Muhammad Anas[3,5], Banzeer Ahsan Abbasi[6], Javed Iqbal[7], Muhammad Nasir Hussain[8], and Shah Fahad[9]

[1] *Department of Agriculture & Forest Science (DAFNE), University of Tuscia, Viterbo, Italy*
[2] *International joint research laboratory for Global Change Ecology, School of Life Science Hennan University Kaifeng, Henan, China*
[3] *Department of Plant Sciences, Faculty of Biological Sciences, Quaid-i-Azam University, Islamabad, Pakistan*
[4] *UNESCO UNISA Africa Chair in Nanoscience and Nanotechnology, College of Graduate Studies, University of South Africa, Pretoria, South Africa*
[5] *Department of Soil Science & Plant Nutrition, Selçuk University, Konya, Türkiye*
[6] *Department of Botany, Rawalpindi Women University, Rawalpindi, Pakistan*
[7] *Department of Botany, Bacha Khan University, Khyber Pakhtunkhwa, Pakistan*
[8] *Department of Chemistry, Quaid-i-Azam University, Islamabad, Pakistan*
[9] *Department of Agronomy, Abdul Wali Khan University Mardan, Khyber Pakh-tunkhwa, Pakistan*

17.1 Introduction

Nanotechnology, as the science of working at the atomic and molecular level at a range usually within 1 to 100 nm, has gained much ground these few years and holds great potential for radical improvement in many areas, one of which is agriculture (Ahmad et al., 2022; Jamil Abbasi et al., 2025). Compared to normal materials, nanomaterials have a higher surface area-to-volume ratio reaction and biological influence; hence, applying nanotechnology in agriculture provides effective solutions to increasing agricultural productivity, resource management, and sustainable production (Haris et al., 2023; Malik et al., 2022; Pechyen et al., 2024). Applications of nanotechnology in agriculture include increasing crop yields, formulating animal health and pest and disease control, uptake of nutrients by plants, utilization of aeroponics, and innovative food packaging (Alex et al., 2025; Ullah et al., 2023a). This promising technology can contribute to solving many of the problems that contemporary agriculture faces: climate change, population increase, and the effectiveness of the utilization of resources through nanoparticles in fertilizers, pesticides, herbicides, and sensors.

Nonetheless, these benefits incumbent in the integration of nanotechnology in agriculture have crucial concerns inherent in them. However, significant emphasis on the increase of innovation in the development of nanotechnology must take time to factor in the ethical,

Nanotechnology Innovations for Food Security and Sustainable Agriculture, First Edition. Edited by Zia Ur Rehman Mashwani, Shah Fahad, Ilyas Ahmad, Shah Saud, Sajid Fiaz, and Taufiq Nawaz.

safety, and regulatory concerns that may hinder the development of possibly useful nanotechnology for mankind without leading to adverse effects on human health, the environment, or society in general (Abbasi et al., 2023; Amutha et al., 2024; Ullah et al., 2023b). When nanomaterials enter agricultural systems, it becomes necessary to examine the impact that they may present, such as off-target effects in ecosystems and the effects on nontarget species, together with effects on the food production chain (Rana et al., 2024). This chapter shall further argue that it is only ethical that these technologies are introduced afresh, provided there is full account taken of fairness, equity, and accountability in how the gains and burdens accruing from such technologies will be shared among the farmers and consumers as well as other members in surrounding communities. However, insufficient regulation to control nanotechnology in the agricultural sector since the current regulations may be inadequate in handling advanced nanotechnology (Kumari et al., 2023). For this reason, ethics, safety, and regulatory concerns that are attached to nanotechnology in the field of agriculture cannot be overemphasized. These are important to avoid misuse and improper use of nanotechnology to the extent that it triggers harm instead of contributing to society's welfare.

Ethical issues in nanotechnology are broader than risks related to health hazards and impacts on civil liberties but legally backed inequality together with a potential for increasing social injustices (Thakur & Thakur, 2024). Some of the issues that can arise include the fact that small-scale farmers in the developing world may be unable to make use of the modern technologies in nanotechnology that may be available in the market today with a view of improving productivity in farming (Dhillon & Moncur, 2023). Additionally, the impacts of nanomaterials on human beings and nature in the long run have not been fully established, which denotes that the uptake of these technologies requires a projecting risk analysis and appraisal (Tawiah et al., 2024). An important aspect of nanoparticles is their safety, and the toxicity of nanoparticles can be considered one of the main issues. One of the main disadvantages of nanoparticles is their propensity to penetrate the biological environment and possibly have negative effects that may not be immediately identifiable (Ahmed et al., 2022). There is also the question of persistence; the functioning of nanomaterials can produce serious long-term effects on environmental bodies such as the soil as well as water (Saleem & Zaidi, 2020). Due to such risks, it is important to analyze risks that might occur in the future outcome of nanotechnology in agriculture and some of the safe practices that could be employed in mitigating them. These protocols should be well researched and analyzed to capture relevant data to ensure that nanomaterials do not pose any risk to human beings or the environment. Such guidelines ensure that nanotechnology products that are produced and applied in agriculture do so appropriately. Still, the modern framework of regulation concerning nanotechnology is rather unclear and frequently trails behind the developments. The present-day policies are orientated mainly towards "conventional" farming practices and may not consider various features of nanomaterials (Chugh et al., 2021). However, in many cases, there are no transparent standard procedures for the testing, labeling, and approval of nanotechnology-based products, which poses a threat to the standardized method of regulation (Ghosh & Kumar, 2024). To this end, there remains the need to develop improved, countrywide codes of practices and international standard guidelines, which can facilitate the enabling use of nanotechnology in agriculture and further the advancement of human civilization. The above indicates that public awareness and the level of trust maintained by society are key determinants in the adoption of nanotechnology in agriculture. The extent of knowledge of and attitudes toward nanotechnology determines the acceptability of these innovative products by the public. Thus, they need to establish communication with the public and offer them all the necessary information concerning the usage of nanotechnology, its advantages and disadvantages, as well as possible security aspects.

Community engagement in nanotechnology decision-making is a way of creating public confidence over technology to establish their preferred direction of development (Pidgeon, 2021). Thus, with the help of nanotechnology, we can say that agriculture will have significant benefits in the future, and this way we can solve some problems that occur in this sector. Yet, its use entails certain risks that have to do with such factors as ethical, safety, and regulatory concerns. The formulation of good policies and, at the same time, practices pertaining to nanotechnology in agriculture should consider several principles that aim at enhancing technology while at the same time enhancing requisite standards concerning use in human health, the environment, and social well-being. It is, therefore, important that the stakeholders tackle these issues in advance so that nanotechnology can benefit agricultural productivity as expected without causing adverse effects on the environment or human health.

17.2 Ethical Concerns in Nanotechnology for Agriculture

Nanotechnology offers strong tools to help agriculture produce more while also creating sustainable improvements. Nanoscience usage brings together different ethical problems that protect both human well-being and environmental sanitation, as well as distribute benefits equally. Although nanotechnology shows strong possibilities for improving food security, it needs effective risk control to make sure it benefits all community segments and preserves natural environments and health (Besha et al., 2020).

17.2.1 Potential Risks to Human Health and the Environment

Agricultural nanotechnology faces its main ethical barrier in its dangerous effects on people along with environmental sustainability (Bhattacharya et al., 2023). Nanomaterials present unique attributes because of the combination of small dimensions and enlarged surface area, which leads to different conduct compared to bulk materials due to intensified performance, enhanced membrane permeability, and toxicological consequences (Sajid, 2022). Nanomaterials demonstrate features that cause either beneficial or harmful reactions based on usage parameters and dosage amounts (Saleem & Zaidi, 2020).

Nanoparticles pose human health risks once they enter the body through skin contact, consumption, or inhalation (Sonwani et al., 2021). Agrochemical operators continue to choose nanoparticle-based fertilizers in combination with pesticides, herbicides, and crop protection tools for their operations (Ali et al., 2023). Processed food consumers as well as field workers dealing with nano-enabled products face the potential for nanomaterial exposure to their bodies (Aguilar-Pérez et al., 2023). Long-term health outcomes have become a concern because nanoparticles show signs of tissue buildup throughout the food system. Some nanoparticles demonstrate the ability to pass through skin tissue until they reach blood circulation, where they induce cellular destruction or inflammatory reactions, as well as potentially leading to cancer development (Bibi et al., 2024). Nanomaterials require additional research evaluation to measure their human health effects because scientists currently lack complete safety protection information for agricultural applications.

Environmental concerns are also significant. Nanoparticles tend to exist longer in the environment after release because of their small size and subsequently build up in the soil and water bodies (Rajput et al., 2020). Nanomaterials in the environment will impact all aspects of soil condition, water purity, and ecosystem balance. The interaction of nanoparticles with different types of microorganisms found in soil could cause both changes to ecosystems and detrimental impacts on natural plant growth-supporting processes (Wahab et al., 2023). The accumulation of nanomaterials inside plants poses environmental risks because these materials are likely to be consumed by eating animals or they may move through the food system (Zhang et al., 2020). The beneficial applications of nanomaterials include boosted nutrient transport to plants, alongside decreased chemical pesticide needs, but scientists require thorough evaluation of their unfounded environmental results. Assessment of environmental implications should remain the highest priority to pursue extensive agricultural use of nanomaterials.

17.2.2 Equity and Accessibility in Agricultural Advancements

The ethical use of nanotechnology in agriculture creates issues of fairness when delivering these advances throughout all farming sectors. Most farmers need nanotechnologyfarm solutions, yet the technology's advantages may reach only developed areas and big operations, not small-scale farmers in both developed and developing nations. Advanced nanotechnology research in developed nations lets large agribusinesses and commercial farms implement new systems first because these companies have ample money and expertise (Khatoon & Velidandi, 2025). Nanotech-based farming supplies give these farms superior outcomes and business gains. Small farming operations, especially in developing nations, require more financial backing, knowledge, and major construction to use modern technology better than large-scale producers (Ashraf et al., 2021). These unequal distribution risks hurt small-scale farmers more because big industrial farms will outperform them both in production and profits. The high development expenses of nanotechnology may prevent poor farmers from using its benefits.

Developing-country smallholder farmers cannot afford to purchase necessary farming supplies such as seed packets and chemical products. Farmers with high incomes will be the only ones able to use and afford new nanotechnology-based products, while poor farmers remain disadvantaged. When this happens, it will cause economic differences between wealthy and poor people to grow, as well as between nations around the world, making improving food security and farming sustainability work less well. The legal protections over nanotechnology products create barriers for small-scale farmers who want to use these technologies (Hofmann et al., 2020). The legal protection for nanotechnology agricultural products prevents farmers from obtaining more affordable product variations (Yang & Duncan, 2021). People question whether patents on agricultural technology are ethically correct, since this technology supports essential food production for billions of small-scale farmers across the globe. To bring nanotechnology benefits to every segment of society, we need to provide small-scale farmers with financial help, while teaching them how to use nanotechnology plus reduce technology price points.

17.2.3 Long-Term Sustainability and Impact on Small-Scale Farmers

Another important ethical issue of nanotechnology in agriculture is its long-term sustainability. Nanotechnology has the potential to attend to immediate impediments, for example, crop yield augmentation, vermin control, and resource management for the executives, yet its long-haul repercussions on farming frameworks, especially for smaller-scale ranchers that depend on customary homestead practices, should be investigated. Small and limited-resource farmers (especially in developing countries) are usually less robust and more vulnerable to economic, social, and environmental conditions (Ijaz et al., 2024). High-tech innovations such as nanotechnology could cause the transformation of existing farming systems, making small-scale farmers dependent on expensive inputs that might put them at risk (Reddy et al., 2022). For example, small farmers that cannot afford the new nanotechnology-based products, like fertilizers or pesticides, will face reduced productivity or even failure to compete in the agricultural market(Marella et al., 2021). At the same time, small-scale farmers depend on the locally adapted seeds and traditional practices of farming and natural resource management, which do not go along with the application of synthetic nanomaterials.

It also has a risk of generating a cycle of dependence on external technologies, where small farmers would be dependent on costly nanotechnology solutions for soil health, crop protection, and pest management instead of promoting sustainable and local agricultural practices (Lowry et al., 2019). While nanotechnology can be an enabler and provide short-term gains, the long-term consequences of relying on external input may seriously weigh on the vulnerability aspects of small-scale farming systems (Singh et al., 2024a). Additionally, it might decrease biodiversity by favoring monocultures or the excess consumption of specific kinds of nanomaterials that, in the long term, may possibly result in soil degradation and ecosystem services loss (Ekka et al., 2023). A dilemma regarding its use lies with respect to the advancement of technology and the conservation of centuries-old, traditional farming systems that have proved to be resilient and sustainable (Sekhar et al., 2024). To be truly sustainable, introducing nanotechnology in agriculture should not only improve productivity, but also concentrate on its implications for the social and economic stability of rural communities. Efforts should be made to realize and encourage the application of these technologies by small-scale farmers in those forms that are compatible with their knowledge and priorities. This could involve bringing together nanotechnology and sustainable farming via farmer training and resources to enable sound technology decision-making and ensuring that the benefits from these advances are distributed fairly.

Consequently, although nanotechnology offers a great opportunity to improve some agricultural processes, the ethics of its use need to be considered to save the favorable merits of this revolution in agriculture (Pramanik et al., 2020). Nevertheless, there is still much that is not known about the toxicity and persistence of nanoparticles in ecosystems, and therefore, great care should be taken when considering the possible risks to human health and the environment. Additionally, the distribution and accessibility of nanotechnology-based agricultural products are likely to be inequitably distributed, and this may be concerning regarding issues of fairness and social justice (Verma et al., 2024). If small-scale farmers in developing countries do not have access to such innovations, nanotechnology could help widen development gaps and imperil efforts to achieve global food security. Additionally, the long-term sustainability of nanotechnology in agriculture should be considered so that dependency on external technologies is not created, or traditional farming systems are not undermined (Altabbaa et al., 2023). Nanotechnology is presented to be an important future space for agriculture; however, nanotechnology in agriculture entails many ethical issues, which are intertwined and need to be addressed holistically while including the pros and cons of the technology (Lombi et al., 2019). Through dealing with these ethical issues, nanotechnology in agriculture can be used for the maximum positive impact, with the limitation of human health risks, damage to environmental quality, and the prevention of social inequity.

17.3 Safety Considerations in Nanotechnology for Agriculture

With the advancement in the use of nanotechnology in agricultural activities, there is a need to protect both humans and the environment in the process (Uddin et al., 2022). Next, properties of nanomaterials that include a large surface and high reactivity can greatly affect fertilizers, pesticides, herbicides, and crop protection products. Even though these materials are relatively good for improving agricultural productivity, sustainability, and efficiency, their use presents some safety risks (Khan et al., 2025a). The main issues include the adverse effects of nanomaterials on the health of the workers and consumers, the effects of such particles on the environment, and potential measures of dealing with such effects. This will be important, as it will contribute to eliminating all the negative effects on human health and the environment as the use of nanotechnology increases in farming.

17.3.1 Health Risks for Workers and Consumers

The use of nanotechnology in agriculture is one of the pressing safety concerns for raising possible health risks to workers and consumers exposed to nanomaterials (Jain et al., 2018). Nanoparticles have special characteristics because they are small and have a large, exposed surface area, which lets them change quickly and penetrate biological barriers (Khan et al., 2025b). Specifically, because of their size, nanoparticles in agricultural products raise major concerns since they can easily pass through human body areas when people take them in orally or through the nose and skin (Sajid et al., 2015). People who work in farming get exposed to direct risks when dealing with nanomaterials at their workplace. Employees in agriculture face nanoparticle exposure through their breath when they work with nanoparticle-based fertilizer and pesticide products (Muthukrishnan, 2022). The body can take these minuscule particles deep into the respiratory system because of their small size, thus leading to diverse dangers to our health. Month-long contact with excessive nanoparticles through breathing methods can create respiratory problems like asthma and bronchitis. Their extreme size permits the particles to penetrate sections of the lungs our bodies cannot shield. Nanomaterial inhalation threats to farmers are growing, and safety measures must be assessed to control respiratory hazards from entering staff areas (Chelliah et al., 2023).

Potential health risks from using nanomaterials in agricultural products also affect the consumers of these products. Nanotechnology is used to produce food in various ways; for example, food products produced with nanotechnology-based fertilizers or treated with nanopesticides may have residual nanoparticles left inside them (Mittal et al., 2020). If swallowed, these particles could be harmful to the digestive system and may accumulate in the tissues or organs over time and cause long-term health problems. Nanomaterials have been studied extensively in labs but still need more research to fully understand the effect on human health; early studies indicate that at least some nanoparticles are toxic to cells, interfere with cellular function, or can induce inflammation (Havelikar et al., 2024). This underscores the imperative of rigorous testing and monitoring to determine whether nanomaterials are safe in food production. In addition, excessive accumulation or magnification (bioaccumulation and biomagnification) of nanomaterials within the food chain is a challenge to consumer safety (Isibor et al., 2024). Under certain conditions, the accumulation of nanoparticles in soil and water systems may lead to an uptake in the food supply by way of plants. This makes it possible for consumers to unknowingly ingest nanoparticles for extended periods without knowing the effects on their long-term health. For example, some nanoparticles could cross the blood–brain barrier or accumulate in organs such as the liver or the kidneys, which will result in such nanoparticles having different effects that are hard to notice till they are fatal (Mohammed et al., 2025). These risks can be moderated with rigorous testing of nanomaterials in agricultural products to ensure the safety of the food supply, food safety protocols, and clear requirements for labeling purposes such that consumers can be informed whether nanoparticles may be a constituent of the food supply.

17.3.2 Environmental Impact of Nanomaterials Used in Agriculture

Nanomaterials used in agriculture present critical safety issues because they create fundamental environmental risks together with human health threats. The agricultural applications of nanotechnology bring two main advantages to farming, including better plant nutrient absorption and limited chemical pesticide usage (Wahab et al., 2024). Current knowledge about how nanoparticles behave in the environment after their release into water bodies, soil, and air remains insufficiently understood. Environmental scientists monitor how nanomaterials stay in existence in natural surroundings and study the effects of their buildup throughout biosystems (Alizadeh et al., 2025). The introduction of nanoparticles into agricultural soils via fertilizers and pesticides leads to unexplained reactions between nanoparticles and soil microorganisms together with organic matter and plants (Munir et al., 2024). Nanoparticles reach biological membranes because of their small dimensions and reactive nature, thus affecting soil ecosystem function (Moulick et al., 2024). Several nanoparticles demonstrate toxicity when they encounter soil-inhabiting life forms such as earthworms, microbes, and beneficial species, thus threatening soil health together with nutrient cycling processes (Sharma et al., 2023).

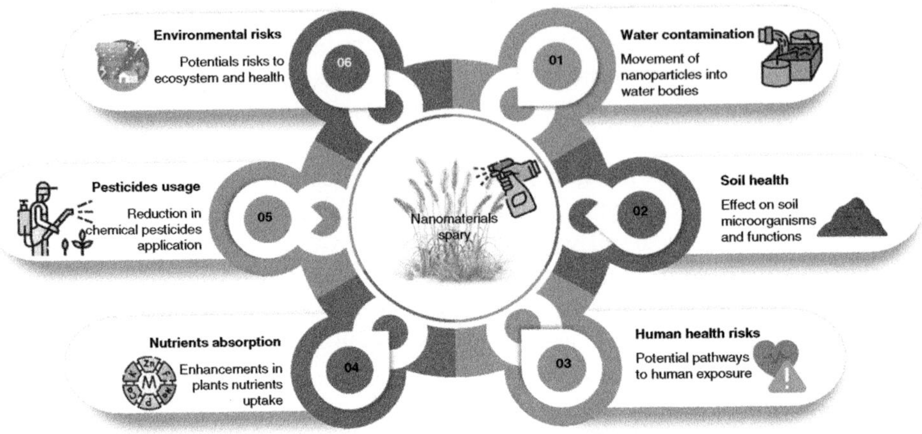

Figure 17.1 Environmental impacts of agricultural nanomaterials.

Nanoparticles' movement through soil and water leads to potential contamination of fundamental water systems used by farmers and consumers for their agricultural operations and drinking supply (Dasgupta et al., 2017). The movement of nanoparticles to water bodies poses risks to aquatic life and the health of organisms such as fish, since they can leak into water sources (Singh et al., 2023). The continuous buildup of nanomaterials in plants, together with animals and microbes, results in unforeseen consequences that get transmitted up through successive feeding levels. Nanoparticles passed to plants from the environment create potential risks for human health through animal consumption until they reach human dietary intake (Committee et al., 2021). The environmental longevity of nanoparticles remains unclear because different studies show both their short-term breakdown and their ability to durably remain in the ecosystem. These materials create lasting concerns about their effects on biodiversity, together with ecosystem services and the entire agricultural ecosystem health status. The environmental monitoring of nanomaterials, along with extensive risk assessment procedures, must be performed to determine their possible long-term environmental impacts (Figure 17.1).

17.3.3 Strategies to Mitigate Risks

The safe application of nanomaterials in agricultural usage requires immediate implementation of strategies to deal with health and environmental dangers that nanotechnology in agriculture has the potential to create. The proper management of risks needs a proactive system because nanotechnology presents both current dangers along with future uncertainties regarding its extended influence.

17.3.4 Risk Assessment and Regulation

Proper risk assessments stand as the primary action to control potential threats from nanotechnology implementation in agricultural practices. Assessments need to evaluate nanomaterial hazards by assessing their toxicity rates as well as measuring their tendency to accumulate in systems and their stability length in the environment. Tests that span the entire risk profile are essential to determine the safety status of nanoparticles regarding human health and agricultural workers and consumers, plus environmental preservation. The regulatory bodies need to create detailed procedures for testing and passing nanomaterials that will be used in agricultural production (Kumari et al., 2023). The current regulatory frameworks show limitations when dealing with nanotechnology challenges since their original focus was conventional agricultural products. New regulatory measures need to be set up because nanotechnology-based products should be safe and present no risks to human health or environmental well-being.

17.3.5 Worker Safety and Protective Measures

Agricultural workers are thus in a position where precautions must be taken to ensure that they are not exposed to nanomaterials. By normal precautions, a practitioner should wear proper protective gear like gloves, masks, and respirators when dealing with nanoparticles in the field (Abbasinia et al., 2018). Therefore, it is also critical to establish the rules and protocols for occupational safety to introduce the workers to the possibility of hazardous impacts of nanomaterials and ways to protect themselves correctly. Nanoparticles can be another source of indoor contamination; therefore, proper ventilation needs to be provided wherever nanoparticles are employed or produced. The techniques of monitoring and exposure should be developed to enable the assessment of the concentration of nanoparticles in the working environment so that no workers are exposed beyond the set limit (Ramachandran et al., 2011).

17.3.6 Consumer Protection and Labeling

For the purpose of consumer protection, there is a need to define the labeling standard for agricultural products with nanomaterials. Such labeling can also help consumers be well-informed when making a purchase decision, whether they wish to purchase any food products that contain nanoparticles (Fantozzi & Garattini, 2025). Therefore, it will also strengthen the continuous testing of food items for the possible presence of residual nanomaterials. It is also necessary for the next steps in the regulation of nanotechnology; there must be set parameters for the allowed levels of nanomaterials in foodstuffs, with the rates of toxicity for the particle sizes, and their probability to bioaccumulate in the human body.

17.3.7 Environmental Monitoring and Cleanup

In relation to the environmental impact of nanomaterials, more random checks and balances on the levels of nanomaterials in soil, water, and air must be conducted (Del Prado-Audelo et al., 2021). There is a belief that nanomaterials could have a large effect on most environments; therefore, impact assessments should be conducted prior to their application in agriculture. In addition, it is necessary to establish certain guidelines and methods that will help to deal with nanomaterial waste to avoid a negative impact on the environment. This also encompasses ways for the proper elimination of nanomaterials and likewise the creation of nontoxic or eco-friendly nanoparticles, as they have adverse effects on the environment.

17.3.8 Research and Public Awareness

Last of all, it is strategic to continue investigating the safety of nanomaterials in agriculture to evaluate the prospect and possibility of the negative impacts of nanomaterials in the agricultural field. In the closing sections, the implication of adopting ethics of ultimate concern in the future uses and development of nanotechnology is that collaborative cooperation between science, policy, industry, and environmental groups will facilitate the creation of a high-quality picture of whether nanotechnology is good or bad for humans, as well as for the environment in the long run (Lo et al., 2020). Awareness will also be crucial to reduce the prospects of nanotechnology products harming the consumers and the workers and encourage them to take safety precautions to reduce risks. In conclusion, it can be stated that nanotechnology has the potential to overcome many barriers associated with agriculture, food production, and resource management. However, it also opens large safety concerns that need to be met to use such technologies appropriately. The adverse effects of being exposed to nanomaterials on the health of the workers and consumers, the effects that the latter have on the environment, and most importantly, the measures that could be taken to minimize these negative consequences are all very important issues that need to be addressed before anything else (Hofmann et al., 2020). While it is crucial to push through with nanotechnology in agriculture to improve crop yield and quality, it is imperative that we do not overlook the fact that there are potential risks attached to its use; it is therefore necessary to work toward minimizing the risks that accompany nanotechnology through carrying out risk assessments, protective measures, establishing nanotechnology-related legislation or acts, and creating awareness.

17.4 Regulatory Landscape in Nanotechnology for Agriculture

This chapter aims to provide an understanding of how nanotechnology has been effectively introduced in agriculture and what specific benefits it has brought into the field. However, as nanomaterials are incorporated into agricultural practices, it becomes quite challenging to guarantee the proper application of these nanomaterials (Usman et al., 2020). The following is a brief look at one of the major approaches that are used to address these risks and the proper use of nanotechnology in agriculture, the regulation approach. These frameworks set out policies and procedures that must be followed when utilizing or designing and implementing new technologies including nanotechnology (He et al., 2019). Although there are regulations in some countries concerning the application of nanomaterials in the agricultural field, the special features of nanotechnology are still not well addressed. In addition, since nanotechnology is an international progressive issue, the key issue is to coordinate with other countries to establish international consistency, as well as synchronize the rules of regulation in nanotechnology for agriculture (Figure 17.2).

17.4.1 Existing Regulations on Nanotechnology in Agriculture

The process of regulating nanotechnology applications for agricultural use remains in its nascent phase (Arshad et al., 2020). Nanotechnology, as a technology, poses no inherent

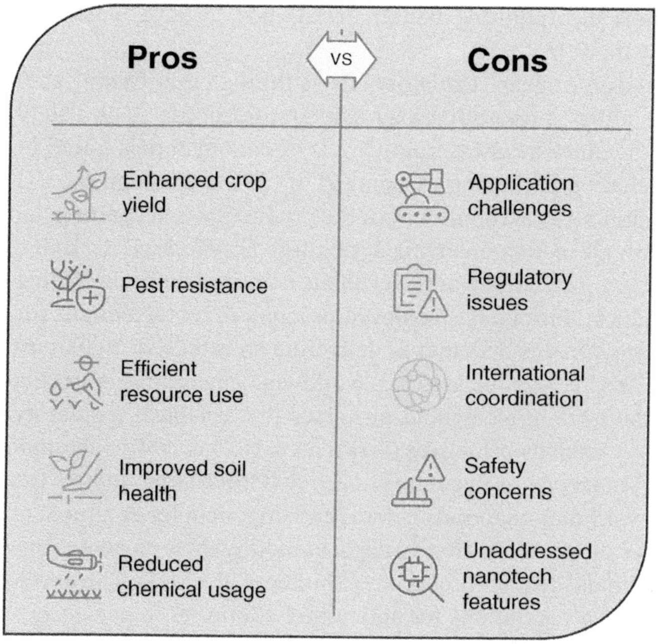

Figure 17.2 Advantages and disadvantages of nanotechnology in agriculture.

harm to products, but implementing it proves to be a significant challenge for existing regulatory systems that focus on conventional agricultural chemicals (Singh & Kumar, 2023). Worldwide regulatory organizations develop appropriate criteria to approve nanotechnology in agriculture as well as its testing procedures and monitoring mechanisms (Mishra et al., 2022). Different locations have already developed these regulations, yet numerous other locations are establishing their preferred application methods. The Environmental Protection Agency, along with the Food and Drug Administration and the Department of Agriculture, acts as a primary regulator of nanotechnology in agriculture within the United States (Cummings et al., 2021; He et al., 2019). The agencies responsible for agricultural product safety supervise both pesticides and food additives alongside fertilizers that incorporate nanomaterials (Pandey, 2018).

The Federal Insecticide, Fungicide, and Rodenticide Act (FIFRA), together with the Federal Food, Drug, and Cosmetic Act (FFDCA), does not possess specific guidelines designed for nanomaterials (Petersen et al., 2021). The agencies now modify current regulatory frameworks because nanomaterials exhibit properties. The Environmental Protection Agency (EPA) uses FIFRA to regulate pesticides regardless of whether nanomaterials are present in these products. Presently, the EPA continues to operate under a pesticide registration system, which they believe lacks proper means to evaluate nanomaterial risks (Bradbury, 2010; Lovell, 2010). The EPA began implementing new procedures to examine nanomaterial pesticides and evaluate their potential harmful effects on human health and the environment (Macko et al., 2022). The FFDCA has authority over food and food additives, but the FDA maintains that traditional safety procedures fail to detect all nanoparticle risks. Food products incorporating nanomaterials now receive individual evaluation for safety purposes by the FDA.

The European Commission leads European regulatory efforts through establishing specific guidelines to manage agricultural nanotechnology use. The European Food Safety Authority (EFSA) established guidelines to assess safety levels of nanomaterials found in food and food-related goods, including packaging (Amenta et al., 2015). Through the EU Nano-Register, the European Union seeks to monitor both the production and commercial activities of nanomaterials within all sectors involving agriculture (Pavlicek et al., 2021). The EU Novel Foods Regulation acts as a framework to evaluate new food items, including nanotechnology-based food products. Europe has achieved progress in its regulations for nanotechnology in agriculture, yet remains deficient in delivering an extensive regulatory structure for this field (Salvi, 2015). Various nations worldwide, including Australia together with Japan and Canada, have implemented processes that establish regulatory requirements surrounding nanotechnology principles (Bowman & Hodge, 2007; Solaiman et al., 2017). The Canadian Pest Management Regulatory Agency (PMRA) functions as the authority to examine pesticides with nanomaterials before releasing them for commercial sale. Canada resides among other jurisdictions that struggle to modify their current pesticide regulations to suit nanomaterials' distinct properties (Bhuller et al., 2021). The development of universal nanotechnology regulations for agriculture continues to persist as a long-term industry challenge, even though diverse regional approaches are underway. Present regulations offer essential direction but fall short when addressing the special dangers that nanomaterials present throughout agricultural applications.

17.4.2 Gaps and Challenges in Current Regulatory Frameworks

Different regulatory organizations have established plans for managing agricultural nanotechnology, yet regulatory obstacles combined with gaps continue to exist in the current oversight framework (Ridge, 2018). Defining and regulating nanomaterials presents difficulties since all groups lack consensus regarding nanomaterials and nanotechnology definitions. The scientific field of nanotechnology lacks clear definitions since it handles diverse materials through many applications, with indistinct transition points between nanotechnology-derived products and other products. The unclear definitions pose regulatory difficulties for conference authorities because they prevent the definition of clear guidelines, yet stakeholders cannot properly decode the requisite regulatory requirements for agricultural nanomaterials. A critical problem for nanomaterials assessment in the industry exists due to the insufficient acceptance of testing standards (Fadeel et al., 2018). Standard testing procedures do not provide adequate evaluation of pesticide safety standards nor fertilizer or food additive safety when examining nanomaterials. The investigation of nanoparticles requires evaluation methodologies to determine human, animal, and plant health threats in conjunction with ecosystem protection because of their dimensions that lead to amplified reactivity, yet biological features (Kumar et al., 2022; López-Serrano et al., 2014). The assessment practices for nanomaterials remain without necessary standardization, and current evaluation methods persistently fail to detect several agricultural risks connected to their utilization (Figure 17.3).

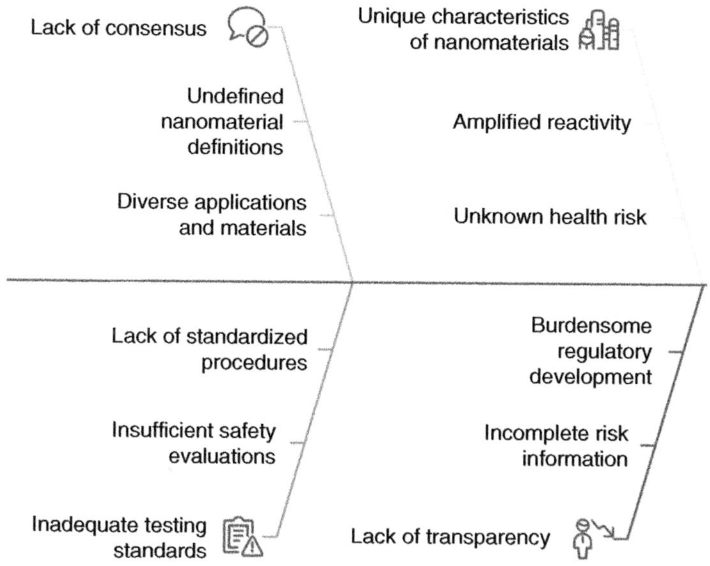

Figure 17.3 Challenges in regulating agriculture nanotechnology.

Several regulatory systems base their operations on the assumption that nanomaterials should display identical behaviors with conventional macro-sized materials. The unique characteristics of nanomaterials generate different biological reactions that produce unknown health and safety dangers (Barhoum et al., 2022). Plants together with animals or humans show better nanoparticle consumption compared to bulk matter materials, but standard environmental health evaluations fail to detect these unexamined risks (Committee et al., 2021). Better assessment models for understanding nanomaterial behavior in agriculture need further research because the current regulatory processes lack transparency and accountability models. The advancement of nanotechnology needs complete information about potential risks and benefits of nanomaterials to be shared between regulatory bodies, industry members, and public entities. The development of regulations requires specific mind-sharing together with transparent decisions and must include multiple stakeholders. The nanotechnology regulatory procedure receives inadequate oversight in numerous agricultural jurisdictions, thus the public loses confidence about nanomaterials' security and regulatory development becomes burdensome.

17.4.3 The Role of International Cooperation in Shaping Regulations

The emerging technology of nanotechnology will affect operations in farming across all nations; therefore, worldwide regulatory structures for nanotechnology need international collaboration. The application of nanotechnology in agriculture does cross national boundaries because its potential traits are associated with potential hazards to every country in the world (Pramanik et al., 2020). Given this, there is still a need to standardize the regulation of such approaches globally because they will ensure the protected and sustainable implementation of agricultural nanomaterials. Important organizations that promote international cooperation in the regulation of nanotechnology are the World Health Organization (WHO) together with the Food and Agriculture Organization (FAO), and the Organization for Economic Co-operation and Development (OECD) (Mbengue & Charles, 2013). As such, it is also seen that intentional organizations form worldwide guidance and security standards for the use of nanotechnology, especially in agricultural operations. The OECD adopted three key components for a standardized safety protocol of nanomaterials that include requirements for evaluation, protocols for testing, and guidelines for participation of the public (Doak et al., 2023). Such guidelines set basic principles to help countries wanting to create new nanotechnology agricultural regulations or to revise existing ones.

International collaboration is a means of appropriate treatment of equity problems related to agricultural nanotechnology usage. Managing the safety-related risks makes way for countries to offer nanotechnology benefits to developing countries' small-scale farmers through a collaborative route. To enable farmers to adopt safe, sustainable, nanotechnology-based practices, they need to have access to resources in the form of financial backing, as well as capacity development through educational programs that are infused with knowledge sharing (Pokrajac et al., 2021). At the same time, many countries and international organizations are elaborating guidelines and regulations to deal with the use of nanomaterials in agricultural applications on emergent regulatory contexts (Mishra et al., 2022). Although the present regulation still provides oversight, there are several problems remaining due to the lack of a definition of what a nanomaterial is, standard tests, or an understanding of its risks. The fragmentation of the current regulatory approach across different regions causes the problem of building an interrelated and successful regulatory system. To create standardized rules for the safe and proper use of nanotechnology in agricultural operations, international cooperation is required. Countries and organizations working in international cooperation will lead to worldwide nanotechnology regulations, ensuring environmental and human safety as well as sustainable world agriculture.

17.5 Public Perception and Trust in Nanotechnology for Agriculture

Nanotechnology will revitalize farming operations because it offers several methods to improve food production while preserving water supplies and protecting natural habitats. Public concerns about nanotechnology impair its development because people have similar doubts about all new technologies (Singh et al., 2024b). Our study looks at both the good things nanotechnology brings to farming plus how people worry about safety while reviewing environmental and ethical results. Farmers determine their nanotechnology use based on what customers think about safety risks even if those beliefs are false or true. Agricultural adoption of nanotechnology requires improved public view and increased trust in nanotechnology. Establishing effective nanotechnology use in farming must combine wide education and awareness efforts with broad scientific openness and proper safety evaluation.

17.5.1 Role of Public Awareness and Education

People's views on the use of nanotechnology in agriculture are based on both how much they know about technology and how confident they are in the science behind it. Since the public lacks awareness of nanotechnology devices, they and others question the technology because they are not told what they are. To understand how to utilize nanotechnology in the process of agricultural sector development, people should learn about nanotechnology and its applications. Nanotechnology is an interdisciplinary subject in which many benefits of agriculture are brought in, such as smarter pest control items, furnishing, and crop protection systems, together with advanced food processing (Kiran et al., 2024). The advantages of technical advancements remain under discussion due to public concern over their safety for people and the environment as well as worries about unforeseen outcomes. Educational programs about nanotechnology science and its possible uses simplify technical information for people to understand how to choose wisely (Mandrikas et al., 2020). Public awareness campaigns need to focus their presentation on the various issues connected to using nanotechnology in farming (Mukherjee et al., 2019). These discussions need to detail health hazards to humans plus ecology while showing results from scientific research. Open sharing of research results and safety monitoring procedures helps the public trust nanotechnology more and use it properly. Multiple organizations support public education on nanotechnology, including educational centers, schools, government bodies, Non-Governmental Organizations (NGOs), and private corporations (Khatoon & Velidandi, 2025). These stakeholders can improve scientific education by creating learning tools for the public and sharing information through meetings and online. People who understand nanotechnology become potential backers of appropriate regulations and its employment in agricultural systems.

17.5.2 Importance of Transparency in Research and Development

Public acceptance and trust formation take root when research and development activities in nanotechnology for agriculture become public. People such as farmers, researchers, policymakers, and society must have full visibility into how nanomaterial technologies are researched and tested to assess their safety and performance both in farms and fields (Gottardo et al., 2021). Society doubts emerging technologies like nanotechnology because their developers and users do not make their work open for people to see or understand. When stakeholders in the nanotechnology sector fail to share reliable facts about nanotechnology safety, people may doubt their dedication to safety and lose faith in the technology. When researchers openly share their findings, industrial impacts get checked against any risks ahead of commercial usage.

Government authorities need to support open nanotechnology research by monitoring its progress and processes. When projects receive public research funds, they must share information from studies by releasing articles, reports, and database content with everyone. Private companies doing nanotechnology research should make their work results public and start open talks with their related stakeholders (Amutha et al., 2024). Nanotechnology stands equally on transparency in both research and policy development (Tawiah et al., 2024). Government officials need to explain the reasons for their choices

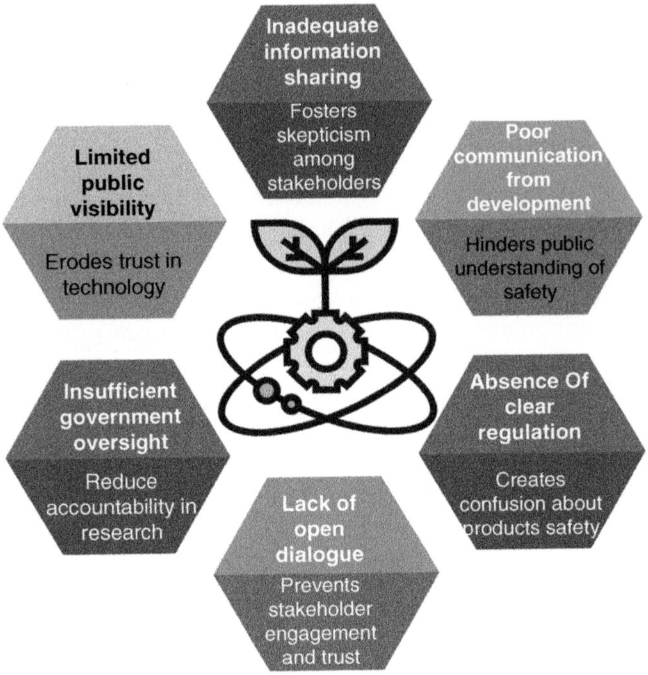

Figure 17.4 Lack of transparency in nanotechnology research.

when monitoring nanotechnology products in farming. In addition, they must indicate at which stages rejection occurs for each nanotechnology product. When authorities release the rules, they realize that people depending on the system trust them better (Figure 17.4).

17.5.3 Balancing Innovation with Caution

However, nanotechnology in agriculture will possibly boast unquestionable benefits that will cause the need for scientists to undertake a thorough understanding before implementation. Therefore, even with nanotechnology development, security and ethics cannot be compromised, and relevant studies should cover the whole range of potential negative influences in each aspect of human well-being, environmental preservation, and societal stability. Achieving an exact stability within new development promotion and risk management is crucial for achieving public trust and optimal nanotechnology benefits (Kamarulzaman et al., 2020). This case of accelerating technological change obligates the need to deploy immediately, which automatically instigates the need for faster agricultural nanotechnology implementation. In terms of innovation development, nanomaterials are pretty much a hurry-up strategy because there is no complete knowledge of what possible dangers nanomaterials pose. Extensive use of nanoparticle-based pesticides or fertilizers would result in contamination of water sources and soil ecosystem and unrevealed environmental issues would happen (Gana et al., 2024). The lack of proper safety assessments of nanomaterials implemented for widespread use can result in health risks to consumers when nanoparticles survive testing for food safety (Wang et al., 2021). The responsible practice of innovation serves as an instrument to achieve equilibrium between innovation and caution. Scientists should create nanomaterials and agricultural products, which demonstrate effectiveness through safety measures, and they must maintain sustainable and ethical boundaries. Nanomaterials should be designed with the main priority of a low environmental footprint and capability to decompose as well as being free from toxic elements to protect ecosystems and human well-being (Corsi et al., 2023). Research should prioritize the discovery of nanomaterials that provide maximum benefits together with minimum adverse effects so agricultural efficiency enhancement can occur without harming biodiversity. Governments should promote adaptive management strategies through which policies and practices receive regular updates based on both newly available scientific information and societal feedback. The ongoing development of nanotechnology remains connected to the most current scientific insights through this approach, so it can respond to new risks and address unresolved challenges.

17.6 Future Directions in Nanotechnology for Agriculture

The potential of nanotechnology in agricultural settings appears to be great and promising, and at the same time, it is associated with certain challenges that require the development of a strategic framework. Over the years, the noble field of nanotechnology opens more opportunities to transform agricultural practices conducive to sustainable protection, management of soil conditions, and even food productivity. However, with these advancements come several critical issues that include regulatory issues, ethics, and the parity of the scientist, policymaker, and the lay public. The goal of this chapter is to regulate nanotechnology for use in agriculture in the best possible way and to avoid negative impacts on human health, the environment, and values.

17.6.1 Advancements in Regulatory Frameworks

The regulation system needs to transform as nanoparticle farming technology continues its growth trend. Countries that have started nanotechnology guidelines need regular updates since scientific progress keeps advancing this technology. The job of health regulators needs to cover both nanomaterial safety and efficiency plus handle technical aspects that come with nanotechnology, such as nanoparticle features. Globally accepted rules for nanotechnology will become standard practice during regulation. Nanotechnology creates worldwide trade barriers when national governments enforce different safety regulations, which makes companies uneasy about their products. Standardization at the global level leads to easier product development and safer use of agricultural nanomaterials through united approval procedures. International health and food organizations, especially WHO and FAO, help nations harmonize their regulations through global discussions about nanoagricultural safety standards. Future regulations should handle matters that concern nanotechnology by design. The regulations should include making fresh tests to see how nanoparticles react with soil, plants, animals, and the environment when used in farming. Traditional chemical risk evaluation methods do not match how nanomaterials behave because their small dimensions and large surface areas create different types of danger to the environment and human health. Scientific and government experts must team up to set proper nanomaterial-testing methods that match their special features in agricultural situations and make sure they remain safe for farmers to use.

17.6.2 Ethical Guidelines for Future Nanotechnology Applications

The main ethical issue is making sure everyone receives equal access to nanotechnology benefits. Technological advancements based on nanoscale research bring more crop yields with less dangerous chemicals while helping to feed the world better. Large farms and wealthy nations may take most of the benefits from nano-based farm improvements even while small-scale farmers and developing areas lose out. Ethical standards need to help all farmers gain from nanotechnology, especially smallholder farmers, who should not pay more for technology adoption. People need to examine how nanomaterials affect our natural environment. Nanoparticles show unexpected movements and reactions in the environment whose outcomes scientists cannot predict entirely. Although nanotechnology helps farming become greener through smarter fertilizers and pest control tools, its benefits depend on keeping nanomaterials from harming nature during use. Environmental impact assessments should completely evaluate how nanoparticles affect soil quality and natural habitats at all stages of their lifecycle. Our ethical rules must include clear statements about how agriculture uses nanotechnology and how it gets customer agreement. The public must learn all advantages and disadvantages before allowing nanotechnology into agricultural production. The public should receive fully detailed information on how nanomaterial food usage affects safety plus environmental and health issues. Nanomaterials pose risks to agricultural workers during their daily tasks, so they must receive proper information and approval before handling them at work. Employers need to educate their staff about job hazards before giving them protective safety equipment.

17.6.3 Collaboration Between Scientists, Policymakers, and the Public

Integrating nanotechnology into agriculture will only be successful with the collaboration of scientists, policymakers, and the public. Each subgroup is bringing a unique set of perspectives and expertise to the table, and working together is required to make sure nanotechnology is developed and deployed in a scientifically as well as socially acceptable manner. Nanotechnology and the development of agricultural knowledge are critical to scientists, as they are areas that are continually growing. By way of research and innovation, they can find new uses for nanomaterials that can help achieve increased crop yields, protect plants from pests and diseases, and make farming practices less harsh on the environment. Nevertheless, scientists should endeavor to work closely with policy influencers and members of the public such that their research is consistent with the needs and values of society. Involving the public is necessary to help decide how agriculture should apply nanotechnology as well as to build trust in technology. On the other hand, policymakers have the task of putting in place the regulatory frameworks that allow for the use of nanotechnology in agriculture. Policymakers must have the latest scientific research at their disposal and be engaged in the discussion on the potential risks and benefits of nanomaterials to make informed decisions. Nanotechnology therefore should be accompanied by policymakers' attention to regulatory frameworks that are flexible and adaptive so that as nanotechnology evolves, challenges can be appropriately addressed. Moreover, they need to make sure that the public concerns and ethical considerations are given attention while formulating policies. Overall, the future of nanotechnology in agriculture looks bright, though success will require work on the part of regulatory authorities in developing responsible regulatory regimes, on the development of ethical frameworks, and on the creation of robust collaborative relationships between scientists, policymakers and the public. However, with expanding nanotechnology, it is necessary to make certain that its use is beneficial with rules and morally correct, characterized by openness, sustainability, and equity. The use of nanotechnology to mitigate risks and maximize benefits to society for tackling global agricultural challenges can be obtained through collaboration and involvement of diverse stakeholders. Designed and developed sensibly, nanotechnology has the potential to reshape agriculture in a way that is sustainable, efficient, and advantageous to all.

17.7 Conclusion

Nanotechnology enables the agricultural industry to produce additional food supplies and guard plant crops in addition to sustaining environmental preservation. All stakeholders must inspect nanotechnology to confirm proper ethical and safe utilization of this technology. The chapter results show that we must answer safety questions regarding responsible nanotechnology use before applying it to agricultural farming methods. The ethical development path of nanotechnology for agriculture is determined by ethical matters. The safe application of nanotechnology requires treating health-related hazards with environmental impacts alongside assessing the advantages that various demographics receive from this technology. The implementation of nanotechnology systems needs to produce helpful results for all farmers but maintain protection for rural communities combined with small-scale farmers and the environment. Society requires straightforward explanations together with balanced rules that will enable it to trust new technological solutions. Before commercial distribution, nanotechnological products require proper emphasis on safety elements. The unique properties together with elevated reactivity of nanomaterials create specific evaluation needs for safety purposes when deployed in agricultural soils. All agricultural applications of nanotechnology need testing that proves their safety for personnel and end-users and examines any environmental impacts. The proper management of nanomaterials together with ongoing monitoring forms the basis for protecting human health and defending environmental conditions. Government needs new regulations to tackle the distinctive threats that stem from nanotechnology. The quick development of nanotechnology requires existing laws to be put into effect. Every nation should follow identical quality requirements during nanomaterial evaluation and permitting procedures. Safety oversight of nanomaterials requires authorities to modify their supervisory procedures based on recently discovered data and potential threats. The proper application of nanotechnology in agriculture necessitates following ethical criteria with proper safety regulations and a legal framework. The strategic consideration of these factors permits nanotechnology to develop agriculturally safe farming solutions that benefit all stakeholder groups. Ethical development, along with appropriate implementation of nanotechnology, determines the future success that this field will achieve in agricultural applications.

References

Abbasi, B. A., Iqbal, J., Yaseen, T., Zahra, S. A., Ali, S., Uddin, S., & Chalgham, W. (2023). Exploring physical characterization and different bio-applications of *Elaeagnus angustifolia* orchestrated nickel oxide nanoparticles. *Molecules*, *28*(2), 654.

Abbasinia, M., Karimie, S., Haghighat, M., & Mohammadfam, I. (2018). Application of nanomaterials in personal respiratory protection equipment: a literature review. *Safety*, *4*(4), 47.

Aguilar-Pérez, K. M., Ruiz-Pulido, G., Medina, D. I., Parra-Saldivar, R., & Iqbal, H. M. (2023). Insight of nanotechnological processing for nano-fortified functional foods and nutraceutical-opportunities, challenges, and future scope in food for better health. *Critical Reviews in Food Science and Nutrition*, *63*(20), 4618–4635.

Ahmad, M., Ali, A., Ullah, Z., Sher, H., Dai, D. Q., Ali, M., & Ali, I. (2022). Biosynthesized silver nanoparticles using *Polygonatum geminiflorum* efficiently control fusarium wilt disease of tomato. *Frontiers in Bioengineering and Biotechnology*, *10*, 988607.

Ahmed, S. F., Mofijur, M., Rafa, N., Chowdhury, A. T., Chowdhury, S., Nahrin, M., Islam, A. B. M. S., & Ong, H. C. (2022). Green approaches in synthesising nanomaterials for environmental nanobioremediation: technological advancements, applications, benefits and challenges. *Environmental Research*, *204*, 111967. https://doi.org/10.1016/j.envres.2021.111967

Alex, R., Suresh, P., Theresa, M., & Radhakrishnan, E. (2025). Introduction to nanotechnology for sustainable agriculture: current perspectives. In *Agricultural sustainability through nanotechnology* (pp. 1–12). CRC Press.

Ali, S., Ahmad, N., Dar, M. A., Manan, S., Rani, A., Alghanem, S. M. S., Khan, K. A., Sethupathy, S., Elboughdiri, N., & Mostafa, Y. S. (2023). Nano-agrochemicals as substitutes for pesticides: prospects and risks. *Plants*, *13*(1), 109.

Alizadeh, M., Qarachal, J. F., & Sheidaee, E. (2025). Understanding the ecological impacts of nanoparticles: risks, monitoring, and mitigation strategies. *Nanotechnology for Environmental Engineering*, *10*(1), 6.

Altabbaa, S., Mann, N. A., Chauhan, N., Utkarsh, K., Thakur, N., & Mahmoud, G. A.-E. (2023). Era connecting nanotechnology with agricultural sustainability: issues and challenges. *Nanotechnology for Environmental Engineering*, *8*(2), 481–498. https://doi.org/10.1007/s41204-022-00289-3

Amenta, V., Aschberger, K., Arena, M., Bouwmeester, H., Moniz, F. B., Brandhoff, P., Gottardo, S., Marvin, H. J., Mech, A., & Pesudo, L. Q. (2015). Regulatory aspects of nanotechnology in the agri/feed/food sector in EU and non-EU countries. *Regulatory Toxicology and Pharmacology*, *73*(1), 463–476.

Amutha, C., Gopan, A., Pushbalatatha, I., Ragavi, M., & Reneese, J. A. (2024). Nanotechnology and governance: regulatory framework for responsible innovation. In S. C. George & B. Tawiah (Eds.), *Nanotechnology in societal development* (pp. 481–503). Springer Nature Singapore. https://doi.org/10.1007/978-981-97-6184-5_14

Arshad, F., Munir, A., Kashif, Q. Q., ul Haq, T., Iqbal, J., Sher, F., & Hussain, I. (2020). Controlled development of higher-dimensional nanostructured copper oxide thin films as binder free electrocatalysts for oxygen evolution reaction. *International Journal of Hydrogen Energy, 45*(33), 16583–16590.

Ashraf, S. A., Siddiqui, A. J., Abd Elmoneim, O. E., Khan, M. I., Patel, M., Alreshidi, M., Moin, A., Singh, R., Snoussi, M., & Adnan, M. (2021). Innovations in nanoscience for the sustainable development of food and agriculture with implications on health and environment. *Science of the Total Environment, 768*, 144990.

Barhoum, A., García-Betancourt, M. L., Jeevanandam, J., Hussien, E. A., Mekkawy, S. A., Mostafa, M., Omran, M. M., Abdalla, M. S., & Bechelany, M. (2022). Review on natural, incidental, bioinspired, and engineered nanomaterials: history, definitions, classifications, synthesis, properties, market, toxicities, risks, and regulations. *Nanomaterials, 12*(2), 177.

Besha, A. T., Liu, Y., Bekele, D. N., Dong, Z., Naidu, R., & Gebremariam, G. N. (2020). Sustainability and environmental ethics for the application of engineered nanoparticles. *Environmental Science & Policy, 103*, 85–98. https://doi.org/10.1016/j.envsci.2019.10.013

Bhattacharya, B., Roy, P., Bhattacharya, S., Prasad, B., & Mandal, A. K. (2023). Chapter 19 - Nanotechnology and sustainable development: overcoming the obstacles by adopting ethical practices for future farming. In A. Husen (Ed.), *Engineered nanomaterials for sustainable agricultural production, soil improvement and stress management* (pp. 431–445). Academic Press. https://doi.org/10.1016/B978-0-323-91933-3.00020-9

Bhuller, Y., Ramsingh, D., Beal, M., Kulkarni, S., Gagne, M., & Barton-Maclaren, T. S. (2021). Canadian regulatory perspective on next generation risk assessments for pest control products and industrial chemicals. *Frontiers in Toxicology, 3*, 748406.

Bibi, F., Khan, A., Iqrar, I., Iqbal, M., & Shinwari, Z. K. (2024). Therapeutic applications and mechanism of action of plant-mediated silver nanoparticles. *Pakistan Journal of Botany, 56*(2), 531–540.

Bowman, D. M., & Hodge, G. A. (2007). A small matter of regulation: an international review of nanotechnology regulation. *Columbia Science and Technology Law Review, 8*(1), 1–45.

Bradbury, S. (2010). FIFRA Scientific Advisory Panel ".

Chelliah, R., Madar, I. L., Sultan, G., Begum, M., Pahi, B., Tayubi, I. A., Selvakumar, V., & Oh, D. H. (2023). Chapter 18 - Risk assessment and regulatory decision-making for nanomaterial use in agriculture. In A. Husen (Ed.), *Engineered nanomaterials for sustainable agricultural production, soil improvement and stress management* (pp. 413–430). Academic Press. https://doi.org/10.1016/B978-0-323-91933-3.00009-X

Chugh, G., Siddique, K. H., & Solaiman, Z. M. (2021). Nanobiotechnology for agriculture: smart technology for combating nutrient deficiencies with nanotoxicity challenges. *Sustainability, 13*(4), 1781.

Committee, E. S., More, S., Bampidis, V., Benford, D., Bragard, C., Halldorsson, T., Hernández-Jerez, A., Hougaard Bennekou, S., Koutsoumanis, K., & Lambré, C. (2021). Guidance on risk assessment of nanomaterials to be applied in the food and feed chain: human and animal health. *EFSA Journal, 19*(8), e06768.

Corsi, I., Venditti, I., Trotta, F., & Punta, C. (2023). Environmental safety of nanotechnologies: the eco-design of manufactured nanomaterials for environmental remediation. *Science of the Total Environment, 864*, 161181.

Cummings, C. L., Kuzma, J., Kokotovich, A., Glas, D., & Grieger, K. (2021). Barriers to responsible innovation of nanotechnology applications in food and agriculture: a study of US experts and developers. *NanoImpact, 23*, 100326.

Dasgupta, N., Ranjan, S., & Ramalingam, C. (2017). Applications of nanotechnology in agriculture and water quality management. *Environmental Chemistry Letters, 15*, 591–605.

Del Prado-Audelo, M., García Kerdan, I., Escutia-Guadarrama, L., Reyna-González, J., Magaña, J., & Leyva-Gómez, G. (2021). Nanoremediation: nanomaterials and nanotechnologies for environmental cleanup. *Frontiers in Environmental Science, 9*, 793765.

Dhillon, R., & Moncur, Q. (2023). Small-scale farming: a review of challenges and potential opportunities offered by technological advancements. *Sustainability, 15*(21), 15478.

Doak, S. H., Andreoli, C., Burgum, M. J., Chaudhry, Q., Bleeker, E. A., Bossa, C., Domenech, J., Drobne, D., Fessard, V., & Jeliazkova, N. (2023). *Current status and future challenges of genotoxicity OECD Test Guidelines for nanomaterials: a workshop report*. Oxford University Press UK.

Ekka, P., Patra, S., Upreti, M., Kumar, G., Kumar, A., & Saikia, P. (2023). Land degradation and its impacts on biodiversity and ecosystem services. In A. Raj, M. K. Jhariya, A. Banerjee, S. Nema, & K. Bargali (Eds.), *Land and environmental management through forestry* (pp. 77–101). Hoboken, NJ: John Wiley & Sons.

Fadeel, B., Farcal, L., Hardy, B., Vázquez-Campos, S., Hristozov, D., Marcomini, A., Lynch, I., Valsami-Jones, E., Alenius, H., & Savolainen, K. (2018). Advanced tools for the safety assessment of nanomaterials. *Nature Nanotechnology, 13*(7), 537–543.

Fantozzi, P., & Garattini, S. (2025). Product labels and advertising: are consumers protected or misled? *Italian Journal of Food Science/Rivista Italiana di Scienza degli Alimenti, 37*(1), p1.

Gana, M., Isibor, P. O., Damisa, J. I., Iseghohi, F., Musa, I. O., & Oyewole, O. A. (2024). Nanoparticle-based remediation and environmental cleanup. In P. O. Isibor, G. Devi, & A. A. Enuneku (Eds.), *Environmental nanotoxicology: Combatting the minute contaminants* (pp. 161–181). Cham: Springer.

Ghosh, M., & Kumar, R. (2024). Regulatory issues in nanotechnology. In M. Prasad, R. Kumar, M. Ghosh, S. M. Syed, & S. Chakravarti (Eds.), *Nanotechnology theranostics in livestock diseases and management* (pp. 765–788). Springer Nature Singapore. https://doi.org/10.1007/978-981-16-1610-5_31

Gottardo, S., Mech, A., Drbohlavová, J., Małyska, A., Bøwadt, S., Sintes, J. R., & Rauscher, H. (2021). Towards safe and sustainable innovation in nanotechnology: state-of-play for smart nanomaterials. *NanoImpact, 21*, 100297.

Haris, M., Fatima, N., Iqbal, J., Chalgham, W., Mumtaz, A. S., El-Sheikh, M. A., & Tavafoghi, M. (2023). *Oscillatoria limnetica* mediated green synthesis of iron oxide (Fe_2O_3) nanoparticles and their diverse in vitro bioactivities. *Molecules, 28*(5), 2091.

Havelikar, U., Ghorpade, K. B., Kumar, A., Patel, A., Singh, M., Banjare, N., & Gupta, P. N. (2024). Comprehensive insights into mechanism of nanotoxicity, assessment methods and regulatory challenges of nanomedicines. *Discover Nano, 19*(1), 165.

He, X., Deng, H., Aker, W. G., & Hwang, H.-m. (2019). Regulation and safety of nanotechnology in the food and agriculture industry. In G. Molina, F. M. P. Inamuddin, & A. M. Asiri (Eds.), *Food applications of nanotechnology* (pp. 525–536). CRC Press.

Hofmann, T., Lowry, G. V., Ghoshal, S., Tufenkji, N., Brambilla, D., Dutcher, J. R., Gilbertson, L. M., Giraldo, J. P., Kinsella, J. M., & Landry, M. P. (2020). Technology readiness and overcoming barriers to sustainably implement nanotechnology-enabled plant agriculture. *Nature Food*, *1*(7), 416–425.

Ijaz, S., Iqbal, J., Abbasi, B. A., Ullah, Z., Ijaz, N., Yaseen, T., Iqbal, R., Murtaza, G., Usman, M., Sampath, S., Hafeez, M. B., Kanwal, S., & Mahmood, T. (2024). Regulatory and ethical concerns of nanotechnology in agriculture. In R. Shahzad, S. Fiaz, A. Qayyum, M. Ul Islam, & I.-J. Lee (Eds.), *Revolutionizing agriculture: a comprehensive exploration of agri-nanotechnology* (pp. 395–427). Springer Nature Switzerland. https://doi.org/10.1007/978-3-031-76000-6_18

Isibor, P. O., Oyegbade, S. A., Oni, J. G., Ahmed, W. W., Abiodun, E. O., & Oyewole, O. A. (2024). Nanoparticles in food chains: bioaccumulation and trophic transfer. In P. O. Isibor, G. Devi, & A. A. Enuneku (Eds.), *Environmental nanotoxicology: combatting the minute contaminants* (pp. 203–233). Springer.

Jain, A., Ranjan, S., Dasgupta, N., & Ramalingam, C. (2018). Nanomaterials in food and agriculture: an overview on their safety concerns and regulatory issues. *Critical Reviews in Food Science and Nutrition*, *58*(2), 297–317.

Jamil Abbasi, A., Anas, M., Elahi, M., Khan, A., Khattak, W. A., Saleem, M. H., Fahad, S., Elansary, H. O., Mahmoud, E. A., Ahmed, T., & Quraishi, U. M. (2025). Restoring wheat productivity and nutrient balance under cadmium stress through reducing toxicity, metal uptake, and oxidative damage using selenium nanoparticles. *Journal of Trace Elements in Medicine and Biology*, *89*, 127644. https://doi.org/10.1016/j.jtemb.2025.127644

Kamarulzaman, N. A., Lee, K. E., Siow, K. S., & Mokhtar, M. (2020). Public benefit and risk perceptions of nanotechnology development: psychological and sociological aspects. *Technology in Society*, *62*, 101329.

Khan, W. S., Asmatulu, E., Zhang, B., & Asmatulu, R. (2025a). Chapter 3 - Safety and ethics of nanotechnology. In R. Asmatulu, W. S. Khan, & E. Asmatulu (Eds.), *Nanotechnology safety* (2nd ed., pp. 43–57). Elsevier. https://doi.org/10.1016/B978-0-443-15904-6.00009-5

Khan, A., Anas, M., Bibi, F., Ali, M., Khalil, A. T., Munawar, K. S., Mohamed, H. E. A., Hkiri, K., Maaza, M., & Shinwari, Z. K. (2025b). Phytochemical-mediated green synthesis of silver, copper, and Ag–Cu bimetallic nanoparticles using *Peganum harmala* demonstrating advanced catalytic, antioxidant, and biomedical applications. *Applied Biochemistry and Biotechnology*. https://doi.org/10.1007/s12010-025-05186-4

Khatoon, U. T., & Velidandi, A. (2025). An overview on the role of government initiatives in nanotechnology innovation for sustainable economic development and research progress. *Sustainability*, *17*(3), 1250.

Kiran, V., Harini, K., Thirumalai, A., Girigoswami, K., & Girigoswami, A. (2024). Nanotechnology's role in ensuring food safety and security. *Biocatalysis and Agricultural Biotechnology*, *58*, 103220.

Kumar, A., Choudhary, A., Kaur, H., Guha, S., Mehta, S., & Husen, A. (2022). Potential applications of engineered nanoparticles in plant disease management: a critical update. *Chemosphere*, *295*, 133798.

Kumari, R., Suman, K., Karmakar, S., Mishra, V., Lakra, S. G., Saurav, G. K., & Mahto, B. K. (2023). Regulation and safety measures for nanotechnology-based agri-products. *Frontiers in Genome Editing*, *5*, 1200987.

Lo, C.-c., Cho, H.-C., & Wang, P.-W. (2020). Global R&D collaboration in the development of nanotechnology: the impact of R&D collaboration patterns on patent quality. *Sustainability*, *12*(15), 6055.

Lombi, E., Donner, E., Dusinska, M., & Wickson, F. (2019). A one health approach to managing the applications and implications of nanotechnologies in agriculture. *Nature Nanotechnology*, *14*(6), 523–531. https://doi.org/10.1038/s41565-019-0460-8

López-Serrano, A., Olivas, R. M., Landaluze, J. S., & Cámara, C. (2014). Nanoparticles: a global vision. Characterization, separation, and quantification methods. Potential environmental and health impact. *Analytical Methods*, *6*(1), 38–56.

Lovell, A. (2010). EPA decision could indicate direction on registering nano pesticides. *Inside EPA's Risk Policy Report*, *17*(36), 1–9.

Lowry, G. V., Avellan, A., & Gilbertson, L. M. (2019). Opportunities and challenges for nanotechnology in the agri-tech revolution. *Nature Nanotechnology*, *14*(6), 517–522.

Macko, M., Antoš, J., Božek, F., Konečný, J., Huzlík, J., Hegrová, J., & Kuřitka, I. (2022). Development of new health risk assessment of nanoparticles: EPA health risk assessment revised. *Nanomaterials*, *13*(1), 20.

Malik, S. B., Saggu, J. I., Gul, A., Abbasi, B. A., Iqbal, J., Waris, S., ... Chalgham, W. (2022). Synthesis and characterization of silver and graphene nanocomposites and their antimicrobial and photocatalytic potentials. *Molecules*, *27*(16), 5184.

Mandrikas, A., Michailidi, E., & Stavrou, D. (2020). Teaching nanotechnology in primary education. *Research in Science & Technological Education*, *38*(4), 377–395.

Marella, S., Nirmal Kumar, A. R., & Tollamadugu, N. V. K. V. P. (2021). Chapter 19 - Nanotechnology-based innovative technologies for high agricultural productivity: opportunities, challenges, and future perspectives. In B. Viswanath (Ed.), *Recent developments in applied microbiology and biochemistry* (pp. 211–220). Academic Press. https://doi.org/10.1016/B978-0-12-821406-0.00019-9

Mbengue, M. M., & Charles, M. (2013). International organizations and nanotechnologies: the challenge of coordination. *Review of European, Comparative & International Environmental Law*, *22*(2), 174–185.

Mishra, A. K., Das, R., Sahoo, S., & Biswal, B. (2022). Global regulations and legislations on nanoparticles usage and application in diverse horizons. In N. B. Turan, O. Engin, & M. S. Bilgili (Eds.), *Comprehensive analytical chemistry* (Vol. 99, pp. 261–290). Elsevier.

Mittal, D., Kaur, G., Singh, P., Yadav, K., & Ali, S. A. (2020). Nanoparticle-based sustainable agriculture and food science: recent advances and future outlook. *Frontiers in Nanotechnology*, *2*, 579954.

Mohammed, P. N., Hussen, N. H., Hasan, A. H., Salh, H. J. H., Jamalis, J., Ahmed, S., Bhat, A. R., & Kamal, M. A. (2025). A review on the role of nanoparticles for targeted brain drug delivery: synthesis, characterization, and applications. *EXCLI Journal*, *24*, 34.

Moulick, D., Majumdar, A., Choudhury, A., Das, A., Chowardhara, B., Pattnaik, B. K., Dash, G. K., Murmu, K., Bhutia, K. L., & Upadhyay, M. K. (2024). Emerging concern of nano-pollution in agro-ecosystem: flip side of nanotechnology. *Plant Physiology and Biochemistry*, *211*, 108704.

Mukherjee, A., Maity, A., Pramanik, P., Shubha, K., Joshi, D., & Wani, S. (2019). Public perception about use of nanotechnology in agriculture. In M. Ghorbanpour & S. H. Wani (Eds.), *Advances in phytonanotechnology: From synthesis to application* (pp. 405–418). Academic Press.

Munir, A., Salah, M. A., Ali, M., Ali, B., Saleem, M. H., Samarasinghe, K., De Silva, S., Ercisli, S., Iqbal, N., & Anas, M. (2024). Advancing agriculture: harnessing smart nanoparticles for precision fertilization. *BioNanoScience, 14*(4), 3846–3863.

Muthukrishnan, L. (2022). An overview on the nanotechnological expansion, toxicity assessment and remediating approaches in agriculture and food industry. *Environmental Technology & Innovation, 25*, 102136. https://doi.org/10.1016/j.eti.2021.10.136

Pandey, G. (2018). Challenges and future prospects of agri-nanotechnology for sustainable agriculture in India. *Environmental Technology & Innovation, 11*, 299–307.

Pavlicek, A., Part, F., Rose, G., Praetorius, A., Miernicki, M., Gazsó, A., & Huber-Humer, M. (2021). A European nano-registry as a reliable database for quantitative risk assessment of nanomaterials? A comparison of national approaches. *NanoImpact, 21*, 100276.

Pechyen, C., Tangnorawich, B., Toommee, S., Marks, R., & Parcharoen, Y. (2024). Green synthesis of metal nanoparticles, characterization, and biosensing applications. *Sensors International, 5*, 100287. https://doi.org/10.1016/j.sintl.2024.100287

Petersen, E. J., Ceger, P., Allen, D. G., Coyle, J., Derk, R., Reyero, N. G., Gordon, J., Kleinstreuer, N., Matheson, J., & McShan, D. (2021). US federal agency interests and key considerations for new approach methodologies for nanomaterials. *Altex, 39*(2), 183–206.

Pidgeon, N. (2021). Engaging publics about environmental and technology risks: frames, values and deliberation. *Journal of Risk Research, 24*(1), 28–46.

Pokrajac, L., Abbas, A., Chrzanowski, W., Dias, G. M., Eggleton, B. J., Maguire, S., Maine, E., Malloy, T., Nathwani, J., & Nazar, L. (2021). *Nanotechnology for a sustainable future: addressing global challenges with the international network4sustainable nanotechnology*. ACS Publications.

Pramanik, P., Krishnan, P., Maity, A., Mridha, N., Mukherjee, A., & Rai, V. (2020). Application of nanotechnology in agriculture. *Environmental Nanotechnology, 4*, 317–348.

Rajput, V., Minkina, T., Ahmed, B., Sushkova, S., Singh, R., Soldatov, M., Laratte, B., Fedorenko, A., Mandzhieva, S., & Blicharska, E. (2020). Interaction of copper-based nanoparticles to soil, terrestrial, and aquatic systems: critical review of the state of the science and future perspectives. *Reviews of Environmental Contamination and Toxicology, 252*, 51–96.

Ramachandran, G., Ostraat, M., Evans, D. E., Methner, M. M., O'Shaughnessy, P., D'Arcy, J., Geraci, C. L., Stevenson, E., Maynard, A., & Rickabaugh, K. (2011). A strategy for assessing workplace exposures to nanomaterials. *Journal of Occupational and Environmental Hygiene, 8*(11), 673–685.

Rana, L., Kumar, M., Rajput, J., Kumar, N., Sow, S., Kumar, S., Kumar, A., Singh, S. N., Jha, C. K., Singh, A. K., Ranjan, S., Sahoo, R., Samanta, D., Nath, D., Panday, R., & Raigar, B. L. (2024). Nexus between nanotechnology and agricultural production systems: challenges and future prospects. *Discover Applied Sciences, 6*(11), 555. https://doi.org/10.1007/s42452-024-06265-7

Reddy, R., Omprasad, J., & Janakiram, T. (2022). Technological innovations in commercial high tech horticulture, vertical farming and landscaping. *International Journal of Innovative Horticulture, 11*(1), 78–91.

Ridge, S. J. (2018). *A regulatory framework for nanotechnology*. Monterey, California: Naval Postgraduate School.

Sajid, M. (2022). Nanomaterials: types, properties, recent advances, and toxicity concerns. *Current Opinion in Environmental Science & Health*, *25*, 100319. https://doi.org/10.1016/j.coesh.2021.100319

Sajid, M., Ilyas, M., Basheer, C., Tariq, M., Daud, M., Baig, N., & Shehzad, F. (2015). Impact of nanoparticles on human and environment: review of toxicity factors, exposures, control strategies, and future prospects. *Environmental Science and Pollution Research*, *22*(6), 4122–4143. https://doi.org/10.1007/s11356-014-3994-1

Saleem, H., & Zaidi, S. J. (2020). Developments in the application of nanomaterials for water treatment and their impact on the environment. *Nanomaterials*, *10*(9), 1764.

Salvi, L. (2015). The EU's 'soft reaction' to nanotechnology regulation in the food sector. *European Food and Feed Law Review*, *10*(3), 186–194.

Sekhar, M., Rastogi, M., Rajesh, C., Saikanth, D., Rout, S., Kumar, S., & Patel, A. K. (2024). Exploring traditional agricultural techniques integrated with modern farming for a sustainable future: a review. *Journal of Scientific Research and Reports*, *30*(3), 185–198.

Sharma, M. M. M., Kapoor, D., Rohilla, R., & Sharma, P. (2023). Nanomaterials and their toxicity to beneficial soil microbiota and fungi associated plants rhizosphere. In A. Husen (Ed.), *Nanomaterials and nanocomposites exposures to plants: response, interaction, phytotoxicity and defense mechanisms* (pp. 353–380). Springer.

Singh, R., & Kumar, S. (2023). Regulatory and safety concerns regarding the use of active nanomaterials in food industry. In R. Singh & S. Kumar (Eds.), *Nanotechnology advancement in agro-food industry* (pp. 269–306). Springer.

Singh, S., Prasad, S. M., & Bashri, G. (2023). Fate and toxicity of nanoparticles in aquatic systems. *Acta Geochimica*, *42*(1), 63–76.

Singh, M., Upadhyay, L., Badekhan, A., Shil, S., Gautam, S. K., & Anand, A. (2024a). Leveraging remote sensing and nanotechnology to overcome barriers to agroforestry adoption by small holder farmers. *Journal of Scientific Research and Reports*, *30*(5), 113–140.

Singh, P., Gupta, P., Verma, V., Yadav, N., & Pandey, N. K. (2024b). Nanotechnology: social acceptance, cultural impact and privacy. In S. C. George & B. Tawiah Nanotechnology in societal development (pp. 65–93). Springer.

Solaiman, S., Yamauchi, Y., Kim, J. H., Horvat, J., Dou, S. X., Alici, G., Ooi, L., Martinac, B., Shiddiky, M. J., & Gopalan, V. (2017). Nanotechnology and its medical applications: revisiting public policies from a regulatory perspective in Australia. *Nanotechnology Reviews*, *6*(3), 255–269.

Sonwani, S., Madaan, S., Arora, J., Suryanarayan, S., Rangra, D., Mongia, N., Vats, T., & Saxena, P. (2021). Inhalation exposure to atmospheric nanoparticles and its associated impacts on human health: a review. *Frontiers in Sustainable Cities*, *3*, 690444.

Tawiah, B., Ofori, E. A., & George, S. C. (2024). Nanotechnology in societal development. In S. C. George & B. Tawiah (Eds.), *Nanotechnology in societal development* (pp. 1–64). Springer Nature Singapore. https://doi.org/10.1007/978-981-97-6184-5_1

Thakur, P., & Thakur, A. (2024). Ethical, legal, and socio-economic impacts of nanotechnology. In A. Thakur, P. Thakur, D. Suhag, & S. M. P. Khurana (Eds.), *Green nanobiotechnology* (pp. 291–309). CRC Press.

Uddin, S., Iqbal, J., Safdar, L. B., Ahmad, S., Abbasi, B. A., Capasso, R., & Quraihi, U. M. (2022). Green synthesis of BPL-NiONPs using leaf extract of *Berberis pachyacantha*: characterization and multiple in vitro biological applications. *Molecules, 27*(7), 2064.

Ullah, Z., Gul, F., Iqbal, J., Abbasi, B. A., Kanwal, S., Chalgham, W., & Mahmood, T. (2023a). Biogenic synthesis of multifunctional silver oxide nanoparticles (Ag_2ONPs) using *Parieteria alsinaefolia* delile aqueous extract and assessment of their diverse biological applications. *Microorganisms, 11*(4), 1069.

Ullah, R., Ullah, Z., Iqbal, J., Chalgham, W., & Ahmad, A. (2023b). Aspartic acid-based nano-copper induces resilience in *Zea mays* to applied lead stress via conserving photosynthetic pigments and triggering the antioxidant biosystem. *Sustainability, 15*(16), 12186.

Usman, M., Farooq, M., Wakeel, A., Nawaz, A., Cheema, S. A., ur Rehman, H., Ashraf, I., & Sanaullah, M. (2020). Nanotechnology in agriculture: current status, challenges and future opportunities. *Science of the Total Environment*, *721*, 137778.

Verma, V., Gupta, P., Singh, P., & Pandey, N. K. (2024). Considerations in the development and deployment of nanotechnology. In S. C. George & B. Tawiah (Eds.), *Nanotechnology in societal development* (pp. 505–540). Springer.

Wahab, A., Munir, A., Saleem, M. H., AbdulRaheem, M. I., Aziz, H., Mfarrej, M. F. B., & Abdi, G. (2023). Interactions of metal-based engineered nanoparticles with plants: an overview of the state of current knowledge, research progress, and prospects. *Journal of Plant Growth Regulation*, *42*(9), 5396–5416. https://doi.org/10.1007/s00344-023-10972-7

Wahab, A., Muhammad, M., Ullah, S., Abdi, G., Shah, G. M., Zaman, W., & Ayaz, A. (2024). Agriculture and environmental management through nanotechnology: eco-friendly nanomaterial synthesis for soil-plant systems, food safety, and sustainability. *Science of the Total Environment, 926*, 171862.

Wang, M., Li, S., Chen, Z., Zhu, J., Hao, W., Jia, G., Chen, W., Zheng, Y., Qu, W., & Liu, Y. (2021). Safety assessment of nanoparticles in food: current status and prospective. *Nano Today, 39*, 101169.

Yang, T., & Duncan, T. V. (2021). Challenges and potential solutions for nanosensors intended for use with foods. *Nature Nanotechnology, 16*(3), 251–265.

Zhang, P., Guo, Z., Zhang, Z., Fu, H., White, J. C., & Lynch, I. (2020). Nanomaterial transformation in the soil–plant system: implications for food safety and application in agriculture. *Small, 16*(21), 2000705.

Index

Nanotechnology Innovations for Food Security and Sustainable Agriculture, First Edition. Edited by Zia Ur Rehman Mashwani, Shah Fahad, Ilyas Ahmad, Shah Saud, Sajid Fiaz, and Taufiq Nawaz.

C

N

O

P

T

Y

Z

Printed and bound by CPI Group (UK) Ltd, Croydon, CR0 4YY
17/09/2026
14964299-0002